Organic Food Systems

Meeting the Needs of Southern Africa

Organic Food Systems

Meeting the Needs of Southern Africa

Edited by

Raymond Auerbach

Nelson Mandela University, George, South Africa

CABI is a trading name of CAB International

CABI
Nosworthy Way
Wallingford
Oxfordshire OX10 8DE
UK

Tel: +44 (0)1491 832111
Fax: +44 (0)1491 833508
E-mail: info@cabi.org
Website: www.cabi.org

CABI
745 Atlantic Avenue
8th Floor
Boston, MA 02111
USA

Tel: +1 (617)682-9015
E-mail: cabi-nao@cabi.org

A catalogue record for this book is available from the British Library, London, UK.

Library of Congress Cataloging-in-Publication Data

Names: Auerbach, Raymond, 1953- editor.
Title: Organic food systems : meeting the needs of Southern Africa / editor: Raymond Auerbach.
Description: Boston, MA : CAB International, 2020. | Includes bibliographical references and index.
Identifiers: LCCN 2019016398 | ISBN 9781786399601 (hbk : alk. paper) | ISBN 9781786399625 (epub)Subjects: LCSH: Organic farming--South Africa. | Organic farming--Zambia. | Organic farming--Uganda. | Organic farming--Tanzania.
Classification: LCC S605.5 .O6795 2019 | DDC 631.5/840968--dc23
LC record available at HYPERLINK “https://protect-eu.mimecast.com/s/aUzGC7p0nHZOnLIWUpjt?domain=lccn.loc.gov” https://lccn.loc.gov/2019016398

ISBN-13: 978 1 78639 960 1 (hardback)

Commissioning editor: Dave Hemming
Editorial assistant: Tabitha Jay
Production editor: Marta Patiño

Typeset by SPi, Pondicherry, India
Printed and bound in the UK by Severn, Gloucester

Dedicated to my teachers

Allexei de Podolinsky
Ingrid Adler
Jeanne Malherbe
and
Robert Mazibuko.

Contents

Contributors

Albert Ackhurst (Nelson Mandela University)

Albert is a conservation scientist, botanist and soil carbon scientist. He holds a BSc (natural sciences), BSc Hons (botany), BCom Hons (training management) and an MTech on the subject of his chapter (the rapid incineration field test for soil organic carbon). He is currently the Head of Biodiversity, with the Department of Environmental Affairs and Development Planning in the Western Cape. He is the Co-chair of the Honeybush Community of Practice, and is also qualified and experienced as a human capital and training specialist, owner/director of SCELETIA (Pty) Ltd, an African herbal tea company, and founder-member/ex-director of THE GREEN TICKET (specializing in agroforestry and biodiversity economies). He is active in the fields of ecological infrastructure, sustainable natural resource management, community development, business intelligence, citizen science and the creation of green-collar jobs. Albert specializes in biodiversity economies, biodiversity policy and strategy development, natural and human resources modelling, project design and development, and learning/training development.

Raymond Auerbach (Nelson Mandela University, co-/supervisor to students listed)

Raymond trained in organic vegetable production at Camphill Hermanus in 1972, and then did an apprenticeship in biodynamic farming with the Australian Institute for BioDynamic Agriculture from 1973 to 1976, with Alex de Podolinsky and Dr Andrew Sargood. He farmed organically until 2002 in KwaZulu-Natal (KZN), trained farmers in organic farming (through his Rainman Landcare Foundation) from 2000 until 2010, and taught soil science and plant production at Nelson Mandela University (George Campus) from 2010 to 2018; he still coordinates their Agroecology Research Group. He runs the African Organic Farming Systems Research project (funded by the South African (SA) National Research Foundation, NRF), and leads projects for the Centre of Excellence in Food Security (with the universities of the Western Cape and Pretoria); he is lead researcher with ecosystem-based solutions for resilient urban agriculture (ECOSOLA) project (with the University of Dar es Salaam and Carl von Ossietzky University, Oldenburg), and teaches 'Systems and technologies for sustainable agriculture' at Stellenbosch University's Sustainability Institute, where he is an extraordinary professor. Raymond is a founder member of the BioDynamic Agricultural Association of Southern Africa, and the Sustainable Organic Agriculture Action Network International Federation of Organic Agricultural Movements (IFOAM). He is on the Steering Committees of the SA Organic Sector Organisation (SAOSO) and Participatory Guarantee Systems – SA (PGS-SA) and is Chair of the Outeniqua Natural and Organic PGS. He is a trustee of the Garden Route Botanical Garden. Having farmed organically for 20 years and trained farmers for 20 years, he is now in the middle of 20 years of research, consulting and policy advocacy.

Jane Battersby (University of Cape Town)
Jane Battersby is a senior researcher at the African Centre for Cities, University of Cape Town, and is the Research Co-ordinator of the 'Consuming Urban Poverty' project and Project Investigator of the 'Nourishing Spaces' project. An urban geographer by training, her work focuses on urban food security, food systems and governance. Jane is currently the Premio Daniel Carasso (prize winner for 2018) and is on the Independent Expert Group of the Global Nutrition Report. She has served in an advisory or consultant role to a number of United Nations (UN) agencies.

Anastasia Caude (Nelson Mandela University intern)
Anastasia has always had an interest in how the world works, and speaks French, Spanish and English; she has lived much of her life in Mauritius. She has a lively understanding of many of the issues around sustainable food systems and agroecology, and completed an agronomy degree at Polytechnique Unilasalle. She is currently doing a master's degree at Cranfield University in Future Food Sustainability.

Stephen Devereux (University of Sussex and University of the Western Cape)
Stephen Devereux is a research fellow at the Institute of Development Studies at the University of Sussex, where he is Co-director of the Centre for Social Protection. He holds the NRF–Newton Fund (SA-UK) Research Chair in Social Protection for Food Security, affiliated to the DST–NRF Centre of Excellence in Food Security and the Institute for Social Development at the University of the Western Cape, SA. He has published nine books and over 70 journal articles, mainly on famine, food security, seasonality and social protection. His current research focuses on trying to explain the 'food security paradox' in SA, where indicators of food security are improving and social grants now reach two-thirds of all children, but indicators of child malnutrition remain unacceptably high.

Catherine Eckert (Nelson Mandela University, lecturer in plant production and agricultural engineering and PhD student on water use efficiency)
Catherine Eckert graduated with a bachelor's degree in agricultural management *cum laude* from Nelson Mandela University in George, SA. She recently completed her master's degree on water use efficiency at the university while lecturing part time, and is now busy with a PhD on the same subject. Catherine has always been passionate about plants, animals and the environment and this in part is why she chose to study agriculture, as she believes farmers can not only feed the world, but also have a positive impact on the environment while doing so. She has previously worked in the SA berry industry, having 5 years' experience in various roles including integrated pest management, trial management, plant tissue culture and nursery management.

Johan Habig (soil microbiologist, MicroLife Research Centre, soil microbiology)
Johan Habig was born and raised on a farm in Heilbron, in SA's Free State Province. He completed his MSc microbiology at Northwest University's Potchefstroom Campus in 2002. He started his career as a researcher at the Agricultural Research Council's (ARC) Plant Protection Research Institute (now Plant Health and Protection) in Pretoria. During his 15 years at the ARC, Johan initially worked on biological nitrogen fixation and trained second economy farmers in Venda, KZN and the Eastern Cape in conservation agriculture and quality maize and legume production in an effort to help them to increase their protein intake. The Soil Microbiology Laboratory was established during 2008 and rendered services to several other ARC institutes, tertiary institutions and organizations in industry. The services provided by the laboratory specialized in studying and determining the impact of various agricultural practices on soil microbial ecology as indicators of soil health. During September 2017, Johan joined the dynamic team of Agri Technovation in Wellington, Western Cape Province, where he was appointed as Senior Researcher and head of the MicroLife Research Centre. This well-equipped laboratory currently provides a wide range of soil microbial assays, statistical analyses on obtained results, and the compiling of easy-to-understand reports to commercial clients who are concerned about the health of the biological component of their soil.

Konrad Hauptfleisch (Manager IFOAM-Organics International Academy)

Konrad Hauptfleisch is Head of Capacity Development at IFOAM-Organics International in Bonn, Germany. Proudly South African by birth, he studied English, linguistics and philosophy at the then Rand Afrikaans University (now University of Johannesburg). While teaching was clearly the focus of such a degree, he spent many years in the performing arts, both behind and in front of the cameras and on stage.

Later in life he became deeply involved in development work and production management. But it was organic agriculture that has been dominant in the last 13 years – first he was chief operations officer of the well-known Bryanston Organic and Natural Market in Johannesburg for several years, during which time he focused on the development of small-scale farmers, market access and appropriate organic guarantee systems for such farmers and farming systems. This work also led him to work alongside fellow activists and experts to develop policy and organic standards for SA, during which time he served on various committees working with the public and private sector, and served as Chair of the SA Bureau of Standards (SABS) Technical Committee developing organic standards. Teaching and training remained his passion, though: he moved to Bonn, Germany in 2012 to take on the task of developing the Organic Academy, and has since worked in more than 42 countries, training and teaching organic leaders, practitioners and activists.

Gareth Haysom (University of Cape Town, co-supervisor, Wim Troosters)

Gareth Haysom's work uses food as a lens to understand the complex nature of the urban transitions currently underway in the Global South. Gareth is based at the African Centre for Cities at the University of Cape Town where he coordinates the 'Hungry Cities Partnership' project. Gareth also works on the 'Consuming Urban Poverty' project, investigating the intersections between food poverty and space in secondary African cities, and the 'Nourishing Spaces' project, a project investigating the connections between food security, nutrition and dietary transitions in large and small African cities.

Jostein Hertwig (BERAS: case studies of organic food systems)

Jostein is an attorney at law with more than 20 years' experience in international business development, cooperation, networking and negotiations. After his business career, Jostein started up (and managed for 12 years) an integrated organic farm in Norway. In 2014 Jostein became head of BERAS (Building Ecological Regenerative Agriculture and Societies) International Foundation, focusing on research, development, education, implementation and communication of results and practical examples within the concepts of ecological regenerative agriculture, learning centres for sustainable food societies and Diet for a Green Planet. He is also one of the coordinators of Organic Food System Programme (OFSP), which was appointed a core initiative of the UN Sustainable Food System Programme in 2017. Since then, he has also been an advisor on sustainability and organic food for REMA 1000 in Norway.

Johannes Kahl (Organic Food Systems project, Kassel University, Germany)

Johannes Kahl worked at the Leibniz Institute for Analytical Sciences and received his doctorate from the University of Dortmund. After his postdoctoral research at the Max Planck Institute for Chemical Ecology in Jena, he researched and taught at Kassel in Germany and Copenhagen in Denmark and was a visiting lecturer at the universities of Barcelona/Spain, Sao Paulo/Brazil and Warsaw/Poland. He coordinates the OFSP, a core initiative of the 10-Year Framework of Programmes on Sustainable Consumption and Production Patterns of the UN. Currently Director of Organic Agriculture at the Faculty of Organic Agricultural Sciences (the only one in Germany, which offers an accredited BSc and MSc (in German and English) in organic agriculture) as well as research and education in the areas of rural development in the subtropics and tropics. The faculty is well known for its research along the food chain including marketing and nutrition culture as well as nature conservation.

Christina Kifunda (PhD student at Carl von Ossietzky University, Oldenburg, Germany)

Christina Kifunda has an MA degree in geography and environmental management on climate change adaptation in coffee in Tanzania. She is currently a PhD student at the Carl von Ossietzky University in Oldenburg, Germany in collaboration with the Centre for Climate Change Studies, University of Dar es Salaam, Tanzania. She is with the ECOSOLA project (ecosystem-based solutions for resilient urban agriculture in Africa) with a research title of: 'The role of gender in supporting livelihoods in a changing climate through urban and peri-urban agriculture: the case of Kinondoni Municipality in Dar es Salaam'. Before joining the PhD programme, Ms Kifunda served as an assistant lecturer and Head of the Department of Geography at Jordan University College (JUCO) located in Morogoro Region. At JUCO, Ms Kifunda was responsible for teaching various courses including physical geography, environmental education and conservation and spatial organization. She has also been a consultant for the Food and Agriculture Organization of the United Nations (FAO) on forest education in primary schools and has a publication on geography and the environment.

Nico Labuschagne (University of Pretoria, supervisor, Mandla Sibiya)

Nico Labuschagne studied plant pathology at the University of Pretoria. He obtained the degrees BSc Agric., MSc Agric. (*cum laude*) and DSc Agric. and started his career as a researcher at the then Department of Agriculture and Fisheries. He later moved to a private agrichemical company as a research plant pathologist. In 1982 he was appointed as a lecturer (and then senior lecturer, later associate professor) in the Department of Microbiology and Plant Pathology at the University of Pretoria. In his research, Professor Labuschagne focused on soilborne plant diseases on a variety of crops. His research on the topic spans more than three decades and he has conducted research on various economically important plant diseases including Phytophthora root rot of avocados, black hull of groundnuts, root rot of citrus caused by *Phytophthora* and *Fusarium* spp., Pythium root rot in hydroponically grown vegetable crops and a number of nematode diseases of citrus, soybeans and tomatoes. For the last 10 years Professor Labuschagne's research focus has been on application of plant growth promoting rhizobacteria (PGPR) as biofertilizers and biocontrol agents of plant diseases. This research programme strikes a balance between fundamental and applied research and has led to the commercialization of a number of biofertilizer and biocontrol products. The PGPR project also intersects with the topic of soil health and has led to Professor Labuschagne becoming involved in a recent project on sustainable crop production as part of the Centre of Excellence in Food Security under the auspices of the SA Department of Science and Technology together with the National Research Foundation. During his career Professor Labuschagne has presented more than 64 papers at national conferences, authored or co-authored 26 papers or posters at international congresses, published more than 60 research articles in peer-reviewed scientific journals and has written three chapters in scholarly textbooks. He has also supervised or co-supervised more than 30 masters and nine doctoral students.

Sandra Lamprecht (ARC, co-supervisor Braam van Niekerk)

Sandra Lamprecht is a specialist researcher and head of the Soilborne Plant Diseases Unit of the ARC-Plant Health and Protection Institute in Stellenbosch, SA. She completed her undergraduate studies at Free State University and obtained her PhD in plant pathology in 1989 on Fusarium diseases of annual *Medicago* spp. from Stellenbosch University. Her research focus is on the epidemiology and management of soilborne diseases of a variety of crops, with special emphasis on the identification and management of soilborne disease complexes. These disease complexes include important pathogens such as *Fusarium*, *Phytophthora*, *Pythium* and *Rhizoctonia* species. She has been involved in developing management strategies for soilborne disease complexes of many crops including lupin, canola, rooibos, soybean, sunflower, wheat and maize with emphasis on integrated management strategies to ensure sustainable management of these diseases. She initiated the annual symposium of the Soilborne Plant Diseases Interest Group of SA in 1990 to promote multidisciplinary collaboration and networking for soilborne plant pathologists with other disciplines and acted as chair of the organizing committee of these annual events for the past 27 years. She received the Applied Plant Pathology Award from the SA Society for Plant Pathology in 2006 and was elected a fellow of this society in 2015.

Andre Leu (International Director of Regeneration International)

Andre is the author of *Poisoning our Children* (2018, Acres USA, Austin, Texas) and the *Myths of Safe Pesticides* (2014, Acres USA). He was the President of IFOAM-Organics International, the world change agent and umbrella body for the organic sector for the past decade. IFOAM-Organics

International has around 850 affiliate organizations in 127 countries. Andre has a degree in communications, with a double major in video/television production and socio-political theory. He has postgraduate qualifications in adult education.

He lectures and teaches at universities, institutions and workshops around the world. He speaks at numerous conferences, seminars, workshops as well as UN events on every continent. He meets with governments, industry, farmers, consumers and non-governmental organizations (NGOs) on the multi-functional benefits of regenerative organic agriculture. He has an extensive knowledge of farming and environmental systems across Asia, Europe, the Americas, Africa and Australasia from over 40 years of visiting and working in over 100 countries. Andre and his wife, Julia, have an organic tropical fruit farm in Daintree, Australia. He has published in magazines, newspapers, journals, conference proceedings, newsletters, websites and other media, as well as doing numerous media interviews for television, radio and online platforms.

Simon Lorentz (University of KwaZulu-Natal, Principal Hydrologist, SRK Consulting Engineers and Scientists, MSc co-supervisor, Catherine Eckert)

Simon Lorentz is a process hydrologist with specialization in vadose zone hydrology and water quality. He obtained a BSc in civil engineering from the University of the Witwatersrand in 1977 and began his career at the Iron and Steel Corporation (ISCOR), with design and project management in mine stormwater protection systems and water supply. He continued his focus on mining hydrology at Steffen, Robertson and Kirsten (SRK) in Johannesburg, where he was also co-author of the first publications of the water resources of SA. Postgraduate research at Colorado State University (CSU) included the theoretical development of flow and solute transport in unsaturated, bimodal porous media and the development of instrumentation for the measurement of unsaturated porous media hydraulic characteristics. He obtained an MSc and PhD in bioresources engineering while at CSU. Over a 20 year period at UKZN (previously University of Natal), Simon developed and led a focus group on hydrological processes observation and modelling. This included hillslope hydrology, vadose zone hydrology, sediment and nutrient transport and unsaturated porous media measurement and simulation. His expertise has been applied to catchment hydrology and local water and solute transport problems including: agricultural water use, sediment and nutrient observation and simulation; forestry water use as well as surface water–groundwater interactions in savannah, agricultural, mining and industrial locations. Simon has been a co-developer of the science of hydropedology in Southern Africa and has pioneered the use of near-surface geophysics and stable isotope use in hydrological and water quality studies in SA. He is currently Principal Hydrologist at SRK Consulting and an honorary associate professor at the Centre for Water Resources Research of UKZN.

Josua Louw (Nelson Mandela University, co-supervisor, Albert Ackhurst)

Josua Louw is a Professor in the School of Natural Resource Management at the Nelson Mandela University's George Campus. He studied at the universities of Stellenbosch, Potchefstroom and Witwatersrand, with postgraduate studies focused on forestry and natural science. His current research interests include soil science and its various applications in natural resource management, landscape ecology and environmental management. Professor Louw's career history includes industry experience in both research and management. Prior to his appointment at the Nelson Mandela University, he specialized in soil surveys, land use evaluations and afforestation assessment, and held project management and research positions at the CSIR (Forestek) and the SA Forestry Research Institute. He has published widely in refereed scientific journals, and also has an extensive track record of student supervision, at master's degree and doctoral level in a variety of research spheres.

N'wa-Jama Mashele (Nelson Mandela University, PhD student)

N'wa-Jama Mashele is a botanist turned agricultural management researcher. She holds BSc and BSc Honours degrees in botany, and an MSc (agricultural management/botany). She is currently registered for a PhD in agricultural management, working on the ECOSOLA project with German and Tanzanian partners, and lectures on a part-time basis at Nelson Mandela University.

Mebelo Mataa (University of Zambia, PhD co-supervisor, Robert Munthali)

Mebelo Mataa is a lecturer in horticulture and plant physiology in the Plant Science Department of the School of Agricultural Sciences, University of Zambia. He holds a PhD in horticulture (physiology of fruit trees) from Kagoshima University, Japan. His interests include: (i) environmental stresses and interaction with plant productivity; (ii) domestication and conservation of under-utilized plants; (iii) plant propagation and postharvest physiology; and (iv) sustainable agriculture. Dr Mataa is co-supervisor of Robert Munthali's PhD study, and has assisted him with conceptualising this work.

Robert Munthali (Klein Karoo Seed Zambia, Lusaka, Zambia, Nelson Mandela University PhD student)

Robert Munthali is currently working for Klein Karoo Seed Zambia, and since February 2014 as Zambia's Production Leader. Robert has a master's degree in sustainable agriculture (University of the Free State), a postgraduate diploma in agricultural development, a postgraduate diploma in lecturing and teaching methodologies for lecturers, and a diploma in technologies for crop production. He has certificates in organic agriculture, food security and globalized agriculture, biodiversity and sustainable agriculture and managing sustainable agricultural enterprises. At present he is studying for a PhD in agronomy in organic agriculture working in collaboration with small-scale farmers in Zambia. Robert has 30 years of work experience in the agricultural sector, and worked for the Organic Producers and Processors Association of Zambia (OPPAZ) as Chief Technical Officer, Audit Control and Expertise, and then worked as Credit Support Services Manager for Seed Company Zambia Limited (SEEDCO). He was later Regional Sales Manager, and worked for the Tobacco Board of Zambia as a tobacco inspector, and for the Zambia Seed Company Limited (ZAMSEED), as Technical Seed Manager. He also worked in public service as a horticulturist at the National Assembly (Parliament Buildings) as well as in the Ministry of Agriculture, Extension Branch as a horticultural officer.

Jane Nalunga (National Organic Agriculture Movement Uganda (NOGAMU); PhD student, Nelson Mandela University)

Jane Nalunga is the Head of Programmes at NOGAMU. She holds an MSc in agricultural economics and a BSc in agriculture, both from Makerere University, Kampala, Uganda. Before becoming the Head of Programmes, she worked as the Senior Training Officer in NOGAMU. She has a good understanding of scientific, economic and socio-economic issues concerning agriculture. She has more than 20 years experience in agricultural management, assessment and training on agricultural production and marketing systems. She has been involved in training agricultural staff, farmers groups and companies in production, postharvest, marketing and quality management, for both local and export markets. She has hands-on experience in performance management, designing quality management systems and producing quality manuals for producer groups and companies. She is also an organic farmer, mainly farming with bananas. In her free time, she enjoys gardening. Her dream is 'making markets work for smallholder farmers'.

Nic Olivier (Professor Extraordinary, North West University; research associate, University of Pretoria)

Nic Olivier was previously a professor at the universities of Pretoria, Potchefstroom and Natal, and, until 2015, Director of the Southern African Development Community (SADC) Centre for Land-related, Regional and Development Law and Policy at the University of Pretoria. He is currently professor extraordinary at North West University. His fields of specialization include governance and development, international and constitutional law, transformative programme and strategic management, policy and law (and linkages to sustainable development and food security). Recently he has collaborated with, among others, the FAO and UN Children's Emergency Fund, as a member of a team of SA and foreign scholars and officials on a number of interdisciplinary projects on food security and nutrition, and was part of the SA team responsible for the background research on which the 2017 South African National Food and Nutrition Security Implementation Strategy is based.

He recently completed the drafting of the Policy Framework for the establishment of a rural development agency and related DFI (rural development finance institution) for the World Bank and the

SA Department of Rural Development and Land Reform, and is helping to develop a Code of Rural Development Law that will consolidate all existing rural development legislation and provide for the establishment and operationalization of the rural development agency and agri-parks. He is co-project leader of a United States Agency for International Development (USAID)-funded project 'Conceptualizing Drivers of Food Security Policy Change: Coordination Mechanisms and Policy Design' in CAADP (Comprehensive Africa Agriculture Development Programme; focus country: Malawi). In addition, he was recently responsible for the execution of an FAO (subregional office: East Africa) multi-country comparative project aimed at determining the level of incorporation of, and compliance with, international, African and subregional food security and nutrition-related obligations and commitments by the nine East African IGAD (Intergovernmental Authority on Development) countries.

Hannelise Piek (Nelson Mandela University)

Hannelise Piek studied consumer science: food and nutrition at the Cape Peninsula University of Technology. After she completed her BTech degree, she worked as a research assistant for the Department of Applied Sciences where she was co-author of several published articles. She finished an MTech degree (on antioxidants in rooibos tea), and now works as a research assistant with the Agroecology Research Group at Nelson Mandela University.

Matthew Purkis (University of Johannesburg and SA Organic Sector Organisation – SAOSO)

Matthew Purkis works with the University of Johannesburg, SAOSO and PGS-SA (Participatory Guarantee Systems – South Africa). Matthew is an organic practitioner and entrepreneur in the green sector. He completed his BA in interior design at the Greenside Design Center. Since his graduation in 2010 he has been developing his skills and knowledge in the fields of agroecology, natural architecture, green events, permaculture land design and project development. Matthew is actively involved in the establishment of organizations and businesses that can drive the development of an alternative food system and pioneer solutions for sustainable human settlements. He is passionate about environmental and social regeneration through innovation and the integration of appropriate technologies into projects and communities.

Matthew is an alumnus of the IFOAM Organic Leadership Academy and has also completed additional studies in the green economy with the University of the North West. He is currently working with the University of Johannesburg as an independent contractor, consulting on green economy projects and training in organic agriculture and PGS. Matthew believes in collaboration and is committed to seeing a strong sector develop that can successfully advocate for an alternative food system and build the foundations of a socially equitable economy in Southern Africa.

Mandla Sibiya (University of Pretoria, MSc student)

Mandla Mcebo Sibiya from Swaziland was a male warder who did his primary, high school and undergraduate studies in his home country between 1994 and 2011 obtaining a 4-year BSc degree in agronomy. He lost his father in 2015, but now lives to make him proud. The move to SA in 2016 to further his studies at the University of Pretoria was inspired by his passion to learn. After 1 year at the university, he graduated with his BSc honours degree in crop science in 2017, after which he enrolled for an MSc degree in plant pathology under the supervision of Professor Nico Labuschagne. He is self-driven and draws his inspiration and motivation from his family back in Swaziland. Being the last born, he plans to become the first PhD graduate of his entire extended family. He is also passionate about sports and never wants to be in the losing team, no matter the goal, either personal or academic. His other passions are his daughter and his mother, both back in Swaziland, forever in his thoughts, he cherishes them dearly.

Bernd Siebenhüner (Carl von Ossietzky University, Oldenburg, Germany, supervisor Maren Wesselow)

Bernd Siebenhüner is a professor of ecological economics at Carl von Ossietzky University, Oldenburg, Germany. He coordinates the Master's programme Sustainability Economics and Management and served as Vice President for Graduate Education and Quality Management from 2010 to 2015. After earning his degrees in economics and political science from the Freie Universität Berlin and his PhD from the University of Halle-Wittenberg, he held positions at the Potsdam Institute for Climate Impact Research, at Harvard University, and at Nelson Mandela University. He headed numerous research undertakings in the fields of social learning, international organizations, global environmental governance, corporate sustainability strategies, climate adaptation and biodiversity governance, and the role of science in global environmental governance.

Charles Ssekyewa (Uganda Martyr's University, PhD co-supervisor, Jane Nalunga)

Charles Ssekyewa studied subtropical horticulture and crop science at London University, obtained a PhD (biological sciences) at Ghent University, and did postdoctoral research in molecular virology at Bangor University (as a Rothamsted Fellow). He is a Professor of Agroecology with Uganda Martyrs University where he facilitates the course on Systems Philosophy and applications of the PhD in agroecology. He is also the Dean of the College of Agriculture and Veterinary Sciences, St Augustine International University, Kampala. Charles is currently the Chairman of the Regional Steering Committee of the African Union/East African Ecological Organic Agriculture Initiative. He developed and ran the first Bachelor of Organic Agriculture degree programme in Africa at Uganda Martyrs University. He was the convenor of the first African Organic Agriculture Conference, and recently the co-convenor of the first Tropical Agroecosystems Conference in Africa. Charles, though a molecular scientist, has understood and embraced ecological organic agriculture. He farms and is the Director of the Centre for Ecosystems Research and Development (CERD-Uganda) which enables communities to optimize their agroecosystems by conserving biodiversity and innovatively adding value to agricultural products.

Sheila Storey (Nemlab, Western Cape)

Sheila Storey is the owner of Nemlab, a plant nematology diagnostic laboratory. She obtained her BSc (Agric.) and MSc (Agric.) in nematology in the 1980s majoring in entomology and plant pathology. She combined her love for the science of nematology and her love for helping others which culminated in the establishment of the laboratory. Sheila has obtained experience in identifying and controlling plant parasitic nematodes on numerous crops all over Southern Africa. She often travels to neighbouring African countries as her nematology expertise is in great demand. In the last few years her interest has turned to soil health with particular reference to the role beneficial nematodes play in soil health. In this regard she has recently started a Soil Health Support Centre to help growers make the transition from conventional farming to more environmentally friendly agricultural practices. Her active role in integrated pest management within the deciduous fruit industry led her to start the company Nema Bio (Pty) Ltd in 2014 to take the laboratory-based results of entomopathogenic nematodes as possible biocontrol agents out into the field.

Carola Strassner (Münster University of Applied Sciences, Organic Food Systems Programme)

Carola Strassner is Professor of Sustainable Food Systems and Nutrition Ecology at the Department of Food, Nutrition and Facilities, at the Münster University of Applied Sciences in Germany and Programme Director for the MSc Sustainability in Service Management and Food Industries, a 2-year collaborative degree programme. She specializes in food systems sustainability, a whole systems approach to sustainability in the sphere of food and nutrition, especially the out-of-home (horeca = hotel-restaurant-catering) context and alternative food networks. Professor Strassner works specifically with the subsystems institutional catering, school meals, and the organic food system. She co-initiated the OFSP, one of eight core initiatives of the UN 10-year Framework of Programmes on Sustainable Consumption and Production Patterns and serves as a coordinator on its board. She has produced a wide range of publications for chefs, kitchen staff and disseminators, and is a regular contributor to textbooks for professionals in the food and organic industry. She initiated and managed an international network 'Organic Out Of Home' for 10 years. Professor Strassner develops teaching and training modules both for undergraduate and postgraduate levels, for theory and practical applications, in both German and English. Train-the-trainer courses and activities form a regular feature of her work. She is managing partner of the business company *a'verdis – Sustainable Foodservice Solutions*. Thus students benefit from a very close proximity to practice and a wide, well-established network.

Marike Swanepoel (Nelson Mandela University; Koppert Biological Systems, integrated pest management)

Marike Swanepoel obtained her MSc (agriculture) *cum laude* from Nelson Mandela University, and holds a BSc (botany, zoology and tourism), BSc Hons (environmental science – plant protection), Advanced Certificate in Project Management (NQF 6), and a postgraduate certificate in education (PGCE). She is currently a technical consultant at Koppert Biological Systems.

Wim Troosters (Nelson Mandela University, MTech graduate)

Wim Troosters is passionate about the environment and sustainable development, from his early days in college in Belgium, and now as a consultant in Africa. He has degrees in chemistry, business management and organic agriculture. From 2006 until 2012, he has worked on a new, innovative development model called Agri-SCIP on the South Coast of KZN in SA. It was this work that inspired him to further his career in agriculture and pro-poor market development. This resulted in a case study of the Agri-SCIP model, which is presented in this book. Since 2015, Wim has lived and worked in Rwanda, where he is involved in the development of a local PGS, in collaboration with the Rwandan Organic Agriculture Movement, and supporting NGOs and small businesses in organizational and sustainable development.

Abraham van Niekerk (Nelson Mandela University; Corteva Agriscience)

Braam van Niekerk holds an MTech degree from Nelson Mandela University in George, SA. He is currently Business Development Manager for Corteva Agriscience, a division of DowDuPont in KZN, SA. He works closely with farmers and distributors of agricultural crop protection products all over KZN and the Eastern Free State provinces, providing training and sustainable solutions in crop protection. He is also the co-founder of ChemSelekt Environmental Science, distributors and exporters of vector-controlling insecticides for combating malaria. Working in agriculture as a technical field representative all of his professional career, he has a passion for crop protection and sustainable solutions in agriculture.

Maren Wesselow (Carl von Ossietzky University, Oldenburg, Germany, PhD student)

Maren Wesselow, research fellow at Carl von Ossietzky University of Oldenburg, Germany, holds a BA in social management (2009) and an MSc in sustainability, geography and regional development (2014). While working in the interdisciplinary SuLaMa project (land management in Madagascar) at the University of Greifswald from 2014 to 2016, she focused on participatory methods to discuss research results with local communities in Madagascar. Since 2017, she investigates formal and informal institutional frameworks of urban agriculture in Tanzania and SA within the ECOSOLA project. Her doctoral research interest is in social processes in sustainable land use in countries of the Global South.

Foreword

Sustainable food production has taken centre stage across the globe in intellectual, technical and policy discussions in recent decades. This is borne out of a series of issues confronting the efficiency of food production and its delivery, as well as climate change. The need to produce more food has been driven by the rapidly growing world population. The annual population growth rate currently stands at 1.07% resulting in an additional 82 million individuals per year. Deducting the death of 55 million per year, the world seems to have an additional 27 million individuals to feed per year. Given that land, which is the principal agricultural production asset, is limited, the task therefore, is to raise productivity from a limited resource base. The efforts of the scientific community in developing a series of intensification methods to ensure increased output per unit area have consistently led to increasing yields through improvement of genetics and agricultural practices, external inputs such as fertilizer and other agrochemicals. This intensification has trade-offs: both rapid degradation of the environment and lately food quality.

Part of the natural response to the prevailing issues is the option of ecological and organic agriculture. It started as an ideological response, which was later supported by scientific evidence. More recently, with the increasing social acceptance of the 'organic food production philosophy', it has mushroomed into an alternative to the chemical-based intensification models, and is becoming mainstream. Organic agriculture relies on naturally occurring inputs, ensuring that seeds, fertilizers and other additives are purely of non-toxic sources and are generated from natural systems. They should also be socially, environmentally and economically sustainable. Evidence shows that organic products are healthy and help sustain the integrity of the production assets. The shortfall of organic production systems is the somewhat reduced yield obtained when compared with the agrochemical-based intensification methods. Research presented in this book shows how scientific interventions can close the yield gap making organic systems comparable with agrochemicals but environment friendly.

Contrary to the common assumption that agriculture in Africa is 'organic' by default rather than 'resource constrained', the real organic production system is knowledge intensive and cannot be treated as a naïve practice because it uses natural materials. Successful organic farmers and practitioners constantly require capacity upgrades to be abreast of more productive systems, emerging trends, market opportunities and science-based production technologies. This is vital to ensure that organic agriculture maximizes the opportunities for promoting food sovereignty.

This book is a vital resource for all stakeholders in organic agriculture. It provides a rich overview of the farming systems' dynamics and the development of the practice of organic agriculture. It provides deep insight on the role of organic farmers in policy issues, as well as other institutional dimensions to foster national development. There is a clear description of the training and other capacity development needs.

Coming from a seasoned voice on the African organic agriculture landscape, I gladly recommend Professor Raymond Auerbach's book for all stakeholders in organic agriculture as a reference work and potent tool to foster further development of organic agriculture as a food production system and enhance its contribution to sustainable agrarian livelihoods and national development.

Yemi Akinbamijo, PhD
Executive Director, FARA

FARA is the Forum for Agricultural Research in Africa (https://faraafrica.org);
With Head Office in Accra, Ghana, it advocates research for development.

Acknowledgements

First, thanks to Professor Janice Jiggins and Dr Gunnar Rundgren for reviewing the book, and providing many constructive suggestions for improving clarity, accuracy and accessibility; their many years of experience in farming, farmer outreach, research and teaching put them in a position to assist the authors and the editor insightfully!

Thanks also to Professor Donald Davis for reviewing Chapters 1 and 7 from a food and nutrition science point of view, and for his useful comments on structure and style, and to Lise Andreasen and Ilse Rasmussen of the International Centre for Research into Organic Farming Systems based at Aarhus University in Denmark, for reviewing Chapter 3, which is largely based on their work and that of Niels Halberg.

Clearly, none of the reviewers are responsible for the opinions expressed in this book.

Second, I must acknowledge the following for funding and research support at various times between 2013 and 2019:

- S. Haddad Agricultural Services (Pty) Ltd in George, Western Cape;
- Hygrotech and BASF for supply of seeds, agrochemicals and biological remedies;
- NRF-RTF for support for the long-term trials and bursary costs;
- CoE-FS for support for the long-term trials, bursary costs and book accessibility;
- AgriSETA, for the 1-year bursary for Abraham van Niekerk;
- AEON for security fence and costs of book accessibility and bursaries;
- the Science Faculty and George Campus Principal for setting up the Mandela Trial site;
- Nelson Mandela University Research and Engagement divisions for ongoing support;
- Niels Halberg, Lise Andreasen and Ilse Rasmussen for supplying research papers for Chapter 3;
- Sheryl Hendriks for supplying draft research papers for Chapter 7;
- GIZ for support for the surveys (Chapter 10 and 11), and assisting with book accessibility;
- International Fund for Agricultural Development (IFAD) through ISASAR, Port Elizabeth for Chapter 12 research costs;
- ECOSOLA (supported by the German government (BMBF and DAAD)) for support with activities for Chapters 8 and 23, stipends and costs of book accessibility;
- For support with statistical analysis: Dr Keith Little and Jeanette Pauw, both of Nelson Mandela University, and Marieta van der Ryst, Agricultural Research Council Stellenbosch;
- For design and construction of the Permaculture Centre: John Turner and Marius Strydom;

- For help with weed control and baboon management: Emily Botha, Leentjie van Wyk and Marius Strydom (also for help with Permaculture Centre management); and
- For support for Chapter 14 by enabling a writing retreat supported by the Nelson Mandela University Department of Engagement, and the facilitation and encouragement of Professor Ruth Albertyn.

Finally, my thanks to Hannelise Piek, who was a most wonderful, patient and thorough research assistant throughout the conceptualization and writing of this book, and without whom the book would have appeared more complex and less consistent!

List of Abbreviations

AEON	Africa Earth Observatory Network
AFASA	African Farmers Association of South Africa
AFRA	Association for Rural Advancement
AfrOnet	African Organic Network
AGRA	Alliance for a Green Revolution in Africa
AGRA-MVP	Alliance for a Green Revolution in Africa's Millennium Villages Project
Agri-SCIP	Agricultural Sustainable Community Investment Project
AHA	American Heart Association
AIDS	acquired immunodeficiency syndrome
ALGOA	Asian Local Governments for Organic Agriculture
ANC	African National Congress
ANOVA	analysis of variance
ARC	Agricultural Research Council
AU	African Union
BD	biodynamic
BERAS	Building Ecological Regenerative Agriculture and Societies
BT	*Bacillus thuringiensis*
CAADP	Comprehensive Africa Agriculture Development Programme
CAN	calcium ammonium nitrate
CBO	community based organization
CBTF	Capacity Building Task Force on Trade, Environment and Development
CCA	canonical correspondence analysis
CEC	cation exchange capacity
CERD	Centre for Ecosystems Research and Development
CFS	Committee on World Food Security
CGIAR	Consultative Group for International Agricultural Research
CIRHEP	Centre for Improved Rural Health and Environment Protection
CoE-FS	Centre of Excellence in Food Security
CSA	community supported agriculture
CSO	Community Skills Officer
CSUP	whole-community substrate utilization profiles
DAFF	Department of Agriculture, Forestry and Fisheries (SA)
DARCOF	Danish Research Centre for Organic Farming
DAWASCO	Dar es Salaam Water and Sewerage Corporation
DBM	diamondback moth
DDAC	didecyldimethylammonium chloride

DDS	dietary diversity score
DDT	dichlorodiphenyltrichloroethane
DGP	Diet for a Green Planet
DM	dry matter
DOK	bioDynamic, Organic, Conventional (Swiss research trials)
EAOPS	East African Organic Products Standard
ECOSOLA	ecosystem-based solutions for resilient urban agriculture in Africa
ECRDA	Eastern Cape Rural Development Agency
EOA	Ecological Organic Agriculture [Initiative]
EPOPA	Export Promotion for Organic Products from Africa
ERA	ecological recycling/regenerative agriculture
EU	European Union
FAO	Food and Agriculture Organization of the United Nations
FARA	Forum for Agricultural Research in Africa
FCCC	Framework Convention on Climate Change
FiBL	Research Institute of Organic Agriculture, Switzerland
FISD	Fostering Innovation for Sustainable Development
FPM	fresh produce market
FSG	Farmer Support Group (University of KwaZulu-Natal)
FSR/E	farming systems research and extension
FYM	farmyard manure
GAP	good agricultural practice
GDP	gross domestic product
GE	genetically engineered
GHG	greenhouse gas
GHS	General Household Survey
GIZ	Deutsche Gesellschaft für Internationale Zusammenarbeit
GM	genetically modified
GMO	genetically modified organism
GSL	glucosinolate
HACCP	hazard analysis critical control points
HCC	Hibiscus Coast Co-operative
HIV	human immunodeficiency virus
HLPE	High Level Panel of Experts
HPCSA	Health Professions Council of SA
IAASTD	International Assessment of Agricultural Knowledge, Science and Technology for Development
ICM	integrated catchment management
ICRISAT	International Crops Research Institute for the Semi-Arid Tropics
ICROFS	International Centre for Research into Organic Farming Systems
ICT	information and communication technologies
IDP	Integrated Development Plan
IFAD	International Fund for Agricultural Development
IFOAM	International Federation of Organic Agriculture Movements
IFP	Inkatha Freedom Party
IGAD	Intergovernmental Authority on Development
INOFO	Intercontinental Network of Organic Farmers Organisations
IPCC	Intergovernmental Panel on Climate Change
IPES	International Panel of Experts on Sustainable Food Systems
IPM	integrated pest management
IQR	inter-quartile range
ISOL	International School of Leadership
ISS	Inba Seva Sangam
ITC	isothiocyanates
ITK	indigenous technical knowledge
IWMI	International Water Management Institute

JUCO	Jordan University College
KIOF	Kenyan Institute for Organic Farming
KOAN	Kenyan Organic Agriculture Network
KZN	KwaZulu-Natal
LAHDC	Ladakh Autonomous Hill Development Council, India
LAPC	Land and Agricultural Policy Centre
LC	learning centre
LCA	life cycle assessments
LEHO	Ladakh Environment and Health Organization
LEISA	low external input sustainable agriculture
LSD	least significant difference
LSM	living standards measure
LWCM	large white cabbage moth
MAAIF	Ministry of Agriculture, Animal Industry and Fisheries [Uganda]
MAR	mean annual rainfall
MB	methyl bromide
MDGs	Millenium Development Goals
MITC	methyl isothiocyanate
MoA	Ministry of Agriculture [Zambia]
MOFI	Manyara Organic Farming Initiative
MS	mean square
MS	metam sodium
MVP	Millennium Village Projects
NABARD	National Bank for Agriculture and Rural Development
NAMC	National Agricultural Marketing Commission (SA)
NARS	National Agricultural Research System
NAU	Natal Agricultural Union (now KwaNALU)
NDP	National Development Plan
NGO	non-governmental organization
NMA	Nutrition in Mountain Agro-ecosystems
NNSDP	National Nutrition and Social Development Programme
NOA	Namibian Organic Association
NOAMs	national organic agricultural movements
NOARA	Network of Organic Agricultural Research in Africa
NOGAMU	National Organic Agriculture Movement of Uganda
NOP	National Organic Program (USA)
NPC	non-profit companies
NQF	National Qualifications Framework
NRF	National Research Foundation
OA	organic agriculture
OAAEA	Organic Agriculture Academy for Extension Agents
OAD	Organic African Development
OASA	Organic Agriculture Association of SA
OECD	Organisation for Economic Co-operation and Development
OFC	Organic Foundation Course
OFSP	Organic Food System Programme
OLC	Organic Leadership Course
OPPAZ	Organic Producers and Processors Association of Zambia
OSA	Organic SA
OSASA	Organic Soil Association of SA
OSSIC	Organic Sector Strategy Implementation Committee
PAW	plant available water
PCA	principal component analysis
PGPR	plant growth promoting rhizobacteria
PGS	Participatory Guarantee System
PHA	Philippi Horticultural Association

PRA	participatory rural appraisal
ProGrOV	Productivity and Growth in Organic Value-chains [project]
R&D	research and development
RDA	Rural Development Administration
RDP	Reconstruction and Development Programme
RIFT	rapid incineration field test
RWH	rainwater harvesting
SA	South Africa/South African
SAAU	South African Agricultural Union
SABS	South African Bureau of Standards
SADC	Southern African Development Community
SADHS	South Africa Demographic and Health Survey
SAFEX	South African Futures Exchange
SANAS	South African National Accreditation Service
SAOSO	South African Organic Sector Organisation
SD	standard deviation
SDC	Siyavuna Abalimi Development Centre, KZN
SDGs	Sustainable Development Goals
SETA	Sectoral Education and Training Authority
SFS	Sustainable Food Societies
SFS	sustainable food system (UN)
SME	small-, micro- and medium-scale enterprises
SNRM	School of Natural Resource Management
SOAAN	Sustainable Organic Agriculture Action Network
SOC	soil organic carbon
SOM	soil organic matter
StatsSA	Statistics South Africa
SWC	soil water content
T&V	Training and Visit [system]
TIPI	Training and Innovation Platform of IFOAM
TOAM	Tanzanian Organic Agriculture Movement
UK	United Kingdom
UN	United Nations
UNCTAD	United Nations Conference on Trade and Development
UNEP	United Nations Environment Programme
UNESCO	United Nations Educational, Scientific and Cultural Organization
US(A)	United States (of America)
USAID	United States Agency for International Development
USDA	United States Department of Agriculture
WB	Walkley-Black [method of soil carbon determination]
WB	water balance
WHO	World Health Organization
WWF	World Wide Fund for Nature
WUE	water use efficiency
YAI	Youth in Agri Initiative

Introduction

Raymond Auerbach*
Nelson Mandela University, George, South Africa

Climate change, food insecurity and ongoing urbanization combined with poor governance in many parts of Africa mean that small-scale farmers are not receiving the support they need. Food quality has also fallen over the past century resulting in increasing obesity, stunting and diabetes, and consumers often do not have access to nourishing food. Much of the support which farmers receive is provided by input suppliers who have a vested interest in selling seeds, fertilizers and agrochemicals. The environmental impacts of industrial agriculture are enormous, with carbon, methane and nitrous oxide emissions, nitrates and phosphates in streams and groundwater, and toxins in food and the environment. The cheapest food on offer is often high in salt, hidden sugars and unhealthy fats. Support for public interest research is at an all-time low in agriculture, with most research funded by companies.

The organic agriculture (OA) movement worldwide helps farmers to produce healthy food with low levels of external inputs, and often shortens value chains, giving farmers a higher share of the consumer dollar; it is backed up by Organic Production and Processing standards, an International Organic Accreditation System, an organic training academy and strong national organic agricultural movements (NOAMs).

Agroecology is a broader approach, including certified and non-certified organic farming, conservation agriculture and a range of other 'almost organic' approaches. The benefit of agroecology is that it is easy for small-scale farmers to practise; the disadvantage is that the consumer cannot be certain whether poisons, chemical fertilizers and/or genetically engineered (GE) seeds have been used by those calling themselves agroecological farmers. Adopting an agroecological approach to sustainable development, without excluding too many farmers would appear to make sense, provided that a balance is maintained between ethical and healthy food production and inclusivity. The idea of regenerative agriculture as a broader framework based on ecological principles is explored in Chapter 2 of this book.

On the other hand, there are controversial calls such as those made by the Swiss Research Institute of Organic Agriculture (FiBL), that organic farming should now selectively include some aspects of GE, and other biotechnologies. This may cause some current farmers and consumers to argue that organics has become too conventional and industrial; there are already calls for 'beyond organic' and 'Organic Plus'. Where should the line be drawn between organic and conventional? How can we do this in such a way that we encourage large-scale farmers to

*raymond.auerbach@mandela.ac.za

move towards greater biodiversity, less poison use and more responsible environmental stewardship? How can we help farmers to take charge of their food production processes, and respond to demands from consumers for health-giving nourishment? How can we educate more consumers about nutrition and responsible food production and processing? How can African food systems promote food sovereignty, rather than corporate interests?

Part of the challenge is understanding that food systems are more than just food production and distribution. Concern about resource use and primary food production, agrochemicals and their residues in food and the environment, food processing, food additives, poor cooking and poor food choices, as well as the increasing impacts of poor nutrition on health has seen a shift in focus from 'enough cheap food' to 'the right kind of food produced sustainably and prepared intelligently'. Consumers, policy makers, researchers and natural resource managers are examining alternatives.

This book reports on long-term comparative organic farming systems' research trials carried out over the last 5 years in the Southern Cape of South Africa (SA), on the George Campus of Nelson Mandela University (the 'Mandela Trials'), as well as research into the successes and failures of the organic sector and the technical tools required for sustainable development in SA, Zambia, Uganda and Tanzania.

The trials compare organic and conventional farming systems, and show how, from an initial situation where organic yields were 20% lower than conventional, this yield gap was closed by the third year, once available soil phosphate levels were attended to in the organic treatments. Soil fertility improved under organic management, and microbial biodiversity was greater. Water use efficiency (WUE) and water retention were also greater in the organic farming system, and pests and diseases were effectively controlled using biological products. The trials examine monocropped cabbage and rotated cabbage, sweet potato and cowpea in a complete randomized block design with four replications, split for organic and conventional farming systems. The trials were intended to run for 10 years, but it seems unlikely that funding for this will be available, with my imminent retirement. At least these preliminary African results have confirmed what other longer-term research found in Europe and the USA: organic yields can exceed conventional yields in dry years, but are likely to be a little lower in wet years. Under climate change, this is important to know, but this yield gap can only be closed if scientific and experiential understanding are combined to develop soil fertility and crop rotations which are ecologically appropriate and economically viable, and which are integrated into local culture and food systems.

The impacts of drought, climate change models, practical analysis of actual climate variability, farmer training approaches, soil carbon analysis, participatory guarantee systems, the Zambian organic farming sector (agronomy) and Ugandan organic farmer (training support) are analysed. Approaches to urban and peri-urban food production in Africa are explored. After the world context for organic and regenerative agriculture has been examined, and the conditions needed for supporting farmer innovation through experiential learning processes have been further explored, a sector plan for Southern African organic farming is developed.

Summary of the Book

The big issues are outlined in the first six chapters which comprise Part 1: a theme running through the book is the importance of Earth's thin, fragile outer layer of soil, and the challenges of producing food on a small planet given climate change in the Anthropocene. Chapter 1 gives a context for participatory and sustainable development, and presents an overview of how farmers can progress from sub-subsistence, through subsistence to semi-commercial and perhaps commercial farming systems using agroecology. The development of organic farming and agroecology in Southern Africa is traced, and a conceptual framework for the book is developed. An overview follows of organic and regenerative farming approaches by one of the world leaders of the organic movement and then a reflection on the importance for policy makers of long-term research in developing sustainable farming systems which can produce healthy food without damaging the environment, drawing on long-term research from Switzerland (the DOK trials), Denmark (the work at Aarhus University) and Rodale

Institute (Pennsylvania), and on policy development work from the United Nations Conference on Trade and Development (UNCTAD) and UNEP (United Nations Environmental Programme). This is followed by a review of international farmer training activities by the head of capacity building and training at the International Federation of Organic Agriculture Movements (IFOAM). Chapters 5 and 6 present the food systems approach, first the concepts and then examples from around the world of how this approach has been applied, drawing on systems theory and integrated approaches to human development as a process of 'eco-development' rather than 'ego development', for the good of the planet!

Progress across the world from a tiny 'organic fringe' at the end of the 20th century to what Chapter 2 describes as 'Organic 3.0' – organic farming as a major part of sustainable food systems – will require a range of adaptations, many of which will be controversial. On the one hand, it has to be possible for farmers, large and small, to produce nourishing food while caring for the environment and the people involved in the whole food system, and still making enough profit to sustain them. On the other hand, we have to accept that growing population pressure and climate change will make this more of a challenge, but that biotechnology will offer an array of emerging tools. In this context, Chapter 3 looks at the links between research and policy, while Chapter 4 looks at farmer training.

Given these developments, we can no longer have agriculturalists looking at food production, food technologists looking at food processing, economists looking at food value chains, nutritionists looking at diets, each in isolation, otherwise we will continue the rapid expansion of the medical community looking at declining health, with pandemics of obesity, diabetes, hypertension, autism and cancer ever more prevalent! This book therefore adopts a 'food systems' approach and examines how African food systems are changing, and how they could become more sustainable and healthy, in line with the work of the Centre of Excellence in Food Security (CoE-FS).

This is described in Chapter 7, at the beginning of Part 2 (in which capacity building is discussed). Holistic systems, inclusive participatory approaches, institution building and experiential learning are examined in subsequent chapters. We also report on research into organic food production, farmer training, value chains in SA, and on climate change, helping farmers to manage drought, building soil carbon and the role of organic farming in sequestering carbon in the soil where it is useful, rather than sending it up into the atmosphere and out into the sea to contribute to the ever-warming greenhouse (Chapters 8–14)!

Part 3 moves to Uganda and Zambia, and Chapter 17 shows how farmers can use simple but accurate soil carbon tests to track the changes in soil carbon; this is a major innovation, and could assist farmers in documenting how OA sequesters carbon in the soil where it is useful, removing greenhouse gases (GHGs) in the process. We describe the long-term Mandela Trials on the George Campus of the Nelson Mandela University, comparing organic and conventional farming systems; we look at changing soil fertility, compare yields and soil microbiology, examine WUE in the two systems and develop biological systems of pest and disease control. We present a number of specialized case studies in various fields. Finally, we present ideas on urban food gardens, we make recommendations for land reform and agricultural transformation in SA, and present a strategy for the organic sector in southern and eastern Africa.

Structure of this Book

The book is structured as follows. In the first section (Chapters 1–6), the historical development of organic farming systems is discussed, global issues which confront us are examined, and some concepts are developed showing a progression in small-scale farmer development and how this can be supported with appropriate training and policy. The difference between national food self-sufficiency and household food security is examined, and the organic sector is introduced.

The first six chapters give a global picture, which is then followed by insights into capacity building in times of climate change, describing the likely future scenarios for SA: Chapter 7 deals with the impacts of the two most recent droughts on SA food prices and consumption patterns, and the concept of 'weather shock'. Of necessity, it will be incomplete, as we are at the time of writing still in the grip of the drought in certain parts of the country. The city of Cape Town is in a situation of critical water shortages. Since, given continuing recurrent droughts Cape Town must address food insecurity through peri-urban food production, Chapter 8 shows how to strengthen community participation in local planning using participatory rural appraisal (PRA) in the case of the Philippi Horticultural Area, situated on the Cape Flats Aquifer, from a workshop funded by the German government agency DAAD in 2018. Chapter 9 looks at value chains in the SA fresh produce sector, and ways of increasing participation by small-scale farmers using a case study from the south coast of KwaZulu-Natal (KZN). Chapters 10 and 11 examine the potential for helping small-scale farmers to access high-end markets through Participatory Guarantee Systems (PGS) and smartphone 'apps' which can help to shorten the value chain, and put research and marketing tools in the hands of small-scale farmers.

Chapter 12 then combines three research papers produced for the International Fund for Agricultural Development (IFAD) on drought prediction models, on long-term rainfall patterns in the Eastern Cape, and on a strategy for supporting farmers in that province, where many areas have experienced a drop in rainfall over the summer rainfall production season. For the 7 months from September to March, rainfed crops require at least 500 mm of rain – many areas of the Eastern Cape used to receive more than this on average, but have experienced a decline in rainfall over the past 20 years. What are the implications for sustainable farming systems, and for farmer livelihoods?

Having looked at rainfall, urban gardens and methods of improving urban water and energy use efficiency are examined in Chapter 13, as well as some strategies for improving household food security in the area around George, where the author is based geographically, on the cusp between the Western Cape winter rainfall region and the Eastern Cape's erratic summer rainfall. Chapter 14 examines approaches to farmer training, looking in detail at two case studies on experiential learning (farmers in KZN at the Rainman Landcare Foundation, and agriculture diploma students at Nelson Mandela University). This concludes Part 2 on approaches to sustainable rural development.

Part 3 examines practical support for organic farmers and organic food systems. It starts with two case studies on the well-developed organic sector in Uganda (Chapter 15), and the developing one in Zambia (Chapter 16), including some of the reasons why some farmers who adopted organic farming and took the trouble to become certified organic producers, have 'dis-adopted' organic certification – the reasons for this are explored, and some lessons for the organic sector are drawn from this experience. A simple and accurate method of determining soil carbon is then explained in Chapter 17, which could allow farmers to test their own soil carbon levels.

The focus then shifts to the results of the Mandela Trials on the George Campus of Nelson Mandela University, with Chapter 18 describing the baseline study carried out in 2014 at the start of the Mandela Trials. Chapter 19 compares changes in WUE between organic and conventional farming systems, while Chapter 20 looks at approaches to pest and disease control, and soil fumigation (biological and chemical). Chapter 21 compares soil microbiology in organic and conventional systems, and Chapter 22 compares soil fertility and crop yields in the Mandela Trials.

Part 4 looks ahead and considers how we can up-scale agroecology. It starts with Chapter 23, a study of the work in Tanzania and SA of the ECOSOLA project (a German government-funded consortium of the University of Dar es Salaam, the Carl von Ossietzky University in Oldenburg, Germany and the Nelson Mandela University). Urban food gardens and their current and potential future role are examined in Dar es Salaam (Tanzania), Giyani (SA's Limpopo Province) and George (Western Cape, SA). The likely impacts of climate change are examined and some preliminary broad strategic ideas for town planning and food security for the region are explored.

The final chapter (Chapter 24), written after consultation through the SA Organic Sector Organisation (SAOSO) and PGS-SA during 2018, becomes much more specific, proposing a strategy for developing the organic sector in SA over the next 10 years as an instrument for transformation of the agricultural sector. The dangers of confrontation between large commercial (mainly white) farmers (unwilling sellers in the context of the newly announced policy of land expropriation), who are currently producing much of the food grown in SA, and the politically empowered but economically disempowered majority of effectively landless South Africans, are examined, and some creative solutions are proposed. Ways of marrying the business, marketing and production skills of commercial farmers with the energy and entrepreneurship of emerging farmers are explored, and possibilities for internships and an apprenticeship system are presented.

Part 1

Conceptual and Global Perspectives

1 The Developing Organic Sector in Southern and Eastern Africa: What Have We Learned About Sustainable Development?

Raymond Auerbach*
Nelson Mandela University, George, South Africa

Abstract

This chapter introduces food systems and organic farming, outlines the development of organic farming in Southern Africa, and grapples with the idea of sustainable development; drawing on earlier work of the author entitled *Sustainable Development: Developing What to Sustain Whom?* a conceptual farmer development process is outlined. The author's personal journey of 48 years including organic gardening, commercial organic farming, farming systems research and extension and farmer training, and concluding with lecturing, mentoring, and managing research and policy development teams, throws light on some of the useful and not-so-useful approaches to African development. Comparative analysis work shows that returns on investments in organic farming systems are more positive than returns on high external input 'Green Revolution' approaches. Anti-organic bias in research funding and policy development is traced back to a particular approach to health, nutrition and agricultural production, which favours research and policy supporting use of inputs produced by pharmaceutical, food and agrochemical companies, and shows how their powerful lobbying efforts have skewed research outputs towards high external input systems. The book outlines some alternative approaches based on organic food systems.

Introduction

How can Africans become food secure, faced with the natural resource, climate and governance challenges we have become familiar with over the past two centuries? How can we de-colonize our mindset and build on indigenous technical knowledge (ITK), while also drawing on the best which biotechnology has to offer organic farming?

This book has been written to address these questions in the light of the challenges facing Africa from climate change, political instability, and poor infrastructure and market development, and in the context of broken food and farming systems. It will provide a basis of evidence for policy change and agroecological support to farmers.

A food system is more than a value chain; it includes production systems, geographic localities and cultural traditions, food processing, food preparation, individual food choices and the preparation and consumption of food. This wide range of activities and human choices requires a deep understanding of social, political, economic, cultural and environmental realities, as well as an appreciation of what nourishment requires and how human health may be established and maintained.

*raymond.auerbach@mandela.ac.za

Part 1 of this volume deals with conceptual and global perspectives, and introduces the broader issues. Part 2 deals with capacity building and climate change, building on the global context, and learning from African experience. Part 3 presents evidence on how to support organic farmers, and Part 4 makes strategic suggestions about how to upscale organic farming and organic food systems in Southern Africa.

The Organic Sector in Africa

Sustainable agriculture developed in many parts of Africa, wherever people took the trouble to understand local ecosystems and learn from careful observation, over many years and generations, what the impacts of food production (given a variety of approaches) would be on the health of the local natural resource base. In many countries of Africa, traditional farmers have been using agroecological approaches for centuries, and rainwater harvesting is part of African culture in many countries (Everson *et al.*, 2011). In introducing his book on institutional dynamics and communal grazing, Cousins (1992) points out that many African agropastoral production systems are managed in arid or semi-arid areas and operate at high stocking rates: 'Very often these are characterised by range scientists and extension personnel as "unsustainable" because "overstocking" is leading to irreversible degradation'. Cousins, in the introduction to his book (and many others in the subsequent chapters of that book) argues that in Zimbabwe these high stocking rates are often sustainable 'because herdowners pursue "opportunistic" strategies which are able to track environmental variability over both space and time'. Local farmers know their ecosystem. Kibue and Auerbach (2013) report the same experience with sustainable transhumance for nomadic Maasai cattle owners in northern Kenya and Ethiopia.

Degradation of some soils in Ethiopia has been continuing for decades, but can be countered by intelligent community conservation activities, provided that local institutional dynamics are understood and respected, as shown by the following quotation:

> The 'Tigray Project', as it is often referred to, demonstrates that ecological agricultural practices such as composting, water and soil harvesting, and crop diversification to mirror the diversity of soil conditions can bring benefits to poor farmers, particularly to women-headed families. Among the benefits demonstrated are increased yields and productivity of crops, an improved hydrological cycle with raised water tables and permanent springs, improved soil fertility, rehabilitated degraded lands, increased incomes, increased biodiversity, and increased mitigation and adaptation to climate change.
>
> (Edwards *et al.*, 2006)

Often, pioneer 'champions' of organic farming emerge, who provide leadership for many years, such as John Njorogo of the Kenyan Institute for Organic Farming (KIOF), Zephaniah Phiri Maseko of Zimbabwe, Robert Mazibuko of the Africa Tree Centre near Pietermaritzburg in South Africa (SA) and Dr Ibrahim Abouleish of SEKEM in Egypt, which adopted a biodynamic (BD) approach (SEKEM, 2018). Christoph and Christa Kieckebusch were the BD pioneers north of Windhoek in Namibia in the 1970s and 1980s, and later Manjo Smith brought farmers together and helped to start the Namibian Organic Association (NOA), one of many national organic agricultural movements (NOAMs).

In several African countries, NOAMs have emerged, such as: (i) Kenyan Organic Agriculture Network (KOAN); (ii) National Organic Agriculture Movement of Uganda (NOGAMU), the largest NOAM in Africa; (iii) Tanzanian Organic Agriculture Movement (TOAM); and (iv) Organic Producers and Processors Association of Zambia (OPPAZ). The emergence of AfrOnet (see www.afronet.bio), with at least 16 African NOAMs as members, marks a development of continental networking, which has been accompanied by research networks such as the Network of Organic Agricultural Research in Africa (NOARA). In Ethiopia, Nigeria, Ghana, Rwanda, Malawi and Zimbabwe, NOAMs are also developing.

In SA, the South African Organic Sector Organisation (SAOSO) has recently become an effective organization, and has assisted in particular with the final chapter of this book (a sector plan for organic agriculture (OA) in Southern Africa). Given SA's apartheid history, the land question is both controversial and emotional, and is currently a topic which is endlessly debated in SA media:

- How can access to land be opened up so that food security can be improved?
- How should land access and land ownership be managed?
- Why has redistribution of land been so spectacularly unsuccessful over the past three decades?
- What support do emerging farmers need to allow them to 'emerge' as sustainable, commercially sound businesses?
- Can collective farms become productive and efficient?
- Can traditional land tenure, stewardship of land and security of tenure be incorporated into SA land reform?

In SA, early interactions between the African National Congress (ANC) and the organic movement saw pioneers working in Soweto and many rural areas, trying to understand where science could serve the needs of traditional farming systems. I worked with Benny Khoapa (Black People's Convention), to develop an agricultural model at Adam's Mission Health Centre (KwaZulu-Natal (KZN)), until the Security Police arrested Benny and destroyed the clinic in 1977. I set up a meeting between the Natal Agricultural Union (NAU) and representatives of the ANC including Derek Hanekom, in 1992.

Subsequently, we developed a report for the Land and Agricultural Policy Centre (LAPC) on the future needs of farmer support in post-apartheid SA (Auerbach, 1994a). The White Paper on rural development flowed out of this report. The report recommended an approach which would allow white commercial farmers to continue production, encouraging them to act as trainers, catalysts and mentors for emerging black commercial farmers. It specified that these farmers did not require financial support from government, but that they were important to continuing national food self-sufficiency. Later, then President Thabo Mbeki, argued that if rural poverty was to be addressed, at least 10% of gross domestic product (GDP) needed to be devoted to rural development; he argued that rural poverty could only be countered by such an investment in infrastructure and human capital development.

Many ignored this recommendation, but years later, I attended a ceremony organized by the Forum for Agricultural Research in Africa (FARA) in Accra. At that meeting, the President of the International Fund for Agricultural Development (IFAD) pointed out that 'To farm successfully, women need agricultural resources and inputs, as well as access to rural finance, education, and knowledge. They also need rights to the land they farm and a voice in the decisions that affect their lives' (IFAD, 2013). Later that week, we presented the President of Ghana with an award, after the ministers of Agriculture and of Education reported to us how Ghana had halved poverty and food insecurity: the key intervention was education of farm women, and this was achieved by doubling of the agricultural education budget in Ghana. The Minister of Education (a qualified social worker) spent time with us, and commented that Thabo Mbeki's insights on rural development had inspired them to invest in rural infrastructure and people. FARA formally recognized this achievement during this Agricultural Science Week in 2013 in Accra 'Africa feeding Africa through Science and Technology', with the acknowledgement of progress towards a food secure Ghana. If we understand and respect local institutional dynamics, much can be achieved.

Chapter 2 of this book gives an overview of the development of OA globally, but for the purposes of this book, it needs to be rooted in the context of African development. We need to understand the importance for African food systems when, in the conclusion of his book, organic pioneer Albert Howard (1940) states that the soil is the true capital of nations, 'real, permanent, and independent':

> To utilise and also to safeguard this important possession, the maintenance of fertility is essential. In consideration of soil fertility many things beside agriculture proper are involved – finance, industry, public health, the efficiency of the population, and the future of civilisation.
>
> (Howard, 1940)

Having explained in great technical detail the importance of composting in developing soil fertility, he concludes that 'The restoration and maintenance of soil fertility has become an universal problem' and 'The connection which exists between a fertile soil and healthy crops, healthy animals and, last but not least, healthy human beings must be made known far and wide'.

I had been inspired in 1968 by Lawrence Hills (founder of Ryton Gardens and Organic Centre in Coventry, UK), and by Richard Rodale of the Rodale Insitute in Pennsylvania, USA; both encouraged me to start practical experimentation. The Swiss Organic Research Institute under Urs Niggli started long-term comparative trials 40 years ago, as did Rodale soon after and Niels Halberg in Denmark 10 years later (see the relevant chapters in Raupp *et al.*, 2006). After my apprenticeship in BD gardening in 1972 under Ingrid Adler in SA, and my BD farming apprenticeship from 1973 to 1976 under Alexei de Podolinsky and Dr Andrew Sargood in Australia, I visited Bo Pettersson in Järna, Sweden, and he showed me his long-term comparative trials (which ran from 1958 to 1990 – see Granstedt and Kjellenberg, 1997). All of these initiatives were characterized by a holistic approach to agriculture, food systems, human development and environmental conservation.

This holistic approach was shared by many African philosophers and politicians, including Jan Smuts, who provided a philosophical legacy which saw the SA colonial government in the early part of the 20th century avoid much of the agricultural specialization which characterized the USA and Australia. His book *Holism and Evolution* (Smuts, 1926) recognizes the changes in our view of the natural world which are the natural consequences of understanding three great discoveries: (i) Darwin's publication of the *Origin of Species* (1859); (ii) Becquerel's discovery of radioactivity (1896); and (iii) Einstein's publication of his General Theory of Relativity in 1915. According to Smuts, this implies that 'the stable results of the 19th century science may once more become unstable and uncertain. But the way will be open for ... the future', and:

> a new interpretation of Nature, including, as it does, Matter, Life, and Mind. Matter, Life, and Mind, so far from being discontinuous and disparate, will appear as a more or less connected progressive series of the same great Process. And this Process will be shown to underline and explain the characters of all three, and to give to Evolution, both inorganic and organic, a fundamental continuity which it does not seem to possess according to current scientific and philosophical ideas.
>
> (Smuts, 1926)

So, even before the Second World War, the importance of holistic healthy food systems was understood by some visionaries; another of these was Dr Rudolf Steiner, who gave the founding course on BD farming at Koberwitz in Silesia in 1924, in response to requests from farmers who felt already at that stage that soil fertility was being abused, and the quality of crops produced was declining. Steiner explained that the farm was indeed an 'organism', characterized by subsystems but nevertheless an individuality of a kind, and if the farmer was sensitive, it would be possible to understand how the minerals, soil, annual and perennial plants, animals and human beings could interact as a whole, which would be found to be greater than the sum of its parts. It is this 'agricultural individuality' or organism, which is at the heart of organic farming, and from which the word 'organic' is derived. In addition, Dr Steiner developed a number of herbal preparations and practices which he claimed would revitalize agriculture (Steiner, 1958). The research and the holistic developments in Sweden, especially at Järna, were based on Steiner's approach to social, educational and agricultural development, and the work at Järna is described in some detail in Chapter 6 of this book.

At the heart of BD farming, however, is excellent compost, the preparation of which is an art and a science, and which is given pride of place on most BD farms. This book will examine the importance of colloidal humus in sustainable agricultural production.

The Organic Sector in Southern Africa

Coming to the history of organics in Southern Africa, a few pioneers should be mentioned: Robert Mazibuko and his deep-trench gardens (Bloch, 1998), and Zephaniah Phiri Maseko (Mabeza, 2015), rainwater harvester of Zimbabwe, are two pioneers, both of whom understood the dynamics of water movement through farms and the importance of retaining water through rainwater harvesting. Marie Roux worked with Mazibuko doing early pioneering work on organic urban food gardens in Soweto; her slogan was: 'Don't feed your dustbin, feed your soil!'

This campaign was supported by the Organic Soil Association of SA (OSASA), which was started by Catherine Parnell, who met Lady Eve Balfour in England in the early 1960s and then changed the name of her gardening association to OSASA, according to Essential Herbal Products (2017); I joined OSASA in 1969, and I remember discussing the establishment of the International Federation of Organic Agriculture Movements (IFOAM) in 1972, and it being agreed that Pauline Raphaelly would be the SA delegate to the Versailles Conference. Not long after this, Jeunesse Park set up Trees for Africa (later 'Food and Trees for Africa'), and encouraged tree planting and vegetable gardens in schools across the country. She was a highly successful communicator and an effective fund-raiser, and continues to inspire thousands of people to plant trees and grow organic vegetables.

Elizabeth Wertheim-Aymes was one of the pioneers of BD farming in SA, and she and her husband Guy helped me to travel to Australia for my 4-year BD apprenticeship; Jeanne Malherbe was the grand old lady of BD farming, and she farmed at Bloublommetjieskloof near Wellington in the Western Cape, SA; she was a teacher to many of us who wanted to combine spiritual mindfulness with a holistic approach to farming. I spent time with Jeanne after my year at Hermanus Camphill School (where Ingrid Adler taught me the basics of BD gardening), and her contribution was not only as a farmer but as a cultured human being, with her Huguenot roots and her love of all aspects of South African culture. After this formative year (1972), I started a 4-year apprenticeship in Australia with Alexei de Podolinsky, Dr Andrew Sargood and Australian biodynamic farmers (1973–1976) after which Elizabeth, Jeanne and I continued with the development of the BD Agricultural Association of Southern Africa.

In the early part of the 20th century, Rudolf Steiner developed 'Anthroposophy', an approach to spiritual mindfulness based on Goethe's phenomenological approach to observing 'the open secrets of nature' through a process of stilling the mind and becoming receptive to the miracles of nature. The Waldorf School movement, the Camphill Schools for Children (and later adults) in Need of Special Care, the alternative health movement and the BD movement were all related to a more mindful and holistic approach to health, education and food systems, mostly based on Steiner's ideas (Steiner, 1958). The market at Michael Mount Waldorf School north of Johannesburg became a centre where alternative lifestyles were explored, as did the Constantia and Michael Oak Waldorf Schools in Cape Town.

I attended several markets with Elizabeth Wertheim-Aymes at Michael Mount School in 1972; we would arrive with milk in her 'bakkie' (light pick-up truck) on a Thursday afternoon after school, and park outside the school with several other farmers in their vehicles. Mothers of children at the school (and a few others who were keen to get organic produce) would buy directly from the farmers. Elizabeth farmed near Magaliesburg with her much-beloved herd of dairy cows, and after the visit of Alexei de Podolinsky in 1976 (he and I did a lecture tour around SA), she started making the BD preparations on the farm. Today's Bryanston Organic and Natural Market (see www.bryanstonorganicmarket.co.za) has developed from these humble beginnings, partly under the direction of Konrad Hauptfleisch who was there in the first 12 years of the century, and then moved to Bonn, to develop training for the IFOAM, where he has been ever since (see Chapter 5, this volume).

In Cape Town, Tenjiwe Christina Kaba and Rob Small developed *Abalimi Bezekhya* (Gardeners of the Home) in the township areas of Cape Town. After the establishment of a packshed, a 'box delivery scheme' developed, to supply Capetonians with organic vegetables; this service was suspended during 2018 due to the serious drought.

Attempts to bring together emerging black commercial organic and existing white commercial organic farmers grew out of the pioneering work which Marie Roux did with GROW in Soweto in the 1970s, as part of the OSASA. Robert Mazibuko taught us the basics of participatory rural development in these years; he was one of the first to combine training in agriculture, health and teaching. He was instructed by Father Bernard Huss of Marianhill: 'Your students must learn to produce food, and you must understand how to build classrooms and make desks, establish school vegetable gardens which can survive under drought conditions, and inspire your students to care for body, mind and spirit'. Here was African Holism, inspired by German practical humanity and common sense, taking root in Zulu culture where it felt quite at home.

Mazibuko often told me during my apprenticeship with him in 1976–1977:

> You must help farmers to see God in the soil;
> only when they have reverence for nature will
> they develop the right approach to farming.
> Where you have 'soil erosion' look at the people;
> the cause is usually 'soul erosion' – they have
> forgotten their traditions, forgotten how their
> grandfathers taught them to respect soil,
> wetlands, and healing plants.
>
> (Robert Mazibuko, Africa Tree Centre near Pietermaritzburg, SA, 1977, personal communication)

Baldwin Mbutho, Ferdi Engel and I spent 6 wonderful months with Mazibuko, and he showed us how all cultures teach respect for elders and for nature. The way he explained working respectfully with farmers is the basis of participatory action research where the knowledge of both parties is valued and built upon, rather than demeaning the approach which local people have developed. 'The missionaries' he would say, 'brought us Churchianity; we are still waiting for Christianity to arrive!'

After OSASA gradually died, the Organic Agriculture Association of SA (OASA) was formed, and later 'Organic SA' (OSA). James Moffett and Alan Rosenberg were steady leaders active in these organizations, but there were also other egos, and petty politics led to bickering and decline. More recently, the emergence of SAOSO marked a commitment to bring together the needs of commercial and emerging organic farmers in a new South African context.

After working with Baba Mazibuko, I began to learn from what was happening in the rest of Africa, notably developments on farming systems research, and the work that Gunnar Rundgren with the United Nations Conference on Trade and Development (UNCTAD) was supporting in East Africa. This work saw the development of the East African Organic Product Standard in 2007, officially launched at the East Africa Organic Conference in Dar es Salaam in July 2013. This led to the development of support materials for developing country governments to assist with the emergence of organic sectors, which I report on in Chapter 3 of this book (UNCTAD, 2008), and to the work of the Export Promotion for Organic Products from Africa (EPOPA) (see EPOPA, 2008). Independent of the organic movement, there was progress with sustainable development and work on the relationship between agricultural research and extension activities.

Farming Systems Research and Extension (FSR/E)

Professors Mike Collinson, Peter Hildebrand and Janice Jiggins were pioneers of the FSR/E approach, which attempted in the last 30 years of the 20th century, to put farming systems at the centre of agricultural development, breaking down reductionist single disciplinary approaches, and to build bridges between farmers, extension and research (Collinson, 2000). Janice Jiggins, in particular, made sure that women farmers and professionals were brought into agricultural research and extension processes, and she and extension specialist Niels Röling encouraged me in my FSR/E work in the 1980s and 1990s. By the time Mike Collinson set up the FSR/E work in Eastern and Southern Africa, Collinson and Jiggins were drawing on the pioneering field studies, methodology development and conceptual insights of many others, based on research from the 1960s onwards in West Africa, Asia and Latin America (Collinson, 2000).

I worked with Dr Ponniah Anandajayasekeram of the International Crops Research Institute for the Semi-Arid Tropics (ICRISAT), and with Ted Stilwell of the SA Development Bank, and, at a conference in Arusha, I was elected as the African Representative on the Board of the International Association for FSR/E, of which Professor Jiggins was President. In 1994, I completed my MSc on FSR/E, taking a systems approach to maize production in southern KwaZulu. This helped to develop my understanding of the importance of what FSR/E calls 'research domains' and 'recommendation domains' (Auerbach, 1995). In subsequent discussions with Rob Small of *Abalimi Bezekhaya*, we agreed on a progression we had both noticed in farmer development, and this led to the evolution of a typology of farming systems, which has helped me to plan farmer support.

A Conceptual Model of South African Small-scale Farmer Development

When developmental agencies attempt to assist food producers, there are often several suppositions, which are often not very accurate. First, there is often an assumption that people with a small piece of land (or even a large one) wish to produce surplus food. Second, it is sometimes assumed that they wish to sell this surplus food. Finally, it is often presumed that they wish to become commercial farmers. While it is true that unemployed, resource-poor rural or urban dwellers do need food, and are often keen to produce some food on the land available to them, this is often a survival or even a desperation strategy, and may not be an aspiration at all.

Even for those who are fairly serious about food production, there may be a desire to use the land optimally, and perhaps a cultural desire to be able to share surplus crops or animals with family or neighbours, but many small-scale gardeners feel reluctant to sell their own produce for money, and it is often seen as part of the bounty which is available for the community. This concept of stewardship and sharing is a deep and wonderful part of many African traditions, and embodies '*ubuntu*' (peopleness). It is also a stark reality of small business development that many people who become able to run a business growing food, having learned trading skills, may decide that there are many less risky ways to trade than primary agricultural production.

In discussion with Rob Small, we agreed that my four research domains (sub-subsistence farming, subsistence farming, semi-commercial farming and commercial farming, see Fig. 1.1), represent a progression in skill and understanding which requires experience. In our work with small-scale farmers, we found that moving from each stage to the next usually requires at least 3 years of praxis (actually doing it).

A person can improve their food production skills by learning basic organic techniques such as crop rotation, mulching and the use of compost.

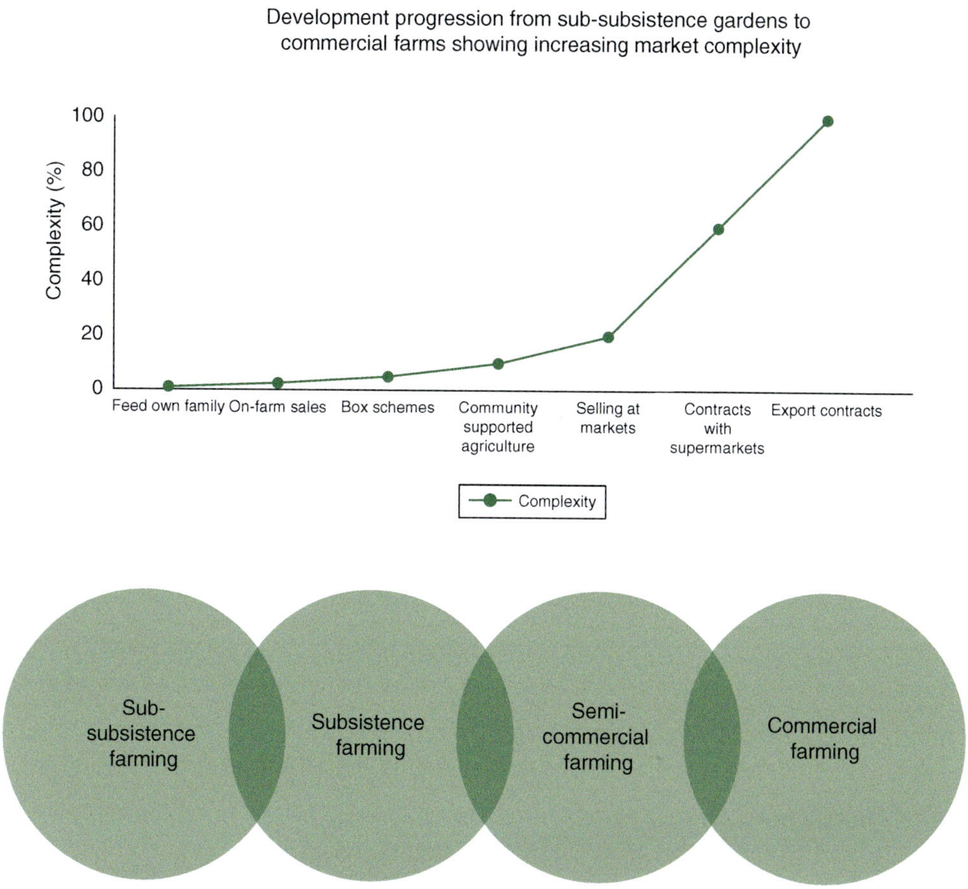

Fig. 1.1. Food production systems, the journey from sub-subsistence to commercial.

They will also need basic husbandry skills for soil preparation, weed control, and the management of pests and diseases. Provision and management of moisture will be critically important. Learning these skills is not technically or intellectually demanding, but it does need a shift from a fatalistic approach (I plant it and it grows if God wills), to an individualistic approach (If I care for the soil and the plant, I am likely to harvest a good crop). This development of a sense of agency is fundamental in helping people to take charge of their lives.

The Intergovernmental Panel on Climate Change (IPCC) agrees:

> Food security in Africa faces multiple threats stemming from entrenched poverty, environmental degradation, rapid urbanization, high population growth rates, and climate change and variability. The intertwined issues of markets and food security have emerged as an important issue in Africa ... Price spikes for globally traded food commodities in 2007–2008 and food price volatility and higher overall food prices in subsequent years have undercut recent gains in food security across Africa.
>
> (IPCC, 2012)

This will be dealt with in Chapter 7 in Part 2 of this volume, and can be counteracted by well-supported small-scale farmer training programmes. After 2 or 3 years of steady application, the small-scale farmer will often find that she has enough fresh vegetables for her children to derive significant nutritional benefits in terms of dietary diversity. After providing for her own family, she begins to find that there is some produce left over to share, and this may become the basis of social exchanges, assistance to those 'less well off', or simply the capacity to provide welcome gifts to friends and family. She may then be ready to move into the subsistence farming system, where she is not quite as desperate, since she is able to provide a lot of food for her family.

Again, she may need a few years to become highly proficient at food production, learning about preserving food, making jams, using her produce in a range of products for her family, and some of which she may find she can sell. This movement from subsistence farming into the market is quite rare, and in each community, one usually finds only a handful of people who become competent subsistence farmers, or where the farming family has developed these skills over generations of farming experience.

The movement into semi-commercial farming requires an expansion of products and the beginning of business planning capacity: what can I produce at a time when there is a market demand, and can I get a good enough price to justify all of my effort? Again, very few subsistence farmers develop sufficiently to become efficient enough to derive major profitable commercial income from farming.

As indicated allegorically in the lower part of Fig. 1.1, if 10,000 sub-subsistence gardeners take part in a farmer development programme, perhaps 1000 will become productive subsistence farmers; of these, perhaps 100 may become semi-commercial farmers with a saleable surplus which brings in some cash.

However, many of these will find less risky outlets for their newly acquired business skills, which is wonderful for the local economy, but not always so good for agricultural development. Of 100 semi-commercial farmers, perhaps one or two may be in a position to intensify, or to access additional land and become commercial farmers.

The upper part of Fig. 1.1 shows how the markets become increasingly complex as gardeners develop into semi-commercial and then commercial farmers. Whereas at the start of this development there is no cash economy involved, gradually business management and marketing become a more major part of the farming system. This increasing complexity requires increasing sophistication of production, as planting schedules must be developed according to what the consumers want, and planning with other suppliers becomes important in order to prevent the market being flooded with too much of certain produce and not enough of others. The farmer is soon running a business which may mean that the food produced is too valuable to give to neighbours!

My experience over the past 45 years has led me to believe that many small-scale farmers, especially women, are reluctant to become commercial farmers partly because of the change in the nature of their relationships with neighbours in the community once they are seen as business people.

Transformation of our rural and peri-urban areas will require sensitive assistance to communities in dealing with feelings of exclusion, and with the commercialization of basic foodstuffs. In 1994 I published an article in the journal *New Ground* entitled 'Sustainable development: developing what to sustain whom?' (Auerbach, 1994b) which I then modified in my doctoral thesis (Auerbach, 1999). As it is relevant in this discussion of conceptual models for SA agriculture, I paraphrase the model below and in Fig. 1.2.

Figure 1.2 summarizes the four common perspectives on rural development which I identified. Agricultural scientists are most comfortable with a production-oriented approach, which is often rather short term and technology centred. This is not to say that national food self-sufficiency is unimportant – it is essential.

However, politicians and social scientists are concerned that the poorer households may not be able to access food if they have to purchase it, and therefore household food security is important if there is to be reasonable equity. Natural resource managers on the other hand, have long been critical of the damage being done to the resource base by industrial agriculture. While their philosophy has always been long term, they were often rather technical in their approach. Over the past 30 years, however, the World Wide Fund for Nature (WWF) has increasingly emphasized the importance of working with communities, if conservation is to become socially sustainable.

So, a tension exists between the technology-centred process of commercial food production, and the people-centred process of household food production; however, both of these perspectives are often short term in nature. Commercial farmers think about 'making a profit this season to repay my production loan', while politicians think about 'helping the people to eat before the next election'!

Natural resource managers think long term, as they work with the conservation of ecosystems, and they have led the way in integrating long-term approaches to the environment, with sensitivity to the contribution of local communities to conservation and a willingness to share resources and economic opportunities with them. As I noted in 1999:

> While most conservationists have learned about people, and social scientists are learning about sustainability, those who research technical

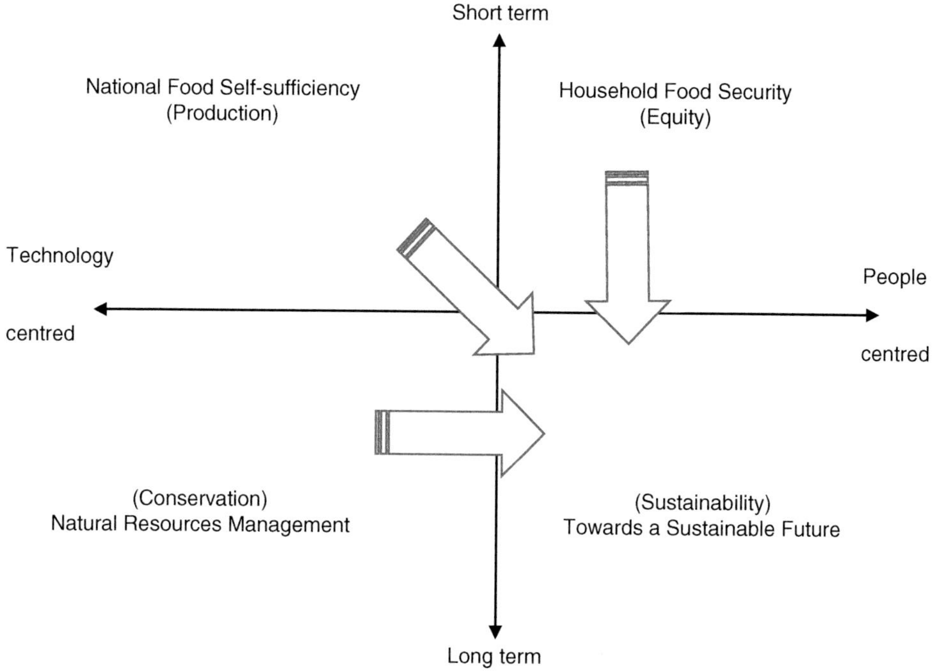

Fig. 1.2. Production, equity, conservation and sustainable development. (From Auerbach, 1999.)

> aspects of agricultural production have to learn about both: their research techniques need to take account of ways of working with people, and of long-term sustainability. This is what is required for 'triple-bottom-line accounting': economics, environment and equity are all important – we cannot sacrifice any of them. We need to find ways to balance this triple equation. We also need to balance short-term individual interests (important to wealth creation and the provision of efficient services) with long-term measures to increase both productivity and equity of access to resources, without damaging our resource base.
>
> (Auerbach, 1999)

Movement towards a sustainable future requires a long-term, people-centred sustainable development perspective, out of which sustainable agroecological farming systems can develop.

Earlier, my MSc research had postulated that interventions to assist farmers will be most successful if they focus on relieving constraints (such as land preparation, access to seed and soil fertility), rather than providing technical advice, and had shown how this can be done practically with maize production in southern KwaZulu (Auerbach, 1995).

In his much-publicized book *The End of Poverty: How We Can Make it Happen in Our Lifetime*, Professor Jeffrey Sachs (2005) postulates that if modern agricultural technology (fertilizer, hybrid seeds, pesticides and mechanization) is combined with interventions on education and health, and made available to African villages, small-scale African farmers will be able to produce a surplus, and by selling this will enter the market economy and improve their livelihoods. This presupposes that there are roads, trucks, agricultural inputs, finance, demand for the crops and a market able to pay for the crops produced. Several critiques of the approach adopted by Sachs claim that it has not worked (Munk, 2013), as the contextual conditions do not simply require technological solutions, but rather human and institutional capacity building. Thus I concluded in my comparative analysis of the work of the Alliance for a Green Revolution on Africa's Millennium Villages Project (AGRA-MVP) and EPOPA, that sustainable development requires a long-term approach to building community participation in agriculture and other aspects of rural development. Resilience, biodiversity, improved productivity and strategies which address soil fertility and water use efficiency (WUE) need to be adapted to local conditions and to robust predictions of the major climate change constraints likely to affect small-scale farming. Capacity building and farmer support are essential in this process.

How this can be done has been the subject of the last 10 years of research of my agroecology group at Nelson Mandela University. Organic farming techniques allow small-scale farmers to use local resources to make compost; this needs to be supplemented by mineral correction of deficient soils, especially where the soil is acidic and low in available phosphate. At the same time, farmer training and capacity building through institutional development is essential for the social and economic dimensions of sustainability.

The dangers of insufficient attention to the building of local institutions and markets was shown by my comparison of the AGRA-MVP and EPOPA projects mentioned above; a histogram from 'Transforming African Agriculture' (Auerbach, 2013a) is reproduced as Fig. 1.3.

Food security, food sovereignty, climate change and food quality: these are four linked topics which all revolve around soil fertility and WUE. However, food systems are also linked to health issues such as diabetes, cancer, obesity and malnutrition, as well as to social justice factors such as household food insecurity, women's access to land, farmers' rights to exchange seed and fair trade access versus dumping of agricultural products; these issues are discussed by my research group in Chapters 18–22 of this book. Food aid, which aims to help those who do not have access to adequate food, often distorts markets in a way which makes it difficult for local farmers to recover from droughts, given the unfair (but well-meaning) competition from free food.

Given this complex international context, helping small-scale farmers in Africa to produce nutritious food while coping with increasingly erratic rainfall and rising temperatures, as well as erratic input supplies and rising prices (especially of energy) becomes challenging (Wilson and Cornell, 2014). Figure 1.2 showed the differences between national food self-sufficiency (enough food for the nation) and household food security (access to sufficient food by resource-poor people) and points out that both are short-term phenomena; conservation takes a longer-term approach, but like national food self-sufficiency,

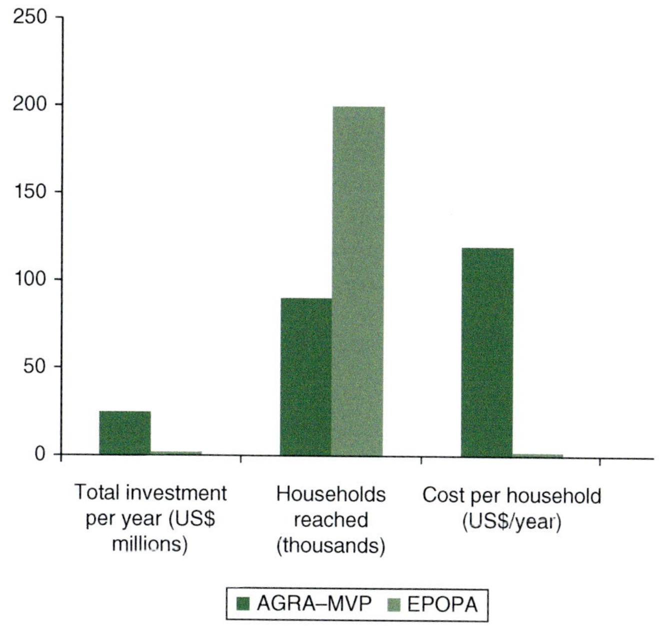

Fig. 1.3. A comparison of the performance of the Alliance for a Green Revolution in Africa's Millennium Village Project (AGRA-MVP) and the Export Promotion for Organic Products from Africa (EPOPA), with regard to scale of investment ($30 million vs $2 million), number of households reached and cost ($120 vs $2) per household per year. (From Auerbach, 2013a.)

it has mostly operated from a technology-centred approach. Recent changes in conservation strategy have seen a more people-centred approach involving local communities in the management of conservation areas, and many short-term household food security practitioners now try to incorporate longer-term resource conservation strategies.

Organic farmers around the world have shown that it is possible to use locally available resources to develop sustainable farming systems which produce nutritious food while improving the soil and environment, as shown in Chapters 5 and 6 of this book. Using less poison, less non-solar energy, less externally purchased inputs, they produce reasonable yields of high quality produce with less impact on the environment. The water- and nutrient-holding capacity of the soil is improved, agrobiodiversity is enhanced, and dependency can be avoided by developing systems which do not rely on bringing in synthetic fertilizers, agrochemicals, genetically modified seeds and levels of industrial processing which negatively affect the quality of the food produced and of the environment.

In Europe and the USA there has been considerable progress in long-term research into organic farming systems. Notably, the long-term comparative trials in Switzerland showed that over a 34-year research period, comparative organic yield levels are about 80% of conventional yields, with lower levels of inputs and higher nutritional quality on 11 parameters (Mäder *et al.*, 2006). Similarly, in Denmark, the ICROFS (International Centre for Research into Organic Farming Systems) trials at Aarhus University obtained similar results over a similar period, also with a yield gap between organic and conventional of about 20% (Rasmussen *et al.*, 2006). In the USA, the Rodale Institute in Pennsylvania has been able to close this yield gap with a number of crops attaining very similar production levels (organic and conventional) in their long-term trials (Hepperly *et al.*, 2006). The Rodale trials found that organic systems tend to out-yield conventional systems in dry seasons, where (as in many parts of Africa) irrigation is not available. All three of these long-running sets of trials (Switzerland, Denmark and the USA) found that organic systems require 3–4 years to build up soil biology to productive levels. Many researchers have pointed to the need to change the way we produce food to reduce environmental impacts, use resources more effectively and develop sustainable food systems. On the other hand, many agricultural scientists dispute the capacity of organic farming systems to 'feed the world', claiming that organic farming has much lower yields and therefore requires much more land; they also argue that more water is needed and that more carbon is emitted by organic systems than conventional systems.

Climate Change and Food Security in Africa

Under climate change, by 2050 nearly 10 billion people will require food security in a world 1.5°C warmer than today; in Africa by this date, probably more than 2 billion people will be competing for increasingly scarce water resources on this continent, and even with the 1.5°C warmer scenario, this will make rainfed crop production highly problematic in many areas.

According to reports of the SR15 meeting of the IPCC in South Korea in *The Guardian* international edition, global temperature increases are likely to reach 2°C by the end of the century (Watts, 2018). Attempts are still underway to limit global warming to 1.5°C, which may avoid the most dramatic effects of climate change, which would include widespread food scarcity and climate-related poverty for hundreds of millions of people.

> At 2°C extremely hot days, such as those experienced in the northern hemisphere this summer, would become more severe and common, increasing heat-related deaths and causing more forest fires. But the greatest difference would be to nature. Insects, which are vital for pollination of crops, and plants are almost twice as likely to lose half their habitat at 2°C compared with 1.5°C. Corals would be 99% lost at the higher of the two temperatures, but more than 10% have a chance of surviving if the lower target is reached.
>
> (Watts, 2018)

The Guardian article reported on the IPCC Report which was commissioned by policy makers at the Paris Climate Talks in 2016. It quoted Dr Debra Roberts (Head of Durban's Environmental Planning and Climate Protection Department, and Co-chair of the IPCC Working Group on Impacts of Climate Change) as saying that scientists who reviewed the 6000 works referenced in the report, said the change caused by just half a degree came as a revelation: 'We can see there is a difference and it's substantial,' Roberts said. The IPCC warns in the same report that carbon pollution would have to be cut by 45% by 2030 to maintain warming at only 1.5°C – if only cut by 20%, the result will be 2°C of warming – and come down to zero by 2050. 'At the current level of commitments, the world is on course for a disastrous 3°C of warming. The report authors are refusing to accept defeat, believing the increasingly visible damage caused by climate change will shift opinion their way' (Watts, 2018).

In acknowledging the recent Climate Research award by the AfriCAN Climate Consortium, Roberts commented on the eThekwini Municipal website:

> This award is further confirmation of the leadership role that eThekwini Municipality plays in the global debate regarding how urban areas – especially those in Africa – should respond to the challenges of climate change. EThekwini Municipality has gone from a passive consumer of climate change data to an effective co-producer of climate change adaptation knowledge. This new type of knowledge is critical as it is capable of driving innovative, local level action that improves the quality of lives of people, ensures the sustainability of ecosystems and increases the long-term resilience of the city.
>
> (Mthethwa, 2018)

Roberts added:

> this is confirmation of my belief that cities around the world are now generating a new breed of local government official, officials that are capable of bridging the gap between science and grass roots action. This emerging cadre of scientifically-inclined practitioners should be encouraged and developed if cities want to remain globally competitive and reap the benefits of a more environmentally aware post 2015 development path.
>
> (Mthethwa, 2018)

African cities, such as Durban and Dar es Salaam, will need to plan for: (i) temperatures well above 40°C; (ii) water scarcity which by 2100 will be worse with temperatures likely to be 2°C warmer; and (iii) the possibly catastrophic scenarios which are predicted to go with feeding over 4 billion people in Africa given these climate changes (M.J. de Wit, Director of the Africa Earth Observatory Network (AEON), Nelson Mandela University, Port Elizabeth, 2018, personal communication). De Wit and Stankiewicz reported on their research findings in the journal *Science* (2006) in an article entitled 'Changes in surface water supply across Africa with predicted climate change', predicting that perennial drainage density (the amount of runoff generating perennial streams) will decrease sharply with climate change. They point out that in higher rainfall regimes (mean annual rainfall (MAR) of about 1000 mm), given a 10% decrease in average precipitation, the decrease in drainage would already be about 17%, but if the same percentage decrease occurs in drier areas (MAR of about 500 mm), the decrease in drainage density would be about 50%.

The AEON database used in this research includes all rivers and lakes in Africa, and actually measured and ordered perennial and non-perennial rivers and streams in Africa, and found that 75% of them are already unstable (de Wit and Stankiewicz, 2006). Their analysis showed that when MAR is less than 400 mm, there is no perennial drainage except in mountainous areas. Data presented in Chapter 12 of this book shows how MAR for many towns in SA's Eastern Cape province decreased over the past 20 years from 550 mm to 430 mm. This means that runoff (or drainage) will be almost nothing in many of these areas in the near future. The implication is that in drier years, some of Africa's major cities may completely run out of water, with all the suffering which accompanies such an event. This can be seen for Cape Town, where this could happen as a result of a combination of climate change, population growth and poor planning and infrastructure management (see Chapter 7, this volume). Sukhmani Mantel's beautiful map (Fig. 1.4) shows the main river systems of SA, illustrating how the Orange River catchment extends all the way through Lesotho, and drains much of the eastern parts of SA. Maarten de Wit points out that the impacts of decreased perennial drainage density will see a dramatic reduction in the amount of irrigation water available in many areas of SA (M.J. de Wit, Director of AEON, Nelson Mandela University, Port Elizabeth, 2018, personal communication).

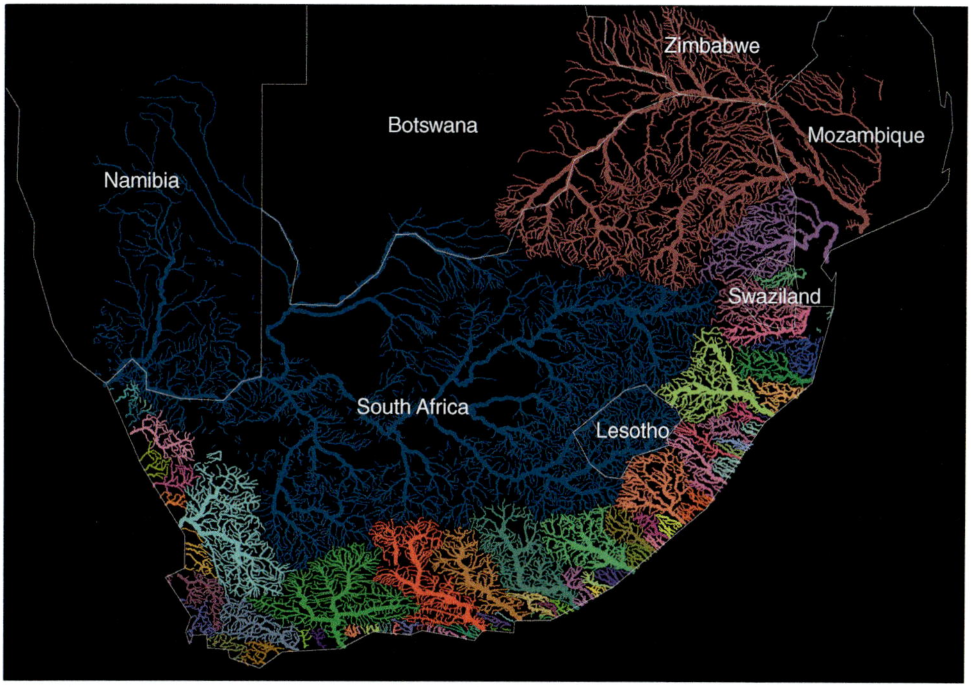

Fig. 1.4. The river systems of South Africa. (Copyright Sukhmani Mantel; used with permission.)

Farmers will need to find ways of using water more efficiently, and of improving the water holding capacity of soil. Most of the freshwater available in SA is used for the irrigation of arable lands, and it is often used rather inefficiently. Improving WUE is a major theme of this book, and research results will show that organic farming systems can help with this much-needed development. To evaluate the role of organic farming systems in helping Southern Africa to meet these challenges, long-term trials were established in 2014. The study commenced with a baseline study (Chapter 18, this volume) to evaluate soil fertility and plant growth on the site where these trials were located (George Campus of Nelson Mandela University, in SA's Southern Cape), after which the first year of the comparative trials was planted in November 2014 (Mashele and Auerbach, 2016). The Mandela Trials have now completed 4 years after the baseline trials, and already the third year data shows good improvement in WUE and soil fertility, and a closing of the yield gap (Chapters 19 and 22, this volume). The trial site is shown in Fig. 1.5.

The research shows (Chapters 18–22, this volume) that organic farming brings about certain improvements (higher soil organic matter, lower soil acidity, better soil biology, aerobic soil conditions, greater WUE, improved soil water retention); while in wet years, low external input organic systems often yield 10–20% less than conventional systems, in dry years these systems often outyield the high-input high-risk conventional systems.

Science Orthodoxy, Food Systems and Farming

We will present evidence from our research over the past 10 years, and make recommendations in the last two chapters of this book, but when considering evidence-based policy related to food systems and farming, it is well to pause for a moment, and learn from history. What is healthy food? What is a healthy diet? What evidence is considered acceptable, and who is heard?

I wish to consider briefly three controversial recommendations related, first to diet, then to seed breeding, and finally back to diet. The first recommendation is that of the American Heart Association (AHA), based on the Seven Countries Study led by Ancel Keys. According to Ilbury (2017), the AHA recommended for many years

Fig. 1.5. The research trial site showing cabbages from the control treatment (left front, no fertilizer), conventional treatment (back left, fertilizer, no mulch) and organic treatment (right front, compost and mulch) with sweet potatoes behind and cowpeas back right (reported in detail in Chapters 18–22, this volume).

that low fat, high carbohydrate diets would help to reduce the incidence of heart disease. He quotes numerous researchers, including Nina Teicholz (2015), writing in the *British Medical Journal* (BMJ), entitled 'The scientific report guiding the US dietary guidelines: is it scientific?' The report questions both the methods and the conclusions of the study, and found the study to be faulty and prejudiced. In response the Center for the Science of Public Interest, based in Washington, DC, issued a letter demanding that BMJ retract the article by Teicholz. After an extensive investigation, the BMJ issued a statement in December 2016 saying that, after further critical evaluation, they stood by her article, and would not retract it.

In similar vein, in 2013, I found myself attacked by various agricultural scientists when I criticized the way Monsanto, Syngenta, DuPont and Bayer are dominating world seed production, and interfering with the rights of farmers to re-plant their seeds, and to exchange seeds with other farmers (Lombard, 2013). A number of unfounded claims about what the scientific literature says about the quality of organic food were made, and my statements were dismissed as 'absolute rubbish'. Eventually, I was able to set the record straight about what the literature does say about organic food, showing that Lombard had misrepresented the findings of the study by the French AFFSA (Agence Française de Sécurité Sanitaire des Aliments – French Agency for Food Safety) (Auerbach, 2013b). Emily Walz (2009) published a note on scientific bias in *Nature*, reporting on several papers which had been published and which were critical of genetic engineering, among others, papers by Chapela and later by Rosi-Marshall; she comments that those who:

> suggest that biotech crops might have harmful environmental effects are learning to expect attacks of a different kind. These strikes are launched from within the scientific community and can sometimes be emotional and personal; heated rhetoric that dismisses papers and can even, as in Rosi-Marshall's case, accuse scientists of misconduct. 'The response we got – it went through your jugular,' says Rosi-Marshall.

Attacks on scientific orthodoxy are not to be made lightly, especially where powerful vested interests are involved!

Returning to dietary recommendations, and the book by Daryl Ilbury (2017): Ilbury reports on A-rated SA scientist Professor Tim Noakes, who was recently hauled before the Health Professions Council of SA (HPCSA), on a charge of unprofessional conduct, arising from his recommendations that babies thrive on high protein diets, and that high fat, high protein, low carbohydrate diets are healthy, and can help those suffering from type 2 diabetes. The HPCSA spent 2 years trying to prove that the recommendations arising from the Seven Countries Study were correct, and that Tim Noakes was being irresponsible in giving different advice. After the extensive hearing, Advocate Joan Adams found Tim Noakes not guilty of unprofessional conduct. Ilbury (2017) quotes the Association of Dietetics (which had lodged the complaint against Noakes) saying 'We accept the verdict and ... We welcome the precedent that this case provides on what we had considered unconventional advice'. Unless unscientific studies are challenged as Nina Teicholz did in the case of the the Seven Countries Study led by Ancel Keys, the majority of scientists will accept the recommendations of industry-funded research, whether it is scientifically valid or not.

The three examples cited above are given to show that scientific orthodoxy may be driven by vested interests, and may take unpleasant steps to support poor science. On the other hand, it is also true to say that unconventional scientific opinions can be based on very flimsy or anecdotal evidence. Robust science should examine evidence, and should test and re-test claims made in a range of contexts. Thomas Kuhn (1962) points out that for a shift in a given conventionally accepted scientific paradigm to take place, considerable evidence may be presented and ignored until the weight of the evidence becomes sufficient to convince a significant proportion of respected scientists that the 'new orthodoxy' is in fact supported by solid evidence!

In presenting evidence in favour of organic food and farming systems, I believe that agricultural and nutrition science has come to just such a point where a paradigm shift is possible and indeed, where it is sorely needed, as the dominant paradigm is contributing to climate change, poor health and dependency in many countries.

Conclusion

Technology alone (such as synthetic fertilizers, agrochemicals, genetically modified seeds and irrigation) will not solve the problems of African food insecurity; it will not (as Jeffrey Sachs claims) 'End poverty in our time' (Sachs, 2005). What is required is systematic capacity building of farmer institutions with farmer training using agroecological or 'low external input sustainable agriculture' (LEISA) approaches. At the same time, it is vital to build market linkages, and to create consumer awareness of the importance of dietary diversity. A LEISA approach will assist in increasing dietary diversity, and agroecology will support improved agrobiodiversity, improved WUE and food sovereignty. Household food security will improve as small-scale farmers move to efficient subsistence farming, and the rural economies will start to develop as they move into semi-commercial farming, and a market economy. Government policies for sustainable rural development will have to understand these inter-related, complex truths if they are to contribute to sustainable development.

The stark realities of climate change in which African farmers find themselves will require a range of approaches to be developed including organic, agroecological, conservation and biological farming. As will be shown in Chapter 7, drought has dramatic impacts on food availability and price; Chapter 12 shows how seasonal rainfall in the Eastern Cape has already decreased in many cases from 550 mm (enough to support most rainfed crops) to 430 mm with a high likelihood of crop failure. Professor Maarten de Wit of AEON, explains that a 10% decrease in precipitation in a high rainfall area would result in a decrease of about 17% in drainage, whereas a 10% decrease in an area with a mean annual rainfall of about 500 mm would cut surface drainage by 50% (de Wit and Stankiewicz, 2006).This would also likely result in decreases in forest species richness, as has happened in north-west Senegal.

What will help African farmers most will be strategies which allow them to use local resources to adapt to the changing conditions without becoming dependent on chemical fertilizers and poisons; in turn, this will help African consumers to rear healthy children even when their financial resources are modest.

References

Auerbach, R. (1994a) *Final Report on Transformation of the South African Department of Agriculture*. Proceedings of national workshop and summaries of provincial workshops and stakeholder group consultations. Land and Agriculture Policy Centre, Johannesburg and Farmer Support Group, Pietermaritzburg, South Africa.

Auerbach, R. (1994b) Sustainable development: developing what to sustain whom? *New Ground* (Johannesburg, South Africa) 15, 3–9.

Auerbach, R. (1995) People, farming and research: how can South African science contribute to sustainable agricultural development? *South African Journal of Science* 91, 3–6.

Auerbach, R. (1999) Design for participation in ecologically sound management of South Africa's Mlazi River catchment. PhD thesis, Wageningen Agricultural University, Wageningen, the Netherlands.

Auerbach, R. (2013a) Organics and AGRA: transforming African agriculture. In: Auerbach, R., Rundgren, G. and Scialabba, N. (eds) *Organic Agriculture: African Experiences in Resilience and Sustainability*. Food and Agricultural Organization of the United Nations (FAO), Rome, pp. 16–37.

Auerbach, R. (2013b) Debate on organic vs GM foods. *George Herald*, 1 August. Available at: https://www.georgeherald.com/News/Article/AgriEco/opinion-debate-on-organic-vs-gm-foods-20170711 (accessed 29 July 2018).

Bloch, J. (1998) The Tree Man: Robert Mazibuko's Story. Available at: https://live.fundza.mobi/home/library/non-fiction-articles-profiles/the-tree-man-robert-mazibukos-story/ (accessed 28 May 2018).

Collinson, M. (ed.) (2000) *A History of Farming Systems Research*. Food and Agricultural Organization of the United Nations (FAO), Rome and CAB International, Wallingford, UK.

Cousins, B. (ed.) (1992) *Institutional Dynamics in Communal Grazing Regimes in Southern Africa*. Proceedings of a workshop held at the University of Zimbabwe, 10–12 December 1990. Centre for Applied Social Sciences, Harare.

De Wit, M.J. and Stankiewicz, J. (2006) Changes in surface water supply across Africa with predicted climate change. *Science* 311, 1917–1921. https://doi.org/10.1126/science.1119929

Edwards, S., Egziabher, T.B.G. and Araya, H. (2006) Successes and Challenges in Ecological Agriculture: Experiences from Tigray in Ethiopia. International Federation of Organic Agriculture Movements (IFOAM). Available at: https://www.ifoam.bio/sites/default/files/page/files/much_greater_resilience_.pdf (accessed 25 May 2018).

EPOPA (Export Promotion of Organic Products from Africa) (2008) *Organic Exports: a Way to a Better Life?* Agro Eco BV and Grolink aB, Hoeje, Sweden.

Essential Herbal Products (2017) Available at: https://essentialherbs.co.za/2017/01/17/organic-agriculture-in-south-africa-and-its-pioneers-part-4-of-whats-all-the-fuss-about-organic/ (accessed 25 May 2018).

Everson, C., Everson, T.M., Modi, A.T., Csiwila, D., Fanadzo, M., Naiken, V., Auerbach, R.M.B. and Moodley, M. (2011) *Sustainable Techniques and Practices for Water Harvesting and Conservation*. Water Research Commission, Pretoria.

Granstedt, A. and Kjellenberg, L. (1997) Long-term field experiment in Sweden: effects of organic and inorganic fertilizers on soil fertility and crop quality. In: Lockerets, W. (ed.) *Proceedings of an International Conference on Agricultural Production and Nutrition*, 19–21 March, Tufts University, Medford, Massachusetts. Tufts University School of Nutrition and Policy, Boston, Massachusetts, pp. 83–111.

IFAD (International Fund for Agricultural Development) and WFP (World Food Programme) (2013) *The State of Food Insecurity in the World 2013. The Multiple Dimensions of Food Security*. Food and Agricultural Organization of the United Nations (FAO), Rome. Available at: http://www.fao.org/docrep/018/i3434e/i3434e.pdf (accessed 9 June 2018).

Hepperly, P.R., Douds, D. and Seidel, R. (2006) The Rodale Institute Farming Systems Trial 1981 to 2005: Long-term analysis of organic and conventional maize and soybean cropping systems. In: Raupp, J., Pekrun, C., Oltmanns, M. and Kopke, U. (eds) *Long-term Field Experiments in Organic Farming*. International Society of Organic Agriculture Research (ISOFAR), Berlin, pp. 15–31.

Howard, A. (1940) *An Agricultural Testament*. Republished 1979. Rodale Press, Emmaus, Pennsylvania.

Ilbury, D. (2017) *Tim Noakes: the Quiet Maverick*. Penguin, London.

IPCC (Intergovernmental Panel on Climate Change) (2012) Chapter 22 of Assessment Report AR5 of 2012. Available at: https://www.ipcc.ch/pdf/assessment-report/ar5/wg2/WGIIAR5-Chap22_FINAL.pdf (accessed 3 June 2018).

Kibue, M. and Auerbach, R. (2013) Keekonyokie: adding value to Maasai beef. In: Auerbach, R., Rundgren, G. and Scialabba, N. (eds) *Organic Agriculture: African Experiences in Resilience and Sustainability*. Food and Agricultural Organization of the United Nations (FAO), Rome, pp. 80–97.

Kuhn, T. (1962) *The Structure of Scientific Revolutions*. University of Chicago, Chicago, Illinois.

Lombard, H. (2013) Organic food not healthier. *AgriEco* 4 July. Available at: http://www.agrieco.net/article.aspx?id=608 (accessed 29 July 2018).

Mabeza, C. (2015) Marrying water and soil: adaptation to climate by a smallholder farmer in Zvishavane, Rural Zimbabwe. PhD thesis on Zephaniah Phiri Maseko, University of Cape Town, Cape Town, South Africa.

Mäder, P., Flieβbach, A., Dubois, D., Gunst, L., Jossi, W., Widmer, F., Oberson, A., Frossard, E., Oehl, F., Wiemken, A., Gattinger, A. and Niggli, U. (2006) The DOK experiment (Switzerland). In: Raupp, J., Pekrun, C., Oltmanns, M. and Kopke, U. (eds) *Long-term Field Experiments in Organic Farming*. International Society of Organic Agriculture Research (ISOFAR), Berlin, pp. 41–58.

Mashele, J. and Auerbach R. (2016) Evaluating crop yields, crop quality and soil fertility from organic and conventional farming systems in South Africa's Southern Cape. *South African Journal of Geology* 119, 25–32.

Mthethwa, T. (2018) Climate Change Award for Dr Debra Roberts. Available at: http://www.durban.gov.za/Resource_Centre/Press_Releases/Pages/Climate-Change-Award-for-Dr-Debra-Roberts.aspx (accessed 20 October 2018).

Munk, N. (2013) The Idealist: Jeffrey Sachs and the Quest to End Poverty. McClelland and Stewart, Toronto, Canada.

Rasmussen, I.A., Askegaard, M. and Olesen, J.E. (2006) The Danish organic crop rotation experiment for cereal production 1997–2004. In: Raupp, J., Pekrun, C., Oltmanns, M. and Kopke, U. (eds) *Long-term Field Experiments in Organic Farming*. International Society of Organic Agriculture Research (ISOFAR), Berlin, pp. 117–134.

Raupp, J., Pekrun, C., Oltmanns, M. and Kopke, U. (eds) (2006) *Long-term Field Experiments in Organic Farming*. ISOFAR (International Society of Organic Agriculture Research), Koster Verlag, Berlin, pp. 41–58.

Sachs, J. (2005) *The End of Poverty: How We Can Make it Happen in Our Lifetime*. Penguin, London.
SEKEM (2018) The Founder of SEKEM, Dr Ibrahim Abouleish. Available at: http://www.sekem.com/en/about/founders/dr-ibrahim-abouleish/ (accessed 3 June 2018).
Smuts, J.C. (1926) *Holism and Evolution*. Republished 1987. N & S Press, Cape Town, South Africa.
Steiner, R. (1958) *Agriculture*. A course of eight lectures given at Koberwitz, Silesia in 1924. English translation by G. Adams. BioDynamic Agricultural Association, London.
Teicholz, N. (2015) The scientific report guiding the US dietary guidelines: is it scientific? *British Medical Journal* 351, 4962.
UNCTAD (United Nations Conference on Trade and Development) (2008) Best Practices for Organic Policy: What Developing Country Governments Can Do to Promote the Organic Agriculture Sector. United Nations Environment Programme. Available at: http://unctad.org/en/Docs/ditcted20073_en.pdf (accessed 12 March 2018).
Walz, E. (2009) Scientific bias. *Nature* 461, 27–32. https://doi.org/10.1038/461027a
Watts, J. (2018) We have 12 years to limit climate change catastrophe, warns UN. Report in *The Guardian*, international edition on 8 October. Available at: https://www.theguardian.com/environment/2018/oct/08/global-warming-must-not-exceed-15c-warns-landmark-un-report (accessed 10 October 2018).
Wilson, F. and Cornell, V. (2014) Overcoming Poverty and Inequality: Guide to Carnegie 3 Conference. SALDRU, University of Cape Town. Available at: http://www.carnegie3.org.za/docs/Carnegie3_April13_WEB.pdf (accessed 12 March 2018).

2 An Overview of Global Organic and Regenerative Agriculture Movements

Andre Leu*
Regeneration International, Minneapolis, Minnesota, USA

Abstract
This chapter gives an historical overview of the organic agriculture (OA) sector and the closely aligned production systems of agroecology, permaculture and regenerative agriculture. It explains how the modern organic movement emerged from concern over the loss of quality in crops with an increase in diseases and pest attacks affecting yields in the later part of the 1800s in Europe and the USA, after the introduction of chemical fertilizers. The researchers and farmers involved in the movements that would lead to modern OA believed that there was a direct relationship between the health of the soil, of the crops that were grown in it and of the animals and people who consumed these crops. The formal international organic movement began in France on 5 November 1972 at a meeting at Versailles and formed the International Federation of Organic Agriculture Movements (IFOAM). It is the global umbrella organization that sets the international standards, policies, definitions and positions around the multi-functionality of organic agriculture through consulting with its members; these cover the whole spectrum of the sector in most countries in the world. Consequently, IFOAM documents are seen as credible source texts for reference material. Most of the world's organic standards and certification systems are based directly or indirectly on IFOAM's Standards. IFOAM put forward the concept of Organic 3.0 in 2013 to enable a widespread uptake of truly sustainable farming systems and markets based on organic principles and imbued with a culture of innovation, of progressive improvement towards best practice, of transparent integrity, of inclusive collaboration, of holistic systems, and of true value pricing. A rapidly emerging sector is regenerative agriculture. It is a concept of agricultural systems improving the resources they use, rather than destroying or depleting them. It is a holistic systems approach to farming that encourages continual innovation for environmental, social, economic and spiritual well-being. It is closely aligned with the concept of Organic 3.0.

Introduction

Organic agriculture (OA) works with ecological systems to produce multi-functional benefits and avoids the use of inputs with adverse effects such as toxic synthetic pesticides. The International Federation of Organic Agriculture Movements (IFOAM)-Organics International, the only global umbrella body for the organic sector, with around 1000 affiliated organizations in 130 countries, developed a consensus definition of OA which was adopted in 2008, clearly showing that organic systems are based on environmental, economic, social and cultural sustainability by working with ecological sciences, natural cycles and people:

> OA is a production system that sustains the health of soils, ecosystems and people. It relies

*andreleu.al@gmail.com

> on ecological processes, biodiversity and cycles adapted to local conditions, rather than the use of inputs with adverse effects. OA combines tradition, innovation and science to benefit the shared environment and promote fair relationships and a good quality of life for all involved.
> (IFOAM, 2008)

The global value of certified organic market sales was estimated to be US$90 billion in 2016 compared to US$62.9 billion globally in 2011 and US$54.9 billion in 2009. The comparison with global organic sales of US$33.2 billion in 2005 and US$15.2 billion in 1999 shows a consistent trend of a high rate of growth (Willer, 2013; Willer and Lernoud, 2018).

The History of Organic Agriculture

The name '*organic*' was popularized by J.I. Rodale in the 1940s.

Jerome Irving Rodale was the first major international author and publisher of books and magazines on organic farming. His primary magazine was called *Organic Farming and Gardening* and was first published in 1942. It was based in the USA; however, this publication was widely read by many thousands of people around the world. He actively promoted the name '*organic farming*' in this and other publications and the name '*organic*' quickly dominated over the numerous other names like natural, permanent and ecological that were being used at the time to describe alternative farming systems.

The word organic has several meanings in the dictionary, with many people taking the dictionary definitions of 'Organised or systematic or co-ordinated' (OUP, 1970) as the context for the use of the word in organic farming. Others take the term to refer to a 'living organism', as mentioned in Chapter 1 of this volume.

Rodale's use of the term 'organic farming' was specific to the farming system's use of organic matter as the primary source of soil health and plant nutrition in contrast to the use of synthetic chemical fertilizers in conventional farming. Rodale repeatedly stated that the fundamental basis of organic farming was to improve soil health and build up colloidal humus through a variety of practices that recycled organic matter (Rodale Institute, 2018).

The OA movement arose over the concern over the loss of quality in crops with an increase in diseases and pest attacks affecting yields in the later part of the 1800s in Europe and the USA, after the introduction of chemical fertilizers.

These chemical fertilizers were based on the published research of Baron Justus von Liebig, in Germany, and John Bennet Lawes, in England, in the 1840s. Von Liebig was the first modern chemist to look at plant growth in a laboratory. He determined that plants needed minerals from the soil and carbon dioxide (CO_2) from the air. He showed that although plants are surrounded by nitrogen in the air, they needed nitrogen in the form of ammonia (which they take up through their roots). Von Liebig stated that nitrogen was the most important mineral and proved that synthetic chemical fertilizers could replace natural ones such as animal manures as the source of nitrogen (R.M.B. Auerbach, 2017, unpublished lecture notes 'Systems and technologies for sustainable agriculture', Stellenbosch University).

Because many of his experiments were done in a laboratory in sand and other soil-less potting media that did not contain humus and organic matter, von Liebig believed that humus did not have a significant role in plant nutrition. He believed that plants only needed minerals in certain types of water-soluble chemical forms and in the correct ratios. Von Liebig did produce a phosphate fertilizer, but it was ineffective.

R.M.B. Auerbach comments:

> It took the work of the practical farmer, Sir John Bennet Lawes, to produce a 'super-phosphate of lime' using sulphuric acid to dissolve bones, and various super phosphates have been used as fertilizers ever since. Lawes experimented with pot-grown plants in the 1830s, and then in the field in 1840–1841, and developed his fertilizer at that time. However, he did not patent it until 1842, as his friends advised him that 'a gentleman does not engage in the manure trade'! However, in 1842 he opened his commercial fertilizer factory, the first in the world. Lawes then appointed the chemist, Joseph Henry Gilbert to help him in 1843, and this partnership resulted in many agricultural innovations, and in assistance to many landowners in prescribing to their tenants what crop rotations they should use, and what practices were needed to maintain soil fertility. Like Justus von Liebig, they argued from research and practical experience, that crop

> rotation was essential to soil health. The work at Rothamsted showed that fields where farmyard manure (FYM) was applied for 160 years became very fertile after about 20 years, even if cropped continuously with barley (the Hoosfield trials). If the fields had not had FYM, yields (and soil organic carbon) declined steadily. After 19 years of treatments, the FYM plots were split, and one half continued to receive FYM, while the other had no fertilizer. A 120 years later, the FYM plots were still yielding twice as much as the plots which never had FYM (although yields had declined in comparison with plots still receiving FYM). Thus, although Rothamsted advocated the use of water-soluble fertilizer, they also advised farmers to look after the soil organic matter, and to practise crop rotation.
>
> (R.M.B. Auerbach, 2017, unpublished lecture notes 'Systems and technologies for sustainable agriculture', Stellenbosch University)

This research fundamentally changed the direction of modern agriculture and became the basis of conventional agriculture as practised around the world.

The people involved in the movements that would lead to modern OA believed that there was a direct relationship between the health of the soil, the crops that were grown in it and the animals and people who consumed these crops.

In the midst of this concern of farmers and researchers, around the turn of the 20th century several key books were published, giving alternatives to chemical fertilizers. These books are still used as reference texts by the organic movement. Two of the critical texts were *Bread from Stones* by Julius Hensel (published in Germany in 1893) and *Farmers of Forty Centuries, Or Permanent Agriculture in Japan, China and Korea*, by F.H. King (written around 1900 and published posthumously in the USA in 1911).

The most significant origins of the formal movement began in Germany in Silesia in 1924 when the philosopher Rudolf Steiner gave a series of eight lectures on agriculture. This was in response to the repeated requests from a group of farmers who were concerned about the declining quality of their soils and crops, since the introduction of synthetic fertilizers and pesticides. The lectures were published later that year in German, and translated in a book titled *Agriculture* (Steiner, 1958).

Steiner tasked Dr Ehrenfried Pfeiffer with developing the specific soil, plant and compost preparations and farming methods based on the broader, philosophical concepts that he used in his lectures.

Dr Pfeiffer developed the preparations and also the name 'biodynamic' (literally 'life force') to describe this new farming approach. He gave numerous lectures throughout Europe and started the biodynamic (BD) movement. This is why many European countries use the words 'bio' or 'biological' to describe organic farming. BD practices and preparations have now spread around the world.

Not long after the beginning of the BD movement a range of other organizations concerned about the link between soil health and human health began to form in the 1930s and 1940s. These organizations were based around the concept of soil health and were called names such as the Soil Association, the Healthy Soil Society, and Soil and Health and were formed primarily in the English-speaking countries or ex-colonies of the UK such as Australia, New Zealand, the USA, India and South Africa (SA).

The most significant of these is the UK Soil Association which still continues to play a leading role in the organic movement in the UK and internationally. Dr Ehrenfried Pfeiffer gave several lectures at key conferences and events that were organized by the founding members of the UK Soil Association. Most of these organizations produced magazines and books that were widely read.

The book from that time that had the most profound influence was *An Agricultural Testament* by Sir Albert Howard (published in 1940, and quoted briefly in Chapter 1, this volume). Howard had spent much of his time in India and had pioneered efficient forms of composting that achieved high yields of healthy plants. Howard had an enormous influence on J.I. Rodale who published the *Organic Farming and Gardening* magazine in 1942 to promote his methods based on the widespread use of recycling organic matter through composting, green manuring and mulching (Rodale Institute, 2018).

The publication of *Silent Spring* in 1962 by Rachel Carson had a significant effect in raising public awareness about the dangers of the pesticides that were being used in farming at the time. *Silent Spring* created a huge controversy and a massive concern about build-up of chemical residues in foods and the environment. Public pressure saw strengthening of pesticide regulations and most importantly the start of the consumer movement that demanded food grown without toxic chemicals. It also saw the beginning of awareness of how farming was impacting on the environment and gave rise to a number of 'whole systems' approaches that fit within the broad organic paradigm.

Examples of these are *The One Straw Revolution* by the Japanese farmer Masanobu Fukuoka. Fukuoka had published earlier books in Japanese, however, *The One Straw Revolution* was published in English in 1978 and quickly became one of the most influential books of that time. His 'natural farming' methods were based on observing how nature works and then designing the system so that nature did the work for you (Fukuoka, 1978). He was one of the pioneers of organic no-till grain systems that did not use herbicides. These systems are easily applied to smallholder farms.

Quite independently of Fukuoka, two Australian researchers, Bill Mollison and David Holmgren published a book called *Permaculture One* in 1978. Permaculture was a shortened word for 'permanent agriculture' (the concept put forward by King, 1911).

Permaculture is a comprehensive whole systems approach which designs completely integrated systems that include cropping and animal systems design. Permaculture works with ecology, horticulture, and vertical stacking of production systems to maximize solar capture, integrating animals, water systems, architecture, energy use efficiency and numerous other concepts. The ideal was to start with a vacant block of land and design the new system based on the specifics of that block, its climate, topography and other attributes. Each permaculture farm would be unique because of this.

In 1979 three books on the concept of agroecology were published. Gliessman (1981) published *Agroecosistemas y tecnologia agricola tradicional*, Cox and Atkins (1979) published *Agricultural Ecology: an Analysis of World Food Production Systems* and Hart (1979) published *Agroecosistemas: Conceptos Básicos*.

This saw the beginning of the agroecology movement. The Laboratory of Agroecology at the University of California, Berkeley offers the following definition:

> Agroecology is both a science and a set of practices. As a science, agroecology consists of the application of ecological science to the study, design and management of sustainable agroecosystems ... This implies the diversification of farms in order to promote beneficial biological interactions and synergies among the components of the agroecosystem so that these may allow for the regeneration of soil fertility, and maintain productivity and crop protection.
> (Altieri, 2002)

It is both a science and a movement. Agroecology has very strong movements in Latin America and Africa and to a lesser extent in Asia, Europe, North America and Australasia. Many of these movements are highly political, focusing not just on farming production systems; they also concentrate on the rights of farmers to have a fair standard of living and the rights of communities to have food sovereignty (Holt-Gimenez, 2006). Agroecology is now taught and researched in many universities on every arable continent and it is gaining considerable credibility as can be seen from the many peer-reviewed scientific publications (see www.orgprints.org).

The Food and Agriculture Organization of the United Nations (FAO) has been promoting agroecology since 2014 when it organized the first United Nations (UN) International Agroecology Conference at its headquarters in Rome. Since then FAO has organized regional agroecology conferences in Africa, Asia and Latin America, and has published several papers and books on the practices and research of agroecology.

The Origins of the Formal International Movement

The formal international movement began in France on 5 November 1972 when at the invitation of Roland Chevriot of Nature et Progrès in France, Lady Eve Balfour a founder of the UK Soil Association, Kjell Arman from the Swedish Biodynamic Association, Pauline Raphaely from the Organic Soil Association of SA, and Jerome Goldstein from the Rodale Institute (and others) held a meeting at Versailles and formed IFOAM (IFOAM, 2018).

IFOAM is the international umbrella movement that has the role to both lead and unite the organic sector around the world. It is the organization that sets the international standards, policies, definitions and positions around the multi-functionality of OA through consulting with its members; these cover the whole spectrum of the sector in most countries in the world. Consequently, IFOAM documents are seen as credible source texts for reference material. Most of the world's organic standards and certifications systems are based directly or indirectly on IFOAM's Standards. The affiliates of IFOAM adopted the name *IFOAM-Organics International* at the General Assembly in 2014. This change makes it easier for people to understand that it is the international change agent and umbrella body for the organic sector.

Organic Agriculture Principles and Definitions

The principles of organic agriculture

The major concerns and concepts that were advocated by the founders and key opinion leaders of the organic movement over the last century, such as soil health, ecology, care, and using the precautionary principle with new technologies have been clearly articulated in IFOAM's four *principles of OA*.

The four principles of OA were developed from current organic practices through extensive worldwide consultation by IFOAM-Organics International. They are the agreed international consensus on the fundamental basis of organic production.

These principles are used by IFOAM-Organics International and other organic organizations to inform the development of practices, positions, programmes and standards. The definition and the four principles may be found on the IFOAM website, which also has many links to training, research and national OA movements (see IFOAM, 2008).

In summary, OA is based on:

- the principle of health;
- the principle of ecology;
- the principle of fairness; and
- the principle of care.

Principle of health

OA should sustain and enhance the health of soil, plant, animal, human and planet as one and indivisible.

This principle points out that the health of individuals and communities cannot be separated from the health of ecosystems – healthy soils produce healthy crops that foster the health of animals and people.

Health is the wholeness and integrity of living systems. It is not simply the absence of illness, but the maintenance of physical, mental, social and ecological well-being. Immunity, resilience and regeneration are key characteristics of health.

The role of OA, whether in farming, processing, distribution or consumption, is to sustain and enhance the health of ecosystems and organisms from the smallest in the soil to human beings. In particular, OA is intended to produce high quality, nutritious food that contributes to preventive healthcare and well-being. In view of this, it should avoid the use of fertilizers, pesticides, animal drugs and food additives that may have adverse health effects.

Principle of ecology

OA should be based on living ecological systems and cycles, work with them, emulate them and help sustain them.

This principle roots OA within living ecological systems. It states that production is to be based on ecological processes, and recycling. Nourishment and well-being are achieved through the ecology of the specific production environment. For example, in the case of crops this is the living soil; for animals it is the farm ecosystem; for fish and marine organisms, the aquatic environment.

Organic farming, pastoral and wild harvest systems should fit the cycles and ecological balances in nature. These cycles are universal but their operation is site specific. Organic management must be adapted to local conditions, ecology, culture and scale. Inputs should be reduced by reuse, recycling and efficient management of materials and energy in order to maintain and improve environmental quality and conserve resources.

OA should attain ecological balance through the design of farming systems, establishment of habitats and maintenance of genetic and agricultural diversity. Those who produce, process, trade or consume organic products should protect and benefit the common environment including landscapes, climate, habitats, biodiversity, air and water.

Principle of fairness

OA should build on relationships that ensure fairness with regard to the common environment and life opportunities.

Fairness is characterized by equity, respect, justice and stewardship of the shared world, both among people and in their relations to other living beings.

This principle emphasizes that those involved in OA should conduct human relationships in a manner that ensures fairness at all levels and to all parties – farmers, workers, processors, distributors, traders and consumers. OA should provide everyone involved with a good quality of life, and contribute to food sovereignty and reduction of poverty. It aims to produce a sufficient supply of good quality food and other products.

This principle insists that animals should be provided with the conditions and opportunities of life that accord with their physiology, natural behaviour and well-being.

Natural and environmental resources that are used for production and consumption should be managed in a way that is socially and ecologically just and should be held in trust for future generations. Fairness requires systems of production, distribution and trade that are open and equitable and account for real environmental and social costs.

Principle of care

OA should be managed in a precautionary and responsible manner to protect the health and well-being of current and future generations and the environment.

OA is a living and dynamic system that responds to internal and external demands and conditions. Practitioners of OA can enhance efficiency and increase productivity, but this should not be at the risk of jeopardizing health and well-being. Consequently, new technologies need to be assessed and existing methods reviewed. Given the incomplete understanding of ecosystems and agriculture, care must be taken.

This principle states that precaution and responsibility are the key concerns in management, development and technology choices in OA. Science is needed to ensure that OA is healthy, safe and ecologically sound. However, scientific knowledge alone is not sufficient. Practical experience, accumulated wisdom and traditional and indigenous knowledge offer valid solutions, tested by time. OA should prevent risks by adopting appropriate technologies and rejecting unpredictable ones, such as genetic engineering. Decisions should reflect the values and needs of all who might be affected, through transparent and participatory processes (IFOAM, 2018).

Organic 3.0

IFOAM-Organics International put forward the concept of Organic 3.0 in 2013. This is the third phase in the development of the organic sector.

> The overall goal of Organic 3.0 is to enable a widespread uptake of truly sustainable farming systems and markets based on organic principles and imbued with a culture of innovation, of progressive improvement towards best practice, of transparent integrity, of inclusive collaboration, of holistic systems, and of true value pricing.
>
> (Arbenz *et al.*, 2016)

The three phases of development in the organic sector are:

- **Organic 1.0** – This was started by our numerous pioneers, who observed the problems with the direction that agriculture was taking at the end of the 19th century and the beginning of the 20th century and saw the need for a radical change.
- **Organic 2.0** – This was started in the 1970s when the writings and agricultural systems developed by our pioneers were codified into standards and then later into legally mandated regulatory systems.
- **Organic 3.0** – This is about bringing organic out of its current niche into the mainstream and positioning organic systems as part of the multiple solutions needed to solve the tremendous challenges faced by our planet and our species.

The aim is to promote OA as a lighthouse for truly sustainable agriculture and agriculture production systems. Organic 3.0 expands the participation options, and positions organic as a modern, innovative farming system that holistically integrates local and regional context including its ecology, economy, society, culture and accountability. Regeneration of resources, responsibility in production, sufficiency in consumption, and the ethical and spiritual development of human values, practices and habits are concepts that guide the building of a new organic culture that can drive societal development. At the heart of Organic 3.0 are the living relationships between consumers and producers, which include information on products, production and the multiple benefits of OA.

Organic 3.0 is not prescriptive but descriptive: instead of enforcing a set of minimum rules to achieve a static result, this model is outcome-based and continuously adaptable to local context. Organic 3.0 is grounded upon clearly defined minimum requirements such as the ones maintained by many government regulations and private schemes around the world, and in the objectives of the IFOAM Standards Requirements. It expands outward from these base requirements. It calls for a culture of continuous improvement through private sector- and stakeholder-driven initiatives towards best practices based on local priorities, and as described in the Best Practices Guidelines of IFOAM-Organics International.

IFOAM-Organics International spent 3 years consulting with a wide cross section of stakeholders to ensure that there was a true consensus over Organic 3.0.

The strategy for Organic 3.0 includes six main features:

1. A culture of innovation, to attract greater farmer adoption of organic practices and to increase yields.
2. Continuous improvement towards best practice, at a localized and regionalized level.
3. Diverse ways to ensure transparent integrity, to broaden the uptake of OA beyond third-party assurance and certification.
4. Inclusiveness of wider sustainability interests, through alliances with the many movements and organizations that have complementary approaches to truly sustainable food and farming.
5. Holistic empowerment from the farm to the final product, to acknowledge the interdependence and real partnerships along the value chain.
6. True value and fair pricing, to internalize costs, encourage transparency for consumers and policy makers and to empower farmers as full partners.

The final document that defines Organic 3.0 (Arbenz *et al.*, 2016) was approved at the IFOAM-Organics International General Assembly in New Delhi, India in November 2017.

Regenerative Agriculture

The Rodale Institute has pioneered regenerative agriculture. Robert Rodale, the son of J.I. Rodale, coined the term 'regenerative OA' to distinguish a kind of farming that goes beyond simply 'sustainable' (Rodale Institute, 2018).

It is a concept in which agricultural systems improve the resources they use, rather than destroying or depleting them. It is a holistic systems approach to farming that encourages continual innovation for environmental, social, economic and spiritual well-being. It is closely aligned with the concept of Organic 3.0.

Regeneration is far more than being sustainable. Sustainable is defined as meeting the needs of the present without compromising the ability of future generations to meet their own needs (Regeneration International, 2018).

However, as pointed out in the Introduction of this book, in this era of the Anthropocene, the world is facing multiple environmental, social and economic crises. These include global warming, climate change, food insecurity, an epidemic of non-contagious chronic diseases, wars, refugees, migration crises, increasing poverty, ocean acidification, the collapse of whole ecosystems, the unsustainable extraction of resources and the greatest extinction event in geological history. Simply being sustainable is not enough.

Do we want to sustain the current status quo or do we want to improve and rejuvenate it? Regeneration improves the current system. We need to do more than ensure that things don't run down any further; we have to repair the extensive damage that our species has caused to our only planet.

Regenerative agriculture improves the land by using technologies that regenerate and revitalize the soil and the environment. The primary aim of regenerative agriculture is to increase the levels of soil organic matter. This leads to multiple positive outcomes such as: (i) better resilience to extreme weather events; (ii) increased efficiency in the soil's water holding capacity;

(iii) fewer diseases due to the beneficial soil biota controlling pathogens; and (iv) increases in the bioavailability of the nutrients that plants, animals and humans need (Regeneration International, 2018).

Regenerative agriculture is dynamic and holistic. It incorporates a mix of best practices that are known to improve soils and agrobiodiversity. These include agroecology, organic farming practices, no-till/low-till, cover crops, crop rotations, holistic grazing, permaculture, composting, mobile animal shelters, pasture cropping, agroforestry, analog forest farming, ecological agriculture and others.

The Formation of the International Regeneration Movement

The international regeneration movement started at a meeting in the Rodale headquarters in New York during the UN Climate Change meeting (Regeneration International, 2018), and the protest march which followed in September 2014.

Regeneration International, an organization that promotes food, farming and land use systems, was formed at a conference in Costa Rica the following year. Its aim is to regenerate and stabilize climate systems, the health of the planet and people, communities, culture and local economies, democracy and peace. At the time of writing (2018) it has over 100 partner organizations in over 30 countries. One of Regeneration International's main aims is to reverse climate change by using photosynthesis processes of agriculture to draw down CO_2 and store it in the soil as soil organic matter. It is working closely with the '4 per 1000 Initiative' (Regeneration International, 2018).

On 1 December 2015, the French Government launched the '4 per 1000 Initiative: Soils for Food Security and Climate', that will use a range of agricultural systems to sequester CO_2 and store it in the ground as soil organic carbon. A total of 31 countries signed on to this initiative along with key international organizations such as the FAO, the Global Environment Facility, the International Fund for Agriculture Development, the World Bank and the Asian Development Bank. In addition, 26 research institutes and universities have signed on along with over 100 non-government organizations (NGOs) and private sector organizations. This initiative is intended to complement the efforts needed to reduce global greenhouse gas (GHG) emissions. The 4 per 1000 Initiative is part of the framework of the Lima–Paris Action Agenda and consequently it is part of the global climate change agreement that was signed in Paris in December 2015. The title comes from research that determined that an annual growth rate of four parts per 1000 in global soil carbon stock would make it possible to stop the present increase in atmospheric CO_2.

> This growth rate is not a normative target for every country but is intended to show that even a small increase in the soil carbon stock (agricultural soils, notably grasslands and pastures, and forest soils) is crucial to improve soil fertility and agricultural production and to contribute to achieving the long-term objective of limiting the temperature increase to +1.5–2°C, the threshold beyond which the IPCC indicates that the effects of climate change are significant.
>
> (Initiative 4 pour 1000, 2018)

Regeneration International is an active part of the 4 per 1000 Initiative consortia. It has been organizing meetings and events to raise awareness that the widespread adoption of regenerative agriculture practices can sequester enough CO_2 to reverse climate change. These events feature published peer-reviewed papers and evidence-based practice (Regeneration International, 2018).

A new Regenerative Organic Certification was launched in 2018. It was developed through a cooperative effort among a coalition of farmers, ranchers, non-profit organizations, scientists and brands, led by the Rodale Institute. Its aim is to reach above the basic organic standards and set guidelines for soil health and land management, animal welfare, and farmer and worker fairness (Rodale Institute, 2018).

Editor's note: To this chapter by Andre Leu, I have added an historical perspective, shown in Box 2.1, consisting of reflections sent to me by Paul Hepperly, who ran the Rodale Institute's long-term trials for many years (Paul Hepperly, Research Director (retired), Rodale Institute, Kutztown, Pennsylvania, 2018, personal communication).

Box 2.1. Paul Hepperly's reflections on the historical perspective of organic agriculture

In the first 40 years of the 20th century advances in genetics, biochemistry and engineering rapidly and profoundly changed farming. The introduction of the internal combustion engine ushered in the tractor era of mechanization, with the allied development of numerous petroleum-driven implements for farm use. Plant breeding research and development led to hybrid seed commercialization. Chemists Haber and Bosch developed a new manufacturing process making nitrogen fertilizer cheap and available. Based on this, nitrogen fertilizers, which rapidly stimulate plant growth at low cost, were widely adopted by farmers in the 20th and 21st centuries.

In 1944 in Mexico the Rockefeller Foundation sponsored the campaign termed the Green Revolution. The Rockefeller Foundation is historically tied to the domestic and global petroleum industry Standard Oil. Green Revolution agriculture depends on heavy use of synthetic fertilizer (which requires fossil fuels), heavy equipment, pesticides, dwarfed grain varieties (which would respond to high input without lodging) and large-scale energy-intensive irrigation.

Machinery diminishes the large need for human hand labour in agriculture: there were almost no tractors in the USA around 1910, but over 3,000,000 by 1950; in 1900, it took one farmer to feed 2.5 people, but currently the ratio is one farmer to well over 100 consumers. As fields grew bigger and cropping more specialized to make more efficient use of machinery the use of animal/human labour was reduced by substituting machinery. This made it possible to farm more intensively and genetics, agrochemistry and mechanization coevolved simultaneously and rapidly and were sold as a systematic approach to modern farming. These technologies, which are petroleum based, have large energy and environmental costs associated with them which, for instance, are shown in agriculture-driven deterioration of soil, air and water quality.

In the USA modern organic farming has grown to respond to Green Revolution agriculture which featured the use of ammoniated fertilizers, pesticides and new narrow genetic proprietary seed resources increasingly dominated by international corporate interests. As more and more sophisticated machinery was developed, large tracts of land were put into production by fewer and fewer farmers. After the Second World War, agricultural research concentrated on new machinery, genetics and agrochemical approaches. The explosive use of chemicals was sold to the public as imperative, in order to improve the ability of producing food for burgeoning human populations. The scare tactic employed was the assertion that high global population growth would lead to massive starvation unless the US' intensified agriculture system was employed globally.

In 1962, Rachel Carson, noted writer, scientist and naturalist, published *Silent Spring*. This transformative work chronicled the effects of DDT (dichlorodiphenyltrichloroethane) on the environment. A bestseller in many countries, including the USA, and widely read around the world, *Silent Spring* is widely considered as being a key factor in the US government's 1972 banning of DDT. The book and its author are often credited with launching the worldwide environmental movement. In the 1970s, global movements concerned with pollution and the environment increased their focus on organic farming. As the distinction between organic and conventional food became clearer, one goal of the organic movement was to encourage consumption of organic and locally grown food. This was promoted through slogans like 'Know Your Farmer, Know Your Food'.

In the USA during the 1970s and 1980s, J.I. Rodale and the Rodale Press led the way in getting Americans to think about the side effects of non-organic methods, and the advantages of organic ones. The Rodale books offered how-to information and advice to Americans interested in trying organic gardening and farming. In 1981 Robert Rodale initiated a biometrically well-designed trial (the Rodale Farming Systems Trial) that compared OA systems to conventional maize and soybean production, as recommended by the state Cooperative Extension Service. Within 3 years, yields of all crops were comparable for organic and conventional systems, but as soil organic matter increased, drought losses decreased and the soil fertility of trial plots under organic management compared favourably with conventional treatments. The Rodale organic systems used no synthetic inputs and produced yields superior to the yields of county conventional farmers.

By 1984, Oregon Tilth and Rodale established an early organic certification service in the USA. In the 1980s, around the world, farming and consumer groups began seriously pressuring for government regulation of organic production. This led to legislation and certification standards being enacted through the 1990s. In the USA, the Organic Foods Production Act of 1990 tasked the United States

Continued

Box 2.1. Continued.

Department of Agriculture (USDA) with developing national standards for organic products, and the final rule establishing the National Organic Program (NOP) was first published in the 2000 Federal Register. From the early 1990s, the retail market for organic farming in developed economies has been growing by about 20% annually due to increasing consumer demand. Concern for the quality and safety of food, and the potential for environmental damage from conventional agriculture, are apparently responsible for this trend. In the 1990s principal field crops starting with soybean, maize and cotton were genetically engineered (GE) by major chemical seed companies to be dependent on glyphosate herbicide and GE *Bacillus thuringiensis* (BT) insecticide.

As concerns about GE technology grew, the acreage of certified organic farms expanded greatly. Throughout this history, the focus of agricultural research and the majority of publicized scientific findings have been on chemical, not organic, farming. This emphasis has continued with biotechnologies in general, and with GE in particular. This imbalance is largely driven by agribusiness, which, through research funding and government lobbying, continues to have a predominating effect on agriculture-related science and policy. Agribusiness has lobbied continually to change the rules of the organic market. Organic farming was driven by small, independent producers and consumers. In recent years, explosive organic market growth has encouraged the participation of agribusiness interests. As the volume and variety of 'organic' products increases, the viability of the small-scale organic farm is at risk, and the meaning of organic farming as an agricultural production system is ever more easily confused with the related but separate areas of organic food and organic certification.

(Paul Hepperly, Research Director (retired), Rodale Institute, Kutztown, Pennsylvania, 2018, personal communication)

References

Altieri, M. (2002) Agroecology: the science of natural resource management for poor farmers in marginal environments. *Agriculture, Ecosystems & Environment* 93, 1–24.

Arbenz, M., Gould, D. and Stopes, C. (2016) *Organic 3.0 – For Truly Sustainable Farming and Consumption*. IFOAM-Organics International, Bonn, Germany and SOAAN, Bonn, Germany. Available at: https://www.ifoam.bio/sites/default/files/organic3.0_v.2_web_0.pdf (accessed 8 May 2019).

Carson, R. (1962) *Silent Spring*. Houghton Mifflin, Boston, Massachusetts.

Cox, G. and Atkins, M. (1979) *Agricultural Ecology: an Analysis of World Food Production Systems*. Freeman (Holzbrink), Stuttgart, Germany.

Fukuoka, M. (1978) *The One Straw Revolution*. Republished, 2009 by New York Review of Books Classics, New York.

Gliessman, S. (1981) *Agroecosistemas y tecnologia agricola tradicional*. Published in English as: The ecological basis for the application of traditional agriculture technology in the management of tropical ecosystems. *Agro-Ecosystems* 7(3), 173–185. https://doi.org/10.1016/0304-3746(81)90001-9

Hart, R. (1979) *Agroecosistemas: Conceptos Básicos*. CATIE, Costa Rica.

Hensel, J. (1893) *Bread from Stones: a New and Rational System of Land Fertilizing and Physical Regeneration*. Translated by John Schacher. Lulu Press, Morrisville, North Carolina.

Holt-Gimenez, E. (2006) *Campesino a Campesino: Voices from Latin America's Farmer to Farmer Movement for Sustainable Agriculture*. Food First Books, New York.

Howard, A. (1940) *An Agricultural Testament*. Republished 1979. Rodale Press, Emmaus, Pennsylvania.

IFOAM (International Federation of Organic Agriculture Movements) (2008) Definition of Organic Agriculture. Available at: https://www.ifoam.bio/en/organic-landmarks/definition-organic-agriculture (accessed 12 November 2018).

IFOAM (International Federation of Organic Agriculture Movements) (2018) History. Available at: https://www.ifoam.bio/en/about-us/history (accessed 6 May 2019).

Initiative 4 pour 1000 (2018) Available at: https://www.4p1000.org (accessed 8 May 2019).

King, F.H. (1911) *Farmers of Forty Centuries, Or Permanent Agriculture in China, Korea and Japan*. Privately published, Madison, Wisconsin. Available at: https://onesearch.library.utoronto.ca/library-info/GERSTEIN (accessed 13 May 2019).

Mollison, W. and Holmgren, D. (1978) *Permaculture One: a Perennial Agriculture for Human Settlements*. Corgi, London.

OUP (Oxford University Press) (1970) Organic. In: *The Concise Oxford Dictionary*, reprint of 5th edn. OUP, London, pp. 854–855.

Regeneration International (2018) About Us. Available at: http://regenerationinternational.org/why-regenerative-agriculture/ (accessed 12 November 2018).

Rodale Institute (2018) History of Rodale Institute. Available at: https://rodaleinstitute.org/about-us/mission-and-history/ (accessed 12 November 2018).

Steiner, R. (1958) *Agriculture. Lectures Delivered at Koberwitz on Biodynamic Farming*. English translation by G. Adams. BioDynamic Association, London.

Willer, H. (2013) *The World of Organic Agriculture, Statistics and Emerging Trends*. Research Institute of Organic Agriculture (FiBL), Frick, Switzerland and IFOAM-Organics International, Bonn, Germany.

Willer, H. and Lernoud, J. (2018) *The World of Organic Agriculture, Statistics and Emerging Trends*. Research Institute of Organic Agriculture (FiBL), Frick, Switzerland and IFOAM-Organics International, Bonn, Germany.

3 Organic Research Contributes to Sector Development and Good Organic Policy: the Danish, Swiss, American and African Case Studies

Raymond Auerbach*
Nelson Mandela University, George, South Africa

Abstract

Research in ten countries (in Africa, America and Europe) has shown that organic research promotes evidence-based policy development. A seven countries study by the United Nations Conference on Trade and Development (UNCTAD) (2008) on how governments can assist organic sectors, provided guidelines about: (i) regulation; (ii) special support for small-scale farmers; and (iii) underpinning the emergence of a market for organic produce without distorting this market. Eight years later, UNCTAD published a further report on financing organic agriculture (OA) in Africa, which concluded that lack of finance hinders the development of OA in Africa. These reports emphasize the need for OA research and this chapter analyses three long-term research projects.

The Swiss long-term research trials showed many benefits of organic farming, but also limitations, with yields 20% lower than conventional farming; they cite many researchers around the world who show the benefits of OA, and argue for the establishment of a global platform for organic farming research, innovation and technology transfer. Long-term research has had a major impact on production, processing, marketing and consumption of organic produce worldwide, as shown by Danish research through four research programmes at Aarhus University (which contributed to sales of organic produce increasing from US$80 million in 1996 to US$821 million in 2010), and this assisted many Danish farmers to expand production and understand the needs of the market. Danish policy makers took note and formulated more supportive organic farming policies. In the USA, the Rodale Institute in Pennsylvania carried out long-term research trials to show that OA can be economically competitive, while benefiting the environment and the health of consumers; they showed that in dry years, organic crops outyield conventional crops. All three studies had close links with agricultural policy, but the Danish and Swiss studies were more sympathetically received and resulted directly in positive changes to agricultural policies in those countries.

Introduction

How can goverments help the organic sector to grow? According to the United Nations Conference on Trade and Development (UNCTAD) seven countries study 'Best Practices for Organic Policy' (2008), organic policy development was studied in seven countries selected to reflect a variety of conditions and stages of development and various levels of government involvement in the sector, from almost none (South Africa (SA)) to deep engagement (Costa Rica and Denmark).

*raymond.auerbach@mandela.ac.za

I wrote the SA case study in 2007, in the middle of a tortuous process to develop national organic standards, which saw 11 years of fruitless engagement with the national department of agriculture, followed by 4 years of frustrating engagement with the South African Bureau of Standards (SABS). Eventually, the South African Organic Sector Organisation (SAOSO) worked with the International Federation of Organic Agriculture Movements (IFOAM) to develop a standard (www.saoso.org) which was accepted into the IFOAM Family of Organic Standards in 2017. Although SABS Organic Standard SANS 1369 was developed, interference made this an internationally unacceptable standard; the two standards may eventually be harmonized. Because the organic sector in SA has been actively opposed by elements of government and agribusiness, it is important to look at what governments should be doing to support the organic sector.

The Foreword to the UNCTAD (2008) study states:

> OA is a production system based on an agro-ecosystem management approach that utilizes both traditional and scientific knowledge... [and]... offers developing countries a wide range of economic, environmental, social and cultural benefits. ... While sales are concentrated in North America and Europe, production is global, with developing countries producing and exporting ever-increasing shares. Due to expanding markets and price premiums, recent studies in Africa, Asia and Latin America indicate that organic farmers generally earn higher incomes than their conventional counterparts. ... Modern organic techniques have the potential to maintain and even increase yields over the long term while improving soil fertility, biodiversity and other ecosystem services that underpin agriculture. Crop rotations in organic farming provide more habitats for biodiversity due to the resulting diversity of housing, breeding and nutritional supply. As synthetic agrochemicals are prohibited in OA, its adoption can help prevent the recurrence of the estimated three million cases of severe pesticide poisoning and 300,000 deaths that result from agrochemical use in conventional agriculture every year. Organic systems have 57% lower nitrate leaching rates compared with other farming systems, and zero risk of surface water contamination. In terms of benefits for climate change, various studies have shown that organic farming uses 20% to 56% less energy per produced unit of crop dry matter than conventional agriculture ... it is estimated that converting the US' 160 million corn and soybean acres (66 million ha) to organic production would sequester enough carbon to meet 73% of that country's Kyoto targets for CO_2 reduction.
>
> (UNCTAD, 2008)

The UNCTAD report (2008) states that organic production helps resource-poor farmers depend less on external resources and enjoy higher and more stable yields and incomes, enhancing food security and supporting farmers' rich heritage of traditional knowledge and traditional agricultural varieties. OA provides employment for youth, and is a powerful tool for achieving the (then) Millennium Development Goals, particularly those related to poverty reduction and the environment. UNCTAD and the United Nations Environment Programme (UNEP) selected it as a priority issue to be addressed in the framework of the UNEP-UNCTAD Capacity Building Task Force on Trade, Environment and Development (CBTF):

> Since 2004, CBTF efforts have focused on promoting production and trading opportunities for organic products in East Africa, including supporting, in cooperation with the IFOAM, the development and adoption in 2007 of the East African Organic Products Standard (EAOPS). The EAOPS is the second regional organic standard after that of the European Union (EU) and the first ever to be developed through a region-wide public-private NGOs partnership process.
>
> (UNCTAD, 2008)

The key challenges for policy makers are: (i) to verify processes for checking compliance with the organic standards (both public and private) of the importing markets in a cost-effective way; (ii) to meet the quality and volume requirements of buyers; (iii) to develop the domestic organic market; and (iv) to build farmers' capacities in organic production techniques and documentation requirements for demonstrating compliance.

The UNCTAD study recommends that developing-country governments should facilitate and integrate rather than try to control OA. They should engage in dialogue with their organic sectors to identify their most pressing needs and consider conducting an integrated assessment of the sector, building OA supply capacities through education, research, extension services, local and regional market development and export facilitation.

Some main points of the Executive Summary of the UNCTAD (2008) report are as follows:

- protection of natural resources (e.g. water and energy) and biodiversity;
- improved quality of soils and thereby a long-term high productivity;
- improved market access;
- improved profitability in farming; and
- improved health or reduced health risks for farmers, farmworkers and consumers.

The report relates experiences from the cases of seven countries: Chile, Costa Rica, Denmark, Egypt, Malaysia, Thailand and SA, as well as from other parts of the world. It shows that OA is developing strongly in all seven countries, despite quite different conditions and very different levels and kinds of government involvement. Most organic production is for export purposes but countries like Egypt, Malaysia and SA have developed substantial domestic markets. Malaysia is even a net importer of organic food. In almost all countries with an organic sector, the early drivers are NGOs and the private sector; governments have rarely played any role in the early stages. 'Countries with a unified organic movement develop the sector quicker ... A starting point for government engagement is to give recognition and encouragement' (UNCTAD, 2008). This also includes the recognition of the relevance of organic sector organizations and the close cooperation between them and governments.'

Summary of UNCTAD (2008 and 2016) Recommendations

There should be integrated assessment of agriculture policies, programmes and plans, to understand how they affect the competitiveness and conditions of the organic sector.

Government involvement in the development of the organic sector should be clarified, and stakeholders should be involved in the policy development and development of plans and programmes.

General and OA policies should support each other where possible to promote effective policy coherence, especially if OA is promoted as a mainstream solution.

An action plan for the organic sector should be developed based on analysis of the state of the sector, participatory consultations and a needs assessment. The action plan should state measurable targets to help agencies and stakeholders focus their efforts.

Governments should actively contribute to awareness raising for OA on all levels, and should set up a formal body for interacting with the sector, establishing organic standards and collecting data.

Producers, especially smallholders, should be supported to comply with standards, certification procedures and regulations. Special considerations should be taken for certification of smallholders. Training programmes for farmer groups to set up internal control systems should be supported. Public procurement of organic products should be encouraged, including featuring organic food in important public events.

Consumer education and awareness should be actively promoted, with a common (national, regional or international) mark for organic products, and support for the organization of farmers through market information systems and export promotion activities, recognizing the special nature of organic markets. Organic exporters should be encouraged to join forces to promote and market their products.

Organic extension services need to be established and the staff trained. Organic extension should be developed and implemented in a participatory manner and have the farm and the farmer as the centre of attention, and should build on traditional knowledge about pest control treatments, etc.

Seed breeding and seed testing should be oriented to organic production. Compulsory seed treatments should be waived for organic farmers and untreated seeds should be made available. Alternative seed treatments should be developed and promoted. Policies for genetically modified organisms (GMOs) need to ensure that GMO seeds do not cause contamination of seeds.

Special research programmes should be established for organic research, and the sector should be involved in priority setting. Research and development in OA should be participatory, build on and integrate traditional knowledge (where relevant) and be based on the needs of the producers.

In 2016, UNCTAD produced a report titled 'Financing Organic Agriculture in Africa', which analyses the financial needs of OA, and barriers and opportunities for funding.

Extracts from the last two sections are summarized as follows. Access to finance in the OA sector remains constrained and survey results do not suggest the situation is improving. The majority (64%) of surveyed stakeholders indicated that, over the last 5 years access to finance had remained the same, and close to a quarter of respondents (23%) even suggested that access to finance has become more restrictive. Limited credit guarantee mechanisms and insufficient capacity of commercial banks to integrate the specificities of OA are major hindrances to the ability of OA stakeholders to finance their activities in Africa. The commitment to support sustainable agriculture expressed in the 2015 Addis Ababa Action Agenda on Financing for Development, and the unanimous approval by the African Union Ministerial Council of the Ecological Organic Agriculture Strategic Plan (2015–2025), are opportunities to bridge the OA funding gap. In this regard, efforts to further embed OA in the Comprehensive Africa Agriculture Development Programme (CAADP) will play a key role in the allocation of funding and the systematic inclusion of OA considerations into national agricultural development plans and strategies.

These two UNCTAD studies show that country organic sectors benefit from certain kinds of government support in training and the setting of organic standards, but that government can also inhibit the development of the sector by a heavy-handed regulatory approach. The same is true of the organic seed sector.

Organic Seed Breeding and its Impacts on Regulatory Frameworks

A doctoral research study by Erica Renaud (2014) examined the regulatory and technical challenges to the organic seed and breeding sectors in the USA, Mexico and the EU. The abstract to this work states:

> The main findings of the regulatory component were: (1) New organisations, procedural arrangements and activities have emerged in the US, EU and Mexico, to support organic seed regulatory development, with both positive and negative results; (2) Official guidance on the interpretation of the regulation in the US has not been sufficiently decisive to prevent divergent interpretation and practice, and in consequence, the needs of a rapidly growing economic sector are not being met; and (3) Growth of the organic seed sector is hindered by regulatory imbalances and trade incompatibilities within and between global markets.
>
> (Renaud, 2014, p. v)

Renaud explains that international organic standards compel organic producers who wish to be certified to use organically produced seeds, seedlings and breeding stock. She points out that regulators are waiting for non-government stakeholders to organize the sector to comply with the organic seed regulations. As there is therefore considerable regulatory ambiguity over what can and cannot be done, this contributes to potential violations of organic integrity, through the use of non-acceptable seed and seed-treatment inputs.

Although the organic seed sector is developing, it is not keeping up with increasing demand, although it is a relatively small sector. As cultivars bred for high-input conventional growing conditions are often not optimal for organic farming systems, it is important to breed seeds under conditions where they are exposed to pests and diseases so that their performance (including pest and disease tolerance) can be assessed realistically. After the defence of Renaud's doctoral thesis on 2 July 2014, several seed breeders told me that they had found her work very interesting and now felt that much of their conventional seed breeding should also be done under conditions where the plant is challenged by pests and diseases, rather than the normal conventional approach, where, during breeding, all pests and diseases are rigorously controlled by chemicals. They felt that breeding seed under more challenging conditions could produce pest and disease tolerant plants which would require fewer chemical treatments.

Long-term Organic Comparative Research: Three Examples

In Europe and the USA there has been considerable progress in long-term research into organic farming systems. Notably, the long-term comparative trials in Switzerland showed that over a 34-year research period, comparative organic yield levels are about 80% of conventional yields, with lower levels of inputs and higher nutritional quality on 11 parameters (Mäder *et al.*, 2006). In Denmark, the International Centre for Research into Organic Farming Systems (ICROFS) trials at Aarhus University obtained similar results over a slightly shorter period, also with a yield gap between organic and conventional of about 20% (Rasmussen *et al.*, 2006). In the USA, the Rodale Institute in Pennsylvania has been able to close this yield gap with a number of crops attaining very similar production levels (organic and conventional) in their long-term trials (Hepperly *et al.*, 2006). The Rodale trials found that organic systems tend to outyield conventional systems in dry seasons, where (as in many parts of Africa) irrigation is not available. All three of these long-running sets of trials (Switzerland, Denmark and the USA) found that organic systems require 3–4 years to build up soil biology to productive levels.

In this chapter, some reflections concerning effects on policy of these three major long-term comparative research trials (which inspired the Mandela Trials in SA) are summarized, starting with the DOK trials (bioDynamic, Organic, Conventional = DOK) run by the Swiss Research Institute for Organic Farming (FiBL), followed by the impacts of Danish organic research, notably the long-term trials at Aarhus University in Denmark run by ICROFS, and finally with some reflections from the Rodale Trials in Pennsylvania (Hepperly *et al.*, 2006). All three of these trials have been running for more than 30 years.

The Swiss long-term research trial

The Swiss perspective is taken from the paper 'Building a global platform for organic farming research, innovation and technology transfer' by Urs Niggli *et al.* (2017).

Under the heading 'Benefits and challenges of organic food and farming systems', they say that empirical evidence shows the benefits and strengths of organic food and farming systems and highlights further challenges and opportunities. Soil quality and health can be improved by organic farming practices, as measured by soil fertility and structure and by biodiversity of soil organisms. Organic farming maintains and increases soil organic matter, sequesters carbon and reduces GHG emissions relative to other forms of agriculture. Soil erosion is less likely in organic soils in the long run, and increased biological activity in the soil helps to suppress pests and diseases and enhance plant immunity to various opportunistic infections. Organic farming has higher nutrient efficiencies by relying on the cycling of nutrients from renewable resources, mainly in the form of organic matter, rather than on synthetic fertilizers that are derived from non-renewable resources.

Swiss organic farming systems have yields about 20% lower than conventional farming systems. Thus, more land may be needed to produce the same amount of food using organic practices, which may diminish the ecological and health benefits of organic relative to conventional farming when measured on a production unit basis. However, the greater biodiversity of organic food and farming systems through cultural practices such as crop rotations, inter- and relay-cropping offsets yield gaps; yields in organic farming systems may also be more stable under environmental stress and adverse weather conditions than in conventional farming systems. Organic farming systems enhance the resilience of agroecosystems by increasing natural pest control and enhancing biodiversity in the soil, as well as at the plot, farm and landscape scale and increase populations of pollinators and other beneficial organisms.

Organic farming is generally more profitable to farmers, particularly when they receive a price premium for their products; yield and gross returns can vary by crop, but the gross margin (after subtraction of the lower production costs) offsets yield reductions in the long run. At least in some cases, organic food and farming systems may also have higher returns to labour; however, organic farming systems can be less profitable, in part because of lower yields. The organic sector's small market share accounts for about 1% of global food sales which is probably the single most limiting factor for farmers in adopting organic practices, although demand has grown steadily over the past 40 years. Organic farming contributes to 'triple bottom line' accounting for social and economic, as well as purely environmental benefits.

Organic farming systems also provide environmental benefits across multiple physical, chemical, biological, economic and social parameters. Life cycle assessments (LCAs) have compared the relative environmental performance of certain aspects of organic and conventional farming, focusing on the inputs used by the different systems. However, LCAs have methodological shortcomings, as they have not been able to capture all of the environmental and social benefits reflected in ecosystem services and the market. The economic and environmental values of the biodiversity conserved by organic farming systems are difficult to estimate given the qualitative differences between extensive organic and intensive conventional production. Given that organic farming systems often require more land to produce the same amount of food, they would theoretically lead to less land being available for unfarmed wildlife habitats. In practice, however, the choices and outcomes are more complex. Land sparing and wildlife friendly agriculture can be complementary.

Because most pesticides are not permitted in organic food, this has significantly lower levels of pesticides than other food. As a result, organic foods pose lower dietary risks from pesticides to human health than conventional foods. Moreover, pesticide risks to the environment are also mitigated by organic production practices. Organic food has lower levels of cadmium, nitrate and nitrite compared with conventionally produced food. Investment in research on organic food and farming systems and other sustainable technologies has increased in recent years but is still marginal compared with research expenditures on agrochemicals, genetic engineering, animal confinement systems and other technologies that are incompatible with organic principles and standards. Most of the research expenditures have been directed at temperate- and Mediterranean-zone agricultures in Europe and North America, while relatively little capacity exists for research on organic food and farming systems in tropical- and subtropical systems, particularly in low-income countries of Africa.

The above is a summary of global progress in establishing evidence-based OA development, based on the Swiss research, and other collaboration.

The Danish long-term trials and consumer impact research

The findings of a report on the measurement of the specific impact of organic research in Denmark (Andreasen *et al.*, 2015) are as follows.

ICROFS analysed the effects of organic research in Denmark (1996–2010) on the Danish organic sector and on society in general. Over these 15 years, three national programmes and one programme with European collaboration were implemented in Denmark (financed through government grants totalling approximately US$80 million). The analysis itself was carried out as a compilation of information from three perspectives, each of which has been independently documented: (i) interviews with end users of results from R&D investigating their assessment of the challenges and solutions in the sector; (ii) assessment of the R&D endeavours in different thematic areas (dairy/milk, pigs, crops, etc.) as they related to end users and the stated challenges at that time; and (iii) documentation of the dissemination of R&D results in relation to themes and challenges in the sector.

The results showed very good correspondence between end-user perceptions of the challenges overcome in the sector, the R&D initiated in the research programmes, and the dissemination of research results and other forms of knowledge transfer. The analysis documented direct effects of the research initiatives targeting the challenges in the sector such as higher yields, weed and pest control, animal health and welfare, the potential for phasing out the use of antibiotics in Danish dairy herds and reducing the problems caused by seedborne diseases. It also described where research did not contribute as much to overcoming challenges. Here the analysis showed that the effects of the research in the organic processing industry and among relevant governmental and non-governmental organizations were of a more indirect character. Research helped stabilize the supply and quality of raw materials at a time of growing demand and sales. Organic research also generated new knowledge and lead to new opportunities which may inspire green conversion, product diversification and growth also in conventional agriculture. The analysis showed that research under the national research programmes overall had been very applied and directed at the barriers in the sector, in order to support the general market and growth conditions for the organic sector. Having laid a solid foundation, the private sector took advantage of commercial opportunities when demand grew, while adhering to the organic policy objectives of market-driven growth in the organic sector.

Since the mid-1980s, organic farming in Denmark has been promoted through political initiatives in order to respond to consumer demand for organic products. The policies of governments included financial support for the conversion of conventional farms, regulation and control, advisory services, information campaigns and education and research in organic farming. At the end of the 1980s and start of the 1990s, Danish research in organic farming was primarily carried out on private farms and in long-term crop rotations at research stations around the country. With the first action plan (Action Plan I) for the promotion of organic food production prepared by the Ministry of Agriculture and Fisheries in 1995 and followed by Action Plan II in 1999, research in organic farming was given a higher priority than earlier times, which resulted in the development of a national research programme and the establishment of the Danish Research Centre for Organic Farming (DARCOF) (now ICROFS) – a 'centre without walls' to coordinate these programmes as research continued within existing research environments throughout the EU.

From 1996 to 2010 Denmark had four research programmes in organic farming and foods financed via special government grants (one of them with European collaboration). While the first programme primarily addressed issues related to primary production, the following programmes also included issues related to industry (including processing), society (including environment and health) and the consumer level (including credibility of the sector). In these programmes, funds were allocated to coordination, communication and dissemination, as well as to knowledge synthesis, research methodology and to research education (PhDs at universities and research centres involved in the research). The centre maintained close contact with the players in the sector via user groups and extensive meeting and dissemination activities in order to ensure the continued relevance of research efforts and applicability of results.

In the same period the organic sector underwent a strong development from niche market to an important part of the Danish food sector. The area under organic farming, including the area under conversion in 2010, was 6.4% of the total farmed area. Of the total food sales in 2010, 7.2% was certified organic after a dramatic increase in sales from approximately US$80 million in 1996 to approximately US$821 million in 2010. The report found that nearly all supermarket chains had a large assortment of organic products and for some product groups, such as eggs and milk, the organic market share was 20–30% of retail sales. Several important factors have contributed to the positive development of the organic sector in Denmark, including: (i) support for marketing and the regulatory framework from public and private sectors; (ii) establishment of strong institutions in organic farming; (iii) entrepreneurs and pioneers in the organic farming, processing and retailing sectors; and (iv) research carried out in universities, research stations and with advisors and farmers at private farms.

Findings arising from dissemination of R&D results

A total of 3173 publications constituted the direct outputs from the projects (Organic Eprints, counted in 2012). About 20% (632) of these were peer-reviewed papers; another 1311 were other publications in English, while 1230 were publications or other forms of dissemination in Danish. Based on a search of the archives of the Knowledge Centre for Agriculture, it was found that there had been dissemination based on R&D in the projects within all the thematic areas. One example is the Danish Crop Rotation Experiment, from which there were 215 publications in Organic Eprints, and at least 50 dissemination articles based on the R&D in the archives of Knowledge Centre for Agriculture. In the interviews, there were many statements about contributions from research. In each case, it was determined that research results were disseminated, so that the statements were justified.

The projects resulted in a high number of peer-reviewed journal articles, in spite of the fact that the research under the four programmes was mainly 'applied research'. The close association between the research scientists and end users in the DARCOF programmes influenced the effects achieved. Both crop production and animal husbandry research projects contributed, with significant new knowledge and methods in response to the considerable challenges in primary production, from the handling of manure and weeds to animal health and feeding. The results have been widely applied partly because many of the projects were designed and selected as a response to challenges formulated by the sector. Organic production would have been much lower today if the research results had not been used, because the production itself is more profitable (higher yields per cow, pigs of higher quality resulting in a higher kilo price, etc.) and because some important problems have been solved, which has reduced the incidence of reconversion to conventional farming (e.g. improved perennial weed control and recycling of nutrients using cover crops and good crop rotations).

The increasing production and the ability to ensure a good and consistent quality and stability has also been a precondition for the establishment of a professional and profitable processing sector. Companies interviewed found that these conditions have had an important effect on their development opportunities. Research had a strong focus on the barriers in the sector and on improving the general market and growth conditions in the sector, and formed the basis for a stronger commercial exploitation of the opportunities. The research focus dealt with the challenges in the commercial sector and also the political ambitions for market-driven growth in the organic sector.

In addition to the direct effects, there are other – more indirect – effects on processing and marketing, such as a better understanding of consumer motives for purchasing organic produce and a higher degree of integrity as a result of research. Integrity – here understood as consumer trust that the organic sector lives up to its declared ideals and added values – has been improved in two ways. First, organic production itself has been improved in areas that are important to the consumer, and second, studies have evaluated organic farming in relation to its principles, consumer expectations and/or the interests of society. In the first instance, research projects have – according to interviews with consultants and representatives of public authorities and organizations – enhanced animal health and welfare on organic farms through the development and description of better farm management, housing, feeding, etc. In the second instance, a series of projects have probed whether organic farming actually confers advantages compared with conventional systems or products.

Some projects have documented positive effects of organic farming on, for example, nutrient balances in livestock farming, conservation of biodiversity in hedgerows, as well as a higher nutritional content of organic produce. However, some results have also been critical regarding specific aspects of organic farming (e.g. when measured either on climate impact per kilo of produce, on flavour, or on general healthiness).

Such results have been used by organizations in the sector to launch campaigns to improve practical aspects of the systems. The sector has focused on improving animal health and welfare on organic farms based on the background of research projects and reviews. It can be assumed that the willingness to admit to weaknesses in the organic systems and the readiness to seek solutions to these has helped maintain integrity in the eyes of the public and ensured continued political backing, although there is no documentation for this.

Some of the research projects have documented that large consumer segments favour organic produce for a variety of personal (health, quality, pesticide-free) and altruistic (animal welfare, environment) reasons. These preferences may also affect conventional food production. In addition to the described effects on the organic sector, the DARCOF projects have also produced results that are relevant for conventional farming and can aid a general green conversion. This applies to replacements for seed treatments, non-chemical weed control, reduction in the consumption of antibiotics and the need for supplementation of synthetic vitamins in animal husbandry. This could save farmers money in the conventional sector if the methods were widely implemented and would further improve the reputation of Danish agriculture as an eco-friendly system supplying high-quality products.

At the international level there is an awareness of the need to improve the relationship between research, extension and agricultural production. In the International Assessment of Agricultural Knowledge, Science and Technology for Development (IAASTD) *Agriculture at a Crossroads – Global Report* (IAASTD, 2009), the conclusions stress that a departure is needed from the traditional model of research and dissemination as separate actions. Instead, there is a need for the farmers' voice to be heard when prioritizing and designing research projects and to integrate their local knowledge and experience into research and policy.

The applied nature and relevance of the projects under the DARCOF programmes has been strengthened via the close and continuous contact with consumer representatives. There has also been contact with the sector via the organic food council and a number of other actors involved in the preparation of the action plans and later in the knowledge synthesis in 2008, on the potential for a market-based development of the organic sector. This influence at programme level has been important for maintaining the relevance of the projects offered and funds granted in relation to the requirements of the sector.

Many of the projects have had contact with advisors and farmers where the acquired knowledge has been continually communicated and discussed. This has had two effects: (i) a rapid application of results, because the users have discussed the results of the research with the scientists and thus achieved a better understanding of how results and knowledge can be adapted to specific practical situations; and (ii) research design and methodology were adapted as a result of practical experience. The scientists were persuaded by dialogue with the users to ensure that treatments are as relevant and practical as possible, without compromising scientific standards.

This shows the complex links between research, development and the application of knowledge in agriculture. The traditional route is one-way communication of scientific results via advisors to producers. Because the project structure and organization of the organic research programmes supported this complexity in knowledge generation and exchange, the findings of the research projects have been used, and many barriers in the sector were overcome. There has also been continuity in many central research activities in terms of long-term experiments at the same localities over many years and in many research programmes.

The long-term trials at Rodale Institute in Pennsylvania, USA

The field trials at Rodale Institute are reported (1981–2004), where conventional and organic 5-year rotations of cereal crops, legumes and forage crops were compared. Both manure and legume-based rotations saw soil carbon rise from 2% to over 2.5%, while conventional soil carbon remained unchanged. Soil carbon and nitrogen accumulation was also significantly greater for manure and legume rotations than for the conventional mineral fertilizer based rotation.

Once soil biology had been improved by the organic farming systems approach, the yield gap was closed. For the 5 drier years (where annual rainfall averaged less than 350 mm, as opposed to about 500 mm in average years), maize grain crops in the organic farming systems outyielded conventional by an average of 31%. In the extreme drought year of 1999, with only 224 mm in the growing season, average organic maize yields were 1511 kg/ha while conventional yields were 1100 kg/ha; for soybeans that year, the mean figures were 1600 kg/ha and 900 kg/ha, respectively.

Conclusions on the Importance of Research for Organic Farming in Africa

This chapter began with extracts from the UNCTAD (2008) seven countries study, which summarized what governments in developing countries can do to support an emerging organic sector, and why this is important. This was followed by an analysis of the failure of governments and aid agencies to support effective organic development in Africa (UNCTAD, 2016). The evidence upon which these two UNCTAD reports was based, included work done in Switzerland by FiBL (the so-called DOK trials and subsequent establishment of a global organic research, training and innovation platform), the work of ICROFS in Denmark, and its impact on government policy and actual organic sector development there, and finally, the work in the USA of the Rodale Institute together with the history leading to the long-term research trials, and their impact. All three of these initiatives have been running for over 30 years, and it is important for African organic farming research and development to draw on the results of this significant mass of research.

We can make five major conclusions based on the results presented in this chapter:

- Without significant government and other support, organic sector development will be slow, and disinformation from suppliers of chemical inputs will continue to dominate the discourse on food sovereignty, food security and nutrition.

- There exists an international body of peer-reviewed evidence to show that organic farming is cost-effective, culturally appropriate and leads to an increase in food security without increasing dependence on expensive external inputs, which are difficult to obtain (and to pay for) in many remote rural areas of Africa.
- Organic farming methods use water and nutrients more efficiently, and build resilience, leading to better performance under drought conditions, and helping farmers to deal with climate change.
- For the organic sector to benefit small-scale farmers, capacity building and institutional development are required. This needs support from National Agricultural Research Systems (NARS), and also needs specially trained organic extension agents to support those farmers wishing to take on organic farming or agroecological farming.
- Finally, organic standards, organic value chains and the support for nutrition education around food quality and healthy food choices are all important to the development of thriving local and export markets for organic products.

References

Andreasen, L., Rasmussen, I.A. and Halberg, N. (2015) Organic research and development in Denmark (1996–2010) – effects on the organic sector and society. *Sustainable Agriculture Research* 4(3), 173–183. Available at: www.orgprints.org/29259 (accessed 3 June 2018).

Hepperly, P., Douds, D. and Seidel, R. (2006) The Rodale Institute Farming Systems Trial 1981 to 2005: Long-term analysis of organic and conventional maize and soybean cropping systems. In: Raupp, J., Pekrun, C., Oltmanns, M. and Kopke, U. (eds) *Long-term Field Experiments in Organic Farming*. International Society of Organic Agriculture Research (ISOFAR), Berlin, pp. 15–31.

IAASTD (International Assessment of Agricultural Knowledge, Science and Technology for Development) (2009) *Agriculture at a Crossroads – Global Report*. Island Press, New York.

Mäder, P., Flieβbach, A., Dubois, D., Gunst, L., Jossi, W., Widmer, F., Oberson, A., Frossard, E., Oehl, F., Wiemken, A., Gattinger, A. and Niggli, U. (2006) The DOK experiment (Switzerland). In: Raupp, J., Pekrun, C., Oltmanns, M. and Kopke, U. (eds) *Long-term Field Experiments in Organic Farming*. International Society of Organic Agriculture Research (ISOFAR), Berlin, pp. 41–58.

Niggli, U., Andres, C., Willer, H. and Baker, B. (2017) Building a global platform for organic farming research, innovation and technology transfer. *Organic Agriculture* 4(3), 173–183. https://doi.org/10.1007/s13165-017-0191-9

Rasmussen, I., Askegaard, M. and Olesen, J. (2006) The Danish organic crop rotation experiment for cereal production 1997–2004. In: Raupp, J., Pekrun, C., Oltmanns, M. and Kopke, U. (eds) *Long-term Field Experiments in Organic Farming*. International Society of Organic Agriculture Research (ISOFAR), Berlin, pp. 117–134.

Renaud, E. (2014) Breeding and regulatory opportunities and constraints for developing broccoli cultivars adapted to organic agriculture. PhD thesis, Wageningen Agricultural University, Wageningen, the Netherlands.

UNCTAD (United Nations Conference on Trade and Development) (2008) Best Practices for Organic Policy. United Nations. Available at: http://unctad.org/en/Docs/ditcted20073_en.pdf (accessed 3 June 2018).

UNCTAD (United Nations Conference on Trade and Development) (2016) Financing Organic Agriculture in Africa: Mapping the Issues. UNCTAD/WEB/DITC/TED/2016/6. Available at: http://unctad.org/en/PublicationsLibrary/webditcted2016d6_en.pdf (accessed 3 June 2018).

4 The Organic Academy of IFOAM-Organics International: Training Multipliers in the Developing World

Konrad Hauptfleisch*
IFOAM-Organics International, Bonn, Germany

Abstract

The Organic Academy of IFOAM-Organics International is 6 years old. It has trained over 600 stakeholders from more than 70 countries in the organic sector, and has conducted its training on four out of five continents. Its main approach is to base its curriculum on the key principles and values of organic agriculture, in order to inspire change and trigger further action, rather than supplying ready-made solutions to students in a classroom environment. Due to its limitations in size and budget, it works actively to develop more multipliers, taking up the concept, methodology and curriculum, to replicate its success globally. Methodology includes experiential learning, with Theory, Action and Reflection making up the complete training experience. The Organic Academy mirrors IFOAM-Organics International's global presence, and works across countries and cultures along the common themes of aspiration, inspiration, network-building and the celebration of diversity.

Introduction

> At the Organic Leadership Course (OLC) I have learned from the wisdom and work of and with others; and experienced at its core what it is to live and breathe the organic movement. Every moment had its nugget of wisdom and I went home bearing with me, 'organic movement' seeds to plant back home.
>
> (Paula Aberasturi, 2012)

It was the 6 April 2012. I was at Bordi Beach, Maharashtra, in north-western India (Fig. 4.1). We were halfway through our first ever Organic Academy training course, and participants were sitting under coconut trees with local farmers, engaged in spirited discussions. They were sharing knowledge, sharing seeds and celebrating life. The sun was setting, but the conversations remained animated and energized. At the time, I do not think I realized the impact this work would have on me, our alumni and our organization.

Six years later I am starting to grasp not only the enormity of our task, but also its true value to our work. We are managing to introduce and link the leaders who will take the organic movement into the future. Organic agriculture (OA) is the one sustainable agriculture system that can offer solutions to global agriculture and global challenges.

In these 6 years, we have managed to train over 600 organic leaders, farmers, extensionists, researchers, rural service providers and activists from more than 70 countries. All of them brought value to us and received value from

*k.hauptfleisch@ifoam.bio

Fig. 4.1. Organic training in Maharashtra. (Photograph by K. Hauptfleisch.)

each other. Three of them are now employed at our Head Office in positions of responsibility and leadership, one is serving on our World Board (three on previous boards), and at least five more are working for IFOAM-Organics International regional structures, sector bodies and standing committees. Some turned their course assignments into doctorates, others published books, and many more are working daily to develop and grow the global organic sector in positions of grass roots or executive leadership.

They bring diversity and energy to a movement that is in need of new and young organic leadership, to contribute to the further growth and development of truly sustainable farming and food systems based on the principles of OA.

Multiplying Organic Capacity: the Method

IFOAM-Organics International describes itself in its new strategy as 'The organic agent of change for true sustainability in agriculture, value chains and consumption; working on behalf of its membership, the global organic movement in over 120 countries.' Adopting a farmer-centred experiential learning approach (Freire, 1970; Holt-Giménez, 2006; Kolb, 2008), the method has been to draw out of local experience the lessons which are suited to the context, and in so doing to support local people in building on local resources and indigenous knowledge.

Leading and uniting the organic world over the last 45 years with a small core of staff, needed a strategic approach. Effecting global change and broader adoption of sustainable food and farming systems based on organic principles even more so. Often, broad adoption goes hand in hand with a technocratic strategy of exponential growth, identifying and scaling-up successful systems. This is the modus operandi of many large corporates and of the industrial food companies. The seminal and definitive critique on Big Agriculture, known as the IAASTD Report (IAASTD, 2009) contains the following inspiring quote: 'If many little people, in many little places, do many little things, we can change the face of the world.' This defines the Organic Academy's approach: we do not scale-up, we replicate. We do not send an army to conduct mass training programmes, but we inspire the multipliers to do many good little things in many little places.

This approach is contained in the second paragraph of the 'Definition of Organic Agriculture' (IFOAM, 2008): '(relying) on cycles adapted to local conditions…'. The Academy does not reinvent or impose content from a 'global up-high' – it brings together the core principles, values, better practices and concepts, into an 'educational seedball' – or, to use a more genetic analogy: similar organisms have similar DNA, but in different regions and under different circumstances, different genetic markers are switched on. The full diversity of solutions and approaches is not only recognized, but actively supported. This is organic learning, in the complex and adaptive ecosystem of adult education.

Similarly, organic solutions are locally appropriate and locally adapted. It therefore makes no sense for the global organization to provide 'ready-made' solutions. Its approach to learning and teaching must honour this principle, and the Academy recognizes that. It aims to develop capacity of local multipliers to co-create and develop solutions (Fig. 4.2), and share and disseminate appropriately.

The Academy training works with the 'common DNA' of organic, namely a well-developed definition, supported by core principles. These principles (IFOAM, 2005), driven by organic values, inspire the content. OA has a proud, centuries-old history of supporting sustainability through agroecologically focused production systems. The pioneers of modern OA recognized this more than a century ago, across all continents and climate zones. From Fukuoka to Steiner, from Balfour to Rodale, from Mazibuko to Phiri to Podolinsky, we see a golden thread of sustainability.

The well-known principles of the organic movement, namely health, ecology, fairness and care, can be used as a yardstick to illustrate how organic systems support a sustainable food system, as explained in Chapters 5 and 6 (this volume).

A food system that considers the well-being of the ecosystem – that supports and emulates it; that considers the health of all the contributors

Fig. 4.2. Group work in Pakistan, 2017. (Photograph by K. Hauptfleisch.)

and beneficiaries; that upholds a principle of fairness in the way that it produces, trades and consumes its output and that takes care to not utilize technologies that could cause untold and unknown damage – such a food system is clearly a sustainable one.

Our current food system is – if not broken – then most definitely severely damaged. In numerous research publications, this becomes clear. *It can no longer be business as usual*. The reasons for failure of the current food system are multiple, and blame is cast in many directions. What is the true *nature* of food? It *is*, and (despite recent technological advances) *remains* a product of a natural ecosystem managed by people. This link between agriculture and food systems seems obvious, but when one observes the way in which our current market economy approaches food as a commodity, its production and consumption as an industrial system, then it is clear that this link is tenuous at best. Nature is an extremely complex adaptive and increasingly vulnerable system – how can it be held captive within an industrial mindset driven by a need to simplify and reduce? Is constraining agriculture in the pipeline of contemporary industry really the most effective way to have sustainable and healthy food systems?

It is here where OA makes a real contribution to changing the food system for the better.

And when we develop any curriculum for organic leaders or practitioners, we can do no better than to base it on the four principles referred to above, and presented in Chapter 2 of this volume: (i) ecology; (ii) health; (iii) fairness; and (iv) care.

The principle of *ecology* features very clearly when we consider farming inputs: they are sourced, developed and improved based on principles of nature rather than an extractive fossil-fuel-based industry. Manures, compost and natural remedies are created by observing nature and natural systems. By using what is available locally, by recycling on-farm nutrients (and in the future, returning off-farm inputs to the farming system). Ecology features further when one considers the growing and production cycle, the sequestration of carbon and responsible use of precious resources like water make real and measurable contributions to the farm ecosystem, and by extension, to the planet.

While organic farming offers an obvious contribution, it goes further than that: organically produced food is free from harmful inputs and chemicals; organic standards also describe which processing agents may or may not be used – there is a clear positive indicator for the principle of *health*. Farmers and farmworkers are healthier because they are not exposed to harmful chemicals. Animals are healthier because they are treated without the use of same, and with respect for their welfare. Consumers of the outputs of this system are healthier both because their food is free from poisons and because it is more nourishing, and the whole ecosystem is healthier as a result.

In order to have a functional food system, *fairness* is key. A farmer's livelihood needs to be ensured by receiving fair compensation for his or her labour of love to provide sustenance for a growing population. Traders, processors, retailers: often these 'middlepersons' are maligned, as purely adding cost without adding true value. OA has a clear focus on fair systems for all involved: most of the world's leading organic standards, supported by farmers' associations across the globe, describe these fairness principles. Here, organic also aligns with its partners in the global movement for change, such as Fairtrade, think-tanks on true cost accounting systems, and the like.

The current buzz-trends in food – transparent, local, seasonal, fresh – these are all principles espoused by organic farmers and traders for decades. Some of the best examples of fair and cooperative farming and trade systems are found within the Organic Movement, and many of the current trends in wholefoods, superfoods and raw food were inspired by the organic pioneers. There is a caveat: the movement needs to ensure that the values of fairness continue to be enshrined even in the regulated organic industry. Governments need to be continually lobbied and engaged to ensure that the regulations echo the values and principles of sustainability.

The first three principles are embraced by and enshrined in the fourth: *care*. Care evokes feelings and perceptions of warmth, safety and support. When we take care, and give care, we celebrate that which is one of the highest aspirations of the human condition: we do not only feed and clothe, but we do it with care for our environment internally and externally. We take care of our health and nourishment, but we also take care of the health and nourishment of the planet. We do this by not supporting technologies that risk harm and destruction of the ecosystem, but by supporting research and technology that gives and takes care. This remains non-negotiable.

The values and principles of OA are clear, and their links to and value for a sustainable food system too. But we cannot build systems only on values and principles. While they give the backbone and the vision, these principles have to be supported by science, technology and economics. It stands to reason that OA will not make a meaningful contribution if it cannot show its contribution clearly.

The Organic Academy therefore bases its curriculum very closely on these fundamentals, in order to inspire and capacitate people, who, in turn, capacitate other people. This approach has certain benefits:

- The Academy is not seen as a prescriptive teacher of recipes, but as an enabler of local solutions.
- As a relatively small global organization, it is able to inspire hundreds, who in turn can teach thousands who can empower millions of farmers to adopt new systems – this is the 'theory of change' of the Academy. With a permanent staff of two, and a small but growing team of regional trainers, a small organization is able to punch above its weight.
- Through the Alumni Programme and Knowledge Network, these multipliers are kept connected through an online platform as well as via events and conferences where alumni meet – sometimes organized, but more often spontaneously.

The Academy focuses on participation as its main approach to adult education: participatory curriculum development, participatory methodology and peer-to-peer learning. Adults approach training differently to younger students: they come with a wealth of experience, and this experience and knowledge needs to be tapped and harvested in a training session. This cannot be done if a trainer is stuck behind a laptop with a laser pointer and a PowerPoint.

While technology is a great tool, and theory is key to learning, it cannot be the only component receiving energy in the participatory classroom. The training programme includes lectures by subject experts and the lead trainer, but this is balanced in equal measure with practical excursions, breakout sessions using different methodologies like 'Fishbowl' and 'World Café', and thirdly ample time to evaluate, discuss and reflect. This Theory-Action-Reflection approach is at the core of the Academy learning cycle (see Fig. 4.3).

The important consideration of the 'T-A-R' cycle is that it is neither sequential, nor hierarchical; in other words, one can start at any of the three corners of the triangle, and move to either of the other two. From there one can return to the first, or move to the third, and cycle between all three in any sequence, depending on the need of the learner, the approach of the facilitator and the topic under scrutiny. Over time, all three will be visited. This is an adaptive, 'organic' and intuitive approach, and has proven itself successful.

Another important component of the Academy's pedagogical approach is the fact that the work is aspirational and inspirational. The organic movement is at its core a social movement: it strives for a *societal* change in the way in which agriculture is discussed; it proposes a different relationship between society and nature, between humans and ecosystems. A strong emphasis is placed on the fact that as a movement, we need to inspire our leaders and multipliers to 'be the change' and, as Antoine de Saint-Exupery may have said (1948) in *The Little Prince*: 'If you want to build a ship, don't drum up the men to gather wood, divide the work and give orders. Instead, teach them to yearn for the vast and endless sea.'

The OLC

Organic pioneers inspired and led the emergence of modern OA in the early 20th century. Actors

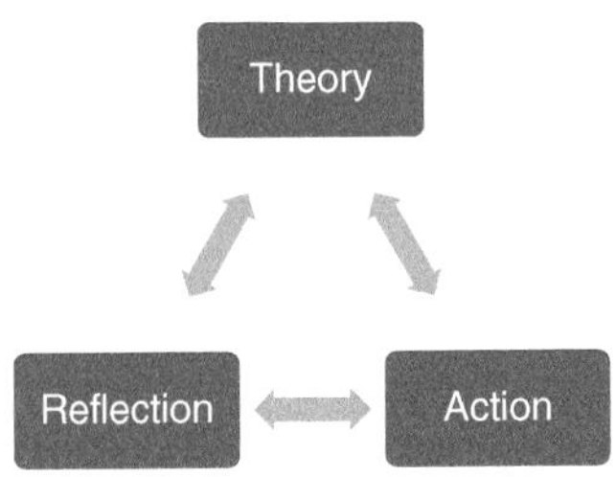

Fig. 4.3. Theory-Action-Reflection approach.

in the organic sector today share their vision of a fairer, healthier and more sustainable world, but sometimes lack the required knowledge, skills and attitude to contribute effectively to organic development.

In the OLC, both present and future leaders come together to learn, share experiences, develop innovative strategies, and build like-minded networks. The Academy has been serving the sector the past 6 years, through effective and comprehensive training programmes, developed to support the strategy of our organization and the needs of our movement.

The OLC provides a unique global learning experience. To date, we have welcomed participants from all continents to courses in India, Korea, Mexico, Switzerland, Sri Lanka, Thailand, the Netherlands, Brazil, Croatia, Germany, Uganda, Tanzania, Ghana and the Philippines.

This flagship programme spans 8–10 months with residential and online learning. The OLC entails a kick-off and closing residential session as well as monthly webinars and assignments where each individual participant actively engages in all residential training sessions (Fig. 4.4), attends monthly webinars consisting of lectures and discussions with subject experts and throughout this period, the group goes on a journey of self-discovery and learning. One of the key ingredients of the OLC is the conception and creation of an individual personal development plan to lead the organic sector in a region, organization or enterprise, with guidance from the lead trainer and peer feedback.

The greatest benefit of the OLC is the network each individual participant connects to – not only during the training, but for the rest of their career in the organic world.

The OLC was inspired by the OA Development course, known as the OAD, a scholarship programme funded by the Swedish Government and conducted for over a decade under the tutelage of Gunnar Rundgren, one of the pioneers of global organic development. The OAD DNA lives on in the OLC of the Organic Academy.

Fig. 4.4. Field excursion with Leadership Group in Kerala. (Photograph by K. Hauptfleisch.)

Short and Tailored Courses

In the further development of the Academy, it was also realized that the need is not only for developing leadership capacity, but also to develop the skills and knowledge of farmers, operators, extensionists, public sector officials and the like. Still with a focus on working with multipliers, the Academy developed additional programmes, including the Organic Foundation Course (OFC).

The OFC

This short, intense 3–4 day course focuses on the fundamental concepts of OA, value chain and certification, policy development and support networks. It is targeted to a diverse audience in public, civil society and private sectors. This course has been conducted with classes as small as 12 in Japan – part of the Asian Local Governments for OA (ALGOA) programme – and a large group of 200 participants at a food and OA-supported training session in Ulaanbaatar, Mongolia. The OFC satisfies the clear need for a general understanding of organic principles, a grasp on the complexities and challenges of organic certification and supply chains, as well as an insight into policy and advocacy. The methodology was inspired by the OLC, but as it is a much shorter programme, content is also adapted accordingly.

The main lesson learnt from such 'bespoke' trainings is that training and education, like any other service provided, has to remain relevant, contextual, adaptive and of value to its target audience. The Academy does not pretend to be a technical university or college – it can better be described as an 'embedded' insititution, working inside the movement rather than beside it. It does not have a campus – its campus is where its students find themselves – in the field, in their regions, in their context.

Both the OLC and the OFC showed the way towards more tailored and focused trainings, developed to serve very specific needs and circumstances of institutions, clients and development projects worldwide.

These trainings range from extensive training programmes with two or more residential sessions and monthly webinars similar to the OLC (Fig. 4.5), to 1-day topical trainings delivered at

Fig. 4.5. The Organic Leadership Course (OLC) in Ghana, 2018. (Photograph by K. Hauptfleisch.)

conferences or workshops. These programmes have been conducted with a diversity of partners and training given to an even greater diversity of audiences – this again illustrates how important the Academy approach is of working with the core principles and adapting content and methodology to the local context. A training in rural Africa on Participatory Guarantee Systems (PGS) looks very different to one given to government officials in North Korea!

The ability of the Academy's trainers to engage in such diverse communities and environments, indicates the success of the approach – while adapting methodology, and tailoring content, the central message of truly sustainable agriculture systems, based on organic values and principles, remains central to the curriculum. One realizes the value of such a principled approach when one sees how that core message rings true in Seoul and Pyong Yang, in Islamabad and Delhi, in Springbok and Stockholm. I had the unique privilege to have Academy participants from the USA, Iran, North and South Korea in one Leadership Group at the same time – this shows the incredible power of OA to bring people from all political, religious and ideological perspectives together to work on a common goal.

The Academy does not only work in so-called 'developing' countries, but has had success in engaging with Organisation for Economic Co-operation and Development (OECD) countries: the Organic Agriculture Academy for Extension Agents (OAAEA) is already a 6-year partnership with the Rural Development Administration (RDA) of South Korea. It consists of two residential trainings every year, developing the capacity of the RDA to advise organic farmers in conversion to OA. Here we work with researchers, scientists, extension officers – all employed in a very well-developed and centralized government institution, with extensive resources and a history of successful, Green Revolution approaches. The whole province of Goesan in South Korea, like Sikkim Province in India and the whole country of Bhutan, plans to convert to 100% organic farming.

Within various projects of IFOAM-Organics International, capacity development is a core component. To this effect, the Academy provided training in the Nutrition in Mountain Agro-ecosystems (NMA) project in five mountainous countries on three continents – Nepal, Pakistan, Kyrgyzstan, Ethiopia and Peru. Here the focus was on nutrition-sensitive agriculture, based on organic and agroecological production solutions.

For the Intercontinental Network of Organic Farmers Organisations (INOFO) project, supported by the International Fund for Agricultural Development (IFAD), the focus was on developing the capacity of grass-roots leaders to represent their constituency from local to international level, sitting beside organizations like Via Campesina (Holt-Giménez, 2006) at the IFAD Farmers' Forum in Rome.

It is still one of my fondest memories of the last 6 years, when young leaders from the Philippines, Costa Rica, Peru, India, Uganda, Zambia, Zimbabwe, SA and Senegal joined with the French-Austrian pioneer and first conveners of INOFO at a conference in Lagos, Nigeria. We broke many preconceptions that night.

Conclusion

The development of experiential training for farmers and leaders in the organic sector worldwide has allowed for the sharing of many organic innovations. I am proud to be part of this unique movement. I am proud to have had the privilege to bring disparate people together, and I am proud to see that we can use training and capacity development as tools to break barriers, to inspire, enlighten and educate.

References

Aberasturi, P. (2012) *Evaluation from Academy Alumna*. International Federation of Organic Agricultural Movements (IFOAM), Dumingag, Mindanao, the Philippines.

De Saint Exupery, A. (1948) Citadelle. Attributed (possibly an abridged quote from an American translation, not referenced in the published originals). Available at: https://tommccallum.com/2018/03/20/writing-i-love-the-little-prince/ (accessed 11 May 2018).

Freire, P. (1970) Pedagogy of the Oppressed. Available at: http://commons.princeton.edu/inclusivepedagogy/wp-content/uploads/sites/17/2016/07/freire_pedagogy_of_the_oppresed_ch2-3.pdf (accessed 11 May 2018).

Holt-Giménez, E. (2006) *Campesino a Campesino: Voices from Latin America's Farmers*. Food First Books, New York.

IAASTD (International Assessment of Agricultural Knowledge, Science and Technology for Development) (2009) *Agriculture at a Crossroads – Global Report*. Island Press, New York.

IFOAM (International Federation of Organic Agricultural Movements) (2005) Principles of Organic Agriculture. Available at: https://www.ifoam.bio/sites/default/files/poa_english_web.pdf (accessed 11 May 2018).

IFOAM (International Federation of Organic Agricultural Movements) (2008) Definition of Organic Agriculture. Available at: https://www.ifoam.bio/en/organic-landmarks/definition-organic-agriculture (accessed 11 May 2018).

Kolb, D. (2008) *Experiential Learning: Experience as the Source of Learning and Development*. Pearson Scientific, Hoboken, New Jersey.

5 Understanding a Food Systems Approach

Carola Strassner[1]* and Johannes Kahl[2]
[1]Münster University of Applied Sciences, Münster, Germany; [2]Kassel University, Kassel, Germany

Abstract
The call to adopt a food systems approach has been made increasingly over the past two decades. A systems perspective contends that the only way to understand a problem fully is to understand the elements (parts) in relation to the system (whole). Systems science and research has been pursued in a very wide range of disciplinary fields. Notwithstanding their differences, most system approaches share some common concepts, which are presented to better understand how a system may be described. Food systems serve the basic human needs of food and drink necessary in order to sustain human life, but also serve wider cultural needs. Systems thinking offers a framework for research on food systems or their subsystems, enabling a wider and deeper contextual analysis. The notion of a food system has received significant attention by key international bodies. However, the literature shows various understandings or use of terms. Used in addition to current approaches, the food systems approach may help organic food and farming by making interdependencies more apparent, examining internal congruence, and providing a real 'living laboratory' to engage in an innovative process towards sustainable food systems. Such studies can: (i) help contribute to finding transformation pathways towards sustainable food systems; (ii) provide the basis for capacity and institution building needed to ensure successful transformation; and (iii) may help to redesign food systems that better support healthy diets in an equitable way.

Introduction

Food security in the context of climate change is a major challenge, not only in Africa but around the world. Chapter 1 of this volume pointed out some of the challenges posed by sustainable development and climate change, and subsequent chapters have illustrated the challenges of supporting sustainable agriculture, and the development of the global organic movement, and more recently, the regenerative agriculture movement, to address these challenges.

Kubik and May (2018) show how weather shocks can contribute to broken food systems, which often do not nourish communities adequately, and this will be explored in Chapter 7 of this volume, while the practical application of a food systems approach will be looked at in detail in Chapter 6 through a number of case studies.

This chapter will explain what is meant by a food systems approach, defining certain terms which will be used throughout the book.

*strassner@fh-muenster.de

The Call to a Food Systems Approach

Over the past two decades the voices developing and applying a food systems approach have been steadily on the rise.

There are a number of factors contributing to this development. Among these are the increased exchanges worldwide arising from globalization, and consequently the higher level of interconnectedness and interdependence (i.e. complexity) very evident in the arena of food, farming, food security and sustainability. Furthermore, there is growing recognition that social, environmental, economic and other challenges and problems facing peoples are common to much, if not all, of humanity. The latter were first declared in the form of the Millenium Development Goals (MDGs), a set of eight goals including the eradication of extreme poverty and hunger, signed as the United Nations (UN) Millenium Declaration by 189 nation states, having been set to expire in 2015. These were followed by the Sustainable Development Goals (SDGs), a set of 17 global goals agreed upon by 193 countries and part of the wider 2030 Agenda for Sustainable Development. Food, being a central issue in daily human life everywhere, is directly or indirectly addressed by every single goal.

The call to adopt a food systems approach for a variety of challenges and issues concerning food has been made by experts in global alliances and recognized authorities such as the International Panel of Experts on Sustainable Food Systems (IPES, 2015, 2017), the Food and Agriculture Organization of the United Nations/World Health Organization (FAO/WHO, 2015) and prepared by the scientific community to apprise the UN Conference on Sustainable Development (also known as Rio+20) (Ingram *et al.*, 2012).

Hailed as a new analytical lens for sustainable food systems (IPES, 2015) it nevertheless suggests strongly that there is a need to understand what exactly a system is.

Understanding Systems

It may be argued that human thought and endeavour are dedicated to furthering our understanding of the world and hence our ability to navigate in it successfully, meeting our needs and developing our capacities. Human thought and learning may take a number of differing routes and all in sum surely bring us forward. Over the centuries we have used moral reasoning, analogy, deduction, induction and many more. Generally, the advent of what we call the scientific method came about with the work of Isaac Newton. The emerging Newtonian worldview is still dominant today, drawing also on the philosopher René Descartes, and rooted in reductionism, materialism and determinism. While it has brought phenomenal discoveries and understanding it excludes or does not address human agency, evolution, values and creativity (Heylighen, 2001). Analysis, as followed here, breaks the object of study into constituent parts and studies one or more of these in isolation. Concurrently (less apparent but not absent) we find (as mentioned in Chapter 1, this volume) what is popularly referred to as holistic thought, more formally known as a precept in Aristotle's Holism, namely that knowledge is derived from understanding of the whole and not (alone) from the understanding of the individual, single parts. This is the providence of a systems approach and of the underlying foundations, systems theories. Capra (1997, cited in Mele *et al.*, 2010) describes systems theory as 'an interdisciplinary theory about every system in nature, in society and in many scientific domains as well as a framework with which we can investigate phenomena from a holistic approach.'

Since Ludwig von Bertalanffy first worked on General System Theory in 1940 many systems theories have been promulgated over the past century over a very wide range of disciplinary fields. Von Bertalanffy developed principles and concepts that are for broad application across many domains of knowledge (von Bertalanffy, 1968).

Systems science and research has been pursued in many areas and there are many more scientists and thinkers who have contributed to building upon this body of knowledge. An indiscriminate catalogue is presented here for illustrative purposes only. These fields include organizational learning and business (Senge *et al.*, 1994), cybernetics and chaos theory (Phelan, 1999), complexity (Diamond, n.d.), biology and sociology (Maturana and Varela, 1980; Bateson, 2000; Luhmann, 2012), health (Gilson, 2012), environment (Meadows, 2009) and psychotherapy (Simon, 2004). On the basis of systems theory, applications to information systems have been developed (Forrester, 2009). Moreover, many elements from various systems theories have been integrated into what the developers term a grand unified systems theory, or 'Biomatrix Theory' (Dostal *et al.*, 2005). It is therefore not surprising that systems theory finds its way into application in the fields of food, farming, food security and sustainability.

Beyond systems science and research (i.e. the epistemological work of dealing with theory, methods and application), the term 'systems thinking' is quite commonly used. Systems thinking includes the study of systems, but many would say that it is more than that, that it is a way of thinking, a way of looking at things, a mindset that can be applied to any field. It can be likened to putting on a pair of glasses and seeing the world in a systems constructed way. It focuses on interactions of elements of a system and looks at the nature of the interaction (relationship) over time. How does energy, material and information flow among the different elements? What are the patterns of interaction? Systems thinking argues that the only way to understand a problem fully is to understand the elements (parts) in relation to the system (whole). Systems thinking in food and farming can be used to construct food systems as a concept. It is also important to draw attention to the distinction between systems (systemic) thinking, which deals with the behaviour of wholes, and systematic thinking, which means having a plan or method, and further, an orderly, methodical or reductionist approach.

Most researchers today distinguish between three traditions of systems thinking: (i) hard systems; (ii) soft systems; and (iii) critical systems (Reynolds and Holwell, 2010; Reynolds, 2011). Notwithstanding their differences, most system approaches share some common key concepts that enable researchers to adopt this perspective. These concepts that underlie most systems literature are the system basics.

System Basics

In order to work with the concept of a system, a number of terms and descriptors are needed to be able to exchange ideas and observations. Indeed, Kim (2000) posits that systems thinking can be seen as a language or a framework. The basic terms are common across systems thinking.

The *Oxford English Dictionary* explains the noun 'system' as 'A set of things working together as parts of a mechanism or an interconnecting network; a complex whole' (OUP, 2018). This definition is an apt staring point for considering systems. What do we mean by 'system'? A **system** is made up of parts, which we call **elements**; a set of interconnected and interdependent elements together forming a whole. The **interconnections** between the elements are the way the elements relate to each other or affect each other (Arnold and Wade, 2015). These interrelationships are internal to the system (F.B. Simon, 2014, unpublished lecture delivered at the Fachtag Sozialmanagement, FH Münster University of Applied Sciences, Münster, Germany).

A system is furthermore an interconnected set of elements that is coherently organized in a way that achieves something. It has a **function** or a **purpose** (Meadows, 2009). The purpose is a property of the system itself, not of its individual elements. In order to achieve the function, all elements are necessary (Kim, 1999).

A system has a **boundary** that delineates what is inside or part of the system and what is outside of and not part of the system. Setting the boundaries of systems depends on or may change with the research question. In living systems the boundaries are permeable – things can come in and pass out; usually this is a regulated or facilitated flow. Through this, a system maintains itself as a distinct entity; it preserves its identity. The metabolism of a system will achieve self-maintenance. A system continually produces, repairs and perpetuates itself. This can be applied to any system. A network is a pattern that is common to all life. The life of a system cannot be ascribed to individual elements but to the entire metabolism of the network of elements together. It produces the boundary itself. Living systems are organizationally closed but energetically and materially open, meaning that energy and matter can flow into and out of a system. The domain outside of a system is referred to as its **environment**. A system encounters physical, chemical and sometimes social constraints of its environment. A system transforms input via throughput to outcomes and is influenced by external and internal drivers. A system can exist within a system: that would be a **subsystem** in a system; this is referred to as being nested. Systems can also overlap with other systems. The relational constellation of the elements gives rise to the **structure** of a system.

However, a system as we use the term here is also **dynamic** in that it is characterized by activity and change. Systems are dynamic, in constant motion, fluid and undergoing change. Notwithstanding the dynamic nature of a system, it will endeavour to maintain stability. It has spatial and temporal dimensions and it exhibits a **behaviour** that can be characterized. Consistent behaviour is managed by **feedback loops**. If the feedback loop reinforces the pattern of behaviour it is a positive feedback loop, if it balances it is a negative feedback loop.

Forms and functions are **emergent properties** (i.e. properties unique to the system and not necessarily expressed or explained by one or more of the system's elements). Together this indicates the **complexity** of a system.

Many things are systems; possibly best known is the term 'ecosystem'. A school, a company, a national economy, a bicycle, a tree or a forest, a farm or a digestive system; all these are examples of systems.

Applying Systems Thinking to Construct Food Systems

To look at a 'food system' the food needs to be connected with the people and/or the people with the food. Food systems serve the basic human needs of food and drink necessary in order to sustain human life. If the function or purpose of a global food system is to feed the global population and the outcome is food insecurity and not food security, then something needs to be optimized (WFP, 2018). As currently millions of people worldwide still suffer food insecurity and hunger, resulting in ill health and premature death, the global food system needs transformation. Taking environmental challenges such as climate change and loss in biodiversity into account as well, the global food system needs to be transformed towards enhanced sustainability in all dimensions (de Schutter, 2014; FAO, 2015a, b, 2017; GloPan, 2016). Further, it seems that making interventions here needs a systems approach (Ostrom, 2009).

Food systems are an interconnected web of activities, resources and people, which involve all domains of the food value chain and more. The organization of food systems reflects and responds to social, cultural, political, economic, health and environmental conditions and can be identified at multiple scales, from individual people to local and regional communities, to nations. To fully understand and develop alternatives to the problems facing global food systems demands a systemic and holistic approach.

Food systems may be entered on different scales and on different levels (Cash *et al.*, 2006). One can look at the full food system on a global scale, or downscale to national food systems (feeding nations) or populations or groups of people. We may ask if a food system can be a regional one. It is a matter of the focal point, of what is in focus. When taking the regional scale, it is about feeding the people of a region, not about the region itself. Feeding a people also means covering all or almost all food and drink (nourishment) needs, so that the study of a regional food system may show how (in)complete it is in this respect (and how much it needs imports into its region in order to achieve comprehensive feeding status).

Scaling down more, a school food system will feed the pupils and perhaps teachers, those members of that (social) organism. The edges of the school food system may be fuzzy so individual school food systems may also feed further members of that community, such as visitors or parents, but its function or purpose is to feed the school pupils. The food systems approach described here is a way of thinking that can be applied to all fields of human endeavour. Systems thinking offers a framework for research on food systems or their subsystems, enabling a wider and deeper, contextual analysis. Systems science may contribute to accelerate the shift towards more sustainable food systems. In other words, one can look at the flows through a system – what flows (matter, energy, information) and the nature of those flows (speed, ease, paths) through it.

Relationships between the elements of the food system and the nature of those relationships can be studied (whether they are equitable, fair, etc.). Furthermore, outcomes of food systems such as health or food literacy or ecological literacy may be assessed. Structures and forms of organization – by design, by evolution, by self – may also be analysed. And also the dynamics of a system, the nature of change and adaptation or the levers that contribute to the greatest change with the least effort can be in focus. Even impacts on other systems (e.g. the textile system) or interactions with other systems (e.g. the energy system) can be described. One of the most challenging ways to apply a systems approach to food and farming is to study transformation of systems from less sustainable to more sustainable food systems and to monitor the transformation processes, assessing how successful they are.

Academic/Scholarly Food Systems Approaches

The notion of a food system has received significant attention by key international bodies, including the FAO, the specialized agency of the UN that leads international efforts to defeat hunger and achieve food security for all. The High Level Panel of Experts (HLPE) on Food Security and Nutrition that advises the Committee on World Food Security (CFS) provides a definition of a food system as follows:

> A food system gathers all the elements (environment, people, inputs, processes, infrastructures, institutions, etc.) and activities that relate to the production, processing, distribution, preparation and consumption of food and the outputs of these activities, including socio-economic and environmental outcomes.
>
> (HLPE, 2014)

Furthermore, it provides a definition for a sustainable food system as follows:

> A sustainable food system is a food system that delivers food security and nutrition for all in such a way that the economic, social and environmental bases to generate food security and nutrition for future generations are not compromised.
>
> (HLPE, 2014)

It may be inferred that a sustainable food system should give rise to sustainable diets (Meybeck and Gitz, 2017). These are defined by FAO as:

> Sustainable diets are those diets with low environmental impacts which contribute to food and nutrition security and to healthy life for present and future generations. Sustainable diets protect and respect biodiversity and ecosystems while being culturally acceptable, accessible, affordable, nutritionally adequate, safe, and healthy.
>
> (Burlingame and Dernini, 2012)

However, the literature shows various understandings or use of terms. Literature is rich in examples on how food systems approaches can be organized and differentiated following a specific purpose and methodology. In the following some examples are given in order to illustrate the broad understanding of a food systems approach.

According to United Nations Environment Programme (UNEP, 2016) taking a food systems approach allows the food chain activities to be linked to their social and environmental context: 'A key element in defining food systems is how the system's activities and actors are linked with each other'. Food systems can be recognized as coupled human and natural systems (Liu *et al.*, 2007a, b) (i.e. complex social-ecological systems) (Ostrom, 2009; Foran *et al.*, 2014; Tendall *et al.*, 2015; Allen and Prosperi, 2016).

For Nesheim *et al.* (2015) food systems as complex adaptive systems are 'composed of many heterogeneous pieces, whose interactions drive system behaviour in ways that cannot easily be understood from considering the components separately'. In this context, Grant (2015) defines food systems as complex adaptive systems giving the meaning:

> that the whole system has properties greater than the sum of its parts, that there are high levels of connectedness and interaction across scales and levels and with diverse agents, and that outcomes may be reached by many possible pathways which are influenced by many factors.

This refers to basic concepts of systems thinking (Meadows, 2009) and system theories (von Bertalanffy, 1968).

Generally coupled human–natural systems 'are not static; they change over time' (Liu *et al.*, 2007a, b). Gladek *et al.* (2016) provide a framework for analysing food systems by the structures, the actors and the impacts. The actors define the behaviour of the system and interact with the structures and they divide the impacts into biophysical and health/well-being impacts (the state of the system).

Ericksen (2008) and Ingram (2011) go further, and link their focus on food security to integrating the different dimensions (food utilization, access and availability). Their framework identifies entry points for changing undesirable outcomes. Ericksen *et al.* (2010) identified external drivers of food systems. They group these drivers into Global Environmental Change drivers as well as socio-economic drivers (Ingram, 2011). Based on this, they suggest working primarily on processes and factors influencing outcomes within and outside a food system as well as looking at trade-offs between these. For them, several approaches and methods may be adopted in which these outcomes can be evaluated, 'depending upon the perspective or objectives of the evaluator, which are shaped by the political and social context' (Ericksen *et al.*, 2010).

Herforth and Ahmed (2015) also focus on food environment and add convenience, and use the broader term 'desirability' which encompasses (but is more than) food quality. Another definition of food environment is given by HLPE (2017): 'Food environment refers to the physical, economic, political and socio-cultural context in which consumers engage with the food system to make their decisions about acquiring, preparing and consuming food.'

Is it Useful to Apply the Food Systems Approach to Organic Food and Farming?

The food systems approach need not replace other ways of seeing or learning such as those mentioned above; instead it adds to the ways usually used to see and learn. In so doing it may help in two ways:

1. Taking a food systems approach to organic food and farming may add something or make something apparent that would not be discovered by looking at single parts (e.g. agricultural practices, consumption patterns or processing methods). An organic food system is about 'feeding people organically'. So additionally, it is an instrument for the organic community to examine whether the purpose is being achieved according to its own precepts (the four principles given in Chapters 2 and 4, this volume) or whether there are areas that need to be addressed.

2. The SDGs show the need to make food systems more sustainable through a shift towards more sustainable farming and processing methods, and healthier food consumption patterns. One may hypothesize that organic food systems act as a model for and, indeed, a living example of a sustainable food system that already has regionally and culturally appropriate diversity. Its approximation of a sustainable food system can be argued on the basis of already existing knowledge, even if it is not yet comprehensive. Hence there is much to learn from observing and studying an organic food system in practice. To become such a role model, the organic food system has to improve where it underperforms, and to develop more comprehensively. This makes the current organic food system a real 'living laboratory' where researchers and stakeholders can engage in an innovative process towards sustainable food systems, integrating all food value chain actors and others involved.

There are some advantages to taking an organic food system as a model. It is globally defined through both ethical principles and assurance criteria (Strassner *et al.*, 2015). This allows comparability and can help address the question of transferability to other regions and cultures as well as scaleability. Here the International Federation of Organic Agriculture Movements (IFOAM) Norms and Best Practice Guidelines (see www.ifoam.bio) can serve as the theoretical foundation and reference of an organic food system and can be applied as a framework for assessing different case studies everywhere in the world. These cases may be described and analysed according to different scales, such as their history of local applications of global principles in almost all climate zones. They can be assessed based on data across sustainability dimensions on primary and secondary production and (increasingly) consumption; in this regard, sustainability indicators such as those proposed in Chapter 24 of this volume may be used. Furthermore, other activities and dimensions may be taken into account such as policies, regulations, certification and labels, education initiatives, technology assessments, stakeholder engagement, advocacy and mobilization.

Conclusion

Taking a food systems view may help to advance understanding of the sustainability of organic food systems, and identify innovations in food production, processing and consumption that increase food systems sustainability (Kahl *et al.*, 2016). Taking organic food systems as an example, the dynamics of present food systems and the leverage points for promoting changes in these food systems towards increased sustainability may be better understood. Such studies can help contribute to finding transformation pathways or conditions and challenges to establishing sustainable food systems. This may not only contribute towards improving organic food systems themselves but also make lessons learned available for different food system groups such as scientists, decision makers and practitioners. Those lessons learned from applying a food systems approach to organic food and farming can help develop and plan strategies for promoting more diverse and sustainable production systems with a better adaptation to different agroecological, cultural, social and political contexts. They will furthermore provide the basis for capacity and institution building needed as a ready basis to ensure successful transformation. They may help to redesign food systems that better support healthy diets in an equitable way with improved efficiency and sustainability. A food systems approach to organic food and farming can be shared with diverse groups, whichever food system group they belong to. There have been many projects arising over the last few years which apply a food systems lens to sustainable development and food security, as will be shown in the next chapter.

References

Allen, T. and Prosperi, P. (2016) Modeling sustainable food systems. *Environmental Management* 57, 956–975.

Arnold, R.D. and Wade, J.P. (2015) A definition of systems thinking: a systems approach. Conference on Systems Engineering Research. *ScienceDirect Procedia Computer Science* 44, 669–678.

Bateson, G. (2000) *Steps to an Ecology of Mind*. University of Chicago Press, Chicago, Illinois.

Burlingame, B. and Dernini, S. (eds) (2012) *Sustainable Diets and Biodiversity: Directions and Solutions for Policy, Research and Action*. Food and Agriculture Organization of the United Nations (FAO) Nutrition and Consumer Protection Division, Rome.

Capra, F. (1997) *The Web of Life*. Doubleday-Anchor Books, New York.

Cash, D.W., Adger, W., Berkes, F., Garden, P., Lebel, L., Olsson, P., Pritchard, L. and Young, O. (2006) Scale and cross-scale dynamics: governance and information in a multilevel world. *Ecology and Society* 11(2), 8.

De Shutter, O. (2014) Final report: The transformative potential of the right to food. Report presented at the 25th session of the United Nations Human Rights Council. Available at: http://www.srfood.org/images/stories/pdf/officialreports/20140310_finalreport_en.pdf (accessed 5 May 2018).

Diamond, L. (n.d.) Twelve Simple Rules of Systems Thinking for Complex Global Issues. 4 pp. Available at: https://www.heartoftheart.org/?p=3978 (accessed 6 May 2018).

Dostal, E. in collaboration with Cloete, A. and Járos, G. (2005) *Biomatrix: a Systems Approach to Organisational and Societal Change*, 3rd edn. Imaging Data Solutions, Cape Town, South Africa.

Ericksen, P.J. (2008) Conceptualizing food systems for global environmental change research. *Global Environmental Change* 8(18), 234–245.

Ericksen, P.J., Stewart, B., Dixon, J., Barling, D., Loring, P., Anderson, M. and Ingram, J. (2010) The value of a food system approach. In: Ingram, J., Ericksen, P. and Liverman, D. (eds) *Food Security and Global Environmental Change*. Earthscan, London, pp. 25–45.

FAO (Food and Agriculture Organization of the United Nations) (2015a) *Climate Change and Food Systems: Global Assessments and Implications for Food Security and Trade*. FAO, Rome.

FAO (Food and Agriculture Organization of the United Nations) (2015b) *Final Report for the International Symposium on Agroecology for Food Security and Nutrition*. 18–19 September 2014, Rome, Italy. FAO, Rome.

FAO (Food and Agriculture Organization of the United Nations) (2017) *The Future of Food and Agriculture – Trends and Challenges*. FAO, Rome.

FAO/WHO (Food and Agriculture Organization of the United Nations/World Health Organization) (2015) *Second International Conference on Nutrition*. Report of the Joint FAO/WHO Secretariat on the Conference. FAO, Rome.

Foran, T., Butler, J.R.A., Williams, R.J., Wanjura, W.J., Hall, A., Carter, L. and Carverry, P.S. (2014) Taking complexity in food systems seriously: an interdisciplinary analysis. *World Development* 61, 85–101.

Forrester, J.W. (2009) *Some Basic Concepts in System Dynamics*. Sloan School of Management, Massachusetts Institute of Technology. Available at: www.cc.gatech.edu/classes/AY2013/cs7601_spring/papers/Forrester-SystemDynamics.pdf (accessed 23 May 2018).

Gilson, L. (ed.) (2012) *Health Policy and Systems Research: a Methodology Reader*. World Health Organization and Alliance for Health Policy and Systems Research, Geneva, Switzerland.

Gladek, E., Fraser, F., Roemers, G., Muñoz, O.S., Kennedy, E. and Hirsch, P. (2016) *The Global Food System: an Analysis*. WWF, Zeist, the Netherlands.

GloPan (Global Panel on Agriculture and Food Systems for Nutrition) (2016) *Food Systems and Diets: Facing the Challenges of the 21st Century*. GloPan, London. Available at: https://glopan.org/sites/default/files/ForesightReport.pdf (accessed 6 May 2018).

Grant, M. (2015) A food systems approach for food and nutrition security. *Sight and Life* 29(1), 87–90.

Herforth, A. and Ahmed, S. (2015) The food environment, its effects on dietary consumption, and potential for measurement within agriculture-nutrition interventions. *Food Security* 7(3), 505–520.

Heylighen, F. (2001) The science of self-organization and adaptivity. In: Kiel, L.D. (ed.) *Knowledge Management, Organizational Intelligence and Learning, and Complexity*. Volume 5(3) of *The Encyclopedia of Life Support Systems (EOLSS)*. Eolss Publishers, Oxford, pp. 253–280.

HLPE (High Level Panel of Experts on Food Security and Nutrition) (2014) *Food Losses and Waste in the Context of Sustainable Food Systems*. A report by the HLPE of the Committee on World Food Security, Rome.

HLPE (High Level Panel of Experts on Food Security and Nutrition) (2017) *Nutrition and Food Systems*. A report by the HLPE of the Committee on World Food Security, Rome.

Ingram, J. (2011) A food systems approach to researching food security and its interactions with global environmental change. *Food Security* 3(4), 417–431.

Ingram, J., Aggarwal, P., Ericksen, P., Gregory, P., Horn-Phathanothai, L., Misselhorn, A. and Wiebe, K. (2012) *Rio+20 Policy Brief. Food Security for a Planet Under Pressure*. Commissioned by the international conference Planet Under Pressure: New Knowledge Towards Solutions. International Geosphere Biosphere Programme, London.

IPES (International Panel of Experts on Sustainable Food Systems) (2015) The case for a new science of for sustainable food systems. Report No. 01, 2015. IPES Food. Available at: http://www.ipes-food.org/_img/upload/files/NewScienceofSusFood.pdf (accessed 1 October 2017).

IPES (International Panel of Experts on Sustainable Food Systems) (2017) *Unravelling the Food-Health Nexus: Addressing Practices, Political Economy, and Power Relations to Build Healthier Food Systems*. The Global Alliance for the Future of Food and IPES-Food. IPES-Food, Brussels. Available at: http://www.ipes-food.org/_img/upload/files/Health_ExecSummary(1).pdf (accessed 4 January 2019).

Kahl, J., Strassner, C., Hertwig, J., Gould, D., Bügel, S., Paoletti, F. and Lairon, D. (2016) Learning from the organic food system as a model for sustainable food systems – the Organic Food System Program. In: Meybeck, A. and Redfern, S. (eds) *Sustainable Value Chains for Sustainable Food Systems*. Food and Agriculture Organization of the United Nations (FAO), Rome, pp. 295–302.

Kim, D.H. (1999) *Introduction to Systems Thinking*. Pegasus Communications, Westford, Massachusetts.

Kim, D.H. (2000) *Systems Thinking Tools: a User's Reference Guide*. Pegasus Communications, Westford, Massachusetts.

Kubik, Z. (2018) Weather shocks, migration and food security: evidence from sub-Saharan Africa. PhD thesis, Paris-Sorbonne University, Paris.

Liu, J., Dietz, T., Carpenter, S.R., Alberti, M., Folke, C., Moran, E., Pell, A.N., Deadman, P., Kratz, T., Lubchenco, J., Ostrom, E., Ouyang, Z., Provencher, W., Redman, C.L., Schneider, S.H. and Taylor, W.W. (2007a) Complexity of Coupled Human and Natural Systems. *Science* 317(5844), 1513–1516.

Liu, J., Dietz, T., Carpenter, S.R., Folke, C., Alberti, M., Redman, C.L., Schneider, S.H., Ostrom, E., Pell, A.N., Lubchenco, J., Taylor, W.W., Ouyang, Z., Deadman, P., Kratz, T. and Provencher, W. (2007b) Coupled human and natural systems. *AMBIO: a Journal of the Human Environment* 36(8), 639–649.

Luhmann, N. (2012) *Soziale Systeme*, 15th edn. Suhrkamp Taschenbuchverlag, Frankfurt am Main, Germany.

Maturana, H.R. and Varela, F.J. (1980) *Autopoiesis and Cognition. The Realization of the Living*. Series: *Boston Studies in the Philosophy and History of Science* Vol. 42. Springer, Dordrecht, the Netherlands.

Meadows, D.H. (2009) *Thinking in Systems. A Primer*. Sustainability Institute, Earthscan, London. Available at: https://wtf.tw/ref/meadows.pdf (accessed 10 November 2015).

Mele, C., Pels, J. and Polese, F. (2010) A brief review of systems theories and their managerial applications. *Service Science* 2(1/2), 126–135.

Meybeck, A. and Gitz, V. (2017) Sustainable diets within sustainable food systems. *Proceedings of the Nutrition Society* 76, 1–11.

Nesheim, M.C., Oria, M. and Yih, P.T. (eds) (2015) *A Framework for Assessing Effects of the Food System*. Institute of Medicine and National Research Council, The National Academies, Washington, DC.

Ostrom, E. (2009) A general framework for analyzing sustainability of social-ecological systems. *Science* 325, 419.

OUP (Oxford University Press) (2018) *Oxford English Dictionary*. Available at: https://en.oxforddictionaries.com/definition/system (accessed 5 May 2018).

Phelan, S.E. (1999) A note on the correspondence between complexity and systems theory. *Systemic Practice and Action Research* 12(3), 237–246.

Reynolds, M. (2011) Critical thinking and systems thinking: towards a critical literacy for systems thinking in practice. In: Horvath, C.P. and Forte, J.M. (eds) *Critical Thinking*. Nova Science Publishers, New York, pp. 37–68.

Reynolds, M. and Holwell, S. (2010) Introducing systems approaches. In: Reynolds, M. and Holwell, S. (eds) *Systems Approaches to Managing Change: a Practical Guide*. Springer, London, pp. 1–23.

Senge, P.M., Roberts, C., Ross, R.B., Smith, B.J. and Kleiner, A. (1994) *The Fifth Discipline: the Art and Practice of the Learning Organisation*. Nicholas Brealey Publishing, London.

Simon, F.B. (2004) *Patterns of War. Systemic Aspects of Deadly Conflicts*. Translated by Sally und Bernd Hofmeister Carl-Auer. Verlag, Heidelberg, Germany.

Strassner, C., Cavoski, I., Di Cagno, R., Kahl, J., Kesse-Guyot, E., Lairon, D., Lampkin, L., Loes, A.K., Matt, D., Niggli, U., Paoletti, F., Pehme, S., Rembialkowska, E., Schader, C. and Stolze, M. (2015) How the organic food system supports sustainable diets and translates these into practice. *Frontiers in Nutrition | Nutrition and Environmental Sustainability* 2, 19.

Tendall, D.M., Joerin, J., Kopainsky, B., Edwards, P., Shreck, A., Le, Q.B., Kruetli, P., Grant, M. and Six, J. (2015) Food system resilience: defining the concept. *Global Food Security* 6, 17–23.

UNEP (United Nations Environment Programme) (2016) *Food Systems and Natural Resources*. A Report of the Working Group on Food Systems of the International Resource Panel. UNEP, Geneva, Switzerland.

Von Bertalanffy, L. (1968) *General System Theory: Foundations, Development, Applications*. George Braziller, New York.

WFP (World Food Programme) (2018) *Food Systems*. Available at: http://www1.wfp.org/food-systems (accessed 31 May 2018).

6 BERAS – a Global Network of Food Systems with Examples from Sweden, Haiti, Tanzania and India

Jostein Hertwig*
CEO BERAS International Foundation, Järna, Södertälje, Sweden

Abstract

Adopting a food systems approach in practice starts from recognizing that food systems serve the basic human needs of food and drink needed to sustain human life, but also serve wider cultural needs. The food systems approach helps organic food and farming by making interdependencies more apparent, examining internal congruence, and providing a real 'living laboratory' to engage in an innovative process towards sustainable food systems.

The original Baltic project of Building Ecological Regenerative Agriculture and Societies brought together many stakeholders in the large area around the Baltic Sea (with a population of 90 million people) to manage natural resources, reduce pollution and improve recycling systems. The project sought to present a realistic, fully integrated ecological alternative for a systemic shift in the whole food chain from farmers to consumers; and thereby to revitalize agricultural and rural development sectors in an economically, socially, culturally and environmentally sustainable manner. Results included: (i) reduction of nitrogen runoff to the Baltic Sea by 47%; and (ii) an increase in soil carbon sequestration of 0.4 t C/ha/year (from on-farm research, 1991–2007).

The three main concepts used are: (i) ecological recycling/regenerative agriculture (ERA); (ii) Diet for a Green Planet (DGP); and (iii) local Sustainable Food Societies (SFS). These concepts resulted in the establishment of learning centres (LCs) around the Baltic, and these were later extended to Haiti, the Dominican Republic, Tanzania and India (where there are now seven LCs). The project has helped communities to set goals for food security and local ecological management, and to build extensive networks.

Building on holistic philosophies which include local values and spiritual and cultural norms, it has been possible to combine the findings of research into biological systems with the local knowledge of communities, in such a way that the communities have been able to take charge of their development in a way which integrates science, technology, culture and human solidarity, so that responses to climate change, globalization and food insecurity build on best global experience.

These case studies contribute to showing transformation pathways towards sustainable food systems, and provide the basis for capacity and institution building needed to ensure successful transformation. They redesign food systems to support healthy diets in a more equitable way.

*jostein.hertwig@beras.eu

Introduction

Chapter 5 explained systems thinking, and how the elements of a system are more than the sum of the parts. The consideration of production, processing, value chains, markets, food choices and cooking, and their relationship to local culture and natural resource management makes for complex analysis. The case studies in this chapter include some (but not always all) of these elements, and are characterized by diversity and adaptation to local customs, indigenous technical knowledge and resources. Many are based on the original Building Ecological Regenerative Agriculture and Societies (BERAS) project which sought cooperation between many stakeholders on the Baltic Sea; this later grew into the BERAS of today, as explained below.

BERAS and Järna in Sweden

A source of inspiration for BERAS flows from the cluster of biodynamic initiatives in Järna, in Sweden. Here a number of Swedish pioneers of research and practical implementation of the basic organic principles of health, ecology, fairness and care, provided some impulses which can guide the organic movement, and which also illustrate a food systems approach. Started many decades ago, Järna (near Södertälje, west of Stockholm) is still a vibrant community with farms, gardens, food processing, wholesalers, shops, restaurants, research, adult education initiatives, schools, its own bank, a hospital, curative homes, social and cultural activities.

As mentioned in Chapter 1 (this volume), the editor visited the research trials of Bo Pettersson (of the Järna-based Nordic Research Circle for Biodynamic Farming) in Uppsala in 1976, where long-term comparative research trials were conducted from 1958 to 1990 (Granstedt and Kjellenberg, 1997). At that time, activities based on the educational, architectural, agricultural and social ideas of the anthroposophist Dr Rudolf Steiner (see Box 6.1) were already flourishing in Järna.

Value-based and holistic-orientated clusters such as Järna may have an important role to play for the development of sustainable food systems, if they meet and interact meaningfully with the surrounding society. To become a real and practical working impulse for the change needed requires certain skills such as social entrepreneurship, inclusiveness, interaction and the creation of an atmosphere of openness, sharing and listening to each other's ideas, aspirations and goals. Of major importance for BERAS is our inclusion in the Organic Food System Programme (OFSP) as a core initiative of the United Nations (UN) Sustainable Food Systems Programme, and the work with the UN

Box 6.1. Rudolf Steiner

As mentioned in Chapter 2 (this volume), Rudolf Steiner was the architect of biodynamic farming; he also contributed to thinking on holistic medicine, Waldorf education, projective geometry, and the provision of the craft- and agriculture-based Camphill movement for children (and later adults) in need of special care. In founding an 'anthroposophical society' after his early involvement with the Theosophical Society, he explained that theosophy means knowledge of God, but that it is difficult for human beings to know God until they know themselves. His anthroposophy therefore developed a spiritual science, taking further what Jan Smuts had to say about holism and evolution (see Chapter 1, this volume). Similar to Smuts' conviction that matter, life and mind formed a continuum and were all part of what he called 'one great process', Steiner's PhD explored concept and percept, and developed a monistic approach to knowing about the non-physical world, as opposed to the dualism of Descartes and Immanuel Kant, who divided the world into that which we can know about through empirical, physical research, and the world of faith, based on that which we believe. His book *The Philosophy of Freedom* (Steiner, 1894) is based on his doctoral work, and was originally translated by Professor Hoernle (and his wife) of the Witwatersrand University in Johannesburg. Whether one follows Steiner's approach to spiritual science or not, it is important to understand the role of values and beliefs in determining how different cultures view the world, and how they can come together to undertake holistic joint actions, which can help them to move towards sustainable ecosystem management.

Sustainable Development Goals (in particular related to Goal no 12: Sustainable Consumption and Production Patterns), as referred to in Chapter 5 of this volume.

Following the conclusion of two European Union (EU) projects, BERAS was consolidated and further developed where it started (in Södertälje Municipality in Sweden), and then shared with initiatives in the Dominican Republic, Haiti, Tanzania and India and also provided many other inputs into the OFSP. In the initial phase, the acronym BERAS had the meaning: Baltic Ecological Recycling Agriculture and Societies. Now that the consolidated phase is reaching out to initiatives globally, this was changed to: Building Ecological Regenerative Agriculture and Societies. The BERAS cases both in Södertälje and as a global network of food systems are living laboratories for some deeper research into more comprehensive application of organic food systems. We need to move from a limited and short-term 'ego orientated system' to an 'eco system' awareness that emphasizes the well-being of the whole system, while understanding and strengthening the constituent elements making up the system.

Early History: Eutrophication as a Major Environmental Challenge for the Baltic Sea and the Two BERAS EU Projects

The Baltic Sea is one of the largest bodies of brackish (part saline) water in the world. It is relatively shallow and almost completely enclosed. Only 3% of the water (by volume) is exchanged each year, implying that it will take more than 30 years for replacement of the total volume. Rivers drain a land area four times larger than the sea itself with a population of nearly 90 million.

Pollution by nutrients (predominantly nitrates and phosphates) cannot easily be absorbed but has rapid and visible impacts. The increasing algal blooms, covering more of the sea each summer, are the result. These algae consume oxygen at the expense of fish and other forms of life and result in dead sea bottoms. The problem is increasing.

One of the major causes of eutrophication comes from excessive nutrient inputs of nitrogen and phosphatic fertilizers from specialized farming practices where crop and animal production are separated, and there are linear flows of plant nutrients.

Addressing the problem in broad terms related to agriculture the challenges are twofold: (i) reducing the intensive, specialized agriculture in the north and west; and (ii) presenting realistic alternatives to the intensive, specialized agriculture for the development in the east and west of this region.

The BERAS example as a food system has been developed through two transnational projects part-financed by the EU and Norway 'Baltic Sea Region Programme', BERAS (2003–2006) and BERAS Implementation (2010–2013). These make up the common efforts by the partnership from 11 countries in the Baltic Sea region: Sweden, Denmark, Germany, Poland, Belarus, Lithuania, Latvia, Estonia, Finland, Russia and Norway (Fig. 6.1). It includes national and local authorities, universities and research institutes, advisory services, ecological and environmental NGOs, farmers organizations, food chain actors and finance institutions – altogether 24 project partners and 35 associated organizations working with the BERAS Implementation team (Fig. 6.2). It is based on an interdisciplinary research, extension and implementation approach including key actors from society.

As an important starting point we acknowledged that organic agriculture (OA) and advice existed, but that it was too narrow in its approach and did not take a systems view. Moreover, to some extent it was recognized that the practical implementation side of OA did not exploit the full potential of its basic principles. The contribution to the challenge from BERAS was to develop specific advice for effective recycling of nutrients at farm level and at the same time to take a systems approach connecting the whole value chain, policy makers and research. To some extent we entered the path of bringing back the original good ideas forming the basis for the organic movement (Figeczky, 2018).

The main objectives for the BERAS projects were:

- to present a realistic, fully integrated ecological alternative for a systemic shift in the

Fig. 6.1. Baltic Sea region. (From Interreg Baltic Sea Region, 2007–2013.)

whole food chain from farmers to consumers; and thereby

- to revitalize agricultural and rural development sectors in an economically, socially, culturally and environmentally sustainable manner.

As Dr Artur Gransted remarked:

> When we at the Swedish Biodynamic Institute initiated these two BERAS projects we were quite unique in taking a holistic view of the problems of Baltic Sea eutrophication in relation to farming and food systems. The visionary

Fig. 6.2. BERAS Implementation team at a conference in Järna, Södertälje, Sweden in 2010. (From BERAS.)

> combination of research and practical good examples is essential for the transition to sustainable food production.
>
> (Granstedt, pers. comm., 2018)

BERAS 2003–2006 (lead partner: Swedish University of Agricultural Science)

Results from the BERAS project 2003–2006 showed that an ecological intensification approach has the following potential:

- to reduce nitrogen runoff to the Baltic Sea by 47% (Granstedt *et al.*, 2008); and
- to increase soil carbon sequestration by 0.4 t C/ha/year (from on-farm research, 1991–2007; Granstedt and Kjellenberg, 2008).

A number of BERAS scientific reports are available online (listed in the Bibliography and References section).

BERAS 2010–2013 (lead partner: Södertörns University, Sweden)

The BERAS project 2010–2013 focused on concept development and implementation where research, innovation and entrepreneurship flowed from multi-sectoral engagement into realistic fully integrated ecological alternatives for the whole food chain from farmer to consumer, both at the level of individuals and at the collective societal level. The three main concepts (derived from systems thinking as shown in Chapter 5, this volume) are as follows:

- ecological recycling/regenerative agriculture (ERA);
- Diet for a Green Planet (DGP); and
- local Sustainable Food Societies (SFS).

ERA is based on the ecological principles of renewable local resources, recycling and biodiversity which see the farm as part of the ecosystem working with natural cycles, organic crop production and animal husbandry. The number of animals is balanced with the farm's own fodder available on the farm. Combined with the cultivation of legumes and grasses as part of the crop rotation, the farms can reach a high degree of self-sufficiency in fodder and fertilizer.

The consumer engagement concept 'Diet for a Green Planet' offers a sustainable lifestyle with consumption of enough and good food

Fig. 6.3. Linking the farmer and the community (a); integrating activities in the food value chain (b).

without threatening the environment. DGP should be developed based on local/regional conditions such as dietary choices and influenced by culture, nutritional knowledge, price of product, availability, taste and convenience. The basic criteria can be summarized as follows:

- availability of tasty and nutritious food;
- food produced according to principles of ERA;
- local food and according to season;
- balanced in vegetables and meat/fish (80% vegetables and 20% meat/fish as a guideline); and
- reduction in food waste.

Local SFS and Learning Centres (LCs)

Local SFS are examples of local food clusters that are socially just, environmentally friendly and economically viable. They involve all actors in the food chain – from farmers to consumers and with the cooperation between universities/research institutions, authorities, business sector and NGOs (Fig. 6.3).

Within each SFS there may be LCs both in rural and in urban areas. They are meeting places – more or less formal, depending on local preferences – that engage in several aspects of a food system. They can be on a farm, a university, a school, NGO, shop, restaurant, etc. based on local preferences and opportunities.

To support conversion to ERA farming, the concepts of DGP and local SFS the project developed:

- guidelines, handbooks and educational material;
- environmental evaluation studies; and
- economic and social impact studies and policy recommendations.

Following the completion of the project a field of interest is a Farmers University Exchange Programme giving students from Belarus and Lithuania the opportunity to learn about ERA farming and practise on Swedish organic farms.

All papers and documents and resources otherwise are available at www.beras.eu.

'The base case' – Södertälje food system

Contact: Jostein Hertwig (jostein.hertwig@beras.eu; http://foodsociety.se/en/ and www.beras.eu)

The Järna cluster includes many of the actors that constitute a food system. A missing element could be the connection to authorities and policy making. In BERAS Implementation (2010–2013) Södertälje Municipality took a major role in developing the food concept 'Diet for a Clean Baltic' (later 'Diet for a Green Planet' (DGP)). In Södertälje a strategic decision was taken in 2001 'Food – the key to a better future: health, environment, good place to work, a viable community'. So here there were good synergies between the objectives of the BERAS project and the strategy of Södertälje Municipality. It also resulted in fruitful long-term cooperation with the municipality and Järna. Following the BERAS project, Södertälje Municipality has been involved in various projects related to sustainable food and agriculture and always with the basic holistic concepts of agriculture, food and societies developed together. Some highlights from Södertälje are:

- Today Södertälje Municipality has implemented DGP in their 90 kitchens for kindergartens, schools and homes for the elderly. They serve 24,000 meals every day, and with 60% organic ingredients.
- In the local Coop store in Järna, 34% of the total sale of food is organic.
- Saltå Kvarn, which is a local wholesaler, is regularly acknowledged as the 'best sustainability brand' in Sweden.

The diversity of actors as the original BERAS opens up to differing situations worldwide, may be seen in Fig. 6.4.

Sharing our work and concepts with initiatives in Tanzania, Haiti and India

Following the conclusion of the initial EU projects, BERAS was consolidated and further developed in Södertälje Municipality in Sweden, and collaborated with initiatives in the Dominican Republic, Haiti, Tanzania and India. The section below has been prepared with inputs from colleagues in all countries involved.

BERAS in Tanzania, Haiti and India

Tanzania – Manyara Organic Farming Initiative (MOFI)

Contact: Vesa-Matti Loiske, Assistant Professor, Södertörns University (Swedenvesa-matti.loiske@sh.se)

Fig. 6.4. The diversity of actors based on the Södertälje food system. (From BERAS.)

MOFI works to strengthen small-scale organic farming communities in Babati and Hanang districts in Manyara Region, Tanzania to improve production and resilience of organic farming and develop a sustainable organic food system. Farmers here, especially women and young people, are generally poor and lack tenure security of land, with agriculture and cattle rearing as the main livelihoods (Fig. 6.5); they are food insecure and belong to the groups most vulnerable to climate change. As a first concrete target MOFI will strengthen the capacity of ten local small-scale organic farmers groups (200 farmers) in ten villages, with 50% men and 50% women, by connecting them to the global organic movement, by organizing Village Community Banks, by acting as their spokesperson towards authorities and by organizing workshops where global experiences, knowledge, agricultural techniques, value adding activities and markets for organic products will be disseminated. MOFI will focus on organic small-scale agriculture and value adding production. MOFI will also strengthen its internal capacity in rights-based approaches, accounting, leadership skills and marketing.

From 2020 the strategy is that MOFI will reach out to a wider circuit of stakeholders than the primary target group. The demand for knowledge on organic farming is high as the baseline study has reported. Therefore, MOFI will help to build an exhibition farm – called the MOFI Learning Centre – in each of the ten MOFI villages where farmers can come, see for themselves, get information orally, acquire pamphlets, leaflets and manuals on different organic techniques and link up to the MOFI organic network. Two of the LCs (located close to Babati and Katesh towns) will also exhibit greenhouse technology. The idea behind the LCs is that they are self-funded entities on functioning farms where a dedicated farmer is prepared to display the organic farming techniques and strategies to other farmers. An additional function for the LCs is to form a knowledge outreach hub regarding organic innovations and techniques. MOFI workshops will be held on these LC farms giving some income to the farm as well as the sales of local seeds and other organic inputs. A survey in the ten MOFI villages showed that there are farmers in all ten villages interested in running such an LC. The LCs will be open for all farmers in Manyara Region and will be promoted through media by the MOFI board.

MOFI will cooperate with OFSP and BERAS International – core members of the 10-Year Sustainable Food Systems Programme of the UN – in planning an organic food system in the Manyara Region, where sustainable rangeland management will play a large part. The MOFI effort is also supported by a research project 'Fostering Innovation for Sustainable Development' (FISD) run by COSTECH (Tanzanian Commission for Science and Technology), SIDO (Small Industries Development Organization) and SICD (Sustainability Innovations in Cooperation for Development).

Fig. 6.5. Cattle are an integral part of Tanzanian farming activities. (Photograph by Lena Loiske.)

Haiti – Vallières in North-Eastern Haiti

Contact: Eunide Lefevre (eunidelefevre@gmail.com) and Marit Brendbekken (maritjazz@gmail.com)

School gardens

The 2010 earthquake and other natural disasters have added to extreme deforestation and erosion of the natural environment in Haiti. Prior to the earthquake, an estimated 70% of the population depended on slash-and-burn farming on steep hillsides; 86% of the inhabitants in Vallières in Département Nord-Est depend today upon their agricultural activity as their main source of income.

The Haitian group within SOFIA/BERAS International (Swedish NGO) and the International School of Leadership (ISOL – Haitian NGO) launched a rights-based small development project in Vallières, in the north-east of Haiti, focusing on children's rights to education, food and a sustainable environment. ISOL has identified that lack of food and outdated school books in French represent obstacles to efficient learning and prevent many children from progressing to secondary school. The school administrators and teachers, and the schoolchildren's parents (the latter organized with six female agricultural Community Skills Officers (CSOs) and three male agricultural CSOs), have asked ISOL for assistance in the process of developing school gardens as sources of organic and healthy food and as sites for learning and teaching. The project will provide the basic capacity building and follow-up needed in order to grow school gardens organically and organize the work, as well as to assist with the development of the educational programme, by providing expertise, international contacts and the sharing of experiences, for example through SOFIA's cooperation with BERAS International.

According to school administrators and teachers, very few children attend secondary school regularly in Vallières. They estimate a drop-out rate of 75%. According to a focus group of parents drawn from 27 CSOs in the arrondissement Vallières, there are two major reasons for this. First, when children reach secondary school, they are no longer given a meal during long school hours. Children walk on empty stomachs for miles in order to reach the nearest school. Head of ISOL, psychologist Eunide Lefevre, links learning problems and illiteracy to the fact that children may endure until late afternoon without eating a meal. Secondly, the teaching material/school books and all examinations are held in French, a language that many of the Creole-speaking children and their parents may not have mastered. The school books basically reflect French society in Europe visually and thematically. The reality presented is far removed from poverty-stricken Haitian children's everyday life and further estranges children from learning.

Organic food production and marketing

One of the aims of the project is to reduce the exploitation of virgin forest for charcoal production, the latter representing a meagre cash income for peasants in this region and also a major threat to environmental viability. Less than 2% of Haiti is currently covered by trees, and deforestation reduces water sources each day. Organic food is in demand in nearby Cap Haitienne owing to a sharp increase in international tourism. This opens up an opportunity for Vallières peasants to sell their natural products rather than engage in charcoal production. In Vallières, the use of synthetic fertilizers, weed killers and insecticides/pesticides is almost absent. Elsewhere in Haiti, there is little or no control of the uses of agroindustrial products, hence, tourism rests on the importation of food from abroad. The female marketing association wants capacity building and assistance in order to make use of this opportunity. Traditionally, women run the markets in Haiti, walking for as much as 50 km each day to markets. They operate their own micro-credit systems. Women also process coffee and cocoa beans and manufacture products such as jam and pickles. They ask for assistance when developing efficient marketing strategies for organic products and to improve the quality of processing of coffee and cocoa and jam production to reach the required standards.

BERAS India and its seven local food systems

Contact: K. Perumal, Head BERAS India (berasindiaperumal@gmail.com)

BERAS India consists of 13 different stakeholders as founding members and with the support of BERAS International, the BERAS India secretariat has been fully operating since 2014. The objectives are to support small-scale farmers and food businesses following the universal concepts of ERA, DGP and local SFS. The network is spread throughout Tamil Nadu and Pondicherry as well as Ladakh in the north. The population of Tamil Nadu is 60 million, of which 70% are in rural areas, of whom 80% are small-scale farmers. BERAS India currently has 23 active partners from various fields: NGOs, universities, research institutes, associations, academic institutions, biodynamic (BD) farmers, producers, traders, local community representatives and policy makers (see Fig. 6.6).

Ladakh – Ladakh Environment and Health Organization (LEHO)

Contact: Mohammed Deen mohammed (deen@yahoo.com; www.leho.in)

Ladakh is the highest plateau in the state of Jammu and Kashmir with much of it being over 3000 m. It extends from the Himalayan to the Kunlun Ranges and includes the upper Indus River valley. Ladakh is a high altitude desert as the Himalayas create a rain shadow, generally denying entry to monsoon clouds. The main source of water is the winter snowfall on the mountains. At the Takmachik ecovillage, rainwater is harvested and used to drive a small grain mill. Terraces have been constructed where wheat is produced. Systems for harvesting and drying organic apricots have transformed this into a high-value crop. After the Organic World Conference in New Delhi in November 2017, we visited the seven Indian training centres (see Fig. 6.7).

The main area of work for LEHO is sustainable development, ecology and health. This was taken up in response to recent changes in agriculture, food habits, social values and culture of the different Ladakhi societies. The concept of sustainable development is based on the holistic approach to using natural resources such as land, water, vegetation and livestock of the villages, integrating with the age-old traditions and cultures, and trying to hold on to and also restore the ecological balance and social harmony which had been maintained over the past centuries.

LEHO works in 46 villages with organic farming models and education of farmers.

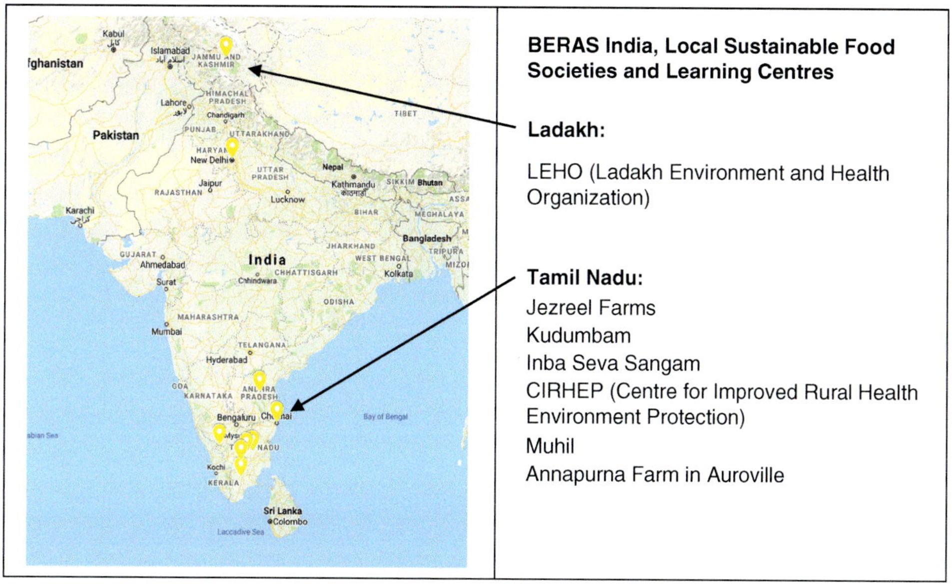

Fig. 6.6. Overview of the BERAS Indian Training Centres. (From Map data ©2019 Google.)

Fig. 6.7. Jostein and Astrid Hertwig and Raymond Auerbach being welcomed with traditional white scarves at the Takmachik Buddhist Organic Village and Training Centre in the Upper Indus Valley. (From left: Phungsog Dolma, Astrid K. Hertwig, Sonam Lhamo, Raymond Auerbach, Jostein Hertwig, Tsering Angmo, Thinles Angmo, Sonam Yangdol and Mohammed Deen, LEHO.)

Particular attention is on seven ecovillages where training and capacity building programmes are organized. In these villages LEHO, together with the local community, has implemented projects for:

- artificial glaciers for storing water during winter;
- drying fruit and other local food processing methods; and
- market development and market availability.

LEHO has a close relationship with the local government LAHDC (Ladakh Autonomous Hill Development Council), in particular to develop a policy for organic farming, organic food production and consumption.

During winters, the people of Ladakh suffer from food shortages as the extreme cold weather does not support any vegetation and farming comes to a halt. Crop farming in Ladakh is limited to 5 months: May–September.

To solve the issue of food shortages and nutritional deficiencies, LEHO proposed the concept of protected cultivation. Protected cultivation is the introduction of 'passive solar greenhouse technology' to produce vegetables in winter in remote areas of mountain regions. The technology is simple, local-resource based, low cost, environment friendly, efficient and is acceptable to the community. It is one of the most suitable technologies to solve nutrition deficiencies and provide food security in Ladakh, as it has the natural advantage of having lots of sunshine in the winter months. The robust greenhouses protect vegetables even during the long, heavy snowy winters (see Fig. 6.8).

The greenhouse technology has helped schools and communities to improve winter food security.

Fig. 6.8. Passive solar greenhouse technology. (a) Detail of greenhouse construction; (b) vegetables growing in the middle of winter; (c) even when covered in snow, the passive solar greenhouse functions. (From LEHO under the Geres project.)

Coonoor, the Nilgiris – Jezreel Farms

Contact: Mahesh L. Melvin (maheshmelvin@gmail.com)

The district of Nilgiris is basically hilly, lying at an elevation of 1000–2600 m. This results in a much cooler and better climate than the surrounding plains. During summer the temperature reaches a maximum of 25°C and a minimum of 10°C. During winter the temperature maximum is 20°C and the minimum 0°C. The district regularly receives rain during both the south-west monsoon and the north-east monsoon. The main cultivation is plantation crops such as tea and coffee, but with some cardamom, pepper and rubber. The area has good conditions for horticulture.

A major task is to find viable organic alternatives to tea plantations. The tea plantation today covers a large area in the Nilgiris. Predominantly the plantations are conventional with intensive use of mineral fertilizer, herbicides and pesticides. This has negative health effects for the workers, the soil and the environment.

The initiative is connected to 11 individual small farms, and four NGOs representing more than 130 farmers. A general challenge for these farms is the inclusion of animals and through this to be able to produce their own manure and apply effective composting methods. Combined farmer training programmes have been conducted in a few tribal villages.

There is a need for capacity building in many other aspects of primary production and also to discuss and develop ideas on value addition for the products. A small wholesaler for organic vegetables and fruits has been established with regular deliveries within the area and to Chennai. There are also plans for setting up an organic restaurant.

Connected to the development of the initiative, there is ongoing work with local schools on ERA farming and DGP. Collaboration with a school for architects has also commenced, both for the practical work with organic farming on their land and also availability of premises for training/meeting local farmers. There are good relationships with the Department of Horticulture and Horticulture Research Centre of Nilgiris (Fig. 6.9).

Kudumbam

Contact: Oswald Quintal and Suresh Kanna (berasindiasureshkannak@gmail.com; http://kudumbam.in/)

Kudumbam was founded in 1982 by Mr Oswald Quintal, Dr Nammalvar and Mr Perianayagasamy. At that time, rural Tamil Nadu was going through an agricultural crisis. The introduction of 'Green Revolution' technology in the 1960s had led to a shift in land use from diverse cropping systems, to monocultures with heavy application of chemical pesticides and fertilizers. Irrigation slowly led to the depletion of groundwater, and small-scale farmers found themselves more marginalized day by day. Kudumbam then started sensitizing rural communities to the rights to community water bodies, and the benefits of drought-tolerant traditional crop varieties. They identified rural leaders and facilitated community participation in deepening and managing community water tanks, raising tree nurseries and re-establishing community forests. Kudumbam also has an organic shop in the Trichy office.

Fig. 6.9. Attendees of a course in natural resource management. (From Kudumbam.)

The vision of Kudumbam is to strengthen vulnerable communities through the building of a multi-stakeholder partnership for the preservation and regeneration of native flora and fauna, in order to ensure a sustainable livelihood.

Kudumbam is currently working in 15 panchayats covering about 45 villages that are undergoing periodical training programmes aimed at the creation of biovillages.

- They organize annual seed fairs and festivals on millets, vegetables and paddy, and encourage farmers to participate in seed distribution and seed sharing and also knowledge sharing.
- Periodical and seasonal long training programmes, practical demonstrations, exposure visits, seed production, training on soil and water conservation techniques and livestock management are some of the education efforts (Fig. 6.10).

KOLUNJI ECOLOGICAL FARM AND TRAINING CENTRE

Since 1990 Kudumbam runs an ecological farm and training centre called 'Kolunji' located in Odugampatty village, Pudukkottai district. The area is dry and the cultivation is dependent on the monsoon rains. The idea was that farmers in the nearby villages would be motivated and inspired to start organic farming from the practical example of Kolunji. For cultivation a number of organic inputs are prepared and used at Kolunji and the nutrients are recycled through vermicompost. Animals such as goats, cattle, hens and ducks are an important part of the nutrient circulation at the farm. Since the farm is situated in a rainfed area the rainwater is conserved and managed by ponds and mounding. Another crucial part of the sustainable farming is the production of tree seedlings for the agroforestry system. In addition to the farm training centre there is also a home for children at Kolunji.

Kudumbam has initiated 'LEISA Safe Food'. This is a market initiative supporting small-scale farmers to get support for collective procurement, processing and marketing of their organic products.

Kudumbam collaborates with officials and authorities from departments of agriculture, animal husbandry, forestry, agriculture engineering, National Bank for Agriculture and Rural Development (NABARD), Department of Co-operatives,

Fig. 6.10. Educational and farmer support activities in Kadavur. (a) Mahesh L. Melvin visiting tribal farmers. (b) The BERAS team in Coonoor.

district collectors, local panchayat presidents, etc., and they participate in its programmes, training and meetings.

Inba Seva Sangam (ISS) – Sevapur, Kadavur district

Contact: K. Perumal (berasindiaperumal@gmail.com; www.inbasevasangam.org)

ISS is an Indian NGO founded in 1968 and based on Gandhian principles of non-violence and welfare of everyone.

According to Dr K. Perumal the 'core values of ISS are truth, non-violence, selfless service, love, brotherhood, self-reliance, mutual understanding, reverence to all faiths and nature, mutual cooperation, dignity of manual labour, devotion, dedication, confidence and discipline.'

ISS is the lead partner of BERAS India and coordinates its activities through a network consisting of 13 different stakeholders as founding members under the supervision of BERAS International Foundation.

ISS has as its main objective to work for the rights of the rural poor, and to enhance their social and economic status. ISS also promotes an ecologically sustainable way of life. ISS target groups are mainly orphans and vulnerable children, vulnerable women and poor peasant families. The organization runs two homes for children (with a total of about 300 children) and one higher secondary school up to standard XII and one community college for rural youths and adults. It works with training and micro-loans (through self-help groups for women), provides literacy programmes and runs 20 'community evening study centres'. ISS also works with water management and the rehabilitation of eroded and abandoned farmlands. ISS owns and runs a BD farm on 13 ha and a herbal garden and nursery area of 1.5 ha for the conservation of endangered species. It has a network of contacts to the organic/BD organizations across the country, and it supports eco-clubs at 140 schools and 300 self-help support groups for women in 70 villages in the area.

ISS has been working in the Kadavur region for the past 12 years empowering farmers, schoolchildren and local government through the Small Farmers Empowerment programme. This programme has led to the formation of a farmer's federation consisting of about 300 certified organic producers or farmers and a scheme for selling their products through the local market. To this end Kadavur Organic Farmers' Association (KOFA) was formed and registered under Tamil Nadu Government Society Act.

During the farmer empowerment programme, ISS has also worked with local schools on promoting environmental well-being. Here there are many reports on challenges including malnutrition and nutrition security in their midday-meal programme. Based upon these findings, ISS plans a project to improve nutrition and food security of rural schoolchildren in the Kadavur valley following the BERAS principles of DGP. In consultation with school authorities and stakeholders the project activities will be implemented in 15 selected rural schools in Kadavur panchayat and one community college – the school of BD farming in Mavathur village panchayat in Tamil Nadu. The project will identify and ensure the participation of about 1400 children/students, 60 teachers and 300 existing certified organic farmers, retailers and market personnel, education department officials, local governments, village youths and public from the region.

ISS collaborates with officials and authorities from departments of agriculture, animal husbandry, forestry, agriculture engineering, NABARD, Department of Co-operatives, district collectors, local panchayat presidents, etc., and they participate in its programmes, training and meetings.

CIRHEP (Natural Resource Management)

Contact: P.M. Mohan and K.A. Chandra (berasindiacirhep@gmail.com)

The Centre for Improved Rural Health and Environment Protection (CIRHEP), a registered NGO established in the year 1994, functions in the districts of Dindigul, Theni and Madurai of Tamil Nadu. CIRHEP is committed to the cause of environment protection especially managing water and soil resources sustainably and its efficient utilization for the benefit of present and future generations. CIRHEP works: (i) with farmers promoting OA and BD inputs in their watershed areas; (ii) with a producer company promoting a sense of responsibility through environmental education; (iii) for women's empowerment; and (iv) with adolescent girls and children as part of an integrated community development approach with the intention of reducing poverty, improving livelihoods and accessing enough quality food for rural people to lead active and healthy lives.

CIRHEP has established the Kadavakurichi Sustainable Agriculture Farmers Producers Company. This market initiative supports small and marginal farmers to get support for marketing their organic products.

As part of soil and water conservation measures, CIRHEP has been doing its pioneering work in watershed development programmes since its inception. The aim of watershed development is to store and conserve water where it falls, within every village, under the direction of the especially constituted village watershed committees. CIRHEP has implemented ten watershed projects so far in Dindigul district covering about 50 villages with the funding support of NABARD and TAWDEVA; there are 3500 beneficiary farmers in an area of about 11,500 ha.

CIRHEP has a centre for natural resource management. This centre serves as a hub to transfer experiences in watershed development, BD/OA, livelihood intervention, and other related concepts to farmers, rural women, youths, children, NGO staff, government officials, students and international visitors.

CIRHEP is promoting a sense of responsibility through environmental education programmes to student communities by sharing their experience. Experienced and caring professionals are committed to positive youth development ensuring this hands-on training programme creates 'episodic memories' for the students. The Environmental Education programme is designed to offer students experiential, hands-on learning opportunities in a rural setting. This is the perfect opportunity to enhance and reinforce what is being taught in the classroom as learning beyond desks and chairs enables them to improve their socialization and curiosity, and is an occasion for immersing themselves in a rural place for a few days.

CIRHEP works closely with local government, panchayats and development departments. They extend its programmes to communities in collaboration with departments of agriculture, animal husbandry, forestry, agricultural engineering, etc.

Muhil – Sustainable Agriculture for Everyone (SAFE)

Contact: J.S.A. Casimir Raj (drcasiraj@gmail.com) and Fr V. Clement Joseph CSsR (clementvincent5@gmail.com)

A diverse group of co-workers is associated with activities at Muhil.

Muhil was established in 1992 inspired by the UN project 'Health for all by 2000'. It is situated in a rural area outside the city of Madurai. Muhil work covers 80 nearby villages where they offer health and other community services. Since 2004 Muhil has been engaged in BD farming among the farming families of

40 villages in the Madurai district. This work is based on a holistic approach for the village people, providing:

- assurance of daily occupation and medical care;
- a sustainable agricultural economy with market facilitation;
- prevention of migration of the rural population to urban areas; and
- protection of the social and natural environment by creating a vibrant rural economy.

A major further achievement was a project to rejuvenate farmland close to the Muhil premises and with the establishment of Karmuhil Organic Farms. This has been extraordinarily successful. With the persistent and skilled use of intercropping, planting of trees, cow manure, different compost methods and watershed management, the soil has been regenerated and is now in full use for growing vegetables, pulses, millets and fodder for the animals, as well as plants and trees for the distillation of essential and aromatic oils.

Karmuhil Organic Farms has launched an organic farming project, which includes training programmes with the following aims:

- to target a total of 600 persons – 200 farmers and 400 farmworkers – to be trained by the end of 2018;
- to consolidate efforts completed on 125 ha and to extend to another 100 ha, totalling 225 ha by the end of 2018; and
- to promote organic food or a recommended diet along with prescribed natural (herbal) remedies or medicine.

In addition, Muhil has started a producer company, a small organic shop and plans to open a restaurant. Muhil has been an active partner in participating in: (i) all the public sector schemes and programmes for village community health services; (ii) planting of trees to protect the environment; (iii) rainwater harvesting plans to conserve water and prevent top soil erosion; and (iv) digging of agriponds and canals to facilitate improvement of rain-fed cultivation.

Annapurna Farm in Auroville

Contact: 'Tomas' (Geert Tomassen), Farm Manager (tomas@auroville.org.in); Lucas Dengel, communication for BERAS India and BERAS International (lucasdl@auroville.org.in; https://www.auroville.org/contents/4154)

In 1966 UNESCO (the United Nations Educational, Scientific and Cultural Organization) commended the 'Auroville Project' in Tamil Nadu, India as a project of great importance for the future of humanity, thereby giving their full encouragement. Today – almost 50 years after its inception – Auroville, with a population of 2550, is recognized as the first and only internationally endorsed ongoing town experiment. Auroville's holistic thinking includes the principles of organic and sustainable agriculture; around 20 organic farms grow food for the community.

The 55 ha Annapurna Farm is the community's largest and only certified organic farm. The farm was started 30 years ago on a barren tract of land. Today Annapurna produces grains, fruits and dairy products for the community. Besides that it functions as the granary of Auroville by storing and processing grains grown on other Auroville farms and some organic farms from the bioregion (see Box 6.2). While growing food for the community, the farm is exploring ways to improve and adjust to the ongoing challenges of the everyday reality. Many questions about sustainability, nutrition, efficiency, economics, wildlife, water scarcity and much more are the daily challenges the farmers have to deal with. For this reason the farm keeps track of ongoing activities; data is being collected and learned from by farm stewards and volunteers/students (see Fig. 6.11).

Besides the ground reality the farm has a spiritual base in which individual growth and development has space to flourish. Working on the farm is more than getting the job done and is a means to develop one's being.

Auroville's economic life includes aspects of capitalist and socialist principles: there is no private ownership of land, housing and immovable assets but on the other hand individual leadership and entrepreneurial qualities are welcome. Basically there are 'service units' and 'income-generating units'. Stewardship is given to the individual by the working group of the particular sector. The 125 income-generating

Fig. 6.11. Farm manager 'Tomas' (Geert Tomassen) explains how the dairy farm is managed (a), and weeds are controlled manually in the rice crop on the Auroville Farm (b). (From Annapurna Farm.)

Box. 6.2. The Auroville Grain Group

Annapurna Farm functions as central granary where grains are dried, cleaned, stored and processed (Fig. 6.12). The working group meets once a month on a grain farm and discusses improvements in grain cultivation, does crop planning, determines prices, reviews crop loan applications and tracks planting and harvesting. Planning is now smoother because grain group members meet regularly and peer review helps all farmers to learn and improve their standards.

Fig. 6.12. Grain storage at Auroville Farm. (From Annapurna Farm.)

units generate profits which help to sustain part of the Auroville economy as a whole. Annapurna Farm is considered a 'service unit', and is not focusing on profit generation but on producing healthy organic food for the community (Fig. 6.11). The farm aims to break even and to generate a small profit to deal with eventualities. This means major infrastructure has to be provided for by fundraising or community support.

Conclusion

The case studies presented in this chapter, from the initial activities at Järna, to the decision to broaden the reach from pollution of the Baltic to food systems in Scandinavia and then to link with Haiti, the Dominican Republic, Tanzania and India, show how community action inspired by shared values and driven by evidence-based scientific results, can have dramatic impacts on the ground.

In each of these case studies, as with the original project in the Baltic, the food system approach has informed development. This has avoided the dangers of 3-year funded projects which often leave very little behind after one or two project cycles, other than some skills with certain members of the community which they may or may not be able to use in the absence of outside project funding.

Resourcing such work, and allowing a broad range of activities while using a food systems lens to understand how various aspects of natural resource management interact, are not simple to achieve! The diversity of activities and localities means that global solidarity can be built without restricting the adaptive capacity of local communities to build on their cultural and spiritual capital. At its core and as a base for truly sustainable development lies the need to move from a limited and short-term 'ego-orientated system' to an 'eco system' awareness that emphasizes the well-being of the whole.

Organic food systems present a living laboratory where the links between various components of a system can be examined. As mentioned in Chapter 5 of this book:

> Food systems are an interconnected web of activities, resources and people, which involve all domains of the food value chain and more. The organization of food systems reflects and responds to social, cultural, political, economic, health and environmental conditions and can be identified at multiple scales, from individual people to local and regional communities, to nations. To fully understand and develop alternatives to the problems facing global food systems demands a systemic and holistic approach.

These case studies have adopted a holistic and integrative food systems approach; we have resisted the temptation to extract definitions of system elements, and pedantically attempt to connect these to system theory. Each case is unique, but shows how individual people and conditions have shaped each system at local level in a global context. Further details are available in the Bibliography to this chapter, with online links which we hope will form a useful resource for leaders wishing to develop food system initiatives. The approach will also inform the more socially applied aspects of Part 2 of this book, and the more technical aspects of Part 3, and will lead towards the final section which looks to the future of African food systems, and food security in times of climate change.

References

Figeczky, G. (2018) IFOAM-Organics International Advocacy. Available at: https://www.ifoam.bio/en/what-we-do/advocacy (accessed 19 December 2018).

Granstedt, A. and Kjellenberg, L. (1997) Long-term field experiment in Sweden: Effects of organic and inorganic fertilizers on soil fertility and crop quality. In: Proceedings of an international conference in Boston (Tufts University, Agricultural Production and Nutrition), Massachusetts, 19–21 March 1997. Available at: http://www.jdb.se/sbfi/publ/boston/boston7.html (accessed 12 July 2018).

Granstedt, A. and Kjellenberg, L. (2008) Organic and biodynamic cultivation – a possible way of increasing humus capital, improving soil fertility and be a significant carbon sink in Nordic conditions. Paper presented at Second Scientific ISOFAR Conference in Modena, Italy, 18–20 June 2008.

Granstedt, A., Seuri, P. and Thomsson, O. (2008) Ecological recycling agriculture to reduce nutrient pollution to the Baltic Sea. *Journal Biological Agriculture and Horticulture* 26(3), 279–307.

Steiner, R. (1894) *Der philosophie der freiheit*. (*The Philosophy of Freedom*, translated by Professor and Mrs A. Hoernle and re-published in 1964). Available at: www.rsarchive.org/Books/Download/Philosophy_of_Freedom-Rudolf_Steiner-4.pdf (accessed 18 December 2018).

Bibliography: Scientific Reports from BERAS Projects

The results from research carried out in BERAS project (2003–2006) form the basis for the BERAS Implementation project (2010–2013) and are available in seven reports. Details of these reports together with an executive summary are provided as follows.

BERAS reports

BERAS Report nr 1

Sepännen, L. (ed.) (July 2004) Local and organic food and farming around the Baltic Sea. *Ekologiskt lantbruk* nr 40.

BERAS Report nr 2

Granstedt, A., Seuri, P. and Thomsson, O. (December 2004) Effective recycling agriculture around the Baltic Sea. *Ekologiskt lantbruk* nr 41.

BERAS Report nr 3

Sumelius, J. (ed.) (2005) Possibilities for and economic consequences of switching to local economical studies within WP3. Ecological Recycling Agriculture. *Ekologiskt lantbruk* nr 43.

BERAS Report nr 4

Kakriainen, S. and von Essen, H. (eds) (August 2005) Obstacles and solutions in use of local and organic food. *Ekologiskt lantbruk* nr 44.

BERAS Report nr 5

Granstedt, A., Thomsson, O. and Schneider, T. (January 2006) Environmental impacts of ecological food systems – Final Report from BERAS. *Ekologiskt lantbruk* nr 46.

BERAS Report nr 6

Sumelius, J. and Vesala, K.M. (eds) (December 2005) Approaches to social sustainability in alternative food systems. *Ekologiskt lantbruk* nr 47.

BERAS Report nr 7

Kahiluoto, H., Berg, P.G., Granstedt, A., Fisher, H. and Thomsson, O. (eds) (June 2006) The power of local – sustainable food systems around the Baltic Sea. *Ekologiskt lantbruk* nr 48.

Some BERAS-related peer-reviewed publications

Granstedt, A. (2000) Increasing the efficiency of plant nutrient recycling within the agricultural system as a way of reducing nutrient pollution to the Baltic Sea. *Agriculture, Ecosystems & Environment* 1570(2000), 1–17.

Granstedt, A., Tyburskij, J. and Stalenga, J. (2007) Nutrient balances in organic farms. Baltic Sea project BERAS (Baltic Ecological Recycling Agriculture and Society), results from Poland. Paper presented at Scientific Agricultural conference Poznan, Poland, August 2007.

Interreg Baltic Sea Region (2007–2013) About the Programme. Available at: https://www.interreg-baltic.eu/about-the-programme/area.html (accessed 28 May 2019).

Part 2

Capacity Building and Climate Change

7 The Likely Impact of the 2015–2018 Drought in South Africa: Lessons From the 2008 Food Price Crisis and Future Implications

Raymond Auerbach,[1]* Hannelise Piek,[1] Jane Battersby,[2] Stephen Devereux[3] and Nic Olivier[4]

[1]*Nelson Mandela University, George, South Africa;* [2]*University of Cape Town, Cape Town, South Africa;* [3]*University of the Western Cape, Cape Town, South Africa;* [4]*University of Pretoria, Pretoria, South Africa*

Abstract

Although still controversial, several reviews of research show that organic foods are more nutritious and contain fewer poisons than conventional foods. Poor food choices lead to low dietary diversity, and refined foods have many nutrients removed; low nutrient density and high sugars, salts and hydrogenated fats lead to the consumption of 'empty calories', and thence to obesity. Access to urban food gardens and to agriculture in rural areas increases dietary diversity scores, but weather shocks lead to price shocks which often decrease availability of food. Climate change will increase the severity of fires, and has contributed (through the 2015–2018 drought) to food prices increasing by 16.1% over this period. Even where child grants and pensions cushioned these shocks, stunting has not decreased. Dietary diversity is poor, often leading to the consumption of empty calories, but urban vegetable gardens can improve dietary diversity scores, as well as community solidarity, though impacts on local ecology only occur when gardens are linked to river health and school activism. A 1% increase in food prices leads to a 2.5% decrease in the number of food items purchased. Although child grants have decreased the experience of hunger, in particular child hunger (in 2011, hunger was less than half of 2001 levels), the percentage of stunted children has not changed much over the past 20 years, with about 27% of children under 5 stunted (10% severely stunted). Food in security is largely related to poverty (cited as the main cause by 86% of respondents). Refining cereals removes a large proportion of protein and protective vitamins, and the situation is aggravated by declining nutrient density in industrial grain and vegetables.

Introduction

Having considered the preceding six chapters, with their global and conceptual overview on farmer development in times of climate change, the evolution of the organic sector globally, organic research and policy, farmer training, food systems and sustainable development, let us pull together some of these ideas in a Southern African context. Household food security in Africa will require adequate food production; commercial agriculture will have to supply much of the food for an increasingly urbanizing continent, but lack of infrastructure and institutional capacity

*raymond.auerbach@mandela.ac.za

will require special attention, both for urban/peri-urban small-scale farming, and if commercial organic farms are to 'emerge' on the continent. Organic food systems are particularly appropriate for Africa, as they allow small-scale farmers to develop drought-tolerant commercial enterprises without becoming dependent on expensive (and often both scarce and poisonous) agricultural inputs (see Chapters 1, 2 and 4, this volume). This requires informed support from NGOs and government (see Chapter 3, this volume), and education of both producers and consumers (see Chapters 5 and 6, this volume).

In this part of the book (Part 2) we will begin to look specifically at capacity building for African farmers in the context of a changing climate. We will look at the impact of drought on food prices and food availability in this chapter, then in the following chapter, examine approaches to develop action research with farmers, using participatory rural appraisal (PRA), and then examine the value chain using a research case study from the south coast of KwaZulu-Natal (KZN), where cooperative development allowed participation of farmers in a short market chain, and facilitated farmer support and training. Participatory Guarantee Systems (PGS) are then examined, showing some of the difficulties involved with meaningful participation, and highlighting the potential of mobile 'apps' for communication and support. Strategies to assist smallholders in accessing high-end markets are examined in Chapter 11 of this volume, while the last three chapters of Part 2 look at likely scenarios under climate change – declining rainfall in the Eastern Cape, techniques of rainwater harvesting, energy reduction and organic vegetable production in urban areas.

This chapter sketches the possible implications of food prices on food availability and on food security, reviewing evidence of consumption behaviour and the likely impact of this on nutrition. National food sufficiency and household food security are differentiated, and analysed through a sustainable food systems lens. Climate change and the impact of weather shock are discussed. The authors draw on a recent draft paper titled 'The likely impact of the 2015/18 drought in South Africa: lessons from the 2008 food price crisis and insight from recent consumption studies' by Professor Sheryl Hendriks and colleagues (S. Hendriks, R. Auerbach, N. Olivier, J. Battersby and S. Devereux, 2016, unpublished draft paper prepared for the Centre of Excellence in Food Security, Cape Town, South Africa). Before looking at drought and food availability, we will look briefly at two contentious aspects of food systems research, which impact on perceptions of what health and nutrition mean in practice.

Scientific nutrition recommendations

Starting with healthy food systems, it is tempting (after Chapters 5 and 6, this volume), to assert that 'healthy food systems are based on dietary recommendations developed over decades by expert nutritionists' and that as we learn to farm better, the quality of the food we produce and process improves – that is what we would like to believe. This should lead to better health, longer life and less need for medical intervention. This does not seem to fit the facts!

Science should gather evidence, formulate hypotheses, test them with double-blind procedures, and analyse results with powerful statistical techniques; at its best, it does exactly that. At its worst, it pushes the agendas of those with vested interests. The case of dietary recommendations of the American Heart Association was examined in Chapter 1 (this volume), and the outcome of the Health Professions Council of South Africa (SA) case against Professor Tim Noakes was reported, as well as the response of the Association of Dietetics (which had lodged the complaint against Noakes): 'We accept the verdict and ... We welcome the precedent that this case provides on what we had considered unconventional advice'.

Nutritional quality and organic farming

The debate about the nutritional superiority of organically versus conventionally grown products has been gathering momentum for some time. Many nutritionists are not prepared to accept that food quality has declined over the past 70 years, and indeed, the evidence for this is contradictory; on the one hand, Davis *et al.* (2004) show how food quality of vegetables in the USA has declined, and Carlo Leifert and colleagues (in an extensive meta-review, Barański *et al.*, 2014), show that for several parameters, the quality of organic food is better than that of conventional food, produced under similar conditions, confirming earlier experimental work (Leifert *et al.*, 2007). On the other hand, Dangour *et al.* (2009) commented that there are no significant differences in quality between organic and conventional foodstuffs.

In the abstract of their paper 'Organic farming, soil health and food quality: considering possible links', Reeve *et al.* (2016) point out that:

> Organic farming systems utilize carbon-based amendments, diverse crop rotations, and cover crops to build soil fertility. These practices increase biologically (active) soil organic matter (SOM) and beneficial soil microbe and invertebrate activities, improve soil physical properties, reduce disease potential, and increase plant health. To date, comparisons of nutrient content between organic and conventional foods have been inconsistent. Recent evidence suggests that organically grown fruits and vegetables contain higher levels of health promoting phytochemicals.
>
> (Reeve *et al.*, 2016)

They quote Hippocrates, the famous ancient Greek doctor, who said 'Let food be thy medicine, and medicine be thy food'. Part 3 of this volume will present evidence showing how soil fertility changed over the first 4 years of organic management in the Mandela Trials, and how the yield gap between organic and conventional systems was closed.

Supplementing his earlier work on vegetables in the USA cited above, Davis (2011) documents increasing yields for 50 fruits and vegetables in the USA from 1970 to 2009, reviews evidence for broadly declining nutrient concentrations in fruits, vegetables and grains, presents an analysis of nutrient declines in 41 botanical fruits and discusses some implications of trade-offs between yield and nutrient concentrations. He traces the decline in nutrients to the introduction of high-yielding hybrids. He concludes that although Darwin (1859) was correct in stating that hybrids possess greater innate constitutional vigour, 'that vigour usually does not extend to the uptake of minerals', and evidence points to 'predominantly decreased nutrient concentrations in hybrids compared to their parents' (Davis, 2011) regardless of the farming system employed.

In a personal communication (July, 2018), Donald Davis explained the complexity of comparisons, and stated: 'Although organic production of plant foods has generally little demonstrated effect on nutrient concentrations (as opposed to some phytochemicals), organic production of animal products has large effects on concentrations of fatty acids'. He backed this up by citing Benbrook *et al.* (2013), who found that:

> Over the last century, intakes of omega-6 (ω-6) fatty acids in Western diets have dramatically increased, while omega-3 (ω-3) intakes have fallen. Resulting ω-6/ω-3 intake ratios have risen to nutritionally undesirable levels, generally 11–15, compared to a possible optimal ratio near 2.3. We report results of the first large-scale, nationwide study of fatty acids in US organic and conventional milk. Averaged over 12 months, organic milk contained 25% less ω-6 fatty acids and 62% more ω-3 fatty acids than conventional milk, yielding a 2.5-fold higher ω-6/ω-3 ratio in conventional compared to organic milk (5.77 vs. 2.28). All individual ω-3 fatty acid concentrations were higher in organic milk—α-linolenic acid (by 60%), eicosapentaenoic acid (32%), and docosapentaenoic acid (19%) – as was the concentration of conjugated linoleic acid (18%).
>
> (Benbrook *et al.*, 2013)

According to Brandt *et al.* (2011), compared with conventional farming, organic produce tends to show higher levels of vitamin C, less nitrate, less total protein and higher levels of phytochemicals. Barański *et al.* (2014) also concluded from their review of 343 published studies that organic crops, on average, have higher concentrations of antioxidants and much lower levels of pesticide residues and heavy metals, especially cadmium, compared with non-organic crops.

Worthington (2001) examined all the available comparisons of crops grown organically with those produced conventionally, using statistical analysis and computerized methods to identify differences and trends. Five frequently consumed vegetables were studied – lettuce, spinach, carrot, potato and cabbage. Results of the study indicated a higher iron and phosphorus content – organic cabbage had a 41% higher iron content and a 22% higher phosphorus content compared with the conventional counterpart (Worthington, 2001). Lairon (2010) and Drake (2009) also reported that organic food usually has a better nutritional quality. As organic agriculture (OA) is a low external input system, and has many environmental and social benefits, the provision of nourishing food from crops and animals produced with locally available inputs makes sense from health, environmental and economic standpoints. The question arises: 'Is it feasible to make organic food generally available in Africa'?

The links between food quality, dietary diversity, food availability and the impacts of drought and climate change on health are diverse and complex.

Food Availability and the Impact of Drought and Climate Change

Food availability is the total amount of food in an area at a specific time (Devereux, 2006). This depends mainly on the total performance of the agricultural sector, but is similarly dependent on the country's ability to import, store, process and distribute food. Domestic production is supplemented by food import options and food consumption patterns prescribe the production and distribution of certain food (DAFF and DSD, 2013). SA has enough food available for all of its population, currently numbering 55 million people (Devereux and Waidler, 2017). However, as pointed out in Chapter 1 of this volume (Fig. 1.1), this is 'national food self-sufficiency', and is not the same as 'household food security', as many people cannot afford to buy the food in the shops. The *Farmer's Weekly* of 11 January 2019 cites Professor Francois Engelbrecht as saying that Southern Africa has been identified as one of the regions in the world where climate change will impact most negatively on economic growth, considering the widespread impact on the agricultural sector, quoting the 'Special Report on Global Warming of 1.5°C'. He says that regional warming could reach levels of 6°C by the end of the century, and this could lead to the total collapse of the maize and livestock sectors (since much of SA livestock is fattened on maize in feedlots). In the same article, fire expert Tiaan Pool says that climate change has resulted in a change in fire regimes, with the frequency, severity, size, type and season being altered: 'With drought events on the increase, and extended periods of below-average rain, there'll be 31% more heatwaves resulting in more veld fires' (*Farmer's Weekly*, 2019).

The SA Department of Agriculture, Forestry and Fisheries (DAFF) concurs, stating that climate change and altered patterns of land use are some of the challenges that pose a threat to domestic production. The African continent is one of the most vulnerable to climate variability and to 'extreme events' such as flooding or droughts. This, together with revitalizing the agricultural sector are two of the challenges that require attention for food supply stability (DAFF and DSD, 2013).

Battersby (2012) identified four key sets of flows households draw upon to ensure food security, which are flows of: (i) food; (ii) cash; (iii) people; and (iv) social networks. These are all directly or indirectly impacted by climate change. It is also important to consider the possible impact of climate change on households' ability to use the food they are able to access, which in turn may shape the foods that households do access and use (Battersby, 2012).

Urban Gardens, Food Sovereignty and Dietary Diversity

The Nyéléni Declaration, made in Mali, defines food sovereignty as:

> the right of peoples to healthy and culturally appropriate food produced through ecologically sound and sustainable methods, and their right to define their own food and agriculture systems. It puts those who produce, distribute and consume food at the heart of food systems and policies rather than the demands of markets and corporations. ... Food sovereignty prioritises local and national economies and markets and empowers peasant and family farmer-driven agriculture ... and food production, distribution and consumption based on environmental, social and economic sustainability. ... It ensures that the rights to use and manage our lands, territories, waters, seeds, livestock and biodiversity are in the hands of those of us who produce food.
>
> (Nyéléni, 2007)

There are six pillars of food sovereignty as shown in Box 7.1.

Food sovereignty as a concept attempts to build locally managed food systems, using

Box 7.1. The six pillars of food sovereignty

The six pillars of food sovereignty developed at Nyéléni are as follows (Nyéléni, 2007):

1. **Focuses on food for people:** Food sovereignty puts people, including those who are hungry, under occupation, in conflict zones and marginalized, at the centre of food, agriculture, livestock and fisheries policies, ensuring sufficient, healthy and culturally appropriate food for all individuals, peoples and communities; and rejects the proposition that food is just another commodity or component for international agribusiness.
2. **Values food providers:** Food sovereignty values and supports the contributions, and respects the rights of women and men, peasants and small-scale family farmers, pastoralists, artisanal fisherfolk, forest dwellers, indigenous peoples and agricultural and fisheries workers, including migrants, who cultivate, grow, harvest and process food; and rejects those policies, actions and programmes that undervalue them, threaten their livelihoods and eliminate them.
3. **Localizes food systems:** Food sovereignty brings food providers and consumers closer together; puts providers and consumers at the centre of decision making on food issues; protects food providers from the dumping of food and food aid in local markets; protects consumers from poor quality and unhealthy food, inappropriate food aid and food tainted with genetically modified organisms; and resists governance structures, agreements and practices that depend on and promote unsustainable and inequitable international trade and give power to remote and unaccountable corporations.
4. **Puts control locally:** Food sovereignty places control over territory, land, grazing, water, seeds, livestock and fish populations on local food providers and respects their rights. They can use and share them in socially and environmentally sustainable ways which conserve diversity; it recognizes that local territories often cross geopolitical borders and ensures the right of local communities to inhabit and use their territories; it promotes positive interaction between food providers in different regions and territories and from different sectors that helps resolve internal conflicts or conflicts with local and national authorities; and rejects the privatization of natural resources through laws, commercial contracts and intellectual property rights regimes.
5. **Builds knowledge and skills:** Food sovereignty builds on the skills and local knowledge of food providers and their local organizations that conserve, develop and manage localized food production and harvesting systems, developing appropriate research systems to support this and passing on this wisdom to future generations; and rejects technologies that undermine, threaten or contaminate these (e.g. genetic engineering).
6. **Works with nature:** Food sovereignty uses the contributions of nature in diverse, low external input agroecological production and harvesting methods that maximize the contribution of ecosystems and improve resilience and adaptation, especially in the face of climate change; it seeks to heal the planet so that the planet may heal us; and, rejects methods that harm beneficial ecosystem functions, that depend on energy-intensive monocultures and livestock factories, destructive fishing practices and other industrialized production methods, which damage the environment and contribute to global warming.

agroecological principles, with an emphasis on stimulating the local economy. Where possible, this means supporting local production in both rural and urban contexts.

Posern (2016) reports that the need for food sovereignty in SA was reported at a food sovereignty assembly in Johannesburg in March 2015 referring to the fact that SA ranks ninth on the Global Hunger Index 2012; 'malnutrition' includes both undernutrition and unhealthy consumption of empty calories. Posern reported that the World Health Organization (WHO) estimates that 59% of men are overweight and 21% obese, while women show a higher percentage (72% overweight and 41% obese), and that 29% of deaths in SA can be attributed to problems of malnutrition.

In her research with gardeners in George, Western Cape Province, SA (mostly from the *Kos en Fynbos* groups in Blanco and Rosedale), Posern (2016) tested the following three hypotheses:

1. The involvement in urban gardening increases food sovereignty with respect to dietary diversity.
2. The involvement in urban gardening increases food sovereignty with respect to ecological sustainability.
3. The involvement in urban gardening increases food sovereignty with respect to community solidarity.

She found that members of *Kos en Fynbos* appreciated the fact that they received training in gardening (both in Blanco and in Rosedale), and they also greatly appreciated the exposure in the local newspaper (*The George Herald*), and the various competitions with accompanying prizes.

She found that dietary diversity is a better indicator of health than quantity of food consumed, and used a dietary diversity score (DDS) to measure dietary diversity. She analysed nutritional diversity with respect to the different subsample groups, and found that a more balanced diet and a better health status were found for gardeners in contrast to non-gardeners. She concludes:

> For the area of dietary diversity H_1, stating that there is a difference between gardeners and non-gardeners in terms of nutrition, is accepted. The hypothesis has been confirmed by the statistically significant difference found for gardeners and non-gardeners within Rosedale [her evidence does not support this conclusion for *Kos en Fynbos* members]. In that context it needs to be added that the consequences of urban gardening on the dietary diversity depend on how urban gardening is performed.
>
> (Posern, 2016)

She concludes that there is no difference between gardeners and non-gardeners in terms of ecological sustainability, for *Kos en Fynbos* in particular. She comes to this conclusion in spite of *Kos en Fynbos* members actively making compost and separating waste. She also accepts that there are differences between gardeners and non-gardeners with respect to community solidarity and comments that three of the four items that measured the solidarity of the respondents revealed a significantly higher solidarity for gardeners. It may be important to link gardening activities to local ecological activities such as river clean-up campaigns, recycling and school environmental action clubs, as was done in the Mlazi River Catchment Management Programme (Auerbach, 1999).

Additionally, she found that the third aspect that investigated the willingness and motivation to invest more time in the community is significantly higher for *Kos en Fynbos* members who form a specific and important part of the group of gardeners in general. She concludes that urban gardening increases food sovereignty with respect to the dietary diversity of gardeners, as well as in terms of solidarity shown between the gardeners. She warns of the dangers of dependence on prizes and inputs from the George Municipality, and comments that the older gardeners at Blanco seem more aware of this danger than the community at Rosedale.

Finally, she states:

> To be specific, urban gardening in this study turned out to enable its activists to realise their right to a healthy and diverse nutrition as well as to enhance the solidarity between gardeners, giving them in this way more power to realise their rights. This detected contribution of urban gardening towards food sovereignty is a particularly important and positive result for two reasons: ... urban food security policies in general are still lacking and ... a positive influence of alternative nutrition strategies is ... needed in SA. This is clearly illustrated from the fact that Sub-Saharan Africa is one of the most rapidly urbanising areas in the world and is confronted with the double burden of disease, namely hunger and undernutrition ... and ... non-communicable diseases.
>
> (Posern, 2016)

School Food Gardens

Gardens at schools can contribute to learner awareness on a range of issues (Auerbach, 1999). Setting up school gardens as part of the Mlazi River Catchment Management Programme was part of a broader environmental education initiative; annual school competitions covered control of invasive alien plants, planting of trees, rainwater harvesting, recycling of rubbish and school gardens. This was part of addressing a research question: 'Can School Environmental Action Clubs effectively involve schoolchildren in caring for their local environment?' The study concluded that:

> [project] activities at schools consitutute an important platform building activity, and involving both teachers and students has far-reaching spin-off effects, as teachers are an important and respected resource in the community, and students carry the message to their homes, and will later impart these values to their own children.
>
> (Auerbach, 1999, p.165)

However, the study found that both gardens and an environmental ethic only developed at schools where there was effective leadership.

A study of the National School Nutrition Programme (NSNP) found that there is great unexploited potential to use school food gardens and nutrition education in schools to improve food security in South Africa. Almost all of the NSNP budget is allocated to school meals, leaving too little for investment in food gardens at most schools (Devereux and Waidler, 2017). Linking school gardens to food sovereignty requires a food systems approach, combining education of scholars concerning healthy food choices, and local organic food production initiatives.

Price Volatility and Drought

Drought has played a significant role in driving agriculture and food policy reform in SA. Drought is a recurrent theme in SA's history. If nothing else, a drought brings people to a point of vulnerability. The 2015–2018 drought raised awareness of our vulnerability to increasingly uncertain climate conditions, including the vulnerability of the city of Cape Town to extreme shortages of drinking water which led to prohibition of *all* crop irrigation in Cape Town's catchments during 2018. We need to change our current production and consumption patterns and to plan water conservation better.

Already in the apartheid era, following a severe drought in SA in the 1980s, significant quantities of maize and other food products were imported at great cost. Because of that drought, various committees were established to investigate elements of national food supply. This included the Ministerial Protein Advisory Council to investigate, advise on and coordinate matters relating to the total demand and supply of protein. Various committees reported concerns regarding the country's natural resources and projected demographic trends. In 1984, the Department of Health and Welfare expressed concern about a possible shortage of locally produced food. In response, the Committee for the Development of a Food and Nutrition Strategy for Southern Africa was appointed by the Ministers of Health and Population Development, and of Agriculture. The findings of this report, as well as the findings and recommendations from the Calitz Committee on Poverty, led to the implementation of the National Nutrition and Social Development Programme (NNSDP, initiated in 1991). The NNSDP aimed to address the nutritional needs of poor communities and households (across SA as well as the *bantustans* (apartheid-era homeland governments)) through the involvement of local communities, NGOs and government institutions by means of feeding schemes and the distribution of food and other commodities. The long-term focus was to help empower communities to become self-reliant and independent through development efforts (Hendriks *et al.*, 2016, unpublished draft paper).

Following another severe drought in the early 1990s, the National Consultative Forum on Drought was established, leading to numerous investigations, commissions and policy changes related to drought monitoring, mitigation and responses. It also influenced and shaped future food security and nutrition policies. A Nutrition Task Force was established under the auspices of the Consultative Forum, mandated to initiate broad public discussion and debate on national nutrition programmes (Hendriks *et al.*, 2016, unpublished draft paper). Policy on drought management and nutrition is not in short supply, nor are committees to discuss implementation; the weak point is implementation of policy in a way which helps ordinary SA citizens to improve their food security and food sovereignty, by helping them to access sufficient, nourishing, preferably locally produced, food in times of climate change. To date, government has not been willing to advocate local seed exchange schemes, organic farming or even an agroecological approach to local food production, preferring to endorse high external input systems backed by companies selling agrochemicals, fertilizers and hybrid and GE seed.

Rainfall and Temperature

In 2015 SA recorded its lowest mean annual rainfall (MAR) on record (South African Weather Service, see Fig 7.1), and with the hottest summer on record and four successive low rainfall seasons in some of the most agriculturally productive areas of the country, harvests for most commodities in 2015 were at record lows. Most of Africa experienced an unusually hot, dry year in 2015.

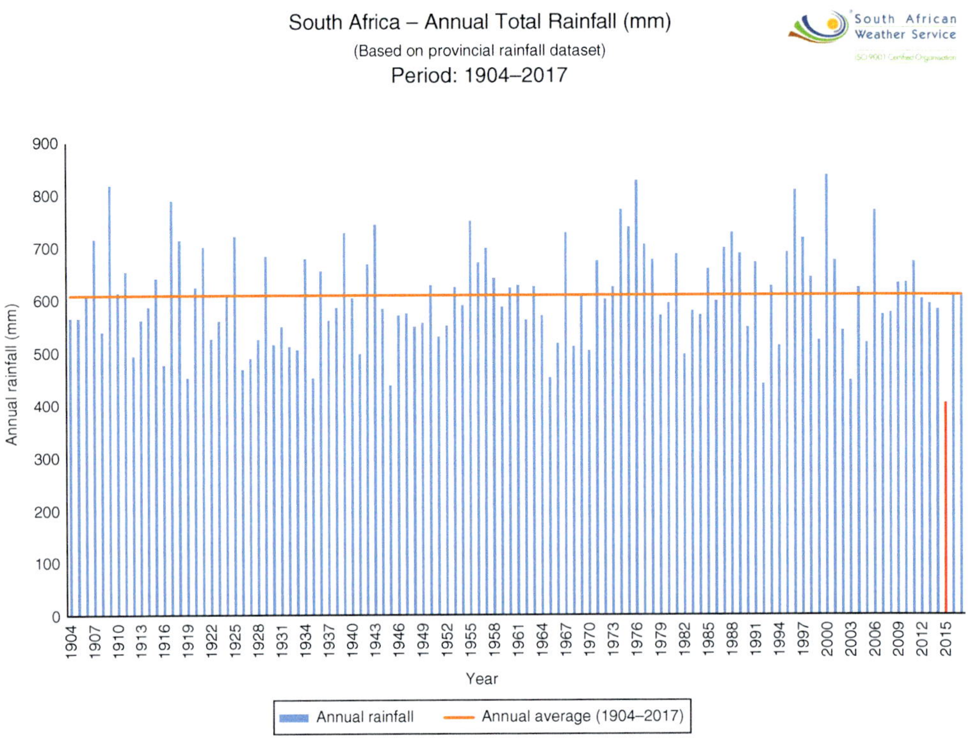

Fig. 7.1. Historical rainfall trends for South Africa 1904–2017. (From South African Weather Service.)

Dam levels dropped and many municipalities experienced severe water shortages. For 2015–2017, above-normal rainfall was restricted to the eastern part of the country for December 2017, and normal to below normal for the rest of SA (DAFF, 2018).

It is important to note that increasingly variable weather patterns and climatic conditions threaten global food security (Chapter 12, this volume). Countries and communities which rely heavily on agricultural production will be the most vulnerable to climate change and variability. Studies on SA farmers' perceptions on climate change report that farmers perceive that there has been an increase in temperatures and a decrease in rainfall over the past two decades. These views are supported by climate data, which shows evidence of an increase in the length of the dry season, frequency of droughts and a later start to the wet season in Limpopo; and increasing inter-annual rainfall variability in North West Province and KZN. In the Eastern Cape Province, an increased vulnerability of households to income losses, poverty and food insecurity has been linked to climate change (Chapter 12, this volume).

Historical Trends in Staple Production and Prices

Food prices are climbing. The prices of all food items increased more steeply than usual in 2015, and during the period between November 2015 and mid-2018 they continued to rise, but more gradually. The main reasons for high inflation levels on food in 2015 were the long period of drought and high temperatures. These factors caused significant increases in prices on food in supermarkets both in 2008 and in 2015. Smith *et al.* (2017) commented that retailers have increased the price on products, aimed for low-income consumers, by 16.1% since November 2015.

Increased food prices, especially for maize and wheat, which are both staple foods of the poor in SA, pose serious difficulties for both urban and rural people as most are net buyers of food (Altman *et al.*, 2009). By the end of January 2016, the year-on-year spot price of white maize had increased by 80%. This raises concern over the affordability of food, especially for lower income earners consuming a largely maize-based diet over the short and medium term. While the country does not have a food security information system to inform current policy debates on appropriate responses to rising prices due to maize shortages, numerous studies have been conducted in SA over the past decade, and provide insight into the possible impacts of domestic food shortages, rising food prices and the potential impact on consumers (Hendriks *et al.*, 2016, unpublished draft paper). Improved crop yields of cereals recently have seen a return to complacency among many policy makers, but problems of food availability and food price volatility are not about to disappear!

Weather Shock

We know from the 2008/2009 global food price crisis that food security and the experience of hunger increased in SA over this period. The current context is somewhat different to the global crisis of 2008/2009. The recent global crisis was caused by a number of global events that led to a global food shortage at the same time as fuel prices increased sharply and biofuels changed the dynamics of maize markets (Hendriks *et al.*, 2016, unpublished draft paper). The 2016 maize shortages in SA were caused by an extreme covariate shock that came at a time when international food prices were the lowest since 2008 but with an exceptionally weak rand at the time. The South African Futures Exchange (SAFEX) prices for white maize rose from a low of R1700/t in August 2014 to an average of about R3000/t a year later, and then rose sharply from November 2015 to over R5000/t in January 2016 (Hendriks *et al.*, 2016, unpublished draft paper).

Kubik (2018) outlines how climate affects food access:

> if weather shocks translate into food price shocks which, in turn, adversely affect food security, then the ongoing climate change should be considered as an important risk. This extends beyond subsistence households relying on own food production, as commonly assumed in the literature, but also, to net food buyers. This question has been addressed mainly in the macroeconomic setting; (with) evidence that rainfall volatility is a factor of food insecurity particularly in sub-Saharan Africa, and that this adverse effect of weather shocks is exacerbated for countries that are vulnerable to food price shocks. [Various sources] predict that as a result of climate change and consecutive price increases, caloric intake in Africa will decrease while the number of malnourished children will increase. ... Weather shocks have been associated with lower child growth ... malnutrition ... and nutrient deficiencies in pregnant women.
>
> (Kubik, 2018)

This author also shows how weather shocks are linked to food price shocks:

> findings suggest that food prices have a significant negative impact on household food security, i.e. a 1% increase in local food prices induced by a weather shock decreases the number of food items consumed by a household by around 2.5%, and the number of food groups by almost 1%. In line with expectations, the low-income households are particularly vulnerable to weather and price shocks; however, their response to shocks seems to depend on the level of poverty.
>
> (Kubik, 2018)

The DDS would thus decrease with weather shocks and food price shocks.

SA is a small player in the global agricultural industry and fluctuations of world prices are therefore key determinants of what happens in the local market. Important international factors included a slowdown in the Chinese economy, favourable production conditions in the current season and rising stock levels, and these contributed to downward pressure on food prices. According to household-level expenditure data, the most popular staple food options (i.e. largest expenditure) among marginalized and lower middle-income consumers (in order of importance) are maize meal, brown bread, rice, white bread and potatoes, representing about 80–85% of these consumers' expenditure on starchy foods (Table 7.1).

According to the National Food Consumption Survey of 2005 (Hendriks *et al.*, 2016, unpublished draft paper), the most important staple food options among individuals aged 10 and older are maize porridge, brown bread, white bread, potato and rice (in order of importance) (Table 7.2). This is the most recent available data, as there has been no national consumption survey since.

Understanding portion and serving sizes of staple foods is important when predicting the adequacy of the food supply. In 2015 portions based on the latest available scientific information for the SA context were defined as: maize meal (100 g raw), potato (70 g raw), rice (140 g cooked, 70 g raw), pasta (85 g raw) and brown or white bread (70 g = two slices). In general two portions of a commodity equal a serving (Hendriks *et al.*, 2016, unpublished draft paper).

According to the analysis the most affordable staple foods are maize meal, potatoes and rice, followed by brown bread. The portion cost of maize meal has been moving closer to the portion cost of potatoes and rice from about mid-2011. Comparing the portion cost of brown bread (being the second most important staple food in SA) with the portion cost of maize meal it is evident that a brown bread portion was around 90% more expensive than amaize meal portion during the period mid-2008 to about mid-2011. However, this ratio has now dropped to around 70% – thus indicating an improvement of the relative affordability of a brown bread portion relative to a maize meal portion, as maize prices rose. In April 2015, the cost of a portion of maize meal was already about R0.75, compared to R1.31 for a portion of brown bread. It should be kept in mind that brown bread has a significant convenience appeal (being ready to eat) and thus saves energy costs and time for the consumer. In April 2016, bread prices increased by 40 cents; this is on top of the 12.7% increase in bread and cereal prices since 2015 due to the drought (Hendriks *et al.*, 2016, unpublished draft paper).

Table 7.1. Dominant staple foods in South Africa according to the StatsSA Income and Expenditure Survey 2010/2011. (From StatsSA, 2012.)

	Share contribution of food type to total expenditure on starchy food category	
Staple food type	Marginalized consumers (%)	Lower middle-income consumers (%)
Maize meal	30.8	27.5
Brown bread	23.2	20.0
Rice	13.2	14.6
White bread	9.6	10.5
Potatoes	8.4	8.2

Potato production has decreased drastically since September 2015 due to the drought and prices doubled between December 2015 and March 2016, described as the highest ever increase in prices. This should be seen against the record plunge of the rand against the dollar, boosting the estimated 7 million t of grains to be imported to make up for the losses due to the drought (Hendriks *et al.*, 2016, unpublished draft paper). The cost of imports of maize, both for staple food and for animal feed, places further pressure on the system.

Food prices and affordability are critical considerations for most consumers in SA. They impact on household food choices and subsequently on aspects such as food insecurity (undernutrition) and obesity. The recent drought

Table 7.2. Dominant staple foods in South Africa according to the National Food Consumption Survey of 2005. (From Hendriks *et al.*, 2016, unpublished draft paper.)

Staple food type	Percentage of group consuming the item (%)	Average consumption of those consuming item (g/person/day)	Average consumption per capita (g/person/day)
Maize porridge and dishes	77.9	848.3	660.7
Brown bread and rolls	55.1	164.7	90.8
White bread and rolls	28.1	161.6	45.5
Potato, cooked	17.1	165.1	28.2
Rice, white/brown, cooked	13.5	163.3	22.1

has also seen increases in the prices and availability of many fruits and vegetables.

The Current Household Food Security Situation and Trends from National Data

In 2014 the South African Government, by means of the National Policy on Food and Nutrition Security gave the following definition of food security and nutrition: 'Access to and control over the physical, social and economic means to ensure sufficient, safe and nutritious food at all times, for all South Africans, in order to meet the dietary requirements for a healthy life' (DAFF and DSD, 2013).

Household food security must be distinguished from national food self-sufficiency; even though there may be food available, if people cannot afford to buy the food, the household is food insecure. Therefore, there is a direct link between income security and food insecurity (Altman *et al.*, 2009).

Auerbach (1999) shows how both national food self-sufficiency and household food security tend to be short-term considerations (see Chapter 1, this volume, Fig. 1.2), while conservation takes a longer-term view, but it tends to be technology centred. However, sustainable development requires that a long-term people-centred approach should be adopted. For conservation, this has meant learning about people and involving communities; for social scientists and policitians, household food security requires a longer-term approach to establishing a more equitable situation with regard to food security and food sovereignty at a household level. For industrial agronomists, a twofold learning needs to take place, if agriculture is to become more sustainable: agriculture must shift towards social inclusivity and also towards long-term environmental responsibility.

Until this happens, hunger and food insecurity remain a reality for many South Africans. This is confirmed in findings from a recent systematic review of 167 food security and nutrition case studies in SA between 1995 and 2014 (Hendriks *et al.*, 2016, unpublished draft paper). Even if only studies that set out to measure food security levels among larger sample sizes (over 200) are included, the picture presented is bleak with an average of 68% of participants in these studies experiencing difficulty in securing a dependable supply of food. Each of these studies is unique in sample type (including age groups) and focus (Table 7.3).

These figures are not at odds with those from national level surveys, with the percentage of households in the country that 'run out of money to buy food' estimated at 70% in the National Food Consumption Survey of 2005, and 46% in the South African Social Attitudes Survey of 2008 (Hendriks *et al.*, 2016, unpublished draft paper).

There is no agreed-on measure of food insecurity, both internationally and nationally (Hendriks *et al.*, 2016, unpublished draft paper). A few nationally representative samples have included food security indicators but the indicator sets are not consistent between surveys in SA. In the case of Statistics South Africa's (StatsSA) General Household Survey (GHS), indicators have not always been consistent over time (StatsSA, 2017).

Data from the GHS (Fig. 7.2) show that generally, the experience of hunger has declined between 2002 and 2011 (StatsSA, 2017). If

Table 7.3. Food security estimates in case studies examining levels of food security in the general population with sample sizes over 200. (Based on Hendriks *et al.* 2016, unpublished draft paper.)

Data cited	Study cited
53% severely food insecure	Rural Limpopo (De Cock *et al.*, 2013)
Nearly 90% food insecure	Urban KwaZulu-Natal (Crush and Caesar, 2014)
73% rural and 87% urban families at high risk of food insecurity	Urban and rural Free State (Walsh and van Rooyen, 2015)
76% worry about not enough money for food	Rural Limpopo (Masekoameng and Maliwichi, 2014)
69% of households severely food insecure	Rural Eastern Cape (Ballantine *et al.*, 2008)
64% of female- and 42% of male-headed households food insecure	Peri-urban Free State (Ndobo and Sekhampu, 2013)

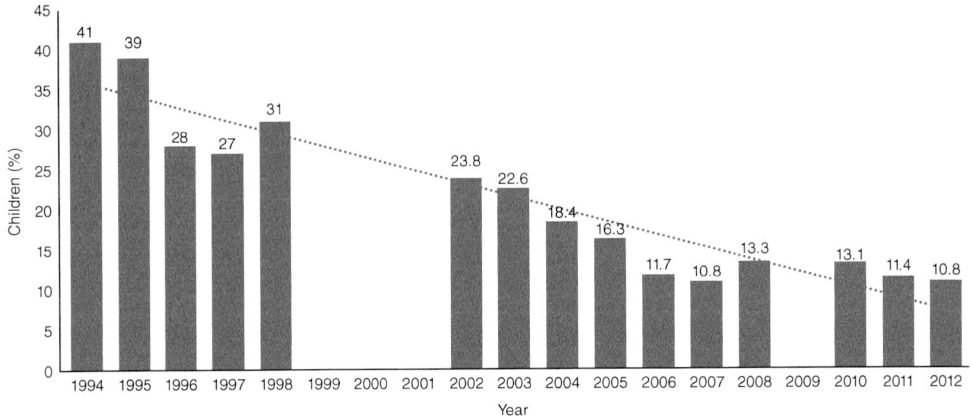

Fig. 7.2. Children experiencing hunger in South Africa, 1994–2012. Note: Questions on hunger were not asked in the 2009 General Household Survey. (From Devereux and Waidler, 2017.)

StatsSA's survey questions regarding the frequency of experiencing hunger are taken as indicators of the depth and severity of food insecurity, the incidence of starvation and acute hunger ('always') has dropped from 2.3% of the population in 2002 to 0.7% in 2011. The proportion of the household experiencing chronic hunger ('often') dropped from 4.4% in 2002 to 1.9% in 2011. But in SA, child hunger remains high, ranging from 9% in the Western Cape to a shocking 43% in the Eastern Cape and Limpopo (DAFF and DSD, 2013).

The 2016 SA Demographic and Health Survey established that 27% of children under the age of 5 years – 30% of boys and 25% of girls – are stunted, of whom 10% are severely stunted (StatsSA, 2017). This indicator of chronic malnutrition has remained above 20% for the past 25 years (see Fig. 7.3). Stunting has immediate and long-term effects on children's health and cognitive development, and contributes to the intergenerational transmission of poverty. Partly in response, social grants have dramatically increased since 1994, and the Child Support Grant now reaches two-thirds of SA's children. Although social grants have significantly raised incomes and children's access to food in low-income families, the persistence of high levels of chronic malnutrition implies that either the value of the grants are not enough, or that children are not accessing the right kinds of food (Devereux and Waidler, 2017).

Once again, these figures are not at odds with national findings, with the National Food Consumption Survey of 2005 suggesting that 18% of South African children were stunted (Labadarios *et al.*, 2011). Even assuming these unwelcome national average statistics are accurate, they mask the significant inter-community and inter-household variations in nutrition security. While many in SA are mired in chronic, debilitating food insecurity marked by retarded development, others may be only mildly insecure, or actually find themselves food secure, and the question of where the next meal is coming from may not be an issue at all.

How does the experience of hunger by children and the availability of food, relate to child health?

The Intergovernmental Panel on Climate Change (IPCC) in Chapter 22 of the Assessment Report 5 of 2012, discusses African food security:

> Africa's food production systems are among the world's most vulnerable because of extensive reliance on rainfed crop production, high intra- and inter-seasonal climate variability, recurrent droughts and floods that affect both crops and livestock, and persistent poverty that limits the capacity to adapt ... However, agriculture in Africa will face significant challenges in adapting to climate changes projected to occur by mid-century, as negative effects of high temperatures become increasingly prominent ... thus increasing the likelihood of diminished yield potential of major crops in

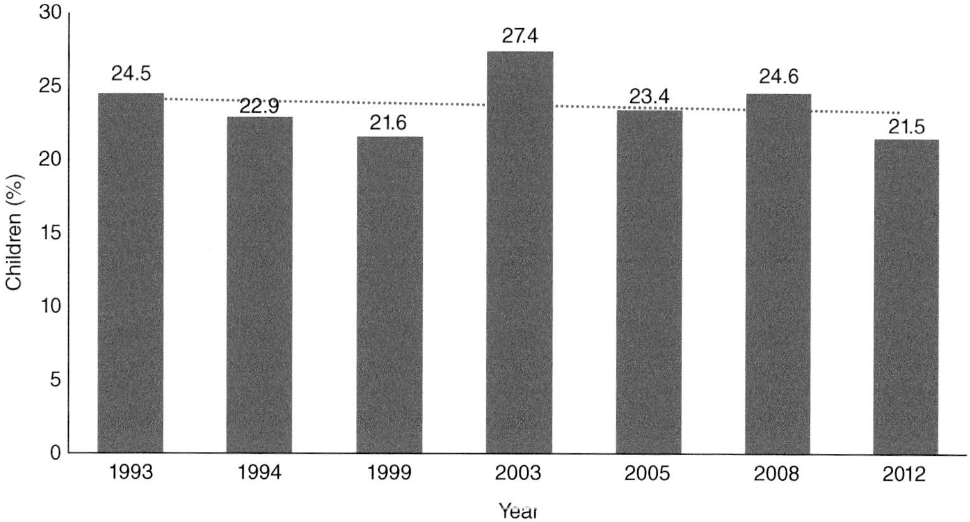

Fig. 7.3. Child stunting rates in South Africa, 1993–2012. (From Devereux and Waidler, 2017.)

> Africa ... The composition of farming systems from mixed crop-livestock to more livestock dominated food production may occur as a result of reduced growing season length for annual crops and increases in the frequency and prevalence of failed seasons ....
>
> (IPCC, 2012)

Professor Jane Battersby (2012) writing a chapter in *Climate Change, Assets and Food Security in Southern African Cities* (Frayne *et al.*, 2012), concludes that the AFSUN study (on which the book reports), found that across southern African cities, about 77% of people living in the poorer township areas are food insecure (with Cape Town marginally higher at 80% of people in poorer areas). Although the study found rural food insecurity even higher in the Eastern Cape (close to 100% of the poorer areas), these results show that food insecurity is not only a rural problem. This discussion is taken further for the Eastern Cape in Chapter 12 of this volume.

Food and nutrition insecurity in SA have context-specific dynamics: drivers and impacts vary across the socio-economic and environmental geographies of SA. This is important for understanding both causes of, and possible responses to, food insecurity. This local-level specificity is illustrated well by there being 91 code-groupings (coded text excerpts) for different 'impacts or outcomes of food insecurity' across the 167 studies. The social dynamics of food insecurity are particularly variable. An investigation of the relationship between food security and attributes of social capital in two villages in Mpumalanga, for example, found no relationship between social capital and food security in one village, but significant relationships between food security and three social capital attributes in the other (namely: collective action and cooperation, social cohesion and self-esteem) (Hendriks *et al.*, 2016, unpublished draft paper).

Despite these dynamics, poverty is a dominant cause of being food insecure in SA (Fig. 7.4). A total of 51% of the studies in the review specifically cited poverty and/or lack of income as a cause of being food insecure, and most, if not all, studies implied it was an underlying causal factor; it is the by far the most-cited driver of food insecurity in case studies over the last 20 years in the country (Hendriks *et al.*, 2016, unpublished draft paper). This has important implications for interventions and decision making. Despite the many factors that can help alleviate food insecurity and malnutrition, being able to afford to purchase food remains a key determinant. This is particularly important when considering food insecurity at the national scale.

The situation is worse both because the cereals are usually highly refined, having had bran and germ (and therefore most of the nutrients and

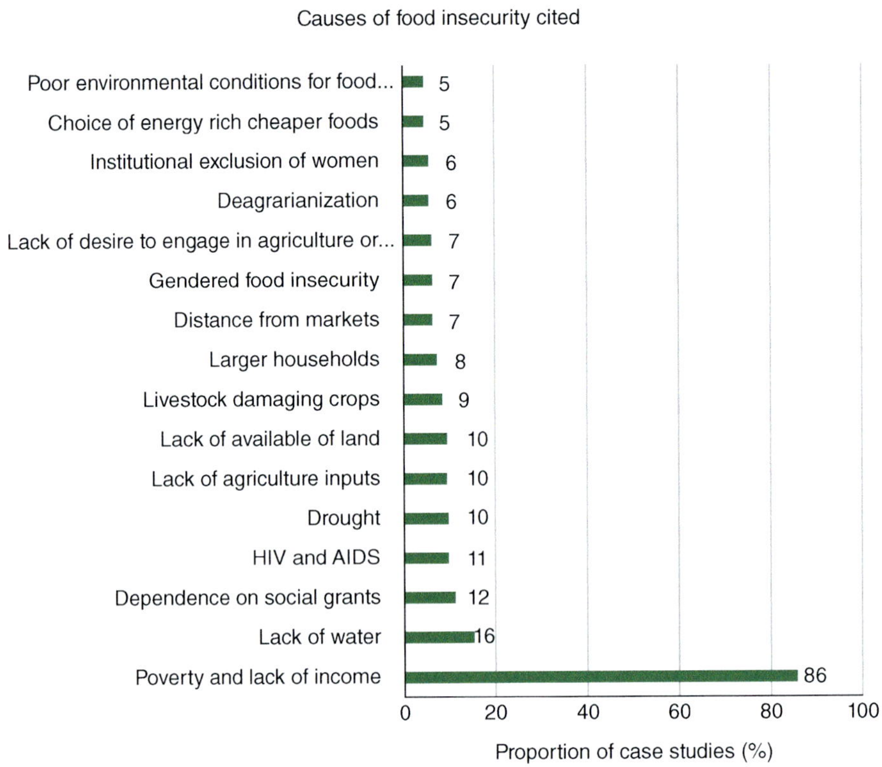

Fig. 7.4. Causes of food insecurity in South Africa cited across the case studies (only causes cited by five or more studies included here). (From Hendriks *et al.*, 2016, unpublished draft paper.)

dietary fibre to be found in wholegrain cereals) removed in processing, and also because the cereals are produced at high plant populations, with high levels of fertilization and intensive herbicide use (mostly glyphosate). The protein content of grains has declined by about 8% over the past 70 years (Davis *et al.*, 2004; Davis, 2009). Although the quantity of nutrients produced per hectare has increased, the nutrient density is often considerably lower than it used to be, contributing to the obesity pandemic.

Recent findings of household consumption patterns from a survey conducted between 2013 and 2015 in four of the poorest rural communities in SA (Ingquza Hill in the Eastern Cape, Jozini in KZN, Maruleng in Limpopo and Ratlou in the North West Province) show that most households are food insecure, with inadequate food available to meet the requirements for a diversified diet (Hendriks *et al.*, 2016, unpublished draft paper). An encouraging number of households engaged in agriculture in Ingquza Hill, Jozini and Maruleng. A very small number of households engaged in agriculture in Ratlou due to the dryness of the area.

Very few households in the sample from these four communities consumed an adequate diversity of fruit and vegetables on a daily basis. Seasonality affected the availability of fresh fruit and vegetables, reducing availability of these foods in winter. A lack of water constrained production of most crops. Participants in focus group discussions reported that drought and climate change have reduced opportunities for diversifying production and the availability of wild foods (Hendriks *et al.*, 2016, unpublished draft paper).

Access to a diversified diet was problematic for households in these four communities. Households reported that a diverse diet was unaffordable. Except for Jozini where dietary diversity was higher, households consumed foods from an average of four food groups each day. The typical diet consisted of maize meal with sugar. Where incomes permitted and production provided ingredients, a relish of onion and tomato or cabbage was added to one meal per day (Hendriks *et al.*, 2016, unpublished draft paper). The largely refined industrially produced maize-based diet is likely the underlying reason both for growth faltering in children and for overweight in women. Most female caregivers were overweight and obese, suggesting that for most households, sufficient dietary energy in the form of the staple food – refined, purchased white maize – is available. The high levels of stunting indicate that the children included in the survey have experienced growth faltering in early life. What is of concern, however, is how many of these stunted children were also overweight (Hendriks *et al.*, 2016, unpublished draft paper).

The study found an encouraging link between engagement in agriculture and improved dietary quality, in common with the earlier reports on urban vegetable gardening and dietary diversity. Engagement in agriculture in these rural areas increased the availability of vegetables and in some cases fruit when in season. This improved household dietary diversity and children's anthropometric scores. Income from farmland production and irrigated agriculture led to increased intakes of vegetables and fruit in general but also meat, eggs, fish, milk, roots and tubers (Hendriks *et al.*, 2016, unpublished draft paper). Hannah Posern found similar relationships between vegetable gardening in peri-urban areas around George in the Western Cape and dietary diversity (Posern, 2016). This contradicts some studies which suggest that agriculture and gardening are not important factors in health.

These findings do not necessarily translate to all urban areas, however. Opportunities for poor households in urban areas to engage in food production are more limited. According to Census 2011 data, the levels of involvement in agricultural activities in SA municipalities vary widely between regions, but in the major cities, the proportion engaged in agricultural production is less than 10%. The analysis of Burger and colleagues of the 2002 and 2007 GHSs comparing urban agriculturalists and urban non-agriculturalists of similar incomes found that households practising urban agriculture were on average actually less secure than non-farming households (Hendriks *et al.*, 2016, unpublished draft paper). This may also indicate that people who lack income are more likely to try to feed themselves through gardening than those who have more secure livelihoods.

Siebert and May (2016) explore the 'Right to the City', and argue that:

> In many cities around the world, individuals, community groups, and NGOs play an increasing role in creating productive spaces. Agriculture can shape the urban habitat, and can therefore contribute to the transformation of the city. Guided by Lefebvre's approach of the 'Right to the City' and the concept of 'Food Sovereignty', this contribution focuses on the activism of inhabitants – the way they produce and manage urban space – and their motivations.
>
> (Siebert and May, 2016)

They conclude that 'By highlighting the importance of bottom-up approaches in the urban food system, the authors discuss how these community efforts can be sustained and integrated on institutional level' (Siebert and May, 2016).

Urban households are highly dependent on incomes and grants in order to meet their food needs and are therefore highly vulnerable to food price increases. The Cape Town AFSUN survey found that 80% of sampled households were either moderately or severely food insecure, according to the Households Food Insecurity Access Prevalence measure. The average household dietary diversity score was six out of 12, with the four most commonly consumed foodstuffs being cereals (92% of households), foods made with oils/fats (consumed by 72% of households), sugar and honey (83%) and 'other' (usually tea and coffee) (88%). This suggests that although the average diet may have caloric adequacy, it is likely to be deficient in vitamins and other micronutrients; 88% of households stated that they had gone without food in the previous 6 months due to unaffordability, while 44% had gone without food once a week or more (Hendriks *et al.*, 2016, unpublished draft paper).

There is a direct connection between the food consumed and the way the body functions. The body needs foods which contain both macronutrients and micronutrients in the correct quantities relative to age, gender and life stage. The result of high food prices and low incomes, is that most South Africans cannot afford the types of food which their bodies actually need to function optimally (Smith *et al.*, 2017). The diet should include a variety of good value whole-foods. Food is a way out of poverty and our economy must be improved to ensure that households enjoy a diverse variety of affordable, good quality nutritious food on the income they are able to secure.

Decline of Agriculture

Although decrease in the role of agriculture is only explicitly identified as a driver of food insecurity in six of the review studies quoted by Hendriks *et al.* (2016, unpublished draft paper), it emerges as a strong theme in the systematic review work and is a feature of food insecurity in SA. 'Deagrarianization' has been defined as a process of reorienting economic activities or livelihoods, changes in occupational activity, and realignment of human settlement away from agrarian patterns. One study in the review in the Eastern Cape Province found that abandonment of farming had peaked following the decreased government support for livestock farming after the political transition in the early 1990s and the resultant decreased animal draft power. Yet, there are a number of other reasons that emerge across the studies included in the review, including: (i) the local availability of high-energy processed foods acting as a disincentive to agricultural production; (ii) grants acting as a disincentive to home farming; and (iii) an ageing farmer population in the area, with the younger generation moving to off-farm ventures which they expect to be more lucrative (Hendriks *et al.*, 2016, unpublished draft paper).

Despite this evidence of deagrarianization, the role of agriculture and food gardening remains important to food security. Its role is less for ensuring absolute security in terms of food quantity than for raising dietary quality and diversity – even if only seasonally – usually by increasing vegetable consumption. Working in George in the Western Cape, Siebert and May (2016) and Posern (2016) found that urban agriculture did contribute to dietary diversity, and to healthy food choices. These findings may be vital if current high levels of stunting are to be addressed.

Food Insecurity in Cape Town and Access to Water

Early in 2018 global attention was turned to Cape Town as the city came perilously close to running out of water. The drought raised a series of important food security questions. One of the key contestations was the relative allocation of water to urban residents and rural agriculture, particularly in the context of the dominant export orientation of the region's agriculture. This raised questions about the orientation of the food system as a whole. The city's relatively limited urban agriculture production was put under considerable pressure, and the drought was a compounding factor in the suspending of the city's largest community-supported agriculture project. The drought has provided an opportunity to consider vulnerabilities across the whole food system and their intersection with other critical urban functions; the drought has affected local, but not national availability. Chapter 8 of this volume reports on the campaign 'Resistance is fertile' in the Philippi Horticultural Area of the Cape Flats, where dispossessed traditional smallholder farmers of the Philippi Horticultural Association (PHA) are fighting against developers and the Cape Town Municipality.

Economic access is determined by both prices and income – the drought has proven ability to disrupt both. Had Cape Town reached Day Zero (the day when city taps were to be switched off) there would have been mass retrenchments which would have fundamentally impacted food access. Additionally, the city is home to substantial food processing, and the forced shut down of these businesses would have reduced access significantly. Furthermore, food utilization would have been severely impacted as households would be unable to prepare and consume food safely. This would have shaped dietary practice and health. The drought in Cape Town should have focused policy makers' minds at national, provincial and local scales on building food system resilience through supported diversified, robust food systems, extending beyond considering just availability and a simplistic framing of accessibility.

Conclusion

Climate change is a new reality to be factored in: we will experience weather shocks and price shocks, and this will decrease access to food for vulnerable households. Training, especially for mothers and for women from rural areas, helped to decrease food insecurity in Ghana (where the government doubled spending on agricultural education, with rural women as the main target beneficiary group).

Food security, food sovereignty and organic farming can go together as the core of interventions which should equip rural women and men, as well as motivating and supporting vulnerable urban households, to produce nutritious and culturally appropriate food locally. Nutrition education and programmes to make unrefined, organically produced food more widely available, can play an important role.

Urban agriculture, and agroecology in general, can mitigate the effects of drought on household food security, and contribute to nourishing the next generation of young South Africans. This is essential if we are serious about addressing the stunting of our children's bodies and the accompanying slowing of mental development. Our children need to be given the opportunity to devlop to their full potential. Child grants have made the financial resources available; an organic food sovereignty movement will educate parents to use these resources productively.

References

Altman, M., Hart, T. and Jacobs, P. (2009) Household food security status in South Africa. *Agrekon* 48(4), 345–361.

Auerbach, R. (1999) Design for participation in ecologically sound management of South Africa's Mlazi River catchment. PhD thesis, Agricultural University, Wageningen, the Netherlands.

Ballantine, N., Rousseau, G.G. and Venter, D.J.L. (2008) Purchasing behaviour as a determinant of food insecurity in Klipplaat. *Journal of Family Ecology and Consumer Sciences* 36.

Barański, M., Średnicka-Tober, D., Volakakis, N., Seal, C., Sanderson, R. *et al.* (2014) Higher antioxidant and lower cadmium concentrations and lower incidence of pesticide residues in organically grown crops: a systemic literature review and meta-analysis. *British Journal of Nutrition* May, 1–18. https://doi.org/10.1017/S0007114514001366

Battersby, J. (2012) Urban food security and climate change: a system of flows. In: Frayne, B., Moser, C. and Ziervogel, G. (eds) *Climate Change, Assets and Food Security in Southern African Cities*. Earthscan, New York, pp. 35–56.

Benbrook, C.M., Butler, G., Latif, M.A., Leifert, C. and Davis, D.R. (2013) Organic production enhances milk nutritional quality by shifting fatty acid composition: a United States-wide, 18-month study. *PLoS One* 8(12), e82429. https://doi.org/10.1371/journal.pone.0082429

Brandt, K., Leifert, C., Sanderson, R. and Seal, C.J. (2011) Agroecosystem management and nutritional quality of plant foods: the case of organic fruits and vegetables. *Critical Reviews in Plant Sciences* 30, 177–197. https://doi.org/10.1080/07352689.2011.554417

Crush, J. and Caesar, M. (2014) City without choice: urban food insecurity in Msunduzi, South Africa. *Urban Forum* 25, 165–175. https://doi.org/10.1007/s12132-014-9218-4

DAFF (Department of Agriculture, Forestry and Fisheries) (2018) Monthly Food Security Bulletin of South Africa: December 2017. Available at: file:///C:/Users/hannelise/Downloads/RSA-Food-Security-Bulletin-Dec-2017.pdf (accessed 7 June 2018).

DAFF and DSD (Department of Agriculture, Forestry and Fisheries and Department for Social Development) (2013) National Food and Nutrition Security Strategy. Available at: www.nda.agric.za/docs/media/NATIONAL%20POLICYon%20food%20and%20nutrition%20security.pdf (accessed 1 September 2019).

Dangour, A.D., Dodhia, S.K., Hayter, A., Allen, E., Lock, K. and Uauy, R. (2009) Nutritional quality of organic foods: a systematic review. *The American Journal of Clinical Nutrition* 90(3), 680–685. https://doi.org/10.3945/ajcn.2009.28041

Darwin, C. (1859) *On the Origin of Species by Means of Natural Selection, or the Preservation of Favoured Races in the Struggle for Life*. John Murray, London.

Davis, D. (2009) Declining fruit and vegetable nutrient composition: what is the evidence? *Hortscience* 44(1).

Davis, D. (2011) Impact of breeding and yield on fruit, vegetable and grain nutrient content. In: Jenks, M.A. and Bebeli, P. (eds) *Breeding for Fruit Quality*. Wiley, Hoboken, New Jersey, pp. 279–305.

Davis, D., Epp, M. and Riordan, H. (2004) Changes in USDA food composition data for 43 garden crops, 1950 to 1999. *Journal of the American College of Nutrition* 23, 669–682.

De Cock, N., D'Haese, M., Vink, N., van Rooyen, C.J., Staelens, L., Schönfeldt, H.C. and D'Haese, L. (2013) Food security in rural areas of Limpopo province, South Africa. *Food Security* 5, 269–282. https://doi.org/10.1007/s12571-013-0247-y

Devereux, S. (2006) *Distinguishing Between Chronic and Transitory Food Insecurity in Emergency Needs Assessments*. World Food Programme, Rome.

Devereux, S. and Waidler, J. (2017) Why does malnutrition persist in South Africa despite social grants? *Food Security SA Working Paper* #001. Department of Science & Technology – National Research Foundation Centre of Excellence in Food Security, Cape Town, South Africa.

Drake, I. (2009) French Research Gives Boost to Organic Food. Available at: http://www.ausfoodnews.com.au/2009/09/14/french-research-gives-boost-to-organic-food.html (accessed 14 June 2018).

Farmer's Weekly (2019) *Farmer's Weekly South Africa: Agriculture & Farming News* (published in Johannesburg) 11–16 January.

Frayne, B., Moser, C. and Ziervogel, G. (2012) *Climate Change, Assets and Food Security in Southern African Cities*. Earthscan, New York.

IPCC (Intergovernmental Panel on Climate Change) (2012) Chapter 22 in Assessment Report 5. Available at https://www.ipcc.ch/pdf/assessment-report/ar5/wg2/WGIIAR5-Chap22_FINAL.pdf (accessed 6 May 2018).

Kubik, Z. (2018) Weather shocks, migration and food security: evidence from sub-Saharan Africa. PhD thesis, Paris-Sorbonne University, Paris.

Labadarios, D., Steyn, N.P. and Nel, J. (2011) How diverse is the diet of adult South Africans? *Nutrition Journal* 10(33).

Lairon, D. (2010) Nutritional quality and safety of organic food. A review. *Agronomy for Sustainable Development* 30, 33–41.

Leifert, C., Nielson, J.H. and Rembialkowska, E. (2007) Third Quality Low Input Food (QLIF) Congress, Hohenheim, Germany, 20–23 March. Available at http://orgprints.org/10417/2/leifert-etal-proceedings-qlif.pdf (accessed 1 September 2019).

Masekoameng, M. and Maliwichi, L.L. (2014) Determinants of food accessibility of the rural households in Sekhukhune District Limpopo Province, South Africa. *Journal of Human Ecology* 47, 275–283. https://doi.org/10.1080/09709274.2014.11906762

Ndobo, F. and Sekhampu, T.J. (2013) Determinants of vulnerability to food insecurity in a South African township: a gender analysis. *Mediterranean Journal of Social Sciences* 4(14). https://doi.org/10.5901/mjss.2013.v4n14p311

Nyéléni (2007) Forum for Food Sovereignty. Sélingué, Mali. Available at: https://nyeleni.org/DOWNLOADS/Nyelni_EN.pdf (accessed 5 January 2019).

Posern, H. (2016) Food sovereignty in urban areas. Promoting marginalized communities through urban gardening. A case study of George, South Africa. MSc thesis, Institute of Development Research and Development Policy, Ruhr University, Bochum, Germany.

Reeve, J., Hoagland, L., Villalba, J., Carr, P., Atucha, A., Cambardella, C., Davis, D.R. and Delate, K. (2016) Organic farming, soil health and food quality: considering possible links. *Advances in Agronomy* 137, 319–368.

Siebert, A. and May, J. (2016) Urbane landwirtschaft und das recht auf stadt: theoretische reflektion und ein praxisbeispiel aus George, Südafrika. In: Engler, S., Stengel, O. and Bommert, W. (eds) *Regional, Innovativ und Gesund: Nachhaltige Ernährung als Teil der Großen Transformation*. Vandenhoeck & Ruprecht, Göttingen, Germany, pp. 153–168.

Smith, J., Abrahams, M. and Chiya, N. (2017) *PACSA Food Price Barometer – Annual Report*. Action Council for Social Awareness, Pietermaritzburg, South Africa.

StatsSA (Statistics South Africa) (2012) Income and Expenditure Survey 2010/2011. StatsSA PO100. Available at: http://www.statssa.gov.za/publications/P0100/P01002011.pdf (accessed 3 January 2019).

StatsSA (Statistics South Africa) (2017) *South Africa Demographic and Health Survey 2016: Key Indicator Report*. Statistics South Africa, Pretoria. Available at: www.statssa.gov.za/publications/Report%2003-00-09/Report%2003-00-092016.pdf (accessed 6 May 2018).

Walsh, C.M. and van Rooyen, F.C. (2015) Household food security and hunger in rural and urban communities in the Free State Province, South Africa. *Ecology of Food and Nutrition* 54(2), 118–137. https://doi.org/10.1080/03670244.2014.964230

Worthington, V. (2001) Nutritional quality of organic versus conventional fruits, vegetables, and grains. *Journal of Alternative and Complementary Medicine* 7(2), 161–173.

8 The Use of Participatory Rural Appraisal (PRA) to Support Organic Food Systems in Africa

Raymond Auerbach*
Nelson Mandela University, George, South Africa

Abstract
The theory of participatory action research recommends finding tools for community participation in analysis of the situation and development of community programmes. Two case studies, from KwaZulu-Natal and from Philippi near Cape Town, are examined. Participatory rural appraisal (PRA) techniques include using: (i) multiple historical perspectives to construct a community timeline; (ii) SWOT (strengths, weaknesses, opportunities, threats) analysis; (iii) issue identification by interest groups; (iv) community voting techniques; (v) transect walks where the actual land is visited and important features are noted and discussed; and (vi) participatory mapping, where such transect walks are converted into rough maps made on a concrete slab or a small piece of land with chalks and various props to represent land features. Such a map can then be 'interviewed' by participants to draw out further understanding of the spatial dynamics. Visioning exercises, Venn diagrams of stakeholder relationships, and analysis of problems and solutions using 'problem' and 'solutions' trees can then be used to clarify problems and draw out solutions. Community participants and young researchers found these techniques useful in developing understanding of community dynamics and clarifying priorities.

Introduction

Having presented some global perspectives in Part 1 of this book, and an overview of the impacts of recent droughts and climate change on South Africa (SA) in Chapter 7, we will look at a number of important rural development tools and approaches, and how they can contribute to capacity building and the development of sustainable institutions in Africa.

One of these tools is participatory rural appraisal (PRA) which is a set of interventions (often called a PRA toolbox) which can help communities to analyse their situation, and to set development objectives and devise strategies (with measurable indicators of success) to achieve them (Auerbach, 2018).

This chapter draws first on a series of PRA exercises carried out in KwaZulu-Natal (KZN) some years ago as part of my doctoral research, and then on a PRA course held in 2018 in the Philippi Horticultural Area of Cape Town (in the well-known Cape Flats). Small-scale farmers have grown vegetables in this part of Cape Town for hundreds of years, and I remember (in 1978) the late Professor Richard van der Ross (former Vice Chancellor of the University of the Western Cape) explaining to me and the late Robert

*raymond.auerbach@mandela.ac.za

Mazibuko (SA's Tree Man) how he grew up on the flats, and he used to accompany his grandfather in a donkey cart to sell the onions which they had produced on their farm at the market in Cape Town. He explained to us how vibrant the area had been and how he owed his appreciation of nature and his work ethic to these early experiences of organic farming, which shaped his view of the world.

History of PRA

The PRA approach was mainly developed in India by the late James Mascarenhas and Robert Chambers (Mascarenhas, 1991; Chambers, 1994). In the late 1980s in Kenya and India, a number of development practitioners and community members found the Rapid Rural Appraisal approach rather extractive, and a number of participatory exercises were developed by communities and practitioners working in collaboration, which promoted community ownership of the process; these then developed into a 'PRA toolkit'. The tools include: (i) direct observation; (ii) participatory mapping and modelling; (iii) seasonal calendars; (iv) semi-structured interviews; (v) timelines; (vi) local histories; (vii) Venn diagrams of relationships of stakeholder organizations; (viii) wealth and well-being rankings; and (ix) folklore, songs, poetry and dance. These techniques are summarized in the concise manual by Luigi Cavestro (Cavestro, 2003). In this chapter I will give two examples from my work to illustrate the usefulness of the approach.

In the 1980s and 1990s, I worked at the University of Natal, first for the Institute of Natural Resources and later for the Farmer Support Group (FSG, as Research Coordinator, with the task of helping colleagues to reconceptualise their work as action research). On behalf of the FSG, the Association for Rural Advancement (AFRA) and the Midlands Rural Development Network (Midnet), Noel Oettle invited Jimmy Mascarenhas to run the first SA PRA Training Course at Stoffelton in KZN in April 1993, which the participants then wrote up as a manual, with major input from Erna Kruger and the late Tessa Cousins, based on a first draft of the workshop notes which I compiled (Auerbach, 2018).

Use of PRA in SA in the 1990s

Jimmy Mascarenhas died soon after he had visited SA, but his approach to PRA has been adopted in many rural development projects around the world. The change in approach in India after the Green Revolution, led to a shift in emphasis from technical training and extension support for farmers, to capacity building and support for local and regional farmer institutions.

The training workshop at Stoffelton (near Bulwer) in KZN saw FSG, Midnet and AFRA make the same shift towards capacity building, and many of the more successful projects run in SA since then have used this approach, with PRAs carried out in a number of projects by participants from the original course, who then trained others.

Debate over integrated catchment management (ICM) in SA also had a major influence on development thinking and led to the Water Act of 1996. A range of projects worldwide based on ICM saw the development of participatory watershed management in India and river basin management in South-east Asia. The Challenge Programme for Water and Food, instituted as a reform programme by the Consultative Group for International Agricultural Research (CGIAR), and driven mainly by the International Water Management Institute (IWMI), was an attempt to make international research more responsive to participatory considerations, and project selection criteria included an emphasis on triple bottom-line development (projects had to include measures to address social, environmental and economic sustainability), as well as participatory processes within the projects.

Having assisted the Land and Agricultural Policy Centre (LAPC) – which was the ANC think-tank on land and agriculture – in a consultation process which included white and black farmers, academics, farmer organizations, trade union representatives and politicians in all four of the (then) provinces of SA, I edited a report for LAPC (Auerbach, 1994) recommending a more participatory approach and new national and provincial departments of agriculture which would focus on capacity building and local institution building. This report contributed directly to the White Paper on SA Land

Policy (Department of Rural Development, 1997), which states (p. 24): 'This work, together with the public comments, was then incorporated into a Draft Statement of Land Policy and Principles that was the basic document discussed at the National Land Policy Conference held on 31 August and 1 September 1995'. The White Paper also advocated facilitating women's access to land by removing legal discrimination (p. 40), and increasing community participation in land reform and land management.

The envisaged transformation of the Department of Land Affairs did not take place as hoped, and land reform has been a slow and contentious process, with many beneficiaries being forced to take on group ownership of land, with all of the management difficulties which this involves. The implications for land reform and agricultural transformation in SA will be discussed in the final chapter of this book.

In the ICM project which I initiated in the Mlazi River catchment (Auerbach, 1997), we made use of PRA in order to deepen our understanding of the history and priorities of the various rural groups we were trying to help. My doctorate summarized these processes in three appendices, and some brief extracts may illustrate the usefulness of PRA processes in involving communities in development planning.

The first PRA plannings in May and June 1996 were as follows:

> The first PRA [helped] to learn about the farming systems of the area, with a view to understanding what support is required to develop sustainable and profitable farming enterprises. It included a time-line study, an activity profile for men and women, interest group perspectives for crop farmers, gardeners and livestock owners, a Venn diagram illustrating what institutions assist people in the area, and a discussion of local approaches to managing conflict which racked the area from 1986 to 1990. Later PRAs gathered specific information about resources in particular micro-catchments.
>
> (Auerbach, 1999)

Appendix 5 (Auerbach, 1999) reported on PRA evaluations in October 1998, of three of the main community gardens, including one where the local Inkatha Freedom Party (IFP) leader heavily criticized the project, and ordered the FSG to withdraw from the garden. Female members later approached project staff, and said that although FSG had been asked to withdraw from the garden by the politicians, nobody had said that we should withdraw from supporting the women's craft groups. At their request, this work thus continued, while the IFP continued to obstruct garden development, especially after the ANC-led local government in Durban's Outer West gave grants to the community gardens to support their work. These examples show that PRA can be a potent instrument for revealing the inner socio-political tensions at work in communities.

My doctoral thesis framed its research question as: 'How can diverse communities often characterized by conflict, be helped to come together to learn more about natural resources and systems, and to manage them collectively in a way which is productive and responsible?' (Auerbach, 1999). On p. 123 of the same work, under the heading 'Design objectives', I touched on the research question and the two design questions:

> a) What strategies are available to help improve food security in a way which addresses aridity, poverty and restricted availability of land?
>
> b) Are there effective methods of producing good yields of crops without resorting to the levels of fertilizers, poisons and other technologies which have produced the environmental problems currently being experienced in Europe?
>
> (Auerbach, 1999)

Earlier, I had commented under the heading 'The socio-political environment' that:

> The process of participatory action research required that there should be some understanding of the governing variables (norms and values in the communities), before developing action strategies, which would have practical consequences.... Right at the start of the programme ....(we brought) together community leaders, resource managers and key informants, in order to develop an overview of the Ntshongweni area and its problems and needs. This was followed by a process of engagement ...., which included .... a PRA process .... and vision building and participatory land use planning processes ....
>
> (Auerbach, 1999)

On the following page, I reported that both the local IFP and ANC leaders had been involved in local violence, and then also 'had played a major role in the peacemaking process which preceded the Ntshongweni Peace Accord of 1989'.

Given the fractured nature of communities in KZN at the time, processes such as those described above were fraught with difficulty, but did contribute to allowing discussion of patriarchal rural governance traditions. Sometimes, this was enough to empower women to insist on changes, and sometimes developers had to bow to the strongly expressed wishes of male-dominated local organizations.

More recently, PRA has been used to assist the Philippi Horticultural Association (PHA) in trying to secure access to peri-urban farming land near Cape Town.

Use of PRA in Peri-urban Cape Town: PHA, 2018

A PRA training course was held with 23 attendants from various parts of the Cape Flats, and including three doctoral students (from Tanzania, Germany and SA). The course was held at the request of PHA, and had been preceded by a series of PHA community training workshops covering soil chemistry, plant nutrition and soil biology, group formation and political protest. The course started with a timeline exercise which is reproduced in full in Fig. 8.1. Timelines help participants to understand the sequence of historical processes, not just from one point of view, but by including older men and women, business people, educators, farmers and the youth; a range of perspectives inform and enrich the written timeline.

As part of the timeline exercise the participants were told the following by members of the PHA:

> Proposed developments include the following, where developers have already purchased the land from farmers:
>
> | | | | |
> |---|---|---|---|
> | Oakland City proposed | 2008 | 570 ha | 30,000 houses |
> | U-vest | 2012 | 280 ha | 10,000 houses, private school, shops |
> | U-vest | 2015 | 5 ha | shopping centre |
> | North Side (Busy Corner) | 2017 | 50 ha | 400 houses |
>
> The national Department of Agriculture, Forestry and Fisheries (DAFF) rejected the 'change of land use' applications (Act 70 of 70), *but* the minister is not taking part actively in the proposed court case, and seems to be allowing the City of Cape Town to do what it wants; what the city wants is more rates from developed property, and the City is not listening to National Policy Objectives. The City is *not* thinking about food, land reform or water conservation through aquifer recharge, but only about short-term increases in the rates base. PHA has been fighting non-agricultural development for 10 years. Urban Edge planners have said '*no*', but province has allowed an Environmental Impact Assessment process and City Rezoning initiatives to go ahead; all appeals thus far have been denied. Now we are going to court to fight these rulings, which will destroy farming in PHA – we want the court to stop the City from flouting national law.
>
> Food security is more important than sand mining and luxury housing development; land reform *must* promote food security! We need to make it easy for urban people to grow food; they should be helped with access to land, water and support.
>
> Part of the problem is that even national government wants us to adopt an industrial farming paradigm. This is impractical and will destroy the environment and the health of people eating industrial food. At present, small-scale farmers, mainly women, are feeding more than half of the world. The industrial model will emphasize wheat, soybeans and maize to supply to food processors, and with hidden sugars, salts and fats; only a few farmers produce broad-acre crops.
>
> The model which PHA is developing would see small-scale commercial organic family farms on the 1300 ha of land which could become accessible in Philippi. More than a 100 farmers could combine with craft producers, agrotourism, and agriprocessing facilities to produce viable home industries; the alternative is more poverty, disempowerment and displacement.

After this timeline exercise, and a discussion on the vision and the challenges to be addressed, participants divided into three geographically based groups: (i) Mitchell's Plain and Strandfontein; (ii) Gugulethu; and (iii) Philippi Horticultural Area. The first group identified the issues shown in Fig. 8.2 in some detail, and then prioritized them, using a participatory voting technique (Table 8.1), while the others prioritized the issues on the same sheet (they have been re-ordered spatially according to ranking) (Tables 8.2 and 8.3).

Pre-1920 The area was used as grazing lands by Xhosa-speaking people. Then some German Protestant farmers arrived and were given some land, but then they were put in concentration camps by the British during the Anglo-Boer war.

1920 The German farmers sold some land to English soldiers.

1930 The Cape Town Fresh Produce Market was at the Good Hope Centre in the 1920s – it then moved to Salt River.

1948 Nationalist Government elected.

1960 Market moved; Epping Municipal Market, still largely horses and carts, some trucks
SA becomes a Republic; Group Areas really applied in the mid-1960s.
Removals from District 6 – huge trucks, many policemen (childhood memories).
Many friends lost; went to Gugulethu (outside toilets – squatting), no electricity.
Mitchell's Plain was declared a 'Coloured Area', so no blacks or Indians were allowed – many were 'removed' under the Group Areas Act.
Some Indian families bought land and stores, even women (Mrs Fatima Sonday told us she had to move to a new area, and bought a store there).

1970 Earthquake at Tulbagh and Ceres – also felt in Philippi and Gugulethu – many people praying in the streets! At this point, most households still had vegetable patches and a flower garden.
People helped each other, especially with surplus vegetables.
The mothers clubbed together – Zenzele Savings Clubs; pooled resources to improve houses, installed toilets and water in houses, and concrete floors.
Spirit of camaraderie – solidarity; some mothers worked in factories.
'I had to leave school, as we had been moved from Bo-Kaap to Rondebosch and Dad could not trade there; he was now unemployed and upset that he could not provide for his family, so children had to leave school and get work; sister refused, insisted on finishing and went into healthcare.'

1976 Many families split up; school riots (not only in Soweto – also Cape Town); many comrades died in Bishop Lavis – it was painful, and many left school (one participant moved school 11 times!).
Removals continued; ghettoes developed, school system deteriorated, DOP Systema saw farmworkers paid partly in booze – there were drunken brawls at the weekends.
As youngsters, we became politically aware of our situation: we heard of Mwalimu Nyerere (Tanzania), of Samora Mashele (Mozambique) – Congo, Mozambique, Zambia – school kids were conscientized, politicized.
Snoek sellers with their bakkies (small trucks) became a feature; donkey carts with fruit and vegetables from local producers; the church was often a unifying centre around which the community organized itself. After 1976, many people blocked out the trauma which they had experienced, and refused to talk about it. We need to write our own history, otherwise others will write it for us, but how do we piece this history together?
At that time, Gugulethu was a classless society; it became a centre of debate.
Some were moved to Ocean View, which is geographically isolated; many then lost their agricultural traditions, and stopped growing vegetables.
A sense of helplessness was carried over to the children.
Some young people questioned the attitude, and asked "Why is the community behaving this way? Why did they allow government to move them?" The adults responded that many worked for government, and felt powerless and hopeless.
The struggle for survival took almost all of their energy. People were made to feel inferior, farmworkers were also displaced – squatter camps, then Belhaar was established, and many squatters moved to town; they were then even more cut off from farming, with three or four generations now alienated from the land.

1985 Kids were making petrol bombs – we learned about political resistance.
We were removed; now we are re-claiming the land; in the 1990s, some Indians had to form companies with white "front" shareholders (a friendly lawyer) to own the store. We had to get permission of neighbours; some were OK, but two white farmers objected; eventually permission was granted for Nazeer to start a bakery, and he was able to get a

Fig. 8.1. Timeline exercise for participants, with focus on Philippi Horticultural Association (PHA).

bond in theory, but the bank delayed for six months. The banks were part of the system, and acted as gate-keepers for *apartheid*.

1994 We formed a Civic Association, and a Housing Association; the community became more organised. We demanded restitution and access to land; many campaigns and good press coverage, but we had to educate people about why land is important, and where their food comes from. We have had support from a range of food consumers who want us to produce healthy, organically grown vegetables. There is still much education and awareness needed.

2000 The developers increased pressure, and some white farmers want to sell their land to make big money from luxury housing and industrial development on their farmland; the national minister of agriculture refused to agree to changes in land use, saying the land is important for food security. Some white farmers agree with us that the land should stay under farming, and we have had support from their Farmers Association, initially. The PHA participated in opposing the development applications, but lost each application, and then appealed and lost again; now it is the end of that process, and will go to court. They need specialists to say "this land is good for farming", which can clearly be seen by the successful crops growing all over the area, even though the developers have said that the soil is too poor for agriculture.
When was this land originally transferred from crown ownership?
We believe many farms were about 20 ha; Mr Terblanche now farms on 200 ha, but does not own all of it – at least he is using the land productively, and is complying with some environmental management requirements (GlobalGAP, etc.); he uses the land under informal leases.

[a]The DOP System was a system of payment used from the beginning of the 19th century up to 2003 in which farmworkers were paid partly in alcohol instead of money.

Fig. 8.1. Continued.

Mitchell's Plain and Strandfontein

<u>**Challenges and issues**</u>

* Security
* Access to land
* Water — Aquifer; Sea water usage through desalinization
* Labour — Skilled; Create jobs (there are people looking for work)
* Sewage
* Transport: Pricing; Distance; Logistics
* Landfill and pollution
* Support of business enterprises
* Climate change: Rainfall; Effects on production
* Education: Training
Educational centre (school)
Agricultural courses in various languages

Fig. 8.2. Mitchell's Plain and Strandfontein interest group's issue identification; priorities were set by discussion and by voting for each area.

Table 8.1. Mitchell's Plain and Strandfontein: issues prioritized.

Rank	Votes	Issue
1	6	Security
1	6	Access to land
1	6	Water
4	5	Transport
5	4	Education
6	3	Business
7	2	Labour
8	2	Waste management
9	2	Climate change
10	0	(Put sewage and landfill together as waste management)

Table 8.2. Gugulethu: issues prioritized.

Rank	Votes	Issue
1	7	Community gatekeepers (politicians) [added afterwards!]
2	6	Land access (agricultural policies)
3	5	Market access
4	4	Access to relevant/appropriate training and education (mentorship, policies)
5	3	Water access
6	2	Policies are not implemented
7	2	Failure of Department of Agriculture to assist us – reliance on NGOs
8	2	Access to tools/inputs (compost, seeds, seedlings)

Table 8.3. Philippi Horticultural Area group: issues prioritized.

Rank	Votes	Issue
1	6	Funding for land access struggle
2	5	Water
2	5	Education (farming)
4	4	Awareness of farmers (farmers invisible – need engagement)
5	4	Markets (access and create new)
6	2	Security
6	2	Lack of endemic species
6	2	Equipment and tools
6	2	Genetics
10	1	Lack of solidarity
10	1	Employees' rights
10	1	Lack of organized structures

Table 8.4. Summary of issues.

Rank	Votes	Issue
1	18	Land access/funds
2	14	Water
3	13	Education/training (farming)

After plenary discussion, three major common issues were identified and prioritized (Table 8.4).

The following day, the group travelled to the area where PHA intends to set up a 1300 ha organic farming project, and we walked over this land, carrying out a transect exercise; notes from the walk and the debriefing discussion (Fig. 8.3) are described in the next section.

Transect walk and participatory mapping exercises

The workshop participants travelled to the farming area which PHA wishes to acquire for commercial organic farming by small-scale black farmers. We walked around the land observing topography, soil, wind direction, vegetation, signs of domestic and wild animals (spoor), and we interviewed two herdsmen who were herding cattle and goats.

On our return to PHA Campaign Headquarters, the following feedback on the transect walk was recorded.

What did we see? What was important?

The walk brought back memories of my childhood – the sand dunes where I had played, the flora and fauna which I remember, the smells, the sounds. When we moved to Gugulethu, the flora and sand dunes were similar; happy memories of childhood, envy for this place.

We saw animals grazing – first cattle of the commercial farmer, which were then taken back to the feedlot, then two herdsmen with cows and goats.

The land is not flat – there are many small hills which provide shelter from the wind in the little valleys; the vegetation varies – the herdsmen told us that they are not allowed to take their animals into the 'government land' (we noted far more vegetation there, but we are not clear who controls this land). The fynbos is resilient

Fig. 8.3. Group discussing future land use. (Photograph by Maren Wesselow.)

and seems to come back quite quickly after disturbance; some have seen it come back when invasive alien plants are controlled. There are still many birds in the bushy ('government') area, fewer were noticed in the cleared areas.

Some areas are protected from wind by the topography – this could help in creating microclimates – permaculture 'Sector' and 'Zone' planning processes could be useful here. There is no evidence of use of poisons or chemical fertilizers on the fynbos areas. The soils are sandy, hot, dry and dead in most places, although there was evidence of some soil microorganisms in certain places (under the cow dung pats).

We noted many productive, successful commercial farms with developed infrastructure (windbreaks, irrigation, roads, sheds, etc.). It is definitely possible to farm on these sandy soils, but the conventional farmers are doing it by pouring fertilizer and water on to the sandy soils – it would be important to build up the soil organic matter on these soils by the use of compost, mulch and crop rotation. There is potential for a training centre here for farmers – the conditions are tough, but possible. Participants noted that it is always 5°C cooler here than in Cape Town. The coastal breeze may allow production of some crops such as potatoes.

We noted several tracks of cattle, goats and small mammals (some dragging a tail, which we could not identify); we found geckos and a snakeskin, as well as many holes (snake holes?) and a wide variety of plants. However, we had the impression of degraded fynbos, and noted some soil erosion, and sand mining with trucks removing the sandy soil. There were many invasive species (Port Jackson willow, *Acacia saligna*) and poor biodiversity. There seemed to be some seasonal wetlands – members will return to check on these, and on water flow, in the wet season as this is a winter rainfall area.

There are footpaths and four-by-four tracks through the area – people are using this place, even where nothing is planted. A gravel road has been made for truck access. Some bush clearing is taking place – perhaps for a road or firebreak. Many trucks are coming and going as part of the sand mining operation. This place has changed a lot compared with 1970!

Our overall impression was that the place is quiet, beautiful and peaceful, the land is elevated, and there is great agrotourism potential, with lovely views of the mountain and the sea. If local government could see the political, social and economic potential of this site, they would fast track appropriate agricultural development!

Visioning exercise: what could/should happen on the land?

Wind and sun can provide alternative sources of energy.

The fynbos should be regenerated, the south-easterly wind could be used to blow rubbish into nets.

We should learn from the local commercial farmers, and ask them: (i) What does it take to farm successfully in this area?; and (ii) Can we farm potatoes, leafy plants, with this coastal breeze?

We should also get to know the herders we met and learn about animal production here; build on the existing local knowledge. We should build on local and other best practice, and also learn from failures! There is agrotourism potential on a farm. It could be a hub of a Cape Town Training Network – 'Know where your food comes from'! With bungalows for visitors to use when they stay over – each farm could potentially be a model: fish, chickens, crops, bees, etc.

Looking at systems thinking, there is a potential for a school for agriculture, with children coming to visit the farm and learn about nutrition; and with links to a national school nutrition programme.

It is a potential research site for urban farming (and a demonstration site, using a working system); with integrated development of a local organic food system.

There may be some conflicts of interest.

We would like to see each farmer having an individual unit, specializing in their chosen operation (livestock, crops, vegetables), but security would be an important consideration.

There is also a potential danger of a squatter camp developing.

Organizing farmers will be a major requirement, and there is potential for some unified elements in the enterprise, at least marketing under a brand. Some rules and management will be required – what will the 'entrance requirements' be?

We will need to develop home-grown models for our conditions – and understand complexity, resilience and equity requirements. Efficient farming systems will need to be developed which use water efficiently and recharge the aquifer (important, as Cape Town water resources are severely limited!), along with capacity, continuity and institution-building processes. Packhouses (or sheds) will be needed, along with brands and market outlets.

Community infrastructure will also be needed including clinics, schools, eco-playschool, alternative health models based on an organic diet and natural medicine, local exchange trading systems, recreational areas, and processing plants to add value to crops.

What will our business model be? Can we access inputs at lower cost?

We need to make the business case to show that when farm production, value adding, agritourism and education are properly costed, as well as ecosystem services, there will be a very strong argument in favour of this development.

We need to gather the following information:

- How does the water move/flood?
- How is the aquifer recharged?
- We need to study the Atlantis water management model, and learn to use water to buffer salinity.
- How can we build soil fertility?
- What are the effects of existing chemical farmers?
- What is the impact of the existing feedlot on the aquifer?
- How should we communicate, make others aware and lobby the community?
- We need to analyse local farmer success – cooperatives, etc.
- What systems from elsewhere (other countries, local) can teach us good practice in ecovillage development and organic farming and processing?
- What skills and capabilities do we have in the group and in support groups?
- Who else can help – human rights, NGOs?
- How do we organize ourselves (civil society is the voice of development)?
- How do we change policy, force government to act, and show up local government failings?

Following the feedback discussion Raymond and Christopher outlined processes for water- and energy-efficient garden development, and well-point development from a technical viewpoint and Ganief shared his pneumatic pump design.

On the final day, we looked at the use of Venn diagrams, seasonal calendars and 'problem' and 'solutions' trees.

Use of Venn diagrams

Venn diagrams are used to describe the relationships between the community project participants and other people, groups or organizations. The method is described in the PRA manuals provided to participants.

Developing a problem tree and a solutions tree

The problem tree is used to understand why you are seeing the effects or consequences which you regard as problematic; look for the causes, and then dig deeper to find the underlying root causes. Then look at the impact of the problems (Fig. 8.4).

Once you have analysed the problems and their causes and impacts, look at possible solutions, starting with the objectives which you feel relate to each problem. Then look at what activities each objective needs, what the outputs will be and what the purpose of each of the objectives (and the accompanying activities) is (Fig. 8.5).

The solutions tree transforms problems into solutions, through purposeful activities.

Once the activities are clear, it is simple to establish costs, timelines and responsibilities.

Three problem tree exercises were looked at: (i) limited access to land; (ii) lack of production; and (iii) ineffective PHA. These are reported in Figs 8.6, 8.7 and 8.8, respectively.

Organizing activities using PRA

Once the PRA exercises were complete the participants were briefed as to how to organize activities using PRA. An outline of the briefing is as follows:

- Start with a participatory planning process using PRA (as we have done).
- Then formulate *goals* and *objectives*, and discuss how they could be achieved. To do this, the causes of the current situation and what the problems are need to be understood – this all forms the *situation analysis*.
- The next step is to move from problems to solutions. Here the problem tree and solutions tree can be useful tools; only then can you decide whether there is one project or several projects. Then you can start looking for help – resources, partners, potential donors; each donor will have their own requirements for approach and structure of a project proposal. A thorough overall logical framework analysis (logframe) can form the basis from which several project proposals may be drawn. There are many forms of logframe; table 8.5 uses the same terms as were used for the problem tree and solutions tree.

Feedback from participants on the use of PRA techniques

Information is power – these are useful techniques for helping the community to agree on

Impact 1	Impact 2	Impact 3	Impact 4	Impact 5
Effect 1	Effect 2	Effect 3	Effect 4	Effect 5
		PROBLEM		
Cause 1	Cause 2	Cause 3	Cause 4	Cause 5
Root cause 1	Root cause 2	Root cause 3	Root cause 4	Root cause 5

Fig. 8.4. What is a 'problem' tree? Start with cause, then dig deeper, then effect, then impact.

Purpose 1	Purpose 2	Purpose 3	Purpose 4	Purpose 5
Output 1	Output 2	Output 3	Output 4	Output 5
		GOAL		
Objective 1	Objective 2	Objective 3	Objective 4	Objective 5
Activities 1.xxx	Activities 2.xxx	Activities 3.xxx	Activities 4.xxx	Activities 5.xxx

Fig. 8.5. What is a 'solutions' tree? Start with objectives, then define activities, then outputs then purpose should become clear; these can then go straight into a logical framework analysis ('logframe').

Exercise 1

Impact 1: Poverty Impact 2: Dependency Impact 3: Corruption (greed)

Effect 1: Landless people Effect 2: Lack of self-sufficiency Effect 3: Poor governance

PROBLEM: No access to land

Cause 1: Stolen land Cause 2: Finances Cause 3: Bureaucracy

Root cause 1: Colonization Root cause 2: Slavery Root cause 3: Politics

Purpose 1: Means of production Purpose 2: Self-sufficiency Purpose 3: The people shall govern (Freedom Charter)

Output 1: Reallocation of land Output 2: Collective of young entrepreneurs Output 3: Collective determination

GOAL: Access to land

Objective 1: Expropriation without compensation Objective 2: Micro-enterprises Objective 3: Civil participation

Activities 1: Decolonize the system Activities 2: Entrepreneur workshops at school Activities 3: Organize/structure community

Fig. 8.6. Problem tree exercise 1: Access to land – problems and solutions.

Exercise 2

Impacts: Inappropriate production Less water No jobs or bargaining power Failures Erosion/poor soils

Effects: Poor and low quality production No rain/waste No sales Directionless Soil degradation

PROBLEM: Lack of Production

Causes: Systems planning Inefficient watering Poor marketing Poor support Poor soil

Root Causes: Poor leadership, disorganized Climate change Don't know consumers Few champions Fertility management

Purposes: Empowered farmers Organized Resilient Increase GDP and happiness Healthy/food secure Resilient/carbon capture

Outputs: Farmers know systems Meet demand Recharge aquifers Job creation Sustained production Healthy

GOAL: Good productivity in our sector

Objectives: Foster awareness of systems Well-planned organic operation Sustainable water usage
Strong markets Continuous quality support Sustainable soil use

Activities: Workshops Crop calendars and PRA Extension
Aquifer care/rainwater harvesting Compost Lobbying

Fig. 8.7. Problem tree exercise 2: Lack of production – problems and solutions. GDP, gross domestic product.

Exercise 3

Impacts: Politicians not held accountable No farmers markets Developers take advantage

Effects: Wrong understanding of democracy No farm for training communities Low participation

PROBLEM: Ineffective PHA

Causes: Lack of awareness/poor organization Not reaching people Hopelessness

Root Causes: No connection between people and food Market distortion: farming and food People disempowered

Purposes 1: Sustainable livelihoods Fresh food/food security National pride

Outputs: Increasing popular involvement People make informed food choices Politicians more responsive

GOAL: How to make PHA more effective

Objectives: Make 50,000 people know about PHA Educate people about food systems Hold local politicians accountable

Activities: Community newspapers and radio, meetings annd pamphlets Demo farm and markets Mobilize local people and get feedback

Fig. 8.8. Problem tree exercise 3: PHA Campaign – problems and solutions.

Table 8.5. An outline for a logical framework analysis (logframe).

	Narrative description	Objectively verifiable indicators	Means of verification	Assumptions
Objectives				
Purpose				
Outputs				
Activities				

complex information and processes; participants were pleased that they actually were able to put the techniques in place using the various perspectives from the group itself on local history and development initiatives. The Venn diagrams were felt to be a useful approach to *stakeholder analysis* and mapping of local resources, which could be used with communities. Participatory mapping was also felt to be a useful technique for visualizing possible future uses of a particular farming area. The issue identification by the different interest groups, and the priority ranking using simple voting techniques was also found to be useful, and to 'stimulate eco-democracy'!

The question arises whether the community will be able to implement these plans. Further action planning will be needed with farmer groups, and resources will need to be mobilized, but this process has helped the community to analyse a large number of issues.

Overview of development approaches

Following the analysis the community participants and researchers had a discussion to conclude the PRA project. At first an overview of the development processes was provided using Figs 1.1 and 1.2 (from Chapter 1, this volume) to describe the development processes. The discussion then followed based on the following prompts:

- Who is involved in land and food security?
- What is the quality of life of common people?
- Economic- versus people-centred development.
- People are designers, and can learn their way into the future.
- SA needs beautiful environments, and we need to use people on the 'inside'; people in government need to take and translate our vision.

Conclusion

PRA techniques helped the community participants to explore the evolution of a compex socio-economic situation, and to form a shared vision of the current situation.

Problem and solutions tree techniques were useful in developing an understanding of the key problems and some possible solutions to some of these problems.

A range of techniques assisted in exploring the situation from multiple perspectives.

Once an in-depth exercise such as this has been carried out, the facts are in place for a logical framework analysis, which can be used for many project proposals.

References

Auerbach, R. (ed.) (1994) *Final Report on Transformation of the South African Department of Agriculture*. Proceedings of national workshop and summaries of provincial workshops and stakeholder group consultations (funded by Dutch Government and Hivos for African National Congress Land Study Group). Land and Agriculture Policy Centre, Johannesburg and Farmer Support Group, Pietermaritzburg, South Africa.

Auerbach, R. (1997) People and South African integrated catchment management – pilot phase report of the Ntshongweni Catchment Management Programme. *Research Report* No. 684/1/97. University of Natal, Natal, South Africa; Water Research Commission, Pretoria, South Africa.

Auerbach, R. (1999) Design for participation in ecologically sound management of South Africa's Mlazi River catchment. PhD thesis, Agricultural University, Wageningen, the Netherlands.

Auerbach, R. (2018) Reader on participatory rural appraisal (PRA), including 'Experiences of using PRA tools in South Africa', written by participants of the workshop facilitated by James Mascarenhas, 19–26 April 1993. Rainman Landcare Foundation, George, South Africa.

Cavestro, L. (2003) *PRA – Participatory Rural Appraisal: Concepts, Methodologies and Techniques* [38 pages]. Faculty of Agriculture, Universita degli Studi di Padova, Padova PD, Italy. Available at: https://docplayer.net/18833298-P-r-a-participatory-rural-appraisal-concepts-methodologies-and-techniques.html (accessed 6 June 2018).

Chambers, R. (1994) The origins and practice of Participatory Rural Appraisal. *World Development* 22(7), 953–969.

Department of Rural Development (1997) White Paper on South African Land Policy April 1997. Department of Land Affairs, Pretoria. Available at: http://www.ruraldevelopment.gov.za/phocadownload/White-Papers/whitepaperlandreform.pdf (accessed 11 July 2018).

Mascarenhas, J. (1991) PRA and participatory learning methods: recent experiences from MYRADA in South India. In: The proceedings of the Bangalore PRA Trainers Workshop, February 1991. *RRA Notes* 13. International Institute for Environment and Development (IIED), London, and MYRADA, Bangalore, India.

9 Strengthening Participation in the Organic Value Chain for Small-scale Farmers in Southern KwaZulu-Natal, South Africa

Wim Troosters,[1]* Raymond Auerbach[1] and Gareth Haysom[2]
[1]*Nelson Mandela University, George, South Africa;* [2]*University of Cape Town, Cape Town, South Africa*

Abstract

While many food security interventions focus on production-related constraints to food security, fewer focus on the integration of smallholder farmers in the supply chain. In the Agricultural Sustainable Community Investment Project (Agri-SCIP), operating on the south coast of the Kwa-Zulu Natal Province, small initiatives at key nodal points allowed for major positive results. As a demand-driven alternative market model, the focus of the project is on the integration of local smallholder farmers into the fresh produce supply chain. This improved their participation in the supply chain by mitigating existing barriers to entry for smallholder farmers. A literature review identified five critical factors for market participation: (i) transport and distance to markets; (ii) product quality; (iii) product quantity; (iv) the buyer–seller relationship; and (v) market information. The impact of the Agri-SCIP project on participation in the fresh produce supply chain was assessed, based on these five critical factors.

The data indicate that many of the existing barriers to entry in the fresh produce supply chain for the smallholder farmers are mitigated, and have been shifted to a collectively owned organic cooperative, with quality management making use of the Participatory Guarantee System (PGS) approach. As an alternative market model, Agri-SCIP can contribute to a sustainable solution for smallholder development in South Africa (SA) with a focus on supply chain participation. The development of a strong smallholder-farmer base and the progression of smallholder farmers into semi-commercial organic farmers are potential long-term results of the Agri-SCIP alternative market model.

Introduction

Although a considerable amount of food is produced by small-scale farmers in South Africa (SA), there is little market participation by smallholder farmers, despite some ad hoc entries mainly in local informal markets (Hauser, 2009). Given the possibilities for progression from sub-subsistence to subsistence farming, and on to semi-commercial and commercial farming outlined in Chapter 1 of this book (see Fig. 1.1), and the skill set which needs to be provided to enable this progression, interventions to facilitate training and market access are required. For the development of smallholder farmers to be successful, alternative marketing models are required. One such model is the Agricultural Sustainable Community Investment Project (Agri-SCIP), operating on the lower south coast of the Kwa-Zulu Natal (KZN) Province in SA. This case provides insights into the extent to which small-scale farmers are able to eliminate or ease off potential barriers to market participation via a demand-driven model, operating under a Participatory Guarantee System (PGS).

There are many agricultural interventions aimed at smallholder farmers in SA, mostly initiated and driven by a diverse group of stakeholders coming from the public sector, civil society and the private sector (Hall, 2009; Senyolo *et al.*, 2009). Many private-sector smallholder-farmer interventions are part of corporate social responsibility programmes. Often, these interventions focus on poverty alleviation via improved household food security (Auerbach, 1994; Kirsten and van Zyl, 1996, 1998; Cousins, 2007). Far fewer apply what could be considered to be a holistic organic food systems approach, aiming to integrate smallholder farmers into the fresh produce supply chain.

The Agri-SCIP aims to achieve exactly that, and is based on an economy development programme conceptualised in the early 2000s by a SA development economist, Dr Norman Reynolds (Geerts, 2014). The programme is aimed at the development of the local economy, whereby local money circulation is stimulated by means of a rights-based approach (N. Reynolds and J. van Zyl, 2002, unpublished report 'Globalisation and localisation: striking a new balance in South Africa'; N. Reynolds, 2006, unpublished report 'South Africa's misunderstood dual economy'). In 2006, it was adopted by the former Department of Provincial and Local Government as one of four local economic development strategies in the 2006–2011 White Paper on local economic development in SA. In the White Paper, it was called SCIP (Republic of South Africa, 2006).

Participatory Guarantee System (PGS)

As explained in Chapter 10 of this volume, PGS is a low-cost certification system run by local consumers and producers, without an external accredited organic certifier, unlike commercial third-party and group certification. It is a locally focused assurance system adapted to the needs of smallholder farmers and accessible in terms of costs and procedures (Katto-Andrighetto and Auerbach, 2009). The International Federation of Organic Agriculture Movements (IFOAM) defines PGS as a locally focused quality assurance system that certifies producers based on active participation of stakeholders. It is built on a foundation of trust, social networks and knowledge exchange (IFOAM, 2013), and is an alternative quality assurance system in relation to third-party and group certification. In SA, PGS-SA is an organization working within the South African Organic Sector Organisation (SAOSO), to assist small-scale farmers (Hauptfleisch *et al.*, 2011). PGS-SA bases its assurance system on the SA Organic Standard; information about SAOSO, PGS-SA and the SA Organic Standard can be found on the SAOSO website (www.saoso.org). Several SA PGS initiatives combine community-supported agriculture with PGS (Hauptfleisch, 2012).

Critical Factors for Smallholder Market Access

A literature review of the fresh produce supply chain showed a number of barriers to entry into the fresh produce market (FPM) in SA, which have an impact on participation of smallholder farmers. Darroch and Mushayanyama (2006) tested 20 key factors by principal components analysis for the Ezemvelo Farmers Organization (a group-certified organic cooperative) based at Umbumbulu on the south coast of KZN. The research found that responses varied between men and women and between older and younger farmers. The farmers listed as top constraints: (i) uncertain climate; (ii) tractor not available when needed; (iii) delays in payment from the packhouse; and (iv) inputs not being available at affordable prices. Some key conclusions were that there is a lack of market information and market influence, crop production expansion constraints, lack of liquidity and cold storage facilities.

The barriers to entry for Agri-SCIP smallholder farmers are diverse in nature and vary for the different supply chain actors. In terms of market participation-related barriers for smallholder farmers, several critical factors are identified in a broader literature review (Troosters, 2014). Each critical factor relates to a specific set of barriers to entry. Grouping them into critical factors allows for a more thematic and concise approach towards the analysis of the challenges regarding market participation of smallholder farmers in the fresh produce supply chain. As a result, the following five critical factors have been identified out of the literature review and project experience.

Critical factor 1: transport and distance to markets

Transport is an important factor in market participation for smallholder farmers. The limited physical access to markets and poor availability of logistical infrastructure were highlighted by Aihoon and Associates (2008) as a barrier for smallholder farmers. In analysing the barriers to entry into FPMs, a report by the National Agricultural Marketing Commission (NAMC) highlighted that smallholder-farmer participation is hampered by a lack of transport facilities as many of them are a long distance away from the market (NAMC, 2002). In order to participate in a local, regional or national market, a farmer needs to have physical access to that particular market, either directly or via an intermediary. This implies either having their own transport or a cost associated with the hiring of a transporter. Transport by foot or bicycle is rather cheap, but very time consuming and labour intensive, with the implication that the market needs to be near to the farmer in order for this mode of transport to be applicable. Walking and cycling generally only allow transportation of low volumes. Motorized transport will allow the farmer to target more distant markets and supply larger volumes, at higher cost. Makhura and Mokoena (2003) argue that poor transport and road infrastructure is a constraint for smallholder farmers to access a market. The traffic network, and especially its condition, has an additional impact on the cost of transport, as well as the time of travel. A poor road network (e.g. dirt roads, bad road connections or traffic congestion) means a longer travel time, more frequent vehicle services and even costly breakdowns. All of these result in a higher cost of transport.

Critical factor 2: product quality

Every market actor demands a certain quality specific to its own needs. Quality requirements can create a barrier to entry for smallholder farmers. Makhura *et al.* (1998) argue that smallholder farmers are often crowded out by local supermarkets and commercial producers as they cannot compete on product quality and supply consistency. Freshness and grading are key drivers for product quality in the fresh produce sector. Size and weight are the predominant factors used to determine the grade. Product-specific factors might be added in the grading process, for example the shape and colour of the particular type of vegetable.

In addition, the growing consumer concern about the quality and safety of fresh produce has led to more stringent product standards beyond mere freshness and grading. Katto-Andrighetto and Auerbach (2009) argue that smallholder farmers have difficulty in complying with product and health standards, which significantly hampers their market participation. Fresh produce standards often refer to the applied production practices and conditions (GlobalGAP (good agricultural practice), hazard analysis critical control points (HACCP), organic), the origin of the produce or the labour conditions during the production process and pricing system (Fairtrade). Most certification systems tend to be rather expensive when organized via third-party certification, and are based on a particular registered set of standards. Often input from the farmer in the form of record keeping and traceability is required, especially for high-value agricultural products such as fresh produce.

Packaging and in certain cases also barcoding, while not directly related to the quality of the product itself, are considered to be quality aspects of fresh produce. Louw *et al.* (2007) argue that supermarkets often tend to shift costs to farmers by insisting that the farmer undertakes packaging of the produce. Therefore packaging requirements can become a barrier to entry in specific markets for smallholder farmers. Often stringent health regulations have to be taken into account when packaging a product.

Critical factor 3: product quantity

It is important for any farmer to produce regular and consistent volumes, often for a variety of products. Aihoon *et al.* (2009) argue that the low production capacity of smallholder farmers limits them in the regular and consistent supply of economic volumes, which results in higher transaction costs when entering markets. In the case of smallholder-farmer participation in supermarkets, Louw *et al.* (2007) argue that the lack of economies of scale (low production capacity) doesn't allow smallholder farmers to supply most supermarkets. When looking at the FPMs, a report by the NAMC (2001) indicated that the FPM system is prone to preferential treatment of producers who supply large consignments of fresh produce. The low production capacity of individual smallholder farmers therefore poses a barrier to entry into specific market channels of the fresh produce supply chain.

Makhura and Mokoena (2003) argue that smallholder farmers also face a challenge due to non-existing or inaccessible storage facilities. Proper storage facilities give farmers a tool to speculate on product prices, enabling them to move away from simply being price-takers during trade negotiations. Smallholder farmers will therefore gain a stronger position in the fresh produce supply chain. This is not possible for all types of fresh produce, but overall, access to proper storage facilities can be an important factor in smallholder-farmer market participation, especially in high volume markets.

Critical factor 4: buyer–seller relationship

This factor emphasizes the relationship between the farmer and other actors in the fresh produce supply chain willing to buy fresh produce. The buyer–seller relationship is, next to the forces of supply and demand, influenced by the power relations in the supply chain. Makhura and Mokoena (2003) note that lower bargaining power and poor power relations in the fresh produce supply chain pose a barrier to entry for smallholder farmers. Not all the supply chain actors have the same position when entering in trade negotiations. Although FPMs are regulated by free-market forces, negotiations never really happen on a level playing field.

Price is a very important factor influencing trade negotiations. The farmer will often be a price-taker, having very little bargaining power during price negotiations. Aihoon *et al.* (2009) argue that smallholder farmers often do not receive the best price at FPMs. This is supported by a report from the NAMC (2001), that agents at the FPM often adopt a 'take it or leave it' attitude towards smallholder farmers during price negotiations.

Other potential barriers to entry for smallholder farmers into specific channels of the fresh produce supply chain include the deduction of a pre-negotiated commission fee for the market agent, as a percentage of sales revenue and a fixed commission for use of the trade facilities at the FPM; these further reduce the farmer's income. A NAMC report (2001) notes that smallholder farmers also bear the financing cost at the FPM as payment is only received when the agent has sold their fresh produce. Louw *et al.* (2007) argue that there is also an issue of delayed payments by supermarkets which puts many smallholder farmers under financial pressure.

Farmers often enter supply contracts, which specify the legal farmer–buyer relationship, strictly controlling and dictating the terms of trade between both parties. Trade in the fresh produce sector ranges from trust-based or 'handshake' agreements, mainly in the informal sector, to legal supply contracts and exclusive trade agreements, mainly in the formal sector. Aihoon *et al.* (2009) argue that the low production capacity of smallholder farmers often results in uncertain contract performance, with overall high transaction costs. This is supported by Louw *et al.* (2007) who report that the lack of economies of scale of smallholder farmers often precludes supply contracts with supermarkets. Uncertain contract performance therefore poses a barrier to entry in certain markets, especially when the buyer favours longer-term or large consignment supply contracts.

Critical factor 5: market information

Access to market information is important for any farmer in order to be able to operate successfully in the FPM space. Makhura and Mokoena (2003) argue that a lack of reliable market information is a market-related challenge for smallholder farmers. Aihoon and Associates (2008) also indicate that inadequate access to market information is a barrier for market participation by smallholder farmers. Market information needs to be accurate, timely, up to date and subjected to a continuous process of feedback, update and analysis. Smallholder farmers need to know what produce existing and potential trade partners and customers want, so that they can accommodate the market. Similarly, knowing how and when buyers and customers want the fresh produce in terms of packaging, delivery times, the operation of the procurement system, the type of trade agreement, etc. is essential information for any farmer who wants to enter into a beneficial trade agreement.

Farmers benefit from having many different trade partners and access to appropriate market information will enable smallholder farmers to track down willing buyers. However, Aihoon and Associates (2008) note that inadequate focus on tapping new markets is an important barrier for smallholder farmers. Access to information is often facilitated by access to telephone and the Internet. A survey by Statistics SA (StatsSA, 2011) highlighted that in rural areas, where most smallholder farmers reside, essential services like telephone and electricity are often lacking. As a result smallholder farmers encounter a barrier to entry in certain markets due to a lack of information.

Research Questions

The main research question is whether smallholder-farmer participation in the fresh produce supply chain improves when potential barriers to entry are shifted to a collectively owned marketing cooperative. The research will show that smallholder-farmer market participation did improve when potential barriers to entry were shifted to a collectively owned marketing cooperative.

Conceptual Framework for the Project

The main aim of Agri-SCIP is to strengthen the position of smallholder farmers in the local fresh produce supply chain by creating a guaranteed demand; therefore Agri-SCIP deploys local cooperatives to guarantee the demand for fresh produce. The cooperatives guarantee that they will buy fresh produce from their local member farmers. The operations of Agri-SCIP are regulated via a PGS, which serves as an overall quality assurance system via which membership of smallholder farmers is arranged.

A number of misconceptions exist about smallholder farmers in SA, mainly in relation to farm size and production efficiency. A small farm size does not necessarily equate to low farm viability, as a 2 ha intensive horticulture farm on prime agricultural land can be more viable than a 500 ha monoculture farm in the dry Karoo. The size of the farm is thus relative in respect to the particular ecological region and soil quality, and to the farming industry involved. Smallholder farms are also not simple scaled-down versions of large commercial farms as they differ significantly, in terms of both input and output intensity. There is often a mistaken perception that smallholder farms are less efficient, because of the absence of assets and industrial equipment. Large, capital-intensive farms are not always more economically efficient than their smallholder counterparts (Kirsten and van Zyl, 1998).

A smallholder farmer is usually one whose scale of operation is too small to attract the provision of the services that he or she needs to be able to increase his or her productivity and market participation (Kirsten and van Zyl, 1996). In comparing the Alliance for a Green Revolution in Africa with the Export Promotion for Organic Products from Africa (EPOPA), Auerbach (2013) comments that while the former used high levels of bought inputs, the latter showed smallholder farmers how to use local natural resources, connected them with markets and strengthened local institutions at a fraction of the cost per farm per year compared with the former (Auerbach, 2013; Chapter 1, this volume). Classifying any farming operation, from a tiny backyard food garden up to a 20 ha farm, as a subsistence farm is therefore a generalization, suggesting that all smallholders form a relatively homogeneous group. On the contrary, there are significant distinguishing features, which allow for a more comprehensive typology to emerge. However, Bernstein (2009) argues that many analysts stress the variability of class identification in concrete social formations, which means that while they may be useful for discerning and analysing broad trajectories of change, they are often difficult to operationalize in the analysis of specific empirical data sets.

Ben Cousins (2010) proposes to distinguish smallholder farmers from subsistence farmers based on the levels of own consumption and marketable surplus. A subsistence farmer produces primarily for their own consumption and has very little market participation beyond some occasional neighbourhood trade. Surplus production is therefore almost non-existent. Distinguishing smallholder farmers from commercial farmers often relies on the size of the farm and the degree of labour intensity, in addition to the level of marketable surplus. In this case, smallholder farming seems to rely mainly on household labour. Based on key differences in the combination of land, labour and capital, smallholder farmers can be classified as semi-commercial or commercially orientated (see Table 9.1 and Chapter 1, Fig. 1.1, this volume). When a surplus is produced in order to enable economic growth, the agricultural activities assume the character of a capital enterprise. A proportion of the surplus value produced by own and/or hired wage labour is reinvested in the production assets in order to expand productive capacity (Cousins, 2010).

In light of the above, and of the progression presented in Chapter 1 of this volume, the classification of subsistence and smallholder farmers in SA shown in Table 9.1 has been adopted, modified from Cousins (2010), and further developed to show the actual development progression among smallholder farmers on the south coast of KZN.

The progression from subsistence to semi-commercial smallholder farmers has taken about 6 years, and there may soon be one or two commercial smallholder farmers. The Siyavuna Abalimi Development Centre (SDC) was established in

Table 9.1. Proposed typology of subsistence and smallholder farmers in South Africa. (From Troosters, 2014.)

Typology	Social production	Expanded production	Criteria
Subsistence farmer	Small contribution from agricultural activity; Heavily reliant on petty enterprise, social grants and/or other wage labour	None via agricultural activity	Mainly own consumption; No significant surplus production; No market participation; Occasional, ad hoc neighbourhood trade or welfare handout
Smallholder farmer	Significant contribution from agricultural activity; Combination with additional forms of income, mainly wage labour	None, to very little, via agricultural activity	Farms on substantial scale; Significant, but rather uncoordinated surplus production; Simple market participation, mainly local; Regular, but mostly unpredictable trade
Semi-commercial smallholder farmer	Fully sustainable from agricultural activity; Only minor combination with additional forms of income; No to very little hired labour	Basic level	Farms on substantial scale; Planned surplus production; Basic market participation, local, regional and/or national; Regular, predictable trade
Commercial smallholder farmer	Fully sustainable from agricultural activity; Only minor combination with additional forms of income; Rely on hired labour	Fully engaged, with capital accumulation	Commercially orientated farm; Highly organized, intensive production; Good market participation; Regular, predictable trade; Potential of vertical integration in the supply chain via value adding

2010 as a not-for-profit organization with the aim of helping subsistence farmers to become smallholder farmers, deriving increasing financial and other benefits from their agricultural activities. By 2013 there were two active agri-marketing cooperatives under SDC. Both are situated on the lower south coast of the KZN, and are part of the Ugu District Municipality (SDC, 2013). The first agri-marketing cooperative, Hibiscus Coast Co-operative (HCC), was set up in 2008 (HCC, 2009, unpublished constitution) and this cooperative operates in the Hibiscus Coast Municipality (Fig. 9.1). In 2011, the second agri-marketing cooperative started operating in the Umdoni Municipality (Fig. 9.2). These two municipalities fall under the Ugu District Municipality.

In the financial year 2013, the Hibiscus Coast agri-marketing cooperative operated in three local communities, namely Nositha, KwaNzimakwe and Gcilima. In 2013 the cooperative started working in Mvutshini. Smallholder farmers form a local Farmers' Association at community level of which every smallholder farmer is a member.

The Umdoni Agri-marketing Co-operative currently operates in three local communities, namely Amahlongwa, Malangeni and Danganya. A registered PGS represents each individual local smallholder farmer.

Gcilima Farmers' Association	KwaNzimakwe Farmers' Association	Nositha Farmers' Association	Mvutshini Farmers' Association
Four collection points	Four collection points	One collection point	Established 2013
131 PGS registered growers	118 PGS registered growers	25 PGS registered growers	Three collection points will be set up in 2013
Sales to cooperative R37,939	Sales to cooperative R23,440	Sales to cooperative R2,707	47 PGS registered growers

Fig. 9.1. Overview of geographical areas of the Hibiscus Coast Co-operative. PGS, Participatory Guarantee System. (From SDC, 2013.)

Amahlongwa Farmers' Association	Malangeni Farmers' Association	Danganya Farmers' Association
Three collection points	Three collection points	New area – established 2013
56 PGS registered growers	51 PGS registered growers	Three collection points will be established in 2013
Sales to cooperative R2,177	Sales to cooperative R8,087	

Fig. 9.2. Overview of geographical areas of the Umdoni Co-operative. (From SDC, 2013.)

Agri-SCIP Fresh Produce Supply Chain

Agri-SCIP is built around an agri-marketing cooperative operating as a social enterprise. It created an enabling environment for the development of smallholder farmers. The main focus is to provide a guaranteed market that is fair, rewarding and sustainable for smallholder farmers in various communities on the lower south coast of KZN (SDC, 2012). An agri-marketing cooperative is a market intermediary in the fresh produce supply chain. In comparison with the established supply chain actors, its position is similar to that of formal market actors, such as FPMs, supermarkets and fresh vegetable shops.

A close relationship between smallholder farmers as chain actors and the agri-marketing cooperative exists, whereby market information is freely exchanged, including product prices. Several services in addition to the marketing of fresh produce are offered by the cooperative to the smallholder farmers such as training, mentoring and the supply of production inputs (SDC, 2012).

The 2012 Siyavuna concept document shows that there was a significant level of vertical integration in the Agri-SCIP supply chain, via the governance of the agri-marketing cooperatives. All the smallholder farmers supplying the agri-marketing cooperative are members and thus co-owners of the cooperative, and are equal co-owners of the body that markets and potentially even processes and adds value to their fresh produce. This approach is also demonstrated, among others, by the Keekenyokie Co-operative in Kenya, which processes Maasai beef (Kibue and Auerbach, 2013). This has a positive impact on the integration and power relations of the smallholder farmers in the general fresh produce supply chain (Delgado, 2006).

In addition, horizontal integration is observed at the level of the smallholder farmers through their membership agreement with the agri-marketing cooperative (Fig. 9.3).

All smallholder farms are not controlled or 'owned' by one (or a few) actor(s), in the way that horizontal integration is generally understood (Riisgaard *et al.*, 2008), but rather through the membership of large numbers of smallholder farmers in specific areas, mainly clustered per community, a sense of general control over the terms of production and consistency in quality and procurement is obtained. This constitutes a potential competitive advantage over other supply chain actors as it contributes to a certain level of horizontal integration. The Agri-SCIP fresh produce supply chain focuses on the short food supply chain concept as well as lean integration (Hobbs and Young, 2000; Riisgaard *et al.*, 2008). Eliminating the so-called 'middle man' as much as possible, in combination with a short distance to the marketplace, are the cornerstones of the Agri-SCIP fresh produce supply chain. Via the PGS a reasonable degree of lean integration is obtained whereby systems and data integration bring the different chain actors closer together. Geographical data and production statistics of individual smallholder farmers are integrated by the PGS, as well as linking them to sales statistics from the agri-marketing cooperative. This level of data and systems integration translates into customer value addition and potentially a competitive advantage for the smallholder farmers. An overview list of the sales outlets of the HCC was analysed according to the nature of sales transaction between direct customer sales and business-to-business sales. Compared to similar market actors in the general fresh produce supply chain, the HCC acts as a direct sales outlet, like a supermarket or vegetable shop, as well as a market intermediary like a wholesaler, focusing primarily on business sales (HCC, 2013, unpublished data). Operating in the direct sales market segment again indicates the application of the short food supply chain concept. Emphasis is placed on local marketing efforts with almost all sales outlets and market intermediaries being within a 30 km radius from the operational base of the HCC.

The agri-marketing cooperative is key to the operations and is in principle the only 'new actor' that the Agri-SCIP introduces into the fresh produce supply chain, despite engaging (new) smallholder farmers in the supply chain. Figure 9.4 illustrates the position of the agri-marketing cooperative in the general fresh produce supply chain with linkages to the formal market actors, informal actors and customers as the cooperative supplies all segments directly.

Key Operations

As members of the cooperative, they gain more control over the supply chain because of the

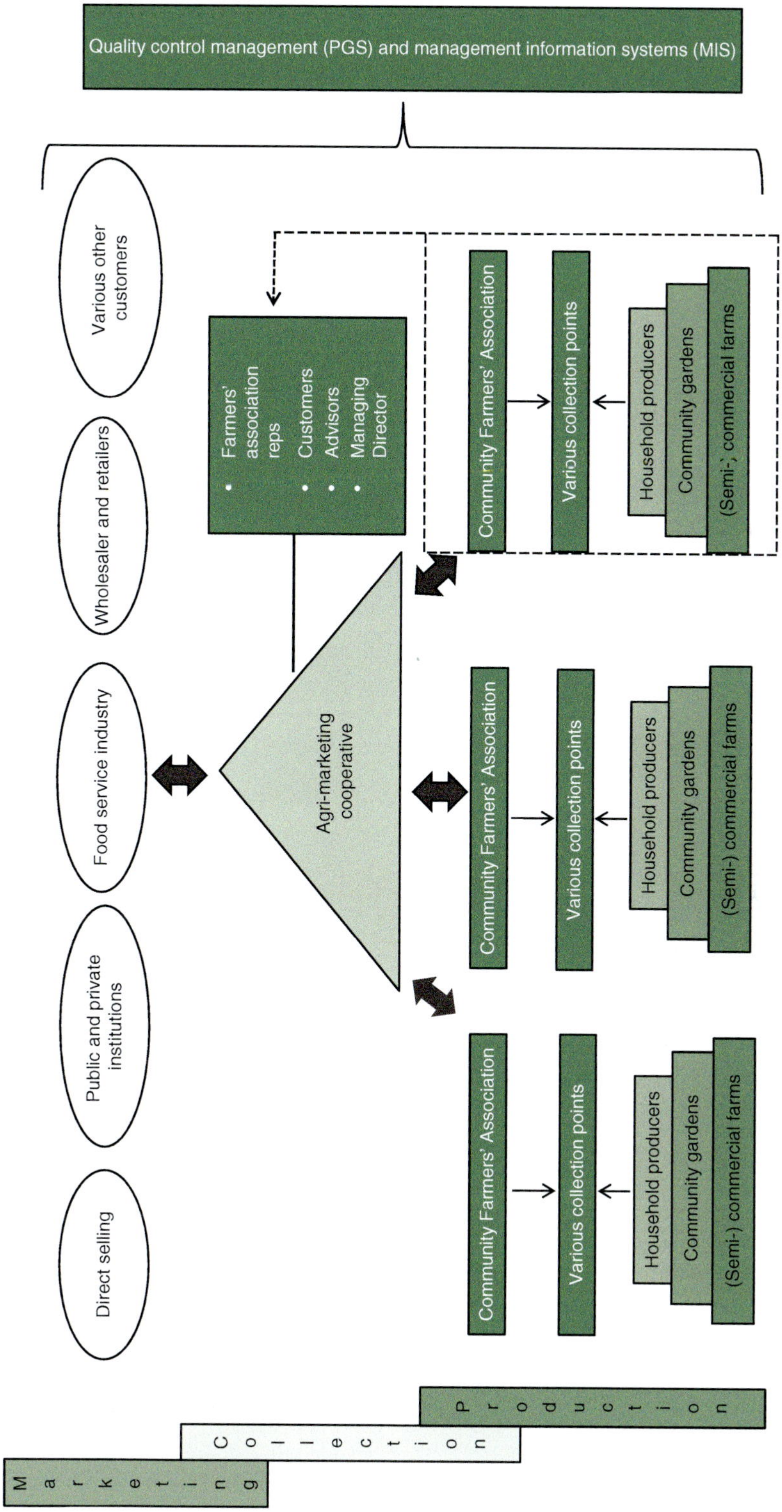

Fig. 9.3. Agri-SCIP fresh produce supply chain. (From SDC, 2012.)

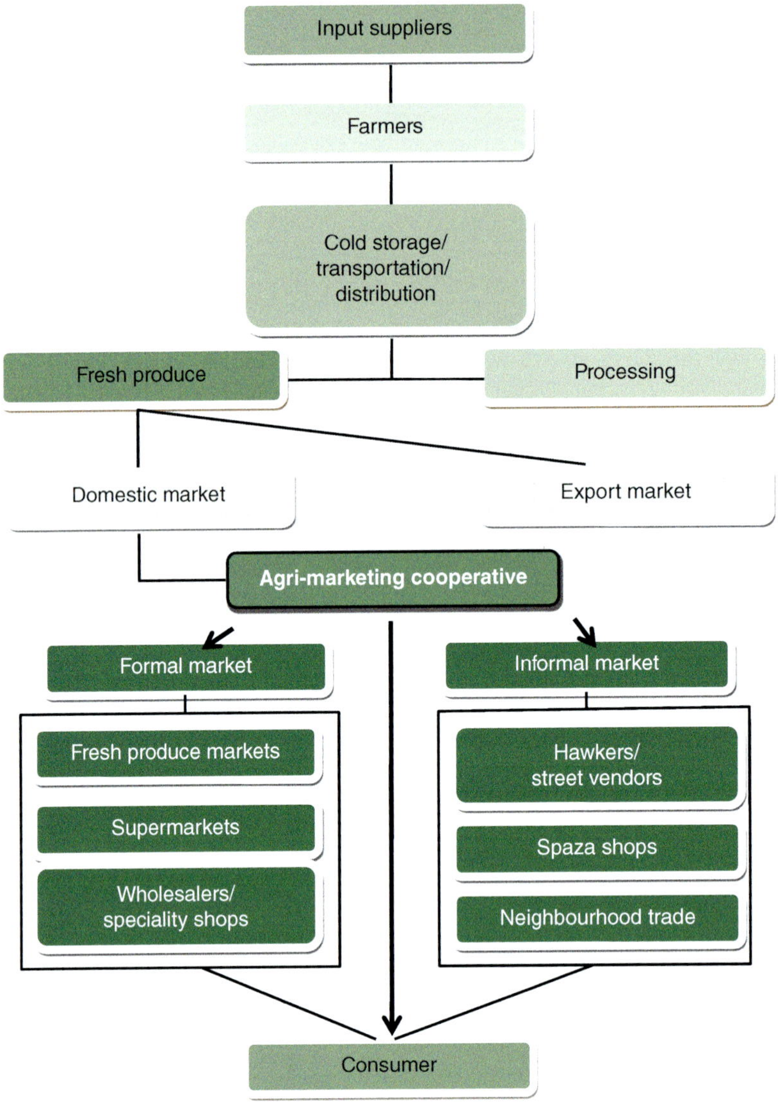

Fig. 9.4. Positioning of the agri-marketing cooperative in the general supply chain analysis of the fresh produce supply chain in South Africa. (From Troosters, 2014.)

packaging, storage, transport and marketing activities of the agri-marketing cooperative. Practically, the vertical integration of smallholder farmers in the fresh produce supply chain, via the PGS, is embedded in the governance and daily management of the agri-marketing cooperative. At local level, smallholder farmers have monthly meetings via their Farmers' Associations where they discuss and share local challenges and success stories. Agricultural extension services are also provided at the Farmers' Association meetings to deal with more complex challenges. Every local Farmers' Association has a direct representation on the board of directors of the agri-marketing cooperative via a democratically elected chairperson and secretary. This ensures that all the smallholder farmers, through their PGS membership are democratically represented in the governance structure of the agri-marketing cooperative.

In addition, a number of external parties (customers and advisors) are included in the governance structure to ensure adherence to good governance practices and viable decision making in the best interest of the cooperative. Customers provide valuable insight on market demands and product quality perceptions, as well as the customer perception of the value proposition in line with the price setting policy of the agri-marketing cooperative. Advisors are skilled in small business and financial management, and have sound bookkeeping skills. The overall diversity and balanced composition of the board of directors of an agri-marketing cooperative, including smallholder farmers, customers and advisors, ensures that the interests of the smallholder farmers, the market (customers) as well as the general business interests of the agri-marketing cooperative are safeguarded (SDC, 2012).

In comparison with a standard primary agricultural cooperative constitution in SA, a more equitable profit-sharing mechanism is adopted in the constitution of the agri-marketing cooperative. Smallholder farmers are paid a share of the profit according to the principle of patronage proportion via their Farmers' Association. This means that each Farmers' Association receives a share of the profit according to the volumes supplied by all farmers from that area proportional to the total quantity (HCC, 2009, unpublished constitution).

When signing up for membership of the agri-marketing cooperative, smallholder farmers pledge to adhere to a set of clear production standards based on organic and natural production principles. As a guideline, a list of production practices and principles that either strictly need to be adhered to or which are simply not allowed in organic agriculture (OA) is provided. In addition, SDC facilitates free courses in OA, as well as providing intensive mentoring services. For SA farmers, PGS-SA provides an organization under which producer groups can receive guidance, and which adheres to the SA Organic Standards (see SAOSO website indicated earlier).

Quality management of organic production standards via the PGS therefore foresees unannounced ad hoc field inspections, as well as obligatory annual field inspections similar to third-party certification. Suspected cases of non-compliance picked up during the field inspection or reported by a smallholder farmer are then subjected to a field inspection which details the reported incident and captures the statement and position of the involved smallholder farmer(s) regarding the incident. The field inspection report is tabled at a meeting of the board of directors, which decides upon the appropriate penalty based on the evidence. Sanctions range from a supply-ban for a number of weeks, accompanied by a remediation plan, to being permanently expelled in the case of a repeat offender (SDC, 2012).

The cooperative operates a rather specific procurement system, mainly because of their aim to attract smallholder farmers, but also because of the nature of the fresh produce supply chain. Procuring from many individual smallholder farmers requires a tailored procurement system. Practically, fresh produce is collected via local collection points at community level. This allows many smallholder farmers to have easy access to the market to sell their fresh produce, as well as having access to valuable inputs such as seeds, seedlings and tools. On a weekly basis, each area or community is serviced by a number of local collection points. They make use of existing local structures such as a church, community hall or even under a tree or on the corner of the street. An information board details the desired quality as well as the procurement price on offer on a weekly basis. Upon delivery at a local collection point, the fresh produce from the smallholder farmers is first subjected to a quality control after which it will be washed, weighed and packaged for the market (SDC, 2012).

There are two important aspects about the procurement system that have an impact on the participation of smallholder farmers in the fresh produce supply chain. First, there is the direct cash payment at collection points and secondly, the unrestricted quantity that can be supplied. All the smallholder farmers supplying produce to a local collection point are paid on the spot for their delivery, no matter how little they supply. The fresh produce supplied by a smallholder farmer is first subjected to a quality check, and when accepted, the smallholder will be paid the full amount in cash, according to the price communicated on the information board at the collection point. This eliminates all cash flow risks of payment on invoice, which in the case of some supply chain actors, such as supermarkets, can be intentionally delayed, causing financial stress situations for resource-poor smallholder farmers. The absence of any minimum supply quantity restriction allows every smallholder farmer, no matter how small or large, to sell their surplus produce to generate an (additional) income. This has a potential positive contribution on the social impact of the agri-marketing cooperative.

A basic assumption is that shifting the risk of disposal of produce from the smallholder farmer to the agri-marketing cooperative will give sufficient encouragement to smallholder farmers, resulting in the willingness to increase production, once risk is shared more equally. The guaranteed demand, with cash payment and no quantity restrictions, does however mean that the risk of sales is transferred to the cooperative. This forces the agri-marketing cooperative to have sufficient market access to ensure that requisite returns are attained, something which is reflected in the diverse range of sales methods via the analysis of the sales outlets of the HCC.

The cooperative adheres to a strict pricing policy. Procurement prices for smallholder farmers are market based. The cooperative values its fresh produce as a high quality and socially equitable product. The pricing mechanism is therefore based on a comparative analysis with the prevailing prices of premium organic products at various retail outlets, supermarkets and speciality shops. This means that a price premium is awarded to the product.

The risk factor included in the pricing system, has a positive impact on the variety of fresh produce being offered by the cooperative. Fresh produce which has a high supply, for example when in season, will get a high risk factor at that point in time, reducing the procurement price and discouraging farmers from producing this type of fresh produce. Low supply of specific types of fresh produce will have the opposite positive effect by reducing the risk factor, because supply is low, raising the procurement price and stimulating farmers to start cultivating that particular crop. In this way variety is promoted, while the risk of over-supply of certain types of fresh produce is mitigated. The same approach has been adopted by Green Road Garden Route, as reported in Chapter 10 of this volume.

Marketing and Branding

The value proposition of the cooperative is based on the values of the fresh produce and is safeguarded by the PGS. The core values are captured in a social brand called 'Kumnandi', which means delicious in isiZulu (Fig. 9.5).

The Kumnandi brand is built on three cornerstones which are the main intrinsic values of the fresh produce supply chain: (i) local; (ii) natural and organic; and (iii) healthy and fresh. First, local, as the fresh produce is grown locally, mainly by previously disadvantaged smallholder farmers. Natural and organic, because all smallholder farmers are trained and mentored to apply natural and organic farming techniques. The fresh produce is free of any kind of chemical pesticides and fertilizers or any other hazardous substance. Healthy and fresh, because the fresh produce is high in nutrients and freshly picked and collected in the morning for the local market.

The primary target market of the cooperative is 'conscious' individual customers. These customers identify themselves with, and thus appreciate, the intrinsic values of a Kumnandi product. This segmentation is mainly based on lifestyle and financial status. Their consumer behaviour is characterized by a strong sense of quality, health and environmental benefits. Additionally, conscious customers value themselves as good citizens and wish to contribute to society at large, and have a high sense of solidarity. Therefore, they are generally prepared to pay a price premium for local, fresh and natural products, such as Kumnandi. The second target market is the lower and middle income community. As the product offering is local and fresh, these customers, while looking for good value for money, are attracted by the superior quality and freshness. However, attracting these customers

Fig. 9.5. Kumnandi social brand. (From SDC, 2012.)

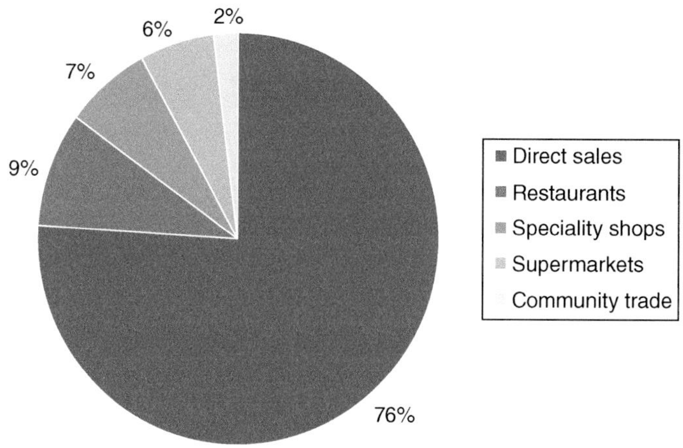

Fig. 9.6. Percentage of sales in different market channels by the Hibiscus Coast Co-operative (HCC, 2013, unpublished data).

requires an expanded distribution system as well as a sensible pricing strategy. The third target market is local community members in the various procurement areas; as part of the social enterprise status, fresh produce is offered at lower prices to these local community members (SDC, 2012).

The cooperative strives to achieve the highest sales profit and therefore focuses on direct sales, restaurants and speciality shops. However, it is important to maintain a diverse sales mix, thus including supermarkets and community trade, to spread the sales risk. Risk is transferred to the cooperative via the specific procurement system which guarantees to the smallholder farmers that all their produce will be purchased. Although certain markets, such as supermarkets for example, may have a lower gross profit margin, they tend to have other benefits in terms of high and varied demand, which other markets might lack. This is important for the financial viability of the cooperative. A diversified range of supply channels allows them to attain a high sales response with very little spoilage. Prioritizing high gross-profit-margin markets supports the financial sustainability.

Analysis of sales figures of the HCC for the financial year 2012/13 indicated that in total 15.5 t of fresh fruit and vegetables had been traded. An overwhelming part of the fresh produce, namely 76%, was sold directly. Lower volumes are traded by the cooperative in other marketing channels as shown in Fig. 9.6. This supports the drive of the agri-marketing cooperative to focus its sales on direct sales, restaurants and speciality shops, which jointly constituted 92% of the total annual sales volume in the 2012/13 financial year (HCC, 2013, unpublished data).

Critical Factors for Smallholder-farmer Participation in Agri-SCIP

A set of five critical factors, grouping barriers to entry into the fresh produce supply chain for smallholder farmers, was identified in the literature review. The status and impact of the Agri-SCIP test case on market participation for smallholder farmers is evaluated according to the same five critical factors.

Critical factor 1: transport and distance to markets

Participation in the FPM by smallholder farmers will improve when barriers to entry related to transport and distance to markets are shifted to a collectively owned marketing cooperative. Having collection points close by, enables the smallholder farmers to bring vegetables on foot, which eliminates the need for long-distance transport, as well as the associated transport cost. While poor road infrastructure is still an issue in most of the communities where local collection points are operative, the risk of high transport costs and long travel times is transferred from the individual smallholder farmer to the agri-marketing cooperative. The cooperative organizes collective transport to the market, which significantly reduces the transaction costs. Furthermore, the vehicle from the cooperative has a high utilization rate, which reduces the fixed costs per kilometre and as a result also lowers the transaction costs (SDC, 2012).

Critical factor 2: product quality

Participation in the FPM by smallholder farmers will improve when barriers to entry related to product quality are shifted to a collectively owned marketing cooperative. In the Agri-SCIP test case, the PGS serves as an overall quality assurance system (SDC, 2012). This alternative assurance scheme allows smallholder farmers to avoid expensive, top-heavy, third-party certification (Katto-Andrighetto and Auerbach, 2009). The PGS sets the quality standards via specifications and guidelines for the production and supply of fresh produce by the smallholder farmers. The production standards are enforced by the PGS via a system of internal control and field inspections, with a penalty system for noncompliance. The quality of the supplied fresh produce is inspected at the local collection point with direct feedback to the smallholder farmer (SDC, 2012). The quality standard and enforcement measures are clearly communicated to the smallholder farmers. Together with the quality inspection and feedback at the collection points, the barrier to entry related to product quality for smallholder farmers is eased off in the Agri-SCIP test case.

A thorough training of farmers in quality management supports them in achieving consistently high quality products. Smallholder farmers are informed on how to grade and check the basic quality of their fresh produce, based on shape, colour, marks, spots, etc. prior to the delivery at the local collection point. This ensures that farmers are fully aware of the quality requirements by the cooperative and therefore lowers the risk of poor quality of produce (SDC, 2012). Freshness is key in the fresh produce sector. In relation to the system of local collection points and the short food supply chain, smallholder farmers harvest on the morning of the collection day by the cooperative. This ensures high levels of freshness, thus tempering potential barriers due to a threat of spoilage for certain specific types of fresh produce (SDC, 2012).

The cooperative organizes collective packaging of the fresh produce supplied by the smallholder farmers at the local collection points (SDC, 2012). The barrier to entry into the FPM for smallholder farmers related to packaging is therefore removed as the full responsibility and risk is shifted from the smallholder farmer to the cooperative.

Critical factor 3: product quantity

Participation in the FPM by smallholder farmers will improve when barriers to entry related to product quantity are shifted to a collectively owned marketing cooperative. In the Agri-SCIP test case, the agri-marketing cooperative is supplied by a group of smallholder farmers, who are all members of the cooperative, via its procurement system of local collection points (SDC, 2012). The collective procurement system raises the confidence level regarding regular and consistent supply by smallholder farmers, which has a positive impact on contract performance and lowers transaction costs. The supply of basic farming inputs at local collection points, such as seeds, seedlings and tools, at bulk market prices, further stimulates regular and consistent supply. Smallholder farmers therefore have a better chance of participating in specific high volume markets such as supermarkets and FPMs with the cooperative as an intermediary.

The pricing strategy of the cooperative incorporates a risk factor. The procurement price by the cooperative is directly influenced by the risk factor which is based on product demand and supply in the market, as well as product-specific risks, such as high perishability or difficult growing conditions (e.g. disease-prone crops). The SDC argues that this has a positive impact on the variety of fresh produce supplied by the smallholder farmers, as it generally raises the price for high demand, low supply products, as well as high risk products, thus stimulating small-scale production (SDC, 2012). The unrestricted quantity that can be supplied at the local collection points has a positive impact on the participation of smallholder farmers in the fresh produce supply chain. The system is transparent with equal prices for small and large consignments of fresh produce. Every smallholder receives similar treatment and the same trade conditions from the cooperative (SDC, 2012).

The focus of Agri-SCIP on a short food supply chain, with weekly local collection points, lowers the need for storage facilities (SDC, 2012). None the less, the agri-marketing cooperative provides the storage for the fresh produce after procurement from the smallholder farmers. The responsibility and risk of fresh produce storage is therefore transferred from the smallholder farmers to the cooperative from the point of delivery at the local collection point.

SDC has observed that in general there is a dual impact on product quantity. First, farmers who produce only very small quantities can sell these through the cooperative. Secondly, the quantities supplied have steadily grown up to a certain point, though it is proving difficult to stimulate further growth.

Critical factor 4: buyer–seller relationship

The hypothesis based on this critical factor is that participation in the FPM by smallholder farmers will improve when barriers to entry related to the buyer–seller relationship are shifted to a collectively owned marketing cooperative. Every smallholder farmer who supplies fresh produce to the agri-marketing cooperative is a member and consequently, co-owner of the cooperative. This improves the level of vertical integration by smallholder farmers in the supply chain. Every smallholder farmer is democratically represented on the board of directors of the agri-marketing cooperative (SDC, 2012). The power relations in the Agri-SCIP supply chain show an improved position and increased bargaining power for smallholder farmers.

SDC argues that the pricing system in the Agri-SCIP test case awards a price premium to the fresh produce, with a procurement price for smallholder farmers between 40% and 60% of the retail price, based on a risk factor. Prices are updated on a monthly basis and communicated via the information board at the local collection points (SDC, 2012). The smallholder farmers have strengthened their bargaining position towards other chain actors in a collective manner, as shown in the first quarterly report by SDC in 2014: the smallholder farmers, represented by a delegation from the agri-marketing cooperative, demanded better services and adaptation to OA practices from local representatives of the Department of Agriculture, Forestry and Fisheries (DAFF) (SDC, 2014). The cooperative operates on a direct cash payment basis. This means that when a smallholder farmer supplies fresh produce at a local collection point, payment is made on the spot in cash (SDC, 2012). This eliminates any barrier to entry into the market related to cash flow risks for smallholder farmers due to delayed payments and financing costs.

The PGS contract or pledge is in many ways different from a general supply contract as it does not include any quantity specifications. In addition, the pricing system is transparent and democratic ownership of the cooperative, with a profit sharing mechanism among its members, are part of the contract.

Critical factor 5: market information

Participation in the FPM by smallholder farmers will improve when barriers to entry related to market information are shifted to a collectively owned marketing cooperative: The Agri-SCIP test case has a number of aspects embedded in its operations which improve the level of market information for smallholder farmers. Successful participation in the Agri-SCIP supply chain is stimulated by a transparent governance and operational system in which an open information flow is promoted. Open information flow is supported by: (i) price boards at local collection points; (ii) democratic governance of the cooperative; (iii) PGS production standards; (iv) a transparent procurement system with weekly local collection points; and (v) collective packaging. Farmers' Association meetings are also an important platform where smallholder farmers can share information of a diverse nature.

Conclusion

An overall assessment of the market participation by smallholder farmers in the Agri-SCIP test case reveals that, despite some remaining barriers, most barriers to entry are either eliminated or at least eased off. By shifting potential barriers to entry on to a collectively owned cooperative, market participation by smallholder farmers in the fresh produce supply chain has improved. However, some barriers are not completely removed, and this indicates that market participation by smallholder farmers is not absolute and unconditional via the introduction of a collectively owned agri-marketing cooperative.

The entry of small-scale farmers into high end markets in Africa is likely to depend on initiatives of this sort, given the ongoing decline in agricultural production due to the changing situation of rural people (Bundy, 1979). In SA, development efforts are needed in this direction as acknowledged in the National Development Plan (South African Government, 2013).

Market participation by smallholder farmers in the SA fresh produce sector is generally poor; introducing agri-marketing cooperatives nationally can address this situation. This assertion is supported by examples drawn from other African countries, demonstrating mixed success in the development of their small-scale farmer sector. These examples include Ghana, Tanzania, Uganda and Zambia (Auerbach *et al.*, 2015). As with the SA Agri-SCIP case, the African examples used alternative production approaches to differentiate their produce, enabling access to a wider market, ensuring greater prosperity and resultant uptake.

References

Aihoon, J.K. and Associates (2008) *Farmer Settlement Support Package (FSSP) Concept and Implementation Document*. Department of Agriculture, Pretoria, South Africa.

Aihoon, J.K., Onumah, G.E. and Mukwene, M. (2009) Empowering smallholder farmers in markets: country background paper – South Africa. European Consortium for Agricultural Research in the Tropics (ECART) and International Federation of Agricultural Producers (IFAP), Pretoria, South Africa.

Auerbach, R. (1994) Sustainable development: developing what to sustain whom? *New Ground* (Johannesburg, South Africa) 15, 3–9.

Auerbach, R. (2013) Organics and AGRA: transforming African agriculture. In: Auerbach, R., Rundgren, G. and Scialabba, N. (eds) *Organic Agriculture: African Experiences in Resilience and Sustainability*. Food and Agricultural Organization of the United Nations (FAO), Rome, pp. 16–37.

Auerbach, R., Amudavi, D., Kadzere, I. and Nicolay, G. (2015) Organic agricultural research in Africa: an overview of new initiatives. Paper presented at Conference: Organic 3.0: the Next Phase of Organic Development – Visions, Trends and Innovations, 9–11 October, Goesan, South Korea.

Bernstein, H. (2009) V.I. Lenin and A.V. Chayanov: looking back, looking forward. *Journal of Peasant Studies* 36(1), 55–81.

Bundy, C. (1979) *The Rise and Fall of the South African Peasantry*. Heinemann, London.

Cousins, B. (2007) *Agrarian Reform and the 'Two Economies': Transforming South Africa's Countryside. The Land Question in South Africa: the Challenge of Transformation and Redistribution*. HSRC Press, Cape Town, South Africa.

Cousins, B. (2010) What is a 'smallholder'? Class-analytical perspectives on small-scale farming and agrarian reform in South Africa. *Institute for Poverty, Land and Agrarian Studies Working Paper* 16. University of Western Cape, Cape Town, South Africa.

Darroch, M.A.G. and Mushanyayama, T. (2006) Smallholder farmers' perceptions of factors that constrain the competitiveness of a formal organic crop supply chain in KwaZulu-Natal, South Africa. *South African Journal of Economic and Management Sciences* NS 9 No. 4.

Delgado, C.L. (2006) *Sources of Growth in Smallholder Agriculture in sub-Saharan Africa: the Role of Vertical Integration of Smallholders with Processors and Marketers of High Value-added Items*. International Food Policy Research Institute, Washington, DC.

Geerts, S. (2014) *A Conceptualisation and Analysis of the Community Investment Programme with Reference to South African Case Studies: Towards a New Model*. University of South Africa, Pretoria, South Africa.

Hall, R. (2009) Land reform how and for whom? Land demand, targeting and acquisition. In: Hall, R. (ed.) *Another Countryside. Policy Options for Land and Agrarian Reform in South Africa*. Institute for Poverty, Land and Agrarian Studies, School of Government, University of the Western Cape, Cape Town, South Africa, pp. 63–91.

Hauptfleisch, K. (2012) Walking the green road: PGS meets CSA in South Africa. *The Global PGS Newsletter* 2(15), 2.

Hauptfleisch, K., Troosters, W. and Auerbach, R. (2011) Participatory Guarantee Systems: a route into local markets for emerging African farmers. Paper presented at 17th IFOAM Organic World Congress, Seoul, South Korea, 26 September–1 October.

Hauser, M. (2009) Gaining control. *Ecology and Farming* August, No. 45. International Federation of Organic Agriculture Movements (IFOAM), Bonn, Germany.

Hobbs, J.E. and Young, L.M. (2000) Closer vertical co-ordination in agri-food supply chains: a conceptual framework and some preliminary evidence. *Supply Chain Management: an International Journal* 5(3), 131–143.

International Federation of Organic Agriculture Movements (IFOAM) (2013) Participatory Guarantee Systems. Available at: https://www.ifoam.bio/en/organic-policy-guarantee/participatory-guarantee-systems-pgs (accessed 4 September 2013).

Katto-Andrighetto, J. and Auerbach, R. (2009) Participatory Guarantee Systems in Africa: farmers' empowerment and local development. *Ecology and Farming* August No.45. International Federation of Organic Agriculture Movements (IFOAM), Bonn, Germany.

Kibue, M. and Auerbach, R. (2013) Keekenyokie: adding value to Maasai beef. In: Auerbach, R., Rundgren, G. and Scialabba, N. (eds) *Organic Agriculture: African Experiences in Resilience and Sustainability*. Food and Agricultural Organization of the United Nations (FAO), Rome, pp. 80–95.

Kirsten, J. and van Zyl, J. (1996) *The Contemporary Agricultural Policy Environment: Undoing the Legacy of the Past*. Oxford University Press, Cape Town, South Africa.

Kirsten, J. and van Zyl, J. (1998) Defining small-scale farmers in the South African context. *Agrekon* 37(4), 560–571.

Louw, A., Vermeulen, H., Kirsten, J. and Madevu, H. (2007) Securing small farmer participation in supermarket supply chains in South Africa. *Development Southern Africa* 24(4), 539–551.

Makhura, M.T. and Mokoena, M. (2003) *Market Access for Small-scale Farmers in South Africa*. University of Natal Press, Pietermaritzburg, South Africa.

Makhura, M.T., Goode, F.M. and Coetzee, G.K. (1998) A cluster analysis of the farmers in developing rural areas of South Africa. *Development Southern Africa* 15(3), 429–448.

NAMC (National Agricultural Marketing Council) (2001) Report on the Investigation into Fresh Produce Marketing: Report 1: Johannesburg Fresh Produce Market. Available at: https://www.namc.co.za/wp-content/uploads/2017/09/FreshProduceStudy1.pdf (accessed 16 March 2013).

NAMC (National Agricultural Marketing Council) (2002) Report on the Investigation into Fresh Produce Marketing: Report 2: Fresh Produce Marketing in South Africa. Available at: https://www.namc.co.za/wp-content/uploads/2017/09/FreshProduceStudy2.pdf (accessed 16 March 2013).

Republic of South Africa (2006) *Stimulating and Developing Sustainable Local Economies: National Framework for Local Economic Development in South Africa*. Department of Provincial and Local Government, Republic of South Africa, Pretoria.

Riisgaard, L., Bolwig, S., Matose, F., Ponte, S., du Toit, A. and Halberg, N. (2008) A strategic framework and toolbox for action research with small producers in value chain. *Danish Institute for International Studies working paper* 2008/17. Danish Institute for International Studies, Copenhagen.

Senyolo, G.M., Chaminuka, P., Makhura, M.N. and Belete, A. (2009) Patterns of access and utilization of output markets by emerging farmers in South Africa: factor analysis approach. *African Journal of Agriculture Research* 4(3), 208–214.

SDC (Siyavuna Abalimi Development Centre NPC) (2012) Concept Document: Agri-SCIP, social enterprise model. SDC, Ramsgate, Kwa-Zulu Natal, South Africa.

SDC (Siyavuna Abalimi Development Centre NPC) (2013) Annual report: 2012/13. SDC, Ramsgate, Kwa-Zulu Natal, South Africa.

SDC (Siyavuna Abalimi Development Centre NPC) (2014) First quarterly report 2014. SDC, Ramsgate, Kwa-Zulu Natal, South Africa.

South African Government (2013) National Development Plan Executive Summary. Available at: http://www.gov.za/sites/www.gov.za/files/Executive%20Summary-NDP%202030%20-%20Our%20future%20-%20make%20it%20work.pdf (accessed 30 October 2015).

StatsSA (Statistics South Africa) (2011) Census 2011 Agricultural Households, Key Highlights. Available at: http://www.statssa.gov.za/census/census_2011/census_products/Agricultural_Households.pdf (accessed 1 September 2019).

Troosters, W. (2014) Demand driven rural agricultural development in South Africa: the case of the agricultural sustainable community investment programme. MTech dissertation, Nelson Mandela University, George, South Africa.

10 Participatory Guarantee Systems as an Organic Market Entry Point for Small-scale Farmers in South Africa

N'wa-Jama Mashele* and Raymond Auerbach
Nelson Mandela University, George, South Africa

Abstract

A survey into the experiences of small-scale commercial farmers in South Africa as well as the development and effectiveness of Participatory Guarantee Systems (PGS) as a tool to connect small-scale organic farmers to the market, showed that the PGS in Gauteng worked well because the coordinator was very knowledgeable, well-resourced and coordinated the system efficiently. The PGS based on the south coast of KwaZulu-Natal was small but functioned well, based around an externally funded project. In the rural areas of Limpopo, the farmers felt let down by the way the PGS was run, as it was more product driven and focused very little on capacitating the farmers to become viable commercially; this was ascribed to lack of resources and a change in leadership. The Western Cape has several PGS groups, all small and struggling. In general, the farmers, retailers and consumers are innovative and well informed about organic farming and have a range of strategies for getting the produce to the consumers. PGS groups in the rural areas are hindered by: (i) logistics; (ii) lack of marketing skills and information; (iii) distance from markets; and (iv) social conflicts. There are considerable benefits to the farmers where PGS groups are well resourced and effectively led.

Introduction

This chapter will draw on the results obtained from a project commissioned by the German government's International Co-operation Department (Deutsche Gesellschaft für Internationale Zusammenarbeit (GIZ)) in South Africa (SA) in 2017 (Mashele and Auerbach, 2017) looking into the experiences of small-scale organic farmers in SA. The research also explored how the Participatory Guarantee System (PGS) has developed in SA, and where participating farmers are located. The project was carried out in conjunction with the South African Organic Sector Organisation (SAOSO) and PGS-SA, and worked together with an assignment by Matthew Purkis on technology and communication (Purkis, 2017; see Chapter 11, this volume), and another by Audrey Wainwright on materials required to start and run a PGS. The materials developed by Audrey Wainwright, and later further developed by Sasha Mentz, are available on the SAOSO website's PGS page (SAOSO, 2018). This chapter includes information from a recently completed PhD by Musa Khapayi looking at the challenges faced by small-scale farmers and agribusiness.

*jama.mashele@mandela.ac.za

Small-scale Farmers in South Africa

The agricultural sector in SA is complex and dualistic in nature (Vink, 2004). It consists of the commercial large-scale farming sector and the comparatively low-productive, struggling smallholder sector. The large-scale commercial farms are industrialized and linked with Western models and technologies while indigenous and small-scale forms of agriculture are marginalized (Kirsten and van Zyl, 1998; Rogerson, 1998). This is a direct result of historical patterns of dispossession and impoverishment, which systematically eroded historically successful land-based production systems and livelihoods in SA (Hebinck *et al.*, 2011).

Currently, a well-integrated, highly capitalized commercial sector which had 60,000 white farmers in 1990 and now has less than 23,000 farmers (see Chapters 1 and 3, this volume) produces around 95% of formal agricultural output on 87% of total agricultural land (Aliber and Hart, 2009). By contrast, the smallholder sector consists of around 4 million black farmers farming in the former homeland areas on 13% of agricultural land of SA (Aliber and Hart, 2009). There is some argument as to who or what a small-scale farmer is (see Table 9.1).

Distinguishing smallholder farmers from commercial farmers often relies on the size of the farm and the degree of labour intensity, in addition to the level of marketable surplus. In this case, smallholder farming seems to rely mainly on household labour. A number of misconceptions exist about smallholder farmers in SA, mainly in relation to farm size and production efficiency. Large, capital-intensive farms are not always more economically efficient than their smallholder counterparts (Kirsten and van Zyl, 1998; Cousins, 2010). Troosters *et al.* report in Chapter 9 (this volume) on the typology of farming which they developed, which seems to present a useful framework. Mashele and Auerbach (2016) showed how long-term research could support efficient production, and they identified reasons for lower production in organic farming systems. Strategies for closing this 'yield gap' are reported in Chapters 18–22 of this volume.

Gradually, but with great difficulty, agricultural support has shifted from support for large-scale commercial farmers to small-scale farmers with the aim of assisting small-scale farmers to become commercial farmers in their own right (van Rooyen *et al.*, 1987; Cousins, 2010). The SA government and the Department of Agriculture, Forestry and Fisheries (DAFF) devised various policies and programmes aimed at supporting small-scale farmers (National Department of Agriculture, 2001; DAFF, 2018), to reduce unemployment and alleviate poverty. However, these policies have not been very successful, have ignored organic agriculture (OA) and agroecology and some have contributed to the downfall of the subsistence agricultural sector (Hall, 2009; Khapayi, 2017). Although most of the research and challenges focused on conventional farmers, the situation is the same with regard to organic farmers (see Chapters 3 and 11, this volume).

Organic Farming and Quality Management

Worldwide, 57.8 million ha of land was certified organic in 2016 (Willer *et al.*, 2018), up 15% from 2015. As reported in Chapter 2 (this volume), the global value of certified organic market sales was estimated to be US$90 billion in 2016 compared to US$62.9 billion globally in 2011 and US$54.9 billion in 2009. The comparison with global organic sales of US$33.2 billion in 2005 and US$15.2 billion in 1999 shows a consistent trend of a high rate of growth (Willer, 2013; Willer *et al.*, 2018).

In Africa, 39 countries are engaged in certified OA, constituting a share of nearly 3% of the total global organic production (Willer and Lernoud, 2019). Willer *et al.* (2018) report that in 2016, the area of certified organic agricultural land in Africa was 1.81 million ha (up from 1.67 million ha in 2015), plus about 12 million ha of certified wild harvest land, and over 700,000 producers. In 2007, there was more than 50,000 ha of OA land under cultivation in SA, but indications are that this figure has fallen since then.

Currently, 167 commercial certified organic farmers are located as follows. The Western Cape has the highest number at 114. This includes, vegetables, livestock, mixed farming and vineyards. The Eastern Cape is the second highest with 14, followed by KwaZulu-Natal (KZN) and the Northern Cape with about a dozen each. Gauteng and Limpopo have a dozen combined, with seven and five organic farmers, respectively, in each province. The Free State, Mpumalanga and North West each have one listed certified organic farmer. There may be more organic farmers but getting information on their whereabouts has proved difficult (Mashele and Auerbach, 2017). New research interviewing farmers in depth identifies some of the problems faced by commercial organic farmers in SA (Du Plessis, 2018).

Quality management is an important part of the credibility and integrity of the organic sector. In a broad sense, three quality management systems are being used: (i) third-party individual certification; (ii) group certification; and more recently (iii) PGS. When organic farmers operate in an anonymous market, certification systems operated by a third, independent party serve as a guarantee to the consumer that a product has been produced organically, according to organic standards (Barrow, 2006). Competent certification leads consumers to trust the organic production system and its products, and has contributed to the steady global growth of the market. Group certification allows for a degree of internal quality management by a group, such as a co-operative managed by an NGO, but the inspection audit is still carried out by an independent third-party inspector working for a certification body. PGS is very different, in that there is no formal inspector, but rather a group of consumers, farmers and distributors which visits farms to assess quality management; if quality is acceptable to consumers, it is 'guaranteed' (not certified) acceptable. PGS groups which are affiliated to PGS-SA agree to abide by the basic provisions of the SA Organic Standard (SAOSO, 2018), and to maintain transparency and abide by PGS good practice.

To better understand the effectiveness of PGS in SA, a survey was conducted which generated rich information highlighting the challenges as well as benefits the farmers have experienced through PGS. The PGS groups are locally focused quality assurance systems. They endorse producers based on active participation of stakeholders and are built on a foundation of trust, social networks and knowledge exchange. PGS is a low-cost guarantee system run by local consumers and producers, without an external accredited organic certifier, unlike third-party and group certification. It is a locally focused assurance system adapted to smallholder farmers and accessible in terms of costs and procedures (Katto-Andrighetto and Auerbach, 2009). The International Federation of Organic Agriculture Movements (IFOAM) defines PGS as a locally focused quality assurance system that certifies producers based on active participation of stakeholders, built on a foundation of trust, social networks and knowledge exchange (IFOAM, 2013).

The PGS approach is designed to fit the needs of each specific group, and is anchored by six core values, or elements which underpin PGS functions. These include:

- participation;
- shared vision;
- transparency;
- horizontality;
- trust; and
- a built-in learning process.

PGS-SA was established in 2011 to support the establishment of PGS in SA and help facilitate simpler market access for local small-scale organic growers and to create an environment where consumers are assured of the integrity of organic products. PGS in SA is a relatively new concept and could help to develop collaboration and increased income-earning potential of smallholder farmers. The value chain is complex and involves many players such as commercial and subsistence farmers, private business, market agents, national and regional government, NGOs, sector bodies, international corporations, other representative bodies and civil society (Purkis, 2017).

SA currently has 358 members in the five registered PGS groups located in four of the nine provinces:

- the Bryanston Organic and Natural Market PGS in Gauteng;
- Siyavuna Abalimi Development Centre in KZN;
- the Giyani PGS in Limpopo Province; and
- the Western Cape Province has three PGS groups: (i) Green Road PGS (along the Garden Route); (ii) the Outeniqua Natural and Organic PGS (George to Plettenberg Bay and Van Wyksdorp); and (iii) the Western Cape PGS, which, however, is not registered with PGS-SA, and operates out of Wolesley.

To facilitate data collection, information was gathered and contact was made with coordinators and the farmers through phone calls and e-mails as far as possible. Some of the PGS groups were difficult to access (due to difficulties communicating with the coordinators and/or poor contact details of farmers). The farm visits covered farmers in George, Johannesburg, Giyani, Stellenbosch and KZN, with a later special visit to Wolesley to try to understand why this PGS did not wish to be part of PGS-SA, and what support they would find useful. The information gathered during farm visits covered topics such as crop and soil management strategies, market access, and challenges and benefits the farmers face.

Bryanston Organic and Natural Market PGS (Gauteng)

Contact: Audrey Wainwright
This PGS consists of eight committee members made up of farmers and retailers and an administrator. There are 11 farmers who are part of the Bryanston PGS. All five farmers visited cultivate vegetables, herbs and fruits (minimal). The market website gives details about the organization of this well-established and effective PGS group, and also has a link to PGS-SA (https://www.bryanstonorganicmarket.co.za).

Siyavuna PGS (south coast of KZN Province)

Contact: Pearl Mqiba
Siyavuna Abalimi Development Centre (KZN): 282 members.

Siyavuna was started in 2009 when Wim Troosters committed to buy produce from the local small-scale farmers (Troosters, 2014). Every Thursday, he showed up at the agreed location and time and bought whatever produce the farmers brought, paying cash. This continued to a point where the farmers became more organized and increased their produce. This reduced the logistical costs and constraints that hindered these small-scale farmers from accessing the market. The collaboration between the farmers and the researcher led to the establishment of a functioning cooperative as well as an organic brand called Kumnandi (see Chapter 9, this volume).

The collaboration grew strong but the departure of the researcher caused the initiative to lose momentum, with a number of changes to the team/coordinators. For the PGS survey, the researcher could not travel to the province due to extreme weather events and had to rely on the collaboration of the coordinator. The coordinator sent the requested information through e-mail communications. According to the coordinator, the farmers are pleased with how the PGS is functioning in their region. They still supply their produce to the cooperative and get a profitable price for it.

Giyani PGS (Limpopo Province)

Coordinator: Butshabelo Mabunda
The Giyani PGS was launched early in 2017. The farmers had been farming organically before the PGS came into effect, through training offered to them by a resident who received training on organic/agroecological systems. The PGS group consists of about 28 members, two of whom are men and the rest are women. The group consists of elderly people, the majority of whom are over 60 years of age. The training was on a voluntary basis and the trainer would conduct demonstrations at the participant's homes. Due to age and the literacy level of the group the trainer had to find innovative and highly participatory methods to train the farmers.

For the conducted survey, the researcher gained access to the farmers through the help of the new coordinator. Arrangements were made weeks in advance with a key informant to assist the researcher in accessing the farmers and navigating to the various farms. Communication hindered effective data collection as the key informant could no longer be reached via phone call despite numerous attempts for planned farm visits.

Eventually a few farmers were reached and agreed to farm visits. The farmers were welcoming and forthcoming with their experience. Each site visit started with an introduction of the researcher and the purpose of the visit. The conversation followed a semi-structured interview set-up with the farmer, followed by a walk through the planted area.

Some of the benefits the farmers have gained from farming organically include:

- They are conscious of soil health and know that their farming is not further degrading the soil.
- They use Moringa powder (*Moringa oleifera*) to clean their water.
- They were taught how to harvest and save their own seeds for future use as opposed to buying from the cooperatives that used to tell them to throw away their own indigenous seeds as they were not viable.
- The assurance of healthy food – because they now grow their own food, they know what is in this food and can rest assured that no harmful chemicals are on it.

These are some of the challenges that came with/surrounded the functioning of the PGS:

- There is very little communication and transparency. The farmers are aware that something went wrong but no one communicated this with them. As a result, the produce they were told to plant is sitting in the fields with no market to sell to. 'The major issue we have is the market. I am not educated enough to approach anyone or even know where to go and sell our produce.'
- The PGS has not been functional since it was launched due to the 'fighting on top'.
- The Department of Agriculture is not impressed with this form of farming and the extension officer avoids them.

The above-mentioned challenges experienced with this PGS in particular are largely due to social conflict issues. The farmers were supplying their produce to the Bryanston Natural and Organic Market through another coordinator. Due to the growth in number of the organic farmers in Giyani, the decision was made to create a local PGS for this area. During the launch of the PGS in Giyani, it was discovered, through farm visits, that some of the farms that were documented as suppliers of some of the products were in fact non-existent. Staying true to one of the core values (transparency) the then coordinator was given the opportunity to explain the discrepancies, however, despite various attempts from other PGS members and coordinators to communicate with the coordinator, she chose to resign and terminate her membership. This caused the production-to-market set-up with the Giyani farmers to come to a standstill. The new coordinator was faced with many challenges when trying to continue the supply to the Bryanston market. There were issues of miscommunication and false information being spread so as to discredit the new coordinator; this also led to a lack of trust from the farmers towards the coordinator. To date this issue still needs to be resolved.

The Giyani situation involves complex social issues that are hindering their ability to actively participate in the supply/value chain. These need to be addressed before other issues such as logistics, market information and other barriers can be addressed.

Green Road PGS (Western Cape Province)

Coordinator: Janet Gracie

Green Road PGS operates along the Garden Route (George, Sedgefield, Hoekwil) and extends to Stellenbosch. There is a shop set up in Sedgefield where produce from local farmers is procured and sold to the public. The coordinator is in contact with farmers from around the Garden Route who supply her with vegetables and in some cases there is produce (mostly fruits) that comes from farmers in the Stellenbosch area. The coordinator has a set schedule that involves a trip once a week to Stellenbosch to meet with the farmers in this area.

The farmers along the Garden Route deliver their produce to the shop or the coordinator collects from the farmers. The produce in the shop is displayed and labelled in a manner so that the consumer knows who, where and how the produce was grown. The label also has stickers of various colours that indicate the type of farming system: (i) conventional (chemicals are used); (ii) organic; and (iii) in transition (moving away from chemicals towards more natural ways of production). All this is done to ensure transparency and encourage consumers to ask questions about their food.

None of the farmers supplying the Green Road shop along the Garden Route were visited. However, after much difficulty in communication via both phone calls and e-mails, a few farmers in Stellenbosch were contacted and farm visits were arranged. Attempts to get in touch with the coordinator, using voice calls and e-mails were not successful. Eventually, through other contacts, a farm visit was arranged with a wine farmer. The farm is a biodynamic farm with the majority of the land cultivated to vineyards and a smaller plot to vegetables. The farmers shared their experiences as well as the potential it offers and want to see it develop.

The challenges experienced by these farmers include:

- 'Green Road is very product driven.' The farmer is seldom taken into account – the only thing that matters is product delivery.'
- Ill-informed market decisions – the farmer was encouraged to plant a crop for which there was no established market. This led to the farmer being stuck with produce that he could not market resulting in a loss in profits.

The benefits are:

- The Green Road shop serves the role of relieving the farmer of the difficulty of finding a market for their products.
- The Green Road shop also offers the farmer 60% of the profit they make from the sales. Currently, the centralized market for procurement and distribution offers the farmer 8–20% of market value. This is not sustainable for the smallholder.
- In cases where the farmer has a surplus and the shop cannot take it/sell it, they repurpose it – some products are chopped and frozen and sold later.

From all the various PGSs the concerns that were consistently being raised by the participants include: (i) market access and information; (ii) transport to the market; (iii) transparency; and (iv) communication. Some of these concerns have been documented by other researchers.

Outeniqua Natural and Organic PGS (Western Cape Province)

Contact: Sasha Mentz
The committee consists of six people. This PGS has been in existence since 2015 (as Eden PGS until 2018) and currently has nine active farmer members who farm with vegetables and fruits, olives and chickens. Three of the farms were visited during the survey. The farm visits started with a structured interview to solicit information including personal details of the farmer and their farming background as well as their farming practices. The visits also included a walk around the farm and any further questions were addressed as the excursion progressed.

The members are generally pleased with how the PGS is running as they as farmers are established and already have market access and connections. The members did, however, raise concerns about the direction the PGS needs to take in order to attract more members and be useful to these members. Some of the challenges/concerns are as follows:

- PGS is still largely unknown to a lot of organic farmers. Focus must be directed towards educating people about what PGS is and how it functions.
- PGS still remains very poorly resourced and cannot work in very rural areas.

'Western Cape PGS' (Wolesley, Western Cape Province)

This PGS claims to have 14 members. It is not affiliated to PGS-SA: the coordinator felt they would rather 'do their own thing'. They saw no personal or commercial advantage for themselves by joining PGS-SA.

Potential Market Access Entry Points for Small-scale Organic Farmers

The challenges of smallholder farmers have been well understood and documented by independent researchers and organizations that are aiming to support smallholder farmers within the agricultural sector.

The five *primary challenges* for smallholder farmers that were identified in the market assessment were:

- the need for a qualified extension service that can support and facilitate knowledge exchange with the farmers;

- transport and distance to markets;
- product quality and quantity;
- the relationship between the seller and buyer through formalized agreements; and
- barriers to market entry due to a lack of, or inaccurate, market information and pricing.

Agriculture can play a leading role in developing SA's gross domestic product (GDP) in the future, if it is supported by programmes that focus on addressing the specific challenges of smallholders. At the heart of the government policy is the redistribution of agricultural land, but this has not been supported with the tools and skills needed to transition into a value chain that is receptive to new entries into the market. There is an apprehension about the capability of smallholder farmers to participate in the market-oriented production due to their lack of access to markets, financial capital, cost-effective inputs, appropriate technologies, and organized extension service.

There has been limited impact with previous projects due to the independent nature of these programmes. These projects are not focusing on the systemic challenges that smallholder farmers are presented with. A collaboration between the various government departments (Rural Development; Health; DAFF) and private/public organization (SAOSO; PGS-SA) could lead to the development of a national food system based on smallholder and micro-farmer production and processing.

There are two 'markets' that can, in theory, be supplied: (i) the formal market; and (ii) the informal market.

The *formal market* has many potential stakeholders and players: (i) restaurants; (ii) box schemes; (iii) government feeding schemes; (iv) corporate canteens; (v) food processors; (vi) complexes, flats and packhouses; and (vii) directly to the end consumer through formalized market places. These are all viable markets for the micro-producers and smallholder farmers to access.

The *informal market* supplies a large percentage of the urban population in SA and opportunities exist such as farm-gate sales, street vendors, informal shops, local restaurants, households, schools and directly to end consumer through street trade. In order for smaller farmers to deal directly with this market, the distribution of produce requires multiple small-scale interventions.

The existing logistics routes have been built around the central markets, and often revolve around one-way transportation of goods between producers and suppliers. The potential of aggregating produce through the PGS, supported by organic standards, will lead to farmers accessing an untapped, lucrative market for which demand exceeds supply and at the same time participating in a regional localized value chain that supplies the local market through high end restaurants, independent retailers and directly into the informal sector through street vendors and innovative businesses (Purkis, 2017).

Conclusion

PGS seems to function very well in some areas where the infrastructure, resources and knowledge base are very well established as opposed to the village areas. The voluntary basis of PGS seems also to be a hindrance as the volunteers (as seen in the case of Giyani) are in need of permanent employment or a source of income and that will be chosen as the priority over volunteering.

The stimulation of local markets will drive the development of a national food system that is coordinated to supply regional markets through the production systems of micro-producers and smallholder farmers, who are currently farming predominantly on 10 ha or less. This will enable many livelihoods to be built on a solid foundation of a participatory agricultural system that provides clean, healthy food to ordinary people, stimulates job creation for women and youth and grows communities with agriculture as the vehicle.

Ideally, a national sector body such as SAOSO with PGS-SA, and in partnership with many of the stakeholders, should develop a programme that supports smallholder farmers with training. This should include best practice, an innovative extension service, integration of innovation and appropriate technology into the supply chain, access to local markets, endorsement, assurance and certification, peer-to-peer information exchange systems, and diverse forms of innovation throughout the value chain.

References

Aliber, M. and Hart, T.G.B. (2009) Should subsistence agriculture be supported as a strategy to address rural food insecurity? *Agrekon* 48(4), 334–458.

Barrow, S. (2006) South African Organic Market Study. Export Promotion for Organic Products from Africa (EPOPA). Available at: http://www.grolink.se/epopa/Publications/Market-studies/SA%20OMS.pdf (accessed 17 July 2018).

Cousins, B. (2010) What is a 'smallholder'? Class-analytical perspectives on small-scale farming and agrarian reform in South Africa. *Institute for Poverty, Land and Agrarian Studies Working Paper* 16. University of Western Cape, Cape Town, South Africa.

DAFF (Department of Agriculture, Forestry and Fisheries) (2018) Draft National Policy on Comprehensive Producer Development Support. Portfolio Committee on Agriculture, Forestry and Fisheries Cape Town, South Africa. Available at: http://pmg-assets.s3-website-eu-west-1.amazonaws.com/1/180508 NPCPDS.pdf (accessed 17 July 2018).

Du Plessis, H.P. (2018) Mapping the contours of organic agriculture: an exploratory study of an underserved population in South Africa. MSc dissertation, Stellenbosch University, Stellenbosch, South Africa.

Hall, R. (2009) *Dynamics in the Commercial Farming Sector: Another Countryside*. School of Government, Institute of Poverty, Land and Agrarian Studies, University of the Western Cape, Bellville, South Africa.

Hebinck, P., Fay, D. and Kondlo, K. (2011) Land and Agrarian reform in South Africa: caught by continuities. *Journal of Agrarian Change* 11(2), 220–240.

IFOAM (International Federation of Organic Agricultural Movements) (2013) Participatory Guarantee Systems. International Federation of Organic Agriculture Movements. Available at: https://www.ifoam.bio/en/organic-policy-guarantee/participatory-guarantee-systems-pgs (accessed 17 July 2018).

Katto-Andrighetto, J. and Auerbach, R. (2009) Participatory Guarantee Systems in Africa: farmers' empowerment and local development. *Ecology and Farming* August No.45. International Federation of Organic Agriculture Movements (IFOAM), Bonn, Germany.

Khapayi, M. (2017) Agribusiness challenges to effectiveness of contract farming in commercialisation of small-scale vegetable farmers. PhD thesis, Nelson Mandela University, Port Elizabeth, South Africa.

Kirsten, J. and van Zyl, J. (1998) Defining small-scale farmers in the South African context. *Agrekon* 37(4), 560–571.

Mashele, J. and Auerbach, R. (2016) Evaluating crop yields, crop quality and soil fertility from organic and conventional farming systems in South Africa's Southern Cape. *South African Journal of Geology* 119, 25–32.

Mashele, J. and Auerbach, R. (2017) *Participatory Guarantee Systems in South Africa*. Report commissioned by the German International Co-operation Department (GIZ). Nelson Mandela University, George, South Africa.

National Department of Agriculture (2001) The Strategic Plan for South African Agriculture. Available at: www.nda.agric.za/docs/sectorplan/sectorplanE.htm (accessed 30 May 2007).

Purkis, M. (2017) *Viability Assessment: Development of an Inclusive Value Chain for Peri-urban Micro Farmers*. Report to German government (GIZ) from South African Organic Sector Organisation (SAOSO), Johannesburg, South Africa.

Rogerson, C.M. (1998) Urban agriculture and urban poverty alleviation. South African debates. *Agrekon* 37, 2.

SAOSO (South African Organic Sector Organisation) (2018) PGS SA – South African Organic Sector Organisation. Available at: https://www.saoso.org/about-pgs (accessed 12 July 2018).

Troosters, W. (2014) Demand driven rural agricultural development in South Africa: the case of the agricultural sustainable community investment programme. MTech dissertation, Nelson Mandela University, George, South Africa.

Van Rooyen, C.J., Fenyes, T. and van Zyl, J. (1987) A comparison of the contribution and relative performance of agriculture in South Africa. *Development Southern Africa* 4(2), 183–198.

Vink, N. (2004) The influence of policy on the roles of agriculture in South Africa. *Development Southern Africa* 21, 155–177.

Willer, H. (2013) *The World of Organic Agriculture. Statistics and Emerging Trends 2012*. FiBL-IFOAM (Research Institute of Organic Agriculture, Switzerland – International Federation of Organic Agriculture Movements) report. FiBL-IFOAM, Bonn, Germany.

Willer, H. and Lernoud, J. (2019) *The World of Organic Agriculture. Statistics and Emerging Trends 2019*. FiBL-IFOAM (Research Institute of Organic Agriculture, Switzerland – International Federation of Organic Agriculture Movements) report. FiBL-IFOAM, Bonn, Germany.

Willer, H., Lernoud, J. and Kemper, L. (2018) *The World of Organic Agriculture. Statistics and Emerging Trends 2017*. FiBL-IFOAM (Research Institute of Organic Agriculture, Switzerland – International Federation of Organic Agriculture Movements) report. FiBL-IFOAM, Bonn, Germany.

11 Development of an Inclusive Value Chain for Peri-urban Micro-farmers

Matthew Purkis*
University of Johannesburg, Johannesburg, South Africa

Abstract

The agricultural sector in South Africa is complex in nature with the historical displacement of its people still impacting on land ownership and farming methods. At the heart of government policy is the redistribution of agricultural land, but this has not been supported with the tools and skills needed to transition into a value chain that is receptive to new entries into the market. Neither does policy recognize the role that smallholder farmers play – internationally – in the creation of a food-secure nation. The growing demand for high-value food commodities (e.g. traditional maize, culinary and medicinal herbs and other niche varieties such as super foods) is opening up opportunities for farmers, especially smallholders to diversify towards commodities that have strong potential for higher returns to land, labour and capital. However, there is an apprehension about the capability of smallholder farmers to participate in the market-oriented production due to their lack of access to markets, financial capital, cost-effective inputs, appropriate technologies, and organized extension services. Many independent practitioners are achieving some success in smaller operations and programmes nationally, often linked to non-governmental organization (NGO) programmes. These programmes often rely on donor funding to continue and are not sustained after funding is withdrawn. Logistics, especially transport, limits small-scale agriculture as most micro-farmers do not have the capital to rent a cold truck as independents. Without cold chain and logistics services, the access to the market is drastically reduced to the immediate community and surrounding traders, and extending shelf life is also a challenge with limited cold chain services, leading to lower prices for small-scale farmers. As central markets operate at the lowest price to retail, this often leaves the farmer with very little margin for profit, once the transaction has taken place. In order to facilitate smooth collaboration between partners, a shared vision is extremely important; this needs to be agreed upon by all participating stakeholders. Ideally a national body such as the South African Organic Sector Organisation (SAOSO) in partnership with many of the stakeholders mentioned in this chapter should develop a programme that supports smallholder farmers with training in best practice, an innovative extension service, integration of innovation and appropriate technology into the supply chain, access to local markets, endorsement and certification, peer-to-peer information exchange systems and a range of innovation in the value chain.

Introduction

The agricultural sector in South Africa (SA) is complex in nature with the historical displacement of its people still impacting on land ownership and farming methods. At the heart of government policy is the redistribution of agricultural land, but this has not been supported with the tools and skills needed to transition into a value chain that is receptive to new entries into the

*matt@saoso.org

market. Neither does policy recognize the role that smallholder farmers play in the creation of a food-secure nation. Existing agricultural policy supports conventional farming and much of the support that is given to farmers is in the form of subsidized inputs and extension training on this method of farming.

Traditional smallholder farmers are reluctant to pursue conventional farming due to the high cost of inputs, and their experiences with the seed's inability to adapt to local conditions. The farmers have been vocal on this matter at the recent provincial seed hearings and some provinces are supporting the concerns that they are expressing about their right to grow food, and save and share seed (NCOP, 2017). Little support is offered to this group of farmers. Smallholder farmers have been very vocal about their rights to save seed without restrictions and many are refusing to take up the Department of Agriculture, Forestry and Fisheries's (DAFF) offer of free commercial genetically modified organism (GMO) seed.

Towards the end of the apartheid era, formerly controlled markets were radically deregulated. At the end of 1996, the Marketing of Agricultural Products Act (Act No. 47 of 1996) was passed and policies developed that were more in line with free market principles (Ledger, 2016). Price controls were removed and single-channel markets disappeared with the abolition of control boards in the 1980s. This ushered in the concept of commodity pricing linked to the international agricultural futures market. Deregulation and liberalization produced 'winners and losers' in the agricultural value chain (Aliber and Hall, 2009). After government reversed the policies of the 1960s and 1970s, the removal of government support produced a 'uniquely hostile' environment for new entrants (Hall, 2004). The effects of deregulation on productivity are discussed in Chapter 12 of this volume; Fig. 12.1 shows how productivity has fluctuated, and the discussion shows how drastically the numbers of commercial farmers have dropped.

In respect of smallholder diversity in SA, a paper delivered at the 29th International Conference of Agricultural Economists, entitled 'Understanding the smallholder farmer in South Africa: towards a sustainable livelihoods classification' (Pienaar and Traub, 2015) highlighted the dualistic nature of the agricultural sector in SA. It consists of a well-integrated, highly capitalized commercial sector with approximately 35,000 white farmers, producing around 95% of agricultural output on 87% of total agricultural land (Aliber and Hall, 2009). Chapter 12 of this volume presents data to show that the number of commercial farmers has now fallen below 22,000. In contrast, the smallholder sector consists of around 4 million black farmers farming in the former homeland areas on 13% of the total land area of SA (Aliber and Hall, 2009). This dualistic nature and division between the commercial, large-scale farming sector and the comparatively low productive, struggling smallholder sector is a direct result of historical patterns of dispossession and impoverishment, which systematically eroded historically successful land-based production systems and livelihoods in SA.

The SA Government has identified smallholder farming as a means to achieve the goals of poverty reduction and rural development and in Chapter Six of the National Development Plan (National Planning Commission, 2012) committed to expanding the number of smallholder farmers to 250,000 by 2014 and 1,000,000 by 2030. However, there appears to be no definition of what constitutes a smallholder farmer. This was discussed in Chapter 9 of this volume (see Table 9.1 for a typology of smallholder farmers). The lack of consistent datasets (StatsSA, 2016) limits the information for policy makers. Descriptive words such as 'small', 'small-scale', 'family', 'subsistence', 'emerging', and even the word 'smallholder' have been used to refer to the group of farmers known as smallholder farmers. In DAFF's recent Policy on Producer Support Programmes 2018 the classification of smallholder farmers is as shown in Table 11.1. In the same report, agricultural household income per province was provided, as shown in Table 11.2.

Support to farmers by DAFF focuses on commercial farmers with smaller farms recognized as subsistence farming. In his working paper, 'What is a smallholder?' Cousins (2017), suggests that a description preceding the word smallholder, might be apt, with the use of adjectives such as 'semi-subsistence', 'semi-commercial' or 'commercially oriented' which might at least indicate the scale of production and extent of marketed surplus. This would assist policy makers who aim to develop farms to be productive small units. The typology proposed by Troosters *et al.*

Table 11.1. Classification of smallholder farmers. (From DAFF Policy on Producer Support Programmes 2018 in DAFF, 2018.)

Classification	Income	Characteristics
Household producer (vulnerable)	No income	Household consumption, vulnerable woman and youth, child-headed households, disabilities, registered indigents
Household producer (subsistence)	R50,000	Household consumption, not classified as indigents, limited surplus production for sale at market
Smallholder producer	R50,000–R5 million	Venture undertaken by individual or business, household consumption, income derived from value chain activities
Medium-scale commercial producer	R5 million–R20 million	Venture undertaken by individual or business to derive an income from value chain activities
Large-scale commercial producer	R20 million +	Venture undertaken by individual or business to derive an income from value chain activities

Table 11.2. Annual household income per province. (From DAFF Policy on Producer Support Programmes 2018 in DAFF, 2018.)

Province	No income	R1–R38,400	R38,401–R307,200	R307,201–R1,228,800	Above R1,228,800	Unspecified	Total
Western Cape	13,922	36,391	24,116	5,685	862	3,591	84,567
Eastern Cape	188,723	349,960	42,962	4,149	831	9,943	596,568
Northern Cape	14,438	27,924	9,798	1,506	305	1,173	55,144
Free State	53,057	116,143	25,877	2,763	570	2,870	201,280
KwaZulu-Natal	221,337	403,715	64,564	7,678	1,135	18,572	717,001
North West	62,904	113,999	29,819	3,059	604	3,658	214,043
Gauteng	68,494	110,224	69,721	18,090	2,793	9,781	279,103
Mpumalanga	80,035	143,503	31,619	2,907	474	4,847	263,385
Limpopo	150,659	260,206	47,237	3,178	670	6,538	468,488
South Africa	853,574	1,562,080	345,726	49,022	8,253	60,977	2,879,632

in Chapter 9 of this volume offers a clear approach to classification of smallholders.

In the context of job creation, and meeting the goals of poverty reduction and rural development, SA's unemployment rate came in at 27.7% in the third quarter of 2017, the same as in the previous two quarters, and remains the highest rate in 13 years (Trading Economics, 2018). The number of unemployed rose by 33,000 to 6.21 million and approximately 56% were classified as youth. During that period 108,000 jobs were shed in the agricultural sector, largely in the commercial sector in the Western Cape.

Households in agriculture also tend to have limited access to basic services compared with the rest of the population. This is illustrated by the fact that whereas 60% of the population has access to a 'flush toilet', for agricultural households only 29% had access to flush toilets. Similarly, 74% of households in general use electricity, compared with 55% for agricultural households (Stats SA, 2011).

It has been shown internationally, that small-scale farming is most productive and cost-effective if farmers are supported in farming according to agroecological principles (Rodale Institute, 2017; Chapter 3, this volume). The agroecological approach is showing positive results when we look to the incorporation of indigenous knowledge systems (Tittonell, 2014). It has also been shown that organic production exceeds conventional production during times of drought. The Mandela Trials undertaken at the Nelson Mandela University are showing yield increases with organic methods after 3 years, comparing favourably with conventional yields, as reported in Chapters 18–22 of this volume.

Agriculture can play a leading role in developing SA's gross domestic product (GDP) in the future, if it is supported by programmes that focus on addressing the specific challenges of smallholders. This will significantly impact the earning potential in the agricultural sector, especially for women and youth who are currently facing unemployment challenges in the urban, peri-urban and rural environments (National Planning Commission, 2012).

Smallholder farmers are challenged to fit into the traditional central markets that were established when deregulation took place in the 1990s and there has been an investigation which has recently revealed price collusion and preferential treatment.

Analysis of Market Access for Smallholder Farmers

The growing demand for high-value food commodities (e.g. traditional maize, culinary and medicinal herbs and other niche items such as super foods) is opening up opportunities for farmers, especially smallholders to diversify towards commodities that have strong potential for higher returns to land, labour and capital. However, there is an apprehension about the capability of smallholder farmers to participate in the market-oriented production due to their lack of access to markets, coordination, financial capital, cost-effective inputs, technology, and extension services. This was reviewed in Chapter 9 of this volume, where Troosters *et al.* show how the relatively small quantities of produce which smallholders typically produce are one of the five critical factors constraining their market access.

As highlighted in the situational analysis of the SA agricultural sector, the small-scale farmer's access to market is also restricted due to the nature of the existing value chain. This is dominated by the large retail and centralized procurement and distribution of produce. The markets that have been identified as 'niche' provide opportunities for smallholder farmers to maximize value.

Recent technological innovations developed for smallholder farmers' access to markets, quality inputs, information and shared services will eventually lead to improvement in productivity and products more aligned to market demand. The South African Organic Sector Organisation (SAOSO) has embarked on developing a suite of existing information and communication technologies (ICT) solutions (applications, or 'apps') that will be used to develop and coordinate the organic value chain for Southern Africa. These are discussed below.

ResearchGo is an app that allows for live data to be gathered by extension officers, field practitioners, non-governmental organizations (NGOs), activists and youth. This live data can assist in accurate research and the design of intervention strategies and support programmes that are relevant to the farmers, surrounding community and the site-specific environmental parameters. This tool has already been extensively used by the University of Johannesburg to gather data on the township economy and in other surveys such as the quality of life survey conducted by government. The tool allows for field workers to get remunerated directly after completing an audited dataset. The University of Johannesburg's 'Youth in Agri Initiative' focuses on mapping of farms, micro-gardens and market access opportunities.

Crop In can be used for specific data on crop production that will allow for the coordination of the supply chain to the distribution hubs, agroprocessing facilities and packhouses, with the ability to aggregate from multiple producers and coordinate the supply chain to meet market demand. This will facilitate an organic value chain that links smallholder producers to the rapidly expanding organic market locally and internationally.

The opportunity for regional African trade will depend on how well the supply chains are coordinated between the participating producers in the different countries. The European market is becoming highly regulated and the opportunity for African countries to collaborate on developing ethical intra-African trade opportunities would enable African capital to circulate on the continent. Most African countries face similar challenges to SA around the scattered and unsupported nature of many of their smallholder producers. Aligning this group of farmers to a coordinated and supported market, would enable the unstable nature of smallholder agriculture to be potentially transformed into a profitable and sustainable African organic market that could compete with the European market in the medium to long term.

On a local level, there are two 'markets' that can, in theory, be supplied. The *formal market* has many potential stakeholders and players: (i) restaurants; (ii) box schemes; (iii) government feeding schemes; (iv) corporate canteens; (v) food processors; (vi) complexes, flats and packhouses; and (vii) directly to the end consumer through formalized market places. These are all viable markets for the micro-producers and smallholder farmers to access.

The *informal market* supplies food to a large percentage of the urban population in SA, and opportunities exist such as farm-gate sales, street vendors, informal shops, local restaurants, households, schools and directly to end consumers through street trade. In order for smaller farmers to deal directly with this market, the distribution of produce requires multiple small-scale interventions. Technologies such as **One App** will allow for the distribution of organic produce to an informal market at an ethical price for the informal consumer. One App is linked directly to a point of sale unit for informal traders allowing the 'spaza shop' owner to sell airtime, electricity and a wide range of pre-ordered products directly off one point-of-sale unit. SAOSO and partners are able to supply PGS-endorsed produce directly to the informal market through this technological innovation. This has been piloted in Gauteng since 2018. Developing a coordinated distribution strategy for the informal market would enable a price point to be found in the informal market for clean, fair organic produce.

Ethical pricing solutions must be practical and based on local economies and markets. Community engagement is key to the development of trust between all stakeholders. The PGS model has been shown globally to be a viable tool that builds community trust and cohesiveness between stakeholders and the groups of farmers. Several case studies (India, South America, parts of East Africa) demonstrate that the PGS approach has successfully developed strong networks of farmers that are accessing markets and receiving an ethical price for their produce at these markets.

As explained in Chapter 10 of this volume, PGS in SA is a relatively new concept and could be a viable vehicle to develop collaboration and increased income-earning potential of smallholder farmers. Given that the value chain is complex and involves many players such as commercial and subsistence farmers, private business, market agents, national and regional government, NGOs, sector bodies, international corporations, other representative bodies and civil society, a support programme that addresses the following challenges is critical for the development of smallholder agriculture in SA.

Government collaboration

Nationally, DAFF is the main contact point in government regarding the agricultural sector but generally represents large commercial farmers. The mandates within the Department of Social Development and the Department of Health are also relevant in the conversation around food resilience. Social Development is responsible for government feeding schemes and the Department of Health is involved in promoting access to nutritious food, but further advocacy is required. The Department of Trade and Industries provides support to farmers through their cooperative support programme, but the level of sophistication required to access this support makes it generally unavailable for smallholder farmers, due to their lack of a track record and a low level of financial literacy. Currently, DAFF promotes 'conservation agriculture' (i.e. minimum tillage with GMOs and pesticides). The Organic Policy, in its tenth draft after 8 years, has still not been operationalized in DAFF, nor has their Agroecology Strategy.

Provincially, the Department of Rural Development and Land Affairs deals with the transfer of land to smallholder farmers and supports these farmers through nationally trained extension officers. These extension officers are spread sparsely and each has to cover more than 500 farms. The extension officers generally promote commercial and input-intensive solutions to small-scale farmers, as they support national policy. Only a few recognize alternative methods of production such as natural/cultural farming or agroecology that do not require costly inputs.

Locally, there are the Departments of Economic Development and Social Development, which operate in the agricultural value chain, and support the mass population through food aid packages and social grants. There has been more engagement with government at local government level through their nutrition programmes and the support that local economic development provides for smallholder farmers.

Policy framework that is currently being reviewed and amended by government departments include the 'Farmer Support Programme', 'Conservation Agriculture', the 'Plant Breeders Rights Bill' (2015) and the 'Plant Improvement Bill' (2015). Many of these policies reflect support for large agribusiness which aims to expand operations more aggressively into African countries through regulation of the seed market and farmers under stricter trade laws embedded in the legalese of these documents. A number of the policies have an emphasis of driving smallholder producers into commodities such as genetically modified (GM) maize, soybean, sugarcane, etc. This policy trend lends itself to conventional farming methodologies and the Green Revolution framework. This has been shown to fail in many countries where this has been rolled out by government policy and extension service (Chapter 1 and 3, this volume). This approach often leads to debt cycles, failing farms and even farmer suicides as documented in India where GM cotton was nationally adopted by government policy. Smallholder agriculture is more suited to the diverse production systems and ecological best practice farming methods advocated by the agroecological and organic community. Internationally there is a massive push back on agribusiness and many grass roots organizations, organic sector bodies and NGOs are strongly advocating for a shift in policy towards agroecological practice. Implementation of policy in the various provinces differs, but the farmers are organized in one or more of the following ways:

- Commercial farmers are organized nationally under the South African Agricultural Union (SAAU), Agri SA, and black emerging and established farmers now under the African Farmers Association of South Africa (AFASA) which was formed in 2011.
- In respect of organics, the SAOSO represents both small-scale and commercial organic and natural farmers. SAOSO was set up to represent the organic sector from the bottom up to the top through the Organic Sector Strategy Implementation Committee (OSSIC).
- Farmers in the rural areas are generally formed under farmers' associations such as the Eastern Cape Rural Development Agency (ECRDA).
- Traditional tribal authorities also represent a large number of farmers in the rural environments.
- Local economic development also plays a role in coordinating the groups of farmers nationally. Generally, many of these groups are informal structures but there is little coordination to support these networks of farmers, resulting in the continued struggle for smallholder producers to find a unified voice representing the needs of the farmers.
- Conventional farmers are also organized into various commodities that represent the interests of the large producers of specific crops. There are organizations such as Grain SA, Cotton SA, Potato SA, etc. These organizations often deal with the bulk trade of specific goods and are fixed to the central market-pricing models. These commodity bodies currently dominate the agricultural value chain, and often have the financial capital to set up large storage, cold chain and processing facilities. This places a lot of pressure on the micro-producers to participate in these centralized value chains, as there is currently no viable coordinated alternative. Many of these institutions have smallholder farmer programmes, but focus is not on the category of producers.

Independent agricultural organizations, nonprofit companies (NPCs) and NGOs also form part of the complex in which farmers are currently represented and supported. Organizations such as SAOSO, PGS-SA, LIMA, Lindros Whole Earth Consultants, Siyavuna, Green Acre Living, Partner Farmer, Biowatch, African Conservation Trust, Save Act, Choice Trust, Food and Trees for Africa, SA Forestry Stewardship Council, Solidaridad, Gender CC, African Centre for Biodiversity, and Southern African Food Lab are just a few of the active stakeholders who provide a service to farmers and advocate for the development of an equitable food system nationally. These NGOs, NPCs and independent organizations play an important role in providing technical support to groups of commercial and smallholder farmers with training, business model development, market access, advocacy, support, logistics, cold chain services, and essential dialogue around food systems reform and policy shift. However, external funding and NGO support must lead to a point of long-term independent sustainability.

What has been discussed in a number of workshops attended by these organizations is a nationally coordinated rollout plan that would catalyse tangible impact on the ground, with the farmers and farmworkers. Finding strategies to source funding collectively through the green economy will lead to the establishment of a supported organic value chain which would allow for the development of projects that could potentially be linked to university research programmes. This approach would grow the opportunities for students and independent researchers to delve into the data that would enable the identification of green economy business models and provide the platform for participatory processes to occur. This could enable a smooth adoption of national programmes with the farmers and in the communities that have been identified through the survey process. Identifying market access opportunities for PGS farmers to supply, will enable market growth to occur and ultimately lead to the establishment of profitable value chains provincially. This initiative must be inclusive of new market entries, linked to business development and job creation in the green economy, but also preserving our natural assets through best practice.

Potential market participants who could represent stakeholders in the market

There are many stakeholders in the inclusive value chain who are potential market participants. Depending on the region or province, the market participants may vary in each region. National stakeholders such as Wellness Warehouse and Fruits Unlimited, Jacksons Real Food Market, Spar and other retailers could potentially play an important role in the establishment of inclusive value chains that would serve as a solid base for retail demand and would accept produce from smallholder producers. Wellness Warehouse and Jacksons Real Food Market are aiming to source as much local produce as possible to supply their national retail outlets. The collaboration between organizations such as SAOSO and PGS-SA could play a lead role in developing these value chains in partnership with other organizations and service providers.

The use of a production standard would be important to secure market participation at a national level. There are currently the SAOSO Standards for Production and Processing that have officially been adopted as part of the IFOAM Family of Standards and the national standards of the South African Bureau of Standards (SABS) that are still currently in draft form. An accepted standard for production will give retailers credibility behind the 'organic claims' at market, as will traceability and transparency for the end consumer and building trust between the producers, suppliers, retailers and consumers. There is much documented international concern around food safety and the issue of contaminated foods that are consumed by the population due to lack of knowledge. Building ethical business relationships between market participants, that offer transparency down the value chain, will allow for trust to develop between the consumers and the producers. They will also be the foundations of a local food system that supports the nutritional requirements of the rapidly growing SA population.

Assessing different micro-farming models

There are many independent practitioners achieving some success in smaller operations and programmes nationally. Such endeavours are often linked to NGO programmes that align to a separate set of key performance indicators. These programmes often rely on donor funding to continue and are not sustained after funding is withdrawn. Sustainability and national impact are often lost as there is limited drive around developing a national focus for smallholder development. This can lead to replication in the NGO space and inefficient use of financial resources. A holistic understanding of project development and support is needed to define sustainable business models within the agricultural value chain. This approach will unlock business development opportunities that could lead to independent businesses contributing to an inclusive value chain that is supportive of business development and innovation. Through this viability assessment several business models were found to have potential to become sustainable with a few small interventions and strategic support. Some examples are:

- A well-documented business model that has gained international recognition is that of *Umgibe Farming Organics and Training Institute*. The Umgibe system is a simple but innovative modular structure that has been designed for the township and refugee market. It uses plastic bags suspended on a wooden structure which assists home growers to farm effectively in the townships with limited space and often very poor soil conditions. The business is providing a successful financially sustainable agricultural service to around 720 farmers in KZN. The business offers agroecological workshops and consultation to smallholder farmers, and it is supported by market linkages. Nonhlanhla Joye is able to trade at market on behalf of the farmers, aggregating the produce and delivering to retail for a fee.
- Another model developed in Limpopo by *Butshabelo Mabunda* is based on providing support to community members wishing to be food secure using their household gardens. This has led to the registration of her own company under which she will offer extension advice to micro- and smallholder farmers for a fee. Additionally, she will offer different training packages within the community focusing on agroecology and food safety.
- *Tim Abaa* has started an agroecological training academy that is demonstrating a viable return. Because Tim is such a passionate facilitator, the people that he is working with see much value in attending the training offered. This model could be scaled up and formalized into a regular training academy. Additional services will involve ongoing farm mentorship. Tim has identified the lack of resources such as transport, tools and other technical equipment, and regular financial remuneration as a few of the key challenges to scaling-up the concept. Perhaps a brand or NGO that he could hang his activities on would gain some traction with local funders.

Business model development with farmers can be supported through existing networks such as the farmer associations like AgriSA, AFASA and SAOSO. Where there are functional PGSs, with their own specific business model relevant to their community (as shown in Chapter 10, this volume) this can provide the institutional framework and the marketing instrument to improve quality and set-up logistics networks to access high-end markets.

The national agrihubs and agri-parks that have been established by government are currently underperforming in many ways. The infrastructure that has been built across SA is not being properly utilized to service the surrounding smallholder farmers. Currently, the agrihubs are managed by the regional municipalities and there is no clear management strategy of these hubs by the provincial agriculture departments. There is a need for public–private partnerships, which would enable a more effective and streamlined management to be developed. Regional service providers are required, who are mandated to support the emerging farmers with an array of services such as organic input supply, training programmes, logistics, market access and off-take agreements. This would enable the support structures to be developed for smallholder farmers in each province and region.

The supply chain is complex, starting with the supply of inputs and seed, production, harvesting, quality grading, cold storage, agroprocessing, etc. The distribution of produce and processed goods into a market can lead to many market participants adding their margins on to the final retail price. This often leads to exorbitant profits by the retailers. In the USA these activities result in only 7% of the consumer dollar going to the farmer; the US organic sector aims for 40% back to the farmer. The value chain needs to be ethical and in SA it seems to be agreed that aiming for 40–50% of retail value back to the farmer is an ethical distribution of value. This will bolster the potential income-generating opportunities for farmers and farmworkers and lead to increased income.

In Uganda, the Export Promotion for Organic Products from Africa (EPOPA) helped smallholders (1–3 ha) to access markets, using local resources, training local trainers and a farming systems approach, and connecting smallholders with quality management systems (organic, PGS) so that they could access a niche quality market (Chapters 1 and 15, this volume). It proved a successful model and has informed some of the PGSs locally. This can also be used to guide the future development of an inclusive agricultural system that provides nourishing food to meet the requirement of the growing population of SA.

Market logistics

Logistics is one of the biggest challenges for micro-farmers, as they often lack the transportation to get their produce to market. Most micro-farmers do not have the capital to rent a cold truck as independents. Without cold chain and logistics services, the access to the market is drastically reduced to the immediate community and surrounding traders; extending shelf life is also a challenge with limited cold chain services. There are currently a number of methods that micro-farmers use to move their product to market. These methods could be broken down into formal and informal methods of transportation.

Formal methods of transportation would consist of logistics partners that may be able to offer a service to the farmers at a reasonable fee. Some trucking companies offer cost-effective solutions, when they are aiming to maximize their loads and fill trucks to and from destinations. Shared cooperative cold trucks or cold trailers allow a group of farmers to share the cost of transport, and this often makes it viable for groups of farmers, as they share the overall cost of transportation as in the case of the Cold Mountain Co-operative in the Western Cape.

Some farmers do have access to vehicles such as 'bakkies' (1 t utility trucks) with no cold storage. This method is viable, but often has to be done in the early morning to prevent the sun from wilting the produce. Businesses like box schemes and online stores can offer logistics solutions to pick up the produce from the farmer, and this will be built into the price that the farmer receives for produce or products from the company. Shared services with commercial farmers have been shown to be a viable method of transportation for micro-farmers.

Online platforms such as Khula are developing logistics solutions for smallholder farmers to add value for the farmers who are participating in the business. Informal methods of transportation can be as simple as on foot with a box or crate, and wheelbarrows are also a popular mode of transport for short distances. The use of public transport has proved to be a successful model of informal transport, especially in coordinating logistics from the rural areas into the urban environment for distribution to market. The requirement for carbon-effective logistics solutions, which can be scaled up will enable a broader market access for the micro-farmers.

Ethical pricing versus conventional pricing

This pricing debate is one of the most important areas of focus within the establishment of an inclusive value chain. The ethical value exchange will effectively support the development of micro-farmers, increasing the income-generating opportunities and micro-business that could be developed on the back of agriculture.

This model has been widely researched internationally, and some key trends have been uncovered. The current conventional prices have several factors that impact the pricing at the market. First, inputs for conventional farming are often subsidized by government schemes. Secondly, agribusiness also plays an influential role as they often distribute synthetic inputs for free or at a significantly reduced cost to the farmer. This distorts the true cost and market value of conventional produce.

The central markets operate at the lowest price to retail, and this often leaves the farmer with very little margin for profit, once transport and commission have been deducted. The market agents and the five main retail outlets absorb most of the profits, to the detriment of the farmer. The consumer also bears the brunt often paying exorbitant prices at retail where, in some cases, retailers are marking up goods and products by 250%. This leaves a lot of room for the development of an inclusive, ethical and localized value chain that ensures profits are distributed to all the role players who are involved in delivering farm produce to the table.

Revealing the true cost accounting of agriculture to participants in the value chain through local research would go some way towards a holistic agroecological approach to sustainable management of resources. Recent work done by Bandel (2017) on true cost accounting for food, farming and finance (TCA-FFF) uses the model of the sustainability flower. Bandel developed a framework with all the impact categories with which the sustainability achievements of organic growers can be evaluated, managed and communicated.

Currently, the centralized market for procurement and distribution offers the farmer 8–20% of market value. This is often not viable for the smallholder farmer and must be replaced by an inclusive value chain with ethical distribution of profits that supports innovation and business development within the agricultural sector.

An inclusive value chain based on sound agroecological practices that sees a fair exchange of value is of high priority to secure the supply chain for smallholder farmers into a broader market, offering a better price to the farmers on a local market level. The stimulation of the local market will drive the development of a national food system that is coordinated to supply regional markets through the production systems of micro-producers and smallholder farmers, who are currently farming predominantly on 10 ha or less. This will enable many livelihoods to be built on a solid foundation of a participatory agricultural system that provides clean, healthy food to the people, stimulates job creation for women and youth and grows communities with agriculture as the vehicle.

As shown in Chapter 7 of this volume, food insecurity in SA is rife, with 14 million people going to bed hungry and the majority consuming cheap calories, processed goods and large quantities of sugar-laden products, often the result of the restrictive nature of the existing value chain that dominates the agricultural space in SA.

The development of an innovative supply chain that links local producers to local markets and shares 40–70% of the market value back to the farmer will require a number of interventions. These include: (i) farmer training support; (ii) youth mentorship programmes; (iii) shared services to drop the cost of production; (iv) start-up capital (private or government); and (v) a shift in policy. All of these areas of intervention will connect, expand and unlock an inclusive value chain and will be the catalyst for many employment opportunities, and will allow people to secure a viable livelihood in the agricultural sector. Agriculture can play a leading role in the transition towards a more socially equitable society, with the long-term objective of SA being a food secure country and trading on a regional and international export market.

The ability of farmers to access agroprocessing opportunities will play a significant role in maximizing the value from smaller farms. Allowing for fresh produce to be processed and preserved would create more value-added opportunities for smallholder farmers. The sales of fresh produce should only be a portion of the total income to the farmer. Setting up independent agroprocessing facilities and independent packhouses will broaden the base of income-generating opportunities for the farmer and many other participants in those specific value chains.

Models based on group decision making such as the Hansalim Cooperative Model in South Korea may be viable options in SA (Chang, 2014). Here, two cooperatives are formed – a farmer cooperative and a buyer cooperative. These cooperatives have a broad representation base and consist of many role players. These forums would have a regular engagement with price setting on a seasonal basis for the different commodities. This model has achieved high levels of social participation and a growth of a healthy agricultural sector in South Korea. This type of hybrid model will build a network of transparency between the farmers and the market. This ultimately will create an environment of trust and transparency and can lead to long-term participation of government and civil society.

Conclusion

Outlined in this chapter is a participatory approach to developing the organic sector in Southern Africa. Many stakeholders and organizations are actively working towards an alternative food system that would begin to address the complexity of the social inequity that is rife within our society. A coordinated organic sector that aims to foster socio-economic growth for the mass population, through innovation in the sector is important. The rollout of effective national organic-farmer support programmes is often overlooked by the green sector and especially government.

Over the last 5 years, the establishment of a functional organic sector in SA founded on ethical grounds has been relatively successful considering the unfunded nature of the this sector; working in collaboration with many key organizations striving for a food sovereign nation is the next logical step.

Currently, with the array of available solutions within the network, SA could lead Africa in demonstrating the potential of agroecology to create sustainable green jobs for millions of people, regenerating soils for the seventh generation, securing our seed to ensure a sovereign food system and most importantly feeding SA with clean, fair, ethically endorsed produce and products.

SA has the potential to align itself with the growing global trend of countries striving for an alternative to the industrialized agriculture model. Parts of Latin America, India, much of Europe and East Africa have all successfully established a growing alternative to the failing model of industrialized conventional agriculture through organic practice. Africa has challenges that only African solutions can address. Collectively we have everything we need to unlock the untapped potential that lies in smallholder organic agriculture to pave the way for the younger generations to rehabilitate our soils, food systems and societies. We need to plan ahead and strategize how best to capacitate the next generation to feed the expanding global population by 2050.

References

Aliber, M. and Hall, R. (2009) Support for smallholder farmers in South Africa: challenges of scale and strategy. *Development Southern Africa* 29(4), 548–562.

Bandel, T. (2017) True Cost Accounting for Food, Farming and Finance (TCA-FFF). EOSTA, Soil & More with Triodos Bank and HIVOS, the Netherlands. Available at: http://www.socila.eu/wp-content/uploads/sites/5/2018/01/True-Cost-Accounting-for-Food-Farming-and-Finance.pdf (accessed 20 December 2017).

Chang, J. (2014) Hansalim Organic Cooperative – A Best Practice Model of Direct Sales Between Farmers and Consumers. Available at: http://orgprints.org/24218/7/24218.pdf (accessed 4 January 2019).

Cousins, B. (2017) What is a smallholder? *PLAAS Working Paper* 16. Available at: http://www.plaas.org.za/sites/default/files/publications-pdf/WP16.pdf (accessed 1 September 2019).

DAFF (Department of Agriculture, Forestry and Fisheries) (2018) Draft National Policy on Comprehensive Producer Development Support. Available at: https://www.daff.gov.za/docs/media/Notice%20to%20gazette%20the%20policy%20(2).pdf (accessed 1 September 2019).

Hall, R. (2004) A Political Economy of Land Reform in South Africa. *Review of African Political Economy* 100, 23–37.

Ledger, T. (2016) *An Empty Plate – Why We Are Losing the Battle for Our Food System, Why It Matters, and How We Can Win it Back*. Jacana Media, Pretoria, South Africa.

National Planning Commission (2012) Our Future – Make It Work. National Development Plan 2030. Available at: http://www.dac.gov.za/sites/default/files/NDP%202030%20-%20Our%20future%20-%20make%20it%20work_0.pdf (accessed 10 December 2018).

NCOP (National Council of Provinces) (2017) Land and Mineral Resources Committee Meeting minutes – 7 November 2017. Available at: https://pmg.org.za/committee-meeting/25412 (accessed 13 February 2017).

Pienaar, L. and Traub, L. (2015) Understanding the smallholder farmer in South Africa: towards a sustainable livelihoods classification. Paper presented at the 29th International Conference of Agricultural Economists, Milan, Italy. Available at: https://ageconsearch.umn.edu/bitstream/212633/2/Pienaar-Understanding%20the%20smallholder%20farmer%20in%20South%20Africa-1233.pdf (accessed 4 January 2019).

Rodale Institute (2017) 30 Year Farming System Trials. Available at: http://rodaleinstitute.org/assets/FSTbookletFINAL.pdf (accessed 19 December 2017).

StatsSA (Statistics South Africa) (2011) Census Agricultural Households. Available at: http://www.statssa.gov.za/?p=1447 (accessed 4 January 2019).

StatsSA (Statistics South Africa) (2016) Agricultural Survey. Available at http://www.statssa.gov.za/publications/P1101/P11012016.pdf (accessed 4 January 2019).

Tittonell, P. (2014) Livelihood strategies, resilience and transformability in African agroecosystems. *Agricultural Systems* 126, 3–14. Available at: https://www.sciencedirect.com/science/article/pii/S0308521X13001340 (accessed 4 January 2019).
Trading Economics (2018) South African Unemployment Rate. Available at: https://tradingeconomics.com/south-africa/unemployment-rate (accessed 10 December 2018).

12 Supporting Vulnerable Communities in the Eastern Cape: Assessing the Rainfall Evidence

Raymond Auerbach*
Nelson Mandela University, George, South Africa

Abstract

Decline in mean annual rainfall of over 10% has been experienced widely in the Eastern Cape over the past 20 years. In many cases, this means that annual rainfall of about 500 mm in the past is now about 400 mm, and the summer seasonal rainfall has declined from 400 mm to 300 mm. This means that drainage density will have decreased by about 50%, and that rainfed cropping is no longer viable in these areas. Strategies for assisting vulnerable households include technical innovations (moisture conservation, soil organic matter management and mulching), as well as a mentorship programme by competent commercial farmers who have been trained in adult education. Attending to capacity building and rural infrastructure development will also be important, as well as agricultural education (especially of young women from the rural areas).

Introduction

> No continent will be struck as severely by the impacts of climate change as Africa. Given its geographical position, the continent will be particularly vulnerable due to the considerably limited adaptive capacity, exacerbated by widespread poverty and the existing low levels of development.
>
> (AMCEN, 2015)

This is true for much of Africa, and certainly true for most of South Africa's (SA) Eastern Cape Province, where rainfall is already highly variable and erratic.

The Guardian of 26 March 2018 reported on the triennial World Water Forum held in Brazil:

> Humans use about 4600 cubic km of water every year, of which 70% goes to agriculture, 20% to industry and 10% to households. Global demand has increased sixfold over the past 100 years and continues to grow at the rate of 1% each year. This is already creating strains that will grow by 2050, when the world population is forecast to reach between 9.4 billion and 10.2 billion (up from 7.7 billion today), with two in every three people living in cities.
>
> (*The Guardian*, 2018)

As reported in Chapter 7 of this book, Cape Town came very close to completely running out of water during 2018 (Friends of the Earth Africa, 2017).

This chapter builds on the science of climate change outlined in Chapter 7 and combines an analytical overview of predicted climate change impacts and approaches to helping vulnerable small-scale farmers to adapt. By analysing rainfall trends in the Eastern Cape and some recommendations for assisting vulnerable

*raymond.auerbach@mandela.ac.za

small-scale farmers, it hopes to show how the resilience of African farmers can be improved. It is based on three papers written for the International Fund for Agricultural Development (IFAD) (Auerbach, 2017a, b, c).

The realities of climate change are now so stark and the data so compelling, that a real effort of burying one's head in the sand is required in order to 'not see' shrinking water bodies, rising world temperatures, unprecedented greenhouse gas (GHG) levels in the atmosphere, ocean acidification and pollution. This has led to the creation of millions of climate change refugees, who are currently prepared to risk their lives in precarious boats headed for Europe. The Intergovernmental Panel on Climate Change (IPCC) has assembled scientifically robust, compelling evidence of man-made climate change (IPCC, 2014).

The IPCC still hopes to limit global warming to 1.5°C (IPCC, 2014), but evidence indicates that this is already unlikely, and even the critical level of 2°C temperature increase, which will see massive impacts on world agriculture, is likely to be exceeded by 2050. Lisa Cox, writing in *The Guardian* on 15 June 2018 says that leaks of the AR6 report from IPCC indicate that:

> Human-induced warming would exceed 1.5°C by about 2040 if emissions continued at their present rate, the report found, but countries could keep warming below that level if they made 'rapid and far-reaching' changes. Under the 2015 Paris climate agreement 2015, almost 200 countries signed up to limit global temperature rises to well below 2°C above pre-industrial levels and to pursue efforts to limit the temperature increase to 1.5°C. Climate scientist and Climate Analytics director Bill Hare said the draft report showed with greater clarity how much faster countries needed to move towards decarbonisation under various temperature situations and that the impacts of climate change greatly increased between 1.5°C and 2°C of warming. Necessary actions include making the transition to renewable energy, powering the transport sector with zero carbon electricity, improving agricultural management and stopping deforestation. 'This IPCC report shows anyone drawing from published papers that there are big differences between 1.5 and 2 degrees warming in both natural and human systems,' Hare said. 'Two degrees warming and the tropical reefs have basically no chance – 1.5 degrees, they have a small to modest chance of survival. There's a range of commentary that comes out of the report that provides a stronger narrative for us to act than ever before'. He said it (AR6 report) showed that if emissions continued on their present pathway, there was no chance of limiting global temperature rises even to 3°C.
>
> (Cox, 2018)

A global temperature rise of 3°C would mean a frightening rise in sea levels.

The links between climate change and food systems, especially in Africa, are fairly obvious, given that most small-scale African food production occurs under rainfed conditions, and as temperatures rise and rainfall becomes more erratic, the already difficult crop and animal production conditions will become ever more precarious.

The Food and Agricultural Organization of the United Nations (FAO) advocates a process (Elbehri, 2015), citing an example from Bangladesh, following a linear sequence of stages:

- raising awareness;
- scientific capacity building;
- generating evidence;
- conducting pilot studies; and
- informing and engaging decision makers on policy planning.

They point out that there is often a mismatch between the approach of those generating data on climate impacts on food and the needs of policy makers; the scientific data is often too detailed, and too careful. Scientists are understandably cautious about making sweeping generalizations, as they have to be honest about the limitations of the particular context of their research, and the many variables which make precision impossible and accuracy difficult. One of my arguments in this chapter will be that precision is relatively unimportant as the changes we are examining are large, and they are (especially in the Eastern Cape) likely to occur in a context which is already highly variable, as existing rainfall data confirm. Precision, however, is very different scientifically to accuracy. We cannot be sure about precisely how much temperatures will increase over the next 20 years, nor exactly how rainfall patterns will change. This need not stop us from analysing the implications of historical data, understanding the trends it illustrates, and preparing for the likely scenarios which common sense accurately tells us we will have to deal with in the two districts where our pilot activities are located.

In the interior of the Eastern Cape, where rainfall already varies from 200 mm/year to 800 mm/year, we do not need precise prediction to within 1% (or even 10%) of what is likely to happen to prepare us for what we should expect over the next 50 years. Rainfed crop production requires 500 mm of rain in the growing season for most crops; we can confidently predict that the current situation, where rainfed crops fail in 1 year out of 3 in many areas, will deteriorate to crop failure every second year by 2020. We need to identify the most vulnerable sub-districts in each of the two districts involved in this study (Sarah Baartman and Amathole), and distinguish between four major agricultural potential classifications, which will each require different support:

- areas where irrigation is available;
- areas where rainwater harvesting is feasible;
- areas where rainfed crop production is currently not risky; and
- areas where there are already regular crop failures.

In looking at support for small-scale farmers in particular, we should take care to learn from local best practice, and develop strategies which make use of readily available, cost-effective resources. The discussion of existing initiatives in Africa will illustrate the dangers of using expensive inputs to produce the wrong crop! Climate change resilience requires strategies which help local people to use locally available resources to produce products which are in demand and can be consumed or sold (not necessarily locally, but if they are to be sold elsewhere, the infrastructure required is a prerequisite for such systems to function, and this is often forgotten by planners).

Climate and Agriculture: Resilience, Biodiversity and Productivity

Climate change science has much to offer policy makers, but it needs to be translated into terms which are general enough and robust enough to be useful to policy makers, and it also should build on decades of developmental work in Africa. There are lessons to learn from history, where millions of dollars have been spent on projects, many of which have not resulted in long-term improvements in African food security, because they often do not improve the resilience of local farmers and farming systems. The lens of resilience is a useful concept through which to examine climate change in terms of likely impacts and possible helpful strategies. What is needed is government policy that will assist small-scale farmers in developing resilience towards climate change.

Resilience

Buzz Holling and Lance Gunderson and their colleagues have worked on ecosystem resilience for many years (Gunderson *et al*., 1995; Gunderson and Holling, 2002; Folke *et al*., 2004). In my doctoral thesis (Auerbach, 1999) I quoted Holling (1995) as follows:

> Holling and (colleagues) examine the tendency of inflexible managers to lose sight of the complexity of the systems they manage once a simple and apparently effective management policy has been developed. Holling (1995) points out that two critical points emerge: firstly, the reduction of ecosystem variability inevitably seems to lead to reduced resilience and increased vulnerability. Secondly, there seems no other way for agencies to manage and to benefit from resource development. He then searches for examples of narrowly controlled systems which are successfully managed in nature. The example of warm-blooded creatures proves interesting. Tightly regulated endotherms are indeed less resilient and more vulnerable in the sense that if their body temperature varies outside a very narrow range, they die. Terrestrial ectotherms ('cold-blooded' animals) can survive a much wider range of temperature variation. However, two interesting aspects appear after careful examination. Endothermy does persist, and is thus a 'revealing metaphor for sustainable development ... First, the kind of regulation is different. Five different mechanisms, from evaporative cooling to metabolic heat generation, control the temperature of endotherms. Each mechanism is not notably efficient by itself. Each operates over a somewhat different range of conditions and with different efficiencies of response. This overlapping of 'soft' redundancy seems to characterize biological regulation of all

kinds. It is not notably efficient or elegant in the engineering sense. But it is robust and continually sensitive to changes in internal body temperature. This is quite unlike the examples of rigid regulation by management where goals of operational efficiency gradually isolated the regulating agency from what it was regulating. (Holling, 1995)

There are some lessons here for resource managers and climate change planners, I suspect!

Earlier in my thesis, I quoted Holling (1995) and commented as follows:

> Managing complex processes such as these means that one has to operate at a variety of scales, from the individual land user's level, to the local level where collective action can lead to effective resource management. At local, provincial and national level, the three spheres of government have a host of regulatory and planning obligations. The interaction between 'bottom-up' processes originating with land users at individual and local level and 'top-down' processes from government at local, provincial and national levels can be conflictual, neutral or synergic. Participatory design processes, through creating platforms for negotiating about resource use, can help to address conflicts and develop a joint vision for a catchment, thus maximising the chance of building synergy among the various stakeholders – this involves designing 'soft systems'. The process of bringing people together, however, is greatly assisted if it is preceded by practical problem-solving work at individual and local level. With increasingly complex systems, a different approach becomes necessary. Even farm, forestry and nature design processes [...] deal with complex systems. When one moves to education and communication design, the shift to 'soft' systems requires an iterative planning process, if it is to avoid over-simplification. Holling (1995) shows in a comprehensive review of large-scale ecological research, policy, design and management efforts, that very often a complex problem is defined too narrowly. This leads to a technical solution, which succeeds in its narrow aim of holding variability at a certain desired level. Thus the design appears effective: uncertainty is reduced, and initially the management intervention seems to have solved the problem. However, Holling points out that it appears to be a characteristic of complex systems that their variability, their very complexity, allows them to absorb a wide range of disturbances. Simplistic interventions often reduce the resilience of ecosystems, because they increase homogeneity. This is often compounded by the switch of the attention of managers from monitoring to managing the 'successful intervention'. He concludes 'The very success in managing a target variable for sustained production of food or fibre apparently leads inevitably to an ultimate pathology of less resilient and more vulnerable ecosystems, more rigid and unresponsive management agencies, and more dependent societies'.
> (Auerbach, 1999)

This conclusion of Holling (1995) is deeply insightful, and highly relevant to useful planning for climate change. How can we understand communities in the context of climate variation? How can we help food and fibre producers to adapt to changes in climate, using robust mechanisms within complex systems? How can we avoid what Holling calls 'an ultimate pathology of less resilient and more vulnerable ecosystems'? How can we modify the responses of management agencies to be more robust and less rigid? Can we help societies to become less dependent?

Ecosystem resilience is defined by Folke *et al.* (2004), citing co-author Holling (1973) as 'the magnitude of disturbance that a system can experience before it shifts into a different state (stability domain) with different controls on structure and function', and he distinguished ecosystem resilience from engineering resilience. They also cite co-author Gunderson (2000) stating that the self-repairing capacity of ecosystems can no longer be taken for granted once resilience has been eroded. Folke *et al.* (2004) then define resilience as 'the capacity of a system to absorb disturbance and reorganize while undergoing change', still maintaining essentially the same function. In examining biodiversity, they go on to speak of 'regime shifts and irreversibility'. Clearly, resilience of ecosystems is closely linked to biodiversity.

Biodiversity

In discussing biodiversity and resilience dynamics, Folke *et al.* (2004) state that 'The diversity of functional groups in a dynamic ecosystem undergoing change, the diversity within species and populations, and the diversity of species in functional groups appear to be critical for resilience and the generation of ecosystem services'. Once functional group diversity is lost, the addition of just one species may have a major effect on the functioning of the ecosystem. More diverse systems are less sensitive to invasion, and thus more resilient. 'Recovery after disturbance has often been measured as return time to the equilibrium state' (Folke *et al.*, 2004). There are clues here for sustainable agriculture: the difference between a monocropping system and a two-crop system is massive; almost equally important is the introduction of a different type of third crop. This is why our long-term research trials use three crops – cabbage (heavy feeder leaf crop), followed by sweet potato (light feeder root crop), followed by cowpea (nitrogen-fixing legume) (see Chapter 18, this volume). Preliminary results of these trials were given in Mashele and Auerbach (2016) and changes in soil fertility (and their impact on water use efficiency (WUE)) are given in Chapters 18–22 of this volume.

When working with the concept 'resilience' in an agroecolological context, it is important to recognize that anthropogenic (human-induced) changes may give rise to regime shifts in the sense that fertile, productive agricultural land becomes infertile and unproductive (Auerbach, 2013). This is often due to monocropping, industrial farming, overuse of chemical fertilizers and poisons and/or poor tillage techniques (Morton, 2007). The loss of biodiversity in the soil and the very low agrobiodiversity in many industrial farming systems reduces the resilience of these systems (Auerbach, 1995); improving productivity requires support which eases the constraints identified.

Productivity

Productivity in times of climate change can only be defined locally, in terms of the resources which are scarce in a given area. In the case of the Eastern Cape, water, energy and tractors are scarce in most areas, and motivated skilled labour is also not as available as many might assume. To understand this, we will examine changes in population, labour migration and changes in SA agricultural productivity.

According to the World Wide Fund for Nature (WWF), SA food consumption since the 1970s has diversified due to the growth of the middle class between 2001 and 2004:

> This has allowed a shift from staple grain crops to a more diverse diet. South Africans have shown a decrease in the consumption of the staples maize and bread, and have massively increased their annual consumption of chicken from 6 kg to 27 kg per person. Per capita egg consumption has also doubled. Interestingly, the per capita consumption of fruit and vegetables has remained constant, while beef, mutton, pork and milk consumption has declined.
>
> (WWF, 2008)

WWF continues:

> Within the region, SA stands out as one of the most water-scarce countries ... characterised by extremely variable rainfall, both geographically and over time. In the 12% of the country that is suitable for the production of rain-fed crops, productivity tracks rainfall, making farming a challenging business. Climate change predictions are that rainfall will be more infrequent but more intense. This will shrink the country's arable land and increase agricultural unpredictability. Farmers will find it increasingly difficult to increase productivity to meet the growing demand for food. This highlights the need for sound cropping and rangeland production practices to retain soil integrity despite these predicted intense rainfall events.
>
> (WWF, 2008)

I reported on research into assisting small-scale maize producers in southern KwaZulu-Natal (KZN) to improve their resilience by building up soil organic matter (SOM), and strategically allocating inputs only as the season unfolded (Auerbach, 1995). I pointed out that while the population of SA had increased by 2.4% per annum in the decade of the 1980s, food production had only increased by 0.6% per annum over the same period (1995). In the 25 years since then, the situation has deteriorated further, although at a somewhat slower rate until 2014. From 2007 to 2013, population increased by 1.3% per annum, but then the rate of growth increased again in 2014 and 2015. Population grew from just over 48 million to almost 55 million people from 2007 to 2015 (Table 12.1).

SA food productivity (defined as production per unit of land) has varied widely, growing at

Table 12.1. South African population and food production, 2007–2015. (Adapted from StatsSA, 2017.)

Year	Population (millions)	Increase (%)	Food production increase (%)
2007	48.91		
2008	49.56	1.3	1.4
2009	50.22	1.3	1.4
2010	50.89	1.3	1.4
2011	51.58	1.3	1.4
2012	52.27	1.3	1.5
2013	52.98	1.3	1.5
2014	54.00	1.9	1.4
2015	54.96	1.7	1.4
Average		1.5	1.4

less than 1% per year from 1900 to 1960, but then seeing a massive increase to 2% and then 4% per year in the 1980s and 1990s; over the past 15 years, productivity has hovered around 1.4% per year increase, after a stagnant period in the year 2000 (Ramaila *et al.*, 2011; Fig. 12.1). More recently, food production is only increasing at about 1.4% (see Table 12.1), while population is now increasing at about 1.7% per annum. We have moved from net food producers to net food consumers, importing 12% of our food. The percentage of our gross domestic product (GDP) derived from agriculture has fallen from 7.1% in 1970, to 1.9% in 2011 (DAFF, 2013). Although population growth was slowed temporarily by HIV/AIDS in the early years of the millennium, it has now risen to an annual level of about 1.7%, and is not likely to decrease, given the relative prosperity of SA. Unless government policy becomes more supportive, agricultural production is likely to fall further.

Ramaila and colleagues (2011) tell us:

> Before 1965 growth of the agricultural productivity was estimated at 0.65% per annum. In 1965, there was no growth of productivity after which (1965–1981) growth increased by 2.15%. This was due to input prices which were rising faster than the output prices farmers received throughout the period in 1965. However, it recovered to 2.15% in 1980s due to a quick adjustment of farmers to the effects of deregulation. Productivity grew rapidly at 3.98% between 1981 and 1989 due to mechanisation and use of fertilizer, herbicides, pesticides, etc. Farmers at this stage were no longer severely constrained by state intervention but had the ability to change the mix of inputs that were less costly after the deregulation phase. From 1989 to 1994 growth of productivity declined to 0.28% due to inflation rates that had reached a peak and negative net farm income.
>
> (Ramaila *et al.*, 2011)

They continue 'After 1994 the growth was positive due to a positive net farm income and then stagnated due to declining output growth and increasing use of inputs around 2008' (Fig. 12.1) (Ramaila *et al.*, 2011).

At present, agriculture has increased the production of high-value crops, and exports have again increased. The sharp decline in commercial farming since the mid-1990s is confirmed by the abstract of agricultural statistics for SA, which reports 60,938 commercial farming units in 1996, 45,818 in 2002, and only 39,966 in 2007 (DAFF, 2017). Estimates are that there are now about 22,000 commercial farmers in the country, of whom about 10,000 produce 80% of the gross agricultural product; half of this (i.e. 40% of the gross agricultural product) is produced by 300 mega-farms. This is a dramatic and worrying change. During the period from 1980 to 2016, the proportion of agricultural products imported against those exported rose from 18% to 75%.

> SA has a dual agricultural economy, with both well-developed commercial farming and smaller-scale communal farming (located in the former homeland areas). Agriculture contributes a relatively small share of the total GDP, but is important in providing employment and earning foreign exchange. The commercial agricultural sector has grown by approximately 14% per year since 1970, while the total economy has grown by 14.5% over the same period, resulting in a decline of agriculture's share of the GDP to 2.5% in 2008. However, there are strong backward and forward linkages into the economy, so that the sector is estimated to actually contribute about 14% of the GDP.
>
> (WWF, 2008)

The mean annual rainfall (MAR) for SA is cyclical, but as Chapter 7 of this volume shows (Fig. 7.1), there has been a distinct downward trend overall, with the driest year on record in 2017. There have been clusters of dry years including 1912–1914, 1930–1933, 1944–1948, 1968–1970, 1982–1984, 2002–2005 and

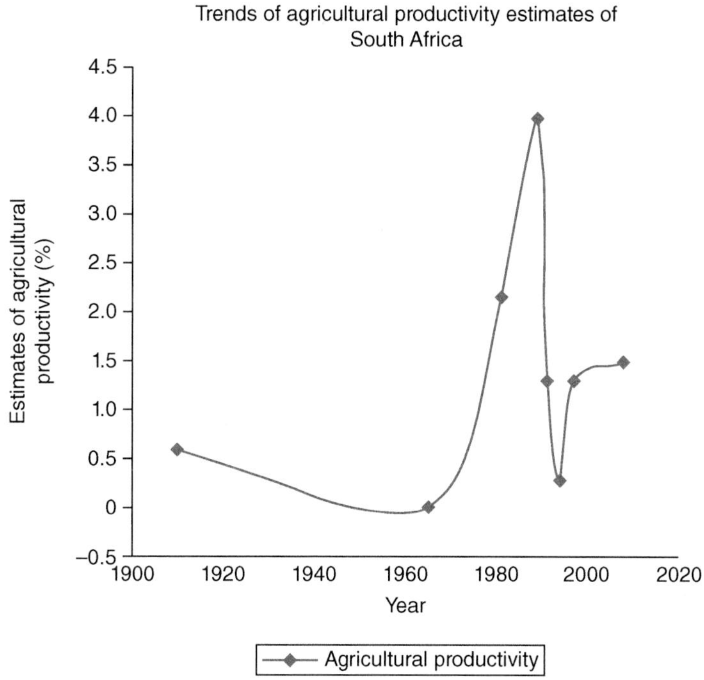

Fig. 12.1. Agricultural productivity estimates from 1910 to 2008. (From Ramaila *et al.*, 2011.)

2012–2017. The intervals are about 14–20 years between clusters of dry years.

So we have a declining number of commercial farmers, decreasing (or stagnant) agricultural productivity, a general failure of land reform policies, a lack of political will to build rural economies, no decrease in child stunting in spite of massive child support grants (Chapter 7, this volume), and a population which now seems to be growing at nearly 2% per year.

I combine these factors in Table 12.2, and although we need to keep in mind that we have an erratically cyclical rainfall pattern and that 2015 was exceptionally dry, looking at the worst aspects of the trend we need to prepare ourselves for some difficult times. This is especially the case in those future cycles where there is a cluster of exceptionally dry years (probably roughly about 4 of every 20 years); this will require some strategic planning.

As SA moves towards larger and more intensive farms, the real costs of agricultural production are not being fully calculated in the cost of production:

> The negative impacts of intensive farming methods on the environment are not being reflected in the input costs. These impacts include pollution of groundwater and surface water, loss of biodiversity, spread of GMOs, loss of soil fertility, erosion, transport costs and climate change, to name a few. It is the individual taxpayer and tomorrow's generations that will pay the real price of these inputs through reduced options.
>
> (WWF, 2008)

Chapter 11 of this volume pointed out that true cost accounting should take note of ecosystem services, and account for the real cost of food production under different farming systems.

The situation on large intensive farms is very different to that in small-scale production, where infrastructure is poor and access to markets is difficult. Both sectors in this dual economy are under severe pressure, but the nature of the pressure is very different. Both sectors are dependent on access to plant-available water, and agriculture is a major user of that available water. Irrigation uses more water than any other sector of the SA economy, and even animal production, traditionally an activity which took place on our veld and pastures (which make up more than 80% of

Table 12.2. Declining commercial farmer numbers and rainfall combined with little growth in productivity, and a steadily growing population will mean that South Africa (SA) is less able to feed itself, and more food insecure. (Compiled from DAFF, 2017 and Table 12.1.)

Year	Number of commercial farmers	Productivity increase (%)	Population increase (%)	SA mean annual rainfall (mm)	Proportion of SA suited to rainfed crop (%)	Food imports versus food exports (%)
1980–1996	60,938	4.4	0.5	800	13	18–55
2002	45,818	1.4	1.4	600	12	60
2012	35,000	1.4	1.8	570	10	65
2015	23,000	1.3	1.9	400	9	75
2030[a]	?	1.3?	2.0?	?	8?	100?
2050[a]	?	?	2.2?	?	?	?

[a]Values in this row are estimated.

our agricultural land), now uses a great deal of water:

> Originally cows grazed on grasslands that were not suitable for crops, converting inedible grass into high-value protein. Today this simple truth has been forgotten and 75% of SA's cattle spend a third of their lives in feedlots, fed by grains grown on the country's scarce arable land. Not only does this practice produce meat with an unhealthy fatty acid profile, it is also a major water issue. Compared to naturally fed beef, it takes about 65 times the quantity of surface water to produce feedlot-finished beef in SA if the feed crops are irrigated – 860 litres for every 500 g grain-fed steak. A sustainable solution is to reduce our daily consumption of red meat and to source natural, range-fed meat.
>
> (WWF, 2008)

What are the implications of this concentration in agriculture, this rapid growth in population, accompanied by slower growth in food production and productivity, and this extravagant use of our water resources? In completing our conceptual framework, let us revisit the idea of agricultural productivity.

Conway (1994), in discussing resilience, distinguishes between production, productivity, stability and equitability; there will have to be trade-offs between these four aspects, he suggests. While agricultural production may increase, productivity only increases if production per unit goes up. The selection of the unit is again a perspective, a lens for examining production efficiencies. Economists will look at return on investment: is my investment productive in terms of return per dollar invested? Fertilizer companies will define improved productivity as 'production per unit area', while advocates of mechanisation will argue for 'production per person-hour'. Water scientists will select 'crop per drop', using WUE as their lens, while renewable energy analysts will look at 'non-solar energy produced as food against non-solar energy used'.

Professor David Pimentel has worked at Cornell University for 50 years, and has shown that organic farming systems are more energy efficient than high-input chemical systems, and he summarizes the situation:

> The total fossil energy used in US conventional crop production is approximately 1,000 liters per ha. Of this, about one-third is for fertilisers, another third is for mechanisation to reduce labour, and about a third covers all other activities and inputs, including pesticides. Past studies on energy use in alternative and sustainable corn [maize] and soybean production systems have pointed to nitrogen fertiliser and pesticides as the inputs leading to the biggest differences in energy use and efficiency, compared to conventional production systems. However, mechanised, irrigated high-input systems are very efficient per unit of land and per person-hour – modern agronomic research has trebled yields per hectare. Mechanisation has allowed a single person to manage 300 ha of land with a big tractor, while in many non-mechanised systems a single hectare is a full-time job, and irrigation is usually a highly effective investment in stability and improved yield, especially in Africa with its erratic rainfall distribution.
>
> (Pimentel, 2006)

Regarding yields in different farming systems, Pimentel (2006) again spells it out:

> Yields of most US crops have increased approximately four-fold since the 1940s ... Steady yield gains have been the result of technological changes rooted in the breeding of higher-yielding plant varieties, increases in the number of seeds planted per acre, more intensive use of fertilizers and pesticides, and more extensive irrigation of cropland. All of these new production technologies depend on the use of significant amounts of fossil energy. The availability of ample quantities of fossil energy and new farm technologies has reduced the human labour required to grow and harvest a hectare of row crops like corn [maize] and soybeans from approximately 1,200 hours per hectare prior to the introduction of farm machinery, to about 11 hours now.
>
> (Pimentel, 2006)

He concludes that for maize, organic agriculture (OA) uses about 31% less fossil fuel and most animal systems use about 28–30% less fuel per unit of product. Bringing together the concepts around climate change, resilience, biodiversity and productivity presented above, sustainable rural development demands that all four of these perspectives should be incorporated into systems which help local people to develop their capacity to use local resources for the production of desired and usable food and fibre.

In 1994 (as part of a study for the new government), I led a process of consultation with farmers, labour unions, farmer trainers, academics and policy makers. I summarized some of the conclusions about the need to build on local capacity, of both commercial and small-scale farmers (Auerbach, 1994). I quoted this work in my doctoral thesis (Auerbach, 1999), and some of the concepts developed were presented in Chapter 1 of this book (Fig. 1.2). In summary, agricultural scientists are most comfortable with a production-oriented approach, which is often rather short term and technology centred. This is not to say that national food self-sufficiency is unimportant – it is essential. However, politicians and social scientists are concerned that the poorer households may not be able to access food if they have to purchase it, and therefore household food security is important if there is to be reasonable equity. Conservationists on the other hand, have long been critical of the damage being done to the resource base by industrial agriculture. While their philosophy has always been long term, they were rather technical in their approach. Over the past 20 years, however, the WWF has increasingly emphasized the importance of working with communities, if conservation is to become socially sustainable.

We will examine the evidence later in this chapter, but thus far will summarize the conceptual framework adopted for this study by concluding that sustainable development requires a long-term approach to building community participation in agriculture and other aspects of rural development. Resilience, biodiversity, improved productivity and strategies which address soil fertility and WUE need to be adapted to local conditions and to robust predictions of the major climate change constraints likely to affect small-scale farming.

Conceptually, we believe that the implications of climate change, resilience and productivity for food security in Africa are that systems using renewable natural resources and based on local culture and local resources are required for Africa.

A Review of Selected African Initiatives

I reviewed two initiatives both aimed at increasing agricultural production in Africa (Auerbach, 2013). I showed how the Alliance for a Green Revolution in Africa (AGRA) developed their AGRA-Millenium Villages Project (AGRA-MVP) based on conventional high external-input maize cropping (Nziguheba *et al.*, 2010), while the Export Promotion of Organic Products from Africa's (EPOPA) approach aimed at broadening agrobiodiversity, stimulating soil biology and thus developing resilience as an emerging property of organic farming systems (EPOPA, 2008).

Earlier in this book, I summarized this work, and compared (Chapter 1, Fig. 1.3) the cost-effectiveness of the two approaches over an initial 5-year period, and found that while AGRA-MVP was spending US$120/farm/year, EPOPA had spent US$2/farm/year. Could there be such a great difference between the two approaches, and if so, what were the reasons for this, and is organic therefore a more viable intervention for improving food security in Africa? Undoubtedly, the AGRA approach was more ambitious, broader in the range of issues it tackled, accurate in identifying the need for infrastructure and honest in trying to address that need. However, the AGRA-MVP project did impose maize as the crop that they would help with, which did not take local variations into account, and therefore had great difficulties with the marketing of crops produced, which were sometimes not locally in demand.

On the contrary, the EPOPA project built on local capacity, concentrated on market linkages and developed local training capacity, based on locally available resources. Aside from the lower cost and larger number of farmers reached, there were claims that while assisting the farmers, EPOPA had significantly benefited the environment in terms of improved resilience, increased agrobiodiversity and reduced dependence on external aid and external agricultural production inputs. If this is so, these factors will likely also be important for improving resilience in SA's Eastern Cape Province. The project will *not* be helpful if it oversimplifies the local economy, ignores local preferences and resources, or attempts to impose interventions on farmers as many past Eastern Cape interventions have done. The comparison of the EPOPA and AGRA-MVP projects should provide several lessons in what to do and how to do it, and also how not to do it.

The major differences in the two approaches relate to how local resources are valued and how they are managed. The conventional paradigm analyses the situation as follows:

> These soils are acidic and highly leached, therefore low in available phosphorus (P) and SOM. Therefore, to increase production quickly, they will require lime and chemical fertilizer.

The organic paradigm takes a different approach:

> These soils are acidic and highly leached, therefore low in available P and SOM. This is a product of climatic and human characteristics which will not disappear. Therefore, if they are to be sustainably managed, systems will have to be developed to counteract the leaching and acidification processes; they will need some lime and then crop rotation, mulching, compost and possibly some rock phosphate. The systems will have to develop local human and institutional capacity, and will need to produce products which are useful to the local people, and to develop networks which will facilitate the sale of produce.

This is much more difficult, but not more expensive, and an approach that can get around the problems of interventions which oversimplify both the problem analysis and the (imposed) solutions, and build real production capacity on the ground! This is at the heart of the agroecological approach to sustainable food system development.

The Agricultural Context of the Eastern Cape in South Africa

The IFAD commissioned a study on climate change variability in the Eastern Cape in 2015. Data indicates that historically (20 years ago) most rainfed cropping areas in the Eastern Cape received more than 500 mm of rain in the growing season for 2 out of 3 years (enough to grow rainfed crops), but many already only have adequate rain for one in three seasons (Auerbach, 2017a). However, an analysis of climate change trends over the past 20 years showed a steady decrease in many areas. These already vulnerable areas should be discouraged from rainfed crop production, and a risk assessment should be carried out for each sub-district based on historic records and a prediction of average temperature increase of 2°C and a decrease in rainfall of about 10% on average, with poorer rainfall distribution (more flooding and more prolonged dry periods), as I reported in three papers for the IFAD-funded study (Auerbach, 2017a, b, c). Added to this, as Chapter 1 of this volume showed, the work of de Wit and Stankiewicz (2006) shows that with annual rainfall under 400 mm there is practically no perennial drainage (or runoff causing river flow); even at 500 mm MAR, a 10% reduction in rainfall means that drainage density decreases by 50%. We will examine changing rainfall in the Eastern Cape on a case-by-case basis, to get an idea of how we might approach support for vulnerable communities.

The previous chapters (in Parts 1 and 2) show that unless the approaches adopted build local human and institutional capacity, they will not prove sustainable, and they are likely to join the ranks of high-budget technology-driven bright ideas which have been implemented throughout Africa, and which show little long-term contribution to food security or food sovereignty in Africa.

While impact modelling globally warns policy makers and agricultural managers about the likely scenarios that farmers and consumers will have to deal with in the next 50 years, it is less useful in telling us what to do about this. Here, practical experience on the ground tells us how we can increase production in the short term where we have easy access to technical expertise and external inputs. Where we wish to develop sustainable systems, improve resilience, and where there is limited access to external inputs, we have shown that agroecological approaches make more sense economically, environmentally and socially. Part 3 of this volume will show that this is also true technically, in terms of soil fertility, WUE, pest and disease management, and also yield and profits. Technical recommendations should be linked to institution-building processes which strengthen the capacity of local farmers to make informed decisions and take joint action.

Technically, climate change adaptation must therefore include strategies to: (i) increase SOM (cover crops and compost); (ii) use mulches wherever possible; (iii) set up small-scale rainwater harvesting systems, especially using the runoff from roads; and (iv) practise crop rotation. Where feasible, compost production should be encouraged. In areas of greatest risk, rainfed crop production should be discouraged, and low-risk strategies for small stock production should be developed.

Socially, little will change in the long term unless infrastructure, institutions and participatory planning processes are put in place.

The two districts Sarah Baartman (formerly Cacadu, north and west of Port Elizabeth) and Amathole (west, north-west and north-east of East London) are shown in Fig. 12.2, with sub-districts 1–9 and 10–16, respectively, and populations of nearly half a million and over three-quarters of a million, respectively, showing the low population density.

The drier areas of these two districts produce sheep, goats (including Angoras for mohair) and some cattle. The moister areas have some rainfed cropping, and the irrigated areas produce citrus, vegetables and a wide range of high-value crops.

Methodology

Nine years of rainfall data for SA and the Eastern Cape are presented, and show that rainfall variability is an inherent characteristic of climate in the Eastern Cape. The major conclusion based on historical rainfall data, is that a bad year in the Eastern Cape is already worse than many 'average' climate change predictions! These figures are presented as a rainfall table for certain areas of the Eastern Cape followed by graphs showing the rainfall trends for each of the identified low rainfall areas.

After I presented these figures to an IFAD workshop in May 2016, it was agreed to use worst scenario historical data as the predictor for conditions under climate change, rather than complex climate models. These figures (with MAR and altitude) have been extracted from SA Weather Service records for 22 weather stations in the Eastern Cape, with records going back on average 30 years.

Rainfall Patterns for the Eastern Cape

Looking at the Eastern Cape in the context of SA, one gets an idea of how variable the climate is. The Eastern Cape Province lies on the south-eastern coast of Africa, with KZN to the north-east, the Western Cape Province to the west, and the Free State Province to the north. The eastern seaboard of SA is much moister than the interior, and rainfall drops off as one moves westward through the Free State and into the arid Northern Cape Province. Large parts of the Eastern Cape regularly receive more than 800 mm MAR, giving them at least 500 mm during the summer growing season (Auerbach, 2017a). However, in many years, some of these areas receive significantly less than 500 mm in the growing season, and this results in crop failure for rainfed agriculture. The high variability of rainfall means that risk of crop failure is a major factor for farmers here. This work shows that 8 years ago there were extensive dry areas

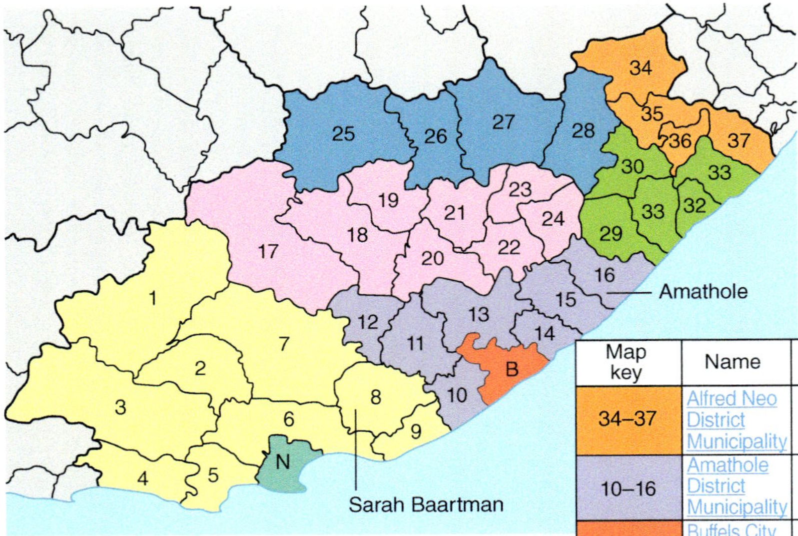

Map key	Name	Code	Seat	Area [km]	Population [2011]	Pop.density [per/km]
34–37	Alfred Neo District Municipality	Code	Mount Frere	6,859	801.344	116.8
10–16	Amathole District Municipality	OC44	East London	21,643	892,397	42.4
B	Buffels City Metropolitian Municipality	BUP	East London	2,536	155,200	287.8
1–9	Sarah Baartman District Municipality	OC10	Port Elizabeth	58,194	450,584	7.7
17–24	Chris Hani District Municipality	OC15	Queenstown	58,555	285,461	21.7
25–28	Jamestown District Municipality	OC14	Barkley East	29,663	343,768	13.6
N	Nelson Mandela bay Metropolitian Municipality	NMA	Port Elizabeth	1,853	1,352,115	588.1
29–33	CB Tambo District Municipality	OC15	Maclear	19,968	1,364,943	85.5

Fig. 12.2. Map of Eastern Cape showing Sarah Baartman and Amathole district boundaries.

in the Eastern Cape and the interior, but that Port Elizabeth, East London and the surrounds had quite a good annual rainfall that year. The research shows that 7 years ago, virtually the entire Eastern Cape Province experienced above average rainfall, even though parts of the interior were extremely dry that year. Six years ago, the coastal areas again did well in terms of rainfall except for a few dry patches, but the interior was again fairly dry. The situation 5 years ago saw good rainfall for the two districts concerned, parts of the Eastern and Western Cape Provinces were again fairly dry.

Things then begin to change as a drier cycle started in 2014/2015, but rainfall was still adequate for most of the Eastern Cape and the eastern seaboard, but there are again drier patches in the interior. Then, by 2015/2016, the moist area on the eastern seaboard is narrower, and the interior is drier.

When we look at 2016/2017 data (Auerbach, 2017a), the relative dryness of the Western and Eastern Cape is apparent, and the moist area on the coast is far smaller. Both Port Elizabeth and Cape Town had severe water shortages from 2015 to 2018.

Historical rainfall records for the Eastern Cape

Rainfall records are available for the past 30 years for many parts of the province. The detailed annual rainfall records were analysed in Auerbach (2017b), where variability between years was examined to assign risk categories to each sub-district. To get some clarity on riskiness of crop production in the various parts of the Eastern Cape, 22 rain stations were examined. Table 12.3 shows the seasonal average rainfall data for these stations, located across the province. The right-hand column in Table 12.3 shows the minimum annual rainfall for the area, as supplied by the SA Weather Service. It is striking that the minimum annual rainfall is often less than half of the MAR.

The altitude is shown (in metres above sea level) in Table 12.3, and in order to understand the significance of the rainfall data, those stations receiving less than 500 mm in the growing period (September–March) are shown with the moister sites at the top, going down to the drier sites. The bottom half of the table shows that Tyefu (north of Bathurst in Amathole District, except where irrigated), Uitenhage, Grootfontein, Graaff-Reinet, De Aar and Hillside Farm (north of Uitenhage in Sarah Baartman District), which all have MAR less than 500 mm, have seasonal rainfall of less than 300 mm per growing season (September–March). These areas are extremely risky for rainfed crop production, and will become even riskier as climate change accelerates. The worst case scenario, which we will use to guide our risk assessment, shows that all experienced at least one growing season with 200 mm of rain or less (minimum annual rainfall).

Increase in drought frequency is likely, but rather than trying to gain precision under conditions of frequent large variability, the current worst case scenario (historically) seems a good place to start. Building resilience into smallholder farming will depend on improving WUE and on harvesting all available moisture carefully. Some techniques (e.g. conservation contours) are well known locally; others (e.g. swales, wetland construction, microclimate development) are relatively unknown. Important agronomic strategies will include organic soil management (increasing colloidal humus using compost and crop rotation, which will improve nutrient- and water-holding capacity) and the use of mulches (to cover the soil and reduce evaporation and soil erosion). Important livestock management strategies will include 'holistic resource management' approaches and other rotational grazing systems, and may also include 'cut and carry' systems for small-scale farmers.

Local institution building is vital for the economic and knowledge economy development. It will be important to identify *at risk* communities. Mapping the relative risk of the various parts of the two districts was the main task of this research study, and the Eastern Cape has been categorized into three sets of attributes: (i) areas with rainfall which is high enough and stable enough that rainfed crop production is likely to be feasible even during times of climate change; (ii) areas where the rainfall is too low already for rainfed crop production; and (iii) areas where rainfall is marginal and/or monthly variability of seasonal rain is so high that rainfed

Table 12.3. A selection of the locations in the Eastern Cape Province where seasonal rainfall (in mm) is less than 500 mm; the growing season is from September to March each year in the summer rainfall areas of South Africa.

Location	Altitude (m)	Mean monthly rainfall (mm)												Mean seasonal rainfall (mm)	Minimum annual rainfall (mm)
		Jul	Aug	Sep	Oct	Nov	Dec	Jan	Feb	Mar	Apr	May	Jun		
Bathurst	259	47	56	59	79	78	54	52	63	82	53	48	46	467	450
Barkley East	1905	7	26	26	43	72	90	89	86	82	51	25	9	488	295
King Williams Tn	400	31	57	37	58	72	61	62	76	63	37	30	22	429	372
Somerset East	717	25	41	30	57	61	58	48	75	84	42	28	21	413	278
Aliwal-Noord	1347	14	19	25	41	56	62	68	98	89	54	25	18	439	355
Buffelsfontein	1783	15	21	30	42	60	65	75	81	82	46	28	16	435	193
Queenstown	1094	9	21	19	48	54	69	65	84	88	41	21	15	427	337
Tyefu	119	30	22	28	60	58	48	35	27	44	38	19	24	300	314
Uitenhage	32	29	36	28	42	46	30	33	35	44	41	30	23	258	207
Grootfontein	1270	13	17	11	27	38	39	47	53	67	33	14	13	282	235
Graaff-Reinet	753	17	21	12	32	36	34	38	45	58	33	21	14	255	160
De Aar	1240	13	15	11	26	30	23	36	60	52	41	16	13	238	155
Hillside Farm	101	10	20	25	39	34	23	32	33	38	37	25	18	224	181

crop production will increasingly be too risky. This classification will help those familiar with farming in the areas to develop recommendations for three sets of strategies:

- higher rainfall, less risky areas;
- moderate likelihood of crop failure; and
- marginal areas where crop failure is likely.

Developing detailed recommendations from a desk-top study would be arrogant and dangerous, and the chapter recommends that a local task team for each district be constituted to do this. However, analysis of the rainfall records of the Eastern Cape for the last 20 years has allowed us to identify certain useful tendencies.

Agriculture: Development Proposals

According to Integrated Development Plan (IDP) 2015–2016, agriculture produces 3% of the Amathole GDP, but accounts for 7% of employment. Figures given for Sarah Baartman also give the contribution of agriculture as 3% of GDP, with the contribution of agriculture for the whole Eastern Cape at less than 2%. On page 31 of the IDP, the agriculture sector in Amathole District Municipality (ADM) is summarized as follows:

> The land use patterns and land ownership in ADM are diverse, varying from communal land ownership, particularly in the former homelands, to private commercial land ownership. Agriculture in most parts of the ADM has not yet developed beyond subsistence because of constraints facing agriculture in rural areas. The prospects of agriculture currently look dim because of the lack of inputs, resources and a lack of interest from the youth.
>
> (ADM IDP Review 2015–2016, version 4 of 5; unpublished report)

The EC Vision 2030 Eastern Cape Provincial Development Plan states under the heading *Ilima labantu* (People's agriculture):

> Multifaceted agriculture-driven development to promote consciousness and participation in agricultural activity and production across scale – households, organised small-scale farmers, large-scale commercial farming and product beneficiation across value-chain. Key to initiative will be – (i) R&D, knowledge and capacity-building; (ii) integrated and coordinated actions by government and partner entities involved, both on supply side (policy, capacity-building and input support), as well demand side (public markets and social economy) and (iii) the promotion of co-operation.
>
> (EC Vision 2030 Eastern Cape Provincial Development Plan; unpublished report)

In the Sarah Baartman District, there have been proposals to extend fruit farming from the Langkloof all the way to Port Elizabeth, and there are many possibilities for intensification of irrigation, concentrating on fruit and dairy farming in the west of the district (ADM IDP Review 2015–2016, version 4 of 5; unpublished report). However, it will be important to focus development appropriately, and to build in an understanding of the likely difficulties which will be posed by increase in rainfall variability in the 'at risk' areas.

What Can We Learn From the Analysis of Long-term Rainfall Data?

Table 12.3 presented a summary of the rainfall data for the Eastern Cape for the last 20 years, and is arranged in descending order of seasonal (not mean annual) rainfall; this is the rainfall in the 7 months from September to March, which is the summer growing season. Towns such as Encobo have been excluded, as they have a higher MAR, even though the standard deviation (SD) is very large. The implications of this large SD are that rainfall may vary from zero to over 100 mm per month from year to year, making farm management decisions very risky. This risk will increase with climate change, and the already highly variable situation in certain areas will become even worse. Each of the towns listed in Table 12.3 will be discussed briefly below. In the discussion, Barkley East, Stutterheim, King Williamstown, Queenstown, Butterworth, Bisho, Grahamstown and Jamestown are grouped together as having a moderate likelihood of crop failure and needing special attention in future, as their rainfall is already low for crop farming. Although these areas have traditionally carried out cropping activities, the low seasonal rainfall and high variability will make rainfed cropping increasingly risky, as the individual trendlines will also show.

Fort Beaufort, Springs, Cradock, Patensie, Uitenhage, Somerset East and Graaff-Reinet, on the other hand, are already so risky for rainfed crop production that few farmers try it without at least some supplementary irrigation capacity. These are classified as marginal areas where crop failure is likely. It is very likely that seasonal (7 month) rainfall will drop below the level of 300 mm, where even drought-tolerant crops will yield very poorly.

The prospects for each town will be discussed in detail below, with reference to rainfall trends and variability.

Barkley East receives most of its seasonal rain during November, December and January, and rainfall in these months is highly variable. Over the past 20 years, rainfall has declined (both annual rainfall, depicted in Fig. 12.3, and seasonal rainfall). The decline in annual rainfall from 750 mm to 550 mm is already drastic and significant, and the average for the last 10 years is close to 500 mm! Only drought-resistant crops should be planted, with as much attention as possible to rainwater harvesting, soil moisture conservation, build up of SOM and where possible the use of mulch to reduce evaporation.

Rainfall for Bisho is spread throughout the year, with high variability, especially in November and February. Over the past 20 years, rainfall has not changed much, with the MAR fairly steady at around 500 mm (Fig. 12.4). The seasonal rainfall of around 361 mm seems fairly stable, but is marginal for rainfed crop production. Only drought-resistant crops should be planted, with as much attention as possible to rainwater harvesting, soil moisture conservation, build up of SOM and where possible the use of mulch to reduce evaporation.

Butterworth receives most of its seasonal rain during November, December and January, and rainfall in these months is highly variable. The MAR is 490 mm, and 7-month rain 371 mm. Over the past 20 years, rainfall has declined (both annual rainfall depicted in Fig. 12.5, and seasonal rainfall). The decline in annual rainfall from 600 mm to under 400 mm is already drastic and significant; however, rain over the past

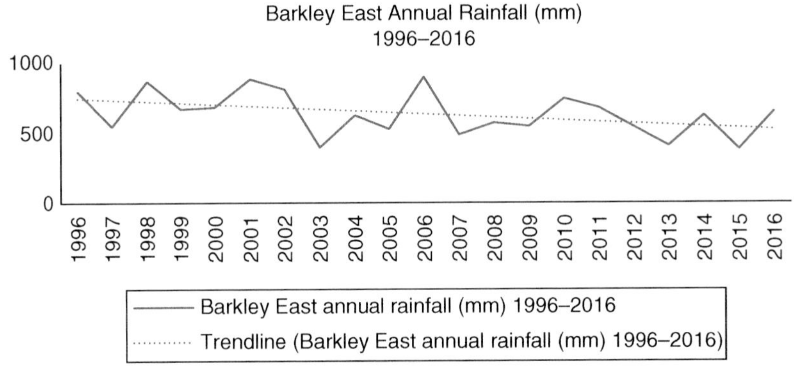

Fig. 12.3. Barkley East annual rainfall (mm) 1996–2016 (mean annual rainfall (MAR) 635 mm).

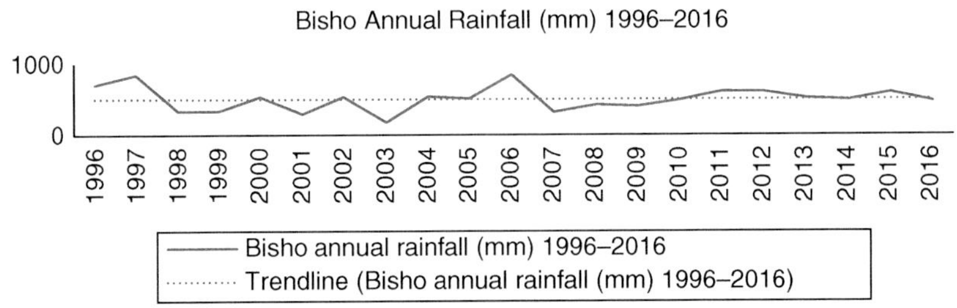

Fig. 12.4. Bisho annual rainfall (mm) 1996–2016 (MAR 503 mm).

2 years has been quite reasonable, but the long-term trend seems to be drier! Only drought-resistant crops should be planted, with as much attention as possible to rainwater harvesting, soil moisture conservation, build up of SOM and where possible the use of mulch to reduce evaporation.

Figure 12.6 shows that rainfall at Cradock is too low to consider rainfed cropping, although there has not been a decline in the annual rainfall or seasonal rainfall over the past 20 years.

Figure 12.7 shows that Engcobo rainfall is higher than the other towns presented here, and it is listed only to show that climate change may well mean that areas with rainfall considered to be adequate for rainfed crop production over the past 20 years may well become marginal in the second half of this century. Although the MAR over the past 20 years is over 1000 mm, the trendline shows a decrease in annual rainfall from 1300 mm to 900 mm. The data show exceptionally high variability for all 7 months of the growing season. An area like this lends itself to water storage in dams and to rainwater harvesting and keyline ploughing to conserve moisture in whatever way is possible. Organic

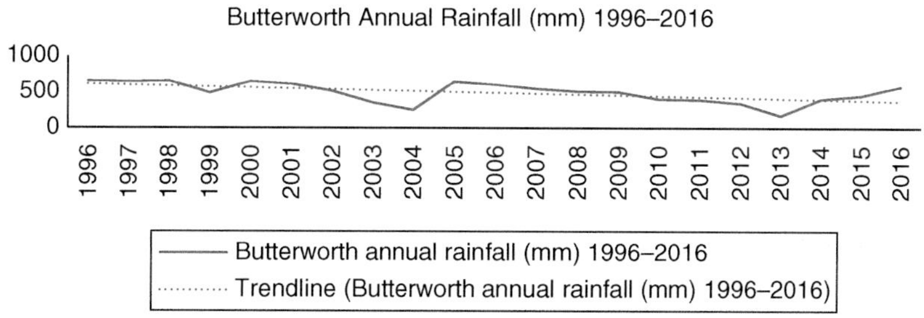

Fig. 12.5. Butterworth annual rainfall (mm) 1996–2016 (MAR 490 mm).

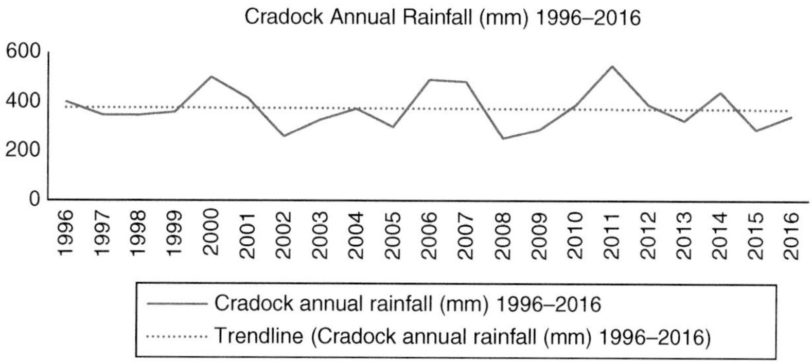

Fig. 12.6. Cradock annual rainfall (mm) 1996–2016 (MAR 375 mm).

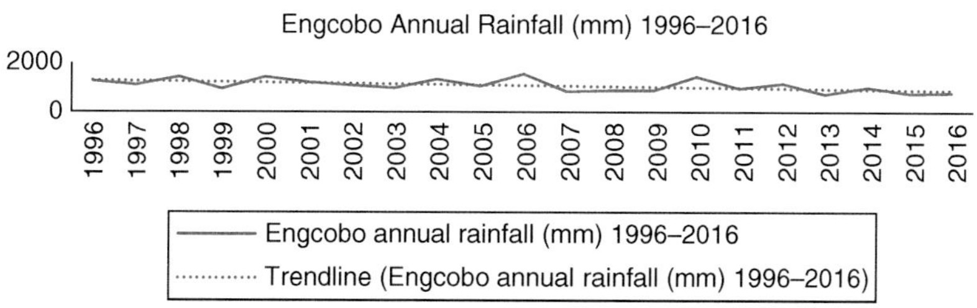

Fig. 12.7. Engcobo annual rainfall (mm) 1996–2016 (MAR 1092 mm).

farming and the use of mulches would be very important in conditions of such high variability.

Figure 12.8 shows that rainfall at Fort Beaufort is too low to consider rainfed cropping, although there has not been a decline in the annual rainfall or seasonal rainfall over the past 20 years; if anything, annual rainfall has increased slightly.

Figure 12.9 shows that rainfall for Graaff-Reinet is similar to that at Fort Beaufort, but at the 300 mm level: too low to consider rainfed cropping, although there has not been a decline in the annual rainfall or seasonal rainfall over the past 20 years; if anything, annual rainfall has increased slightly.

Figure 12.10 shows that rainfall at Grahamstown is also similar to Fort Beaufort, but here, at the 500 mm level, it is not too low to consider rainfed cropping, and there has not been a decline in the annual rainfall or seasonal rainfall over the past 20 years; if anything, annual rainfall has increased slightly. Only drought-resistant crops should be planted, with as much attention as possible to rainwater harvesting, soil moisture conservation, build up of SOM and where possible the use of mulch to reduce evaporation.

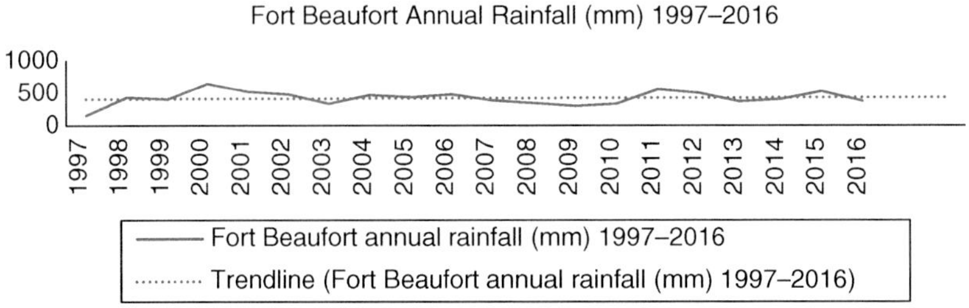

Fig. 12.8. Fort Beaufort annual rainfall (mm) 1997–2016 (MAR 401 mm).

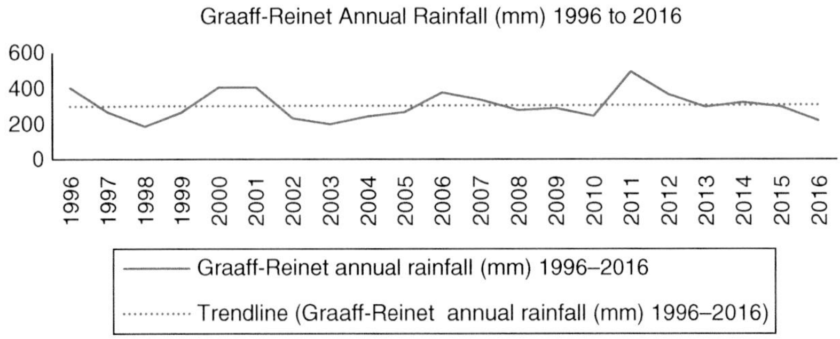

Fig. 12.9. Graaff-Reinet annual rainfall (mm) 1996–2016 (MAR 304 mm).

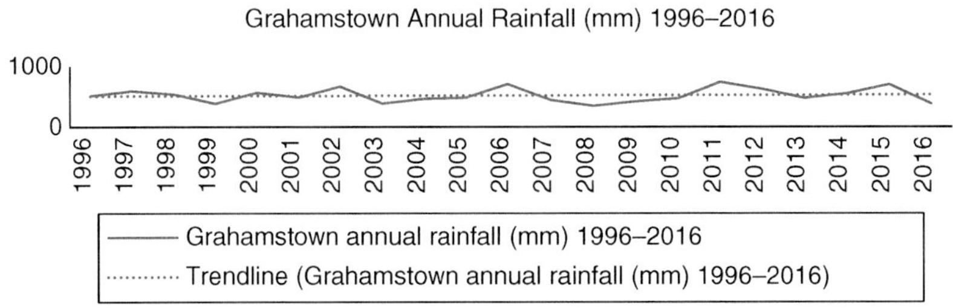

Fig. 12.10. Grahamstown annual rainfall (mm) 1996–2016 (MAR 521 mm).

Jamestown receives most of its seasonal rain during November, December and January, and rainfall in these months is highly variable. Over the past 20 years, rainfall has declined (both annual rainfall, depicted in Fig. 12.11, and seasonal rainfall). The decline in annual rainfall from over 500 mm to below 400 mm is already drastic and significant, and the average for the last 10 years is close to 400 mm. Only drought-resistant crops should be planted, with as much attention as possible to rainwater harvesting, soil moisture conservation, build up of SOM and where possible the use of mulch to reduce evaporation.

King WilliamsTown receives most of its seasonal rain during November, December and January, and rainfall in these months is highly variable; there is often rain at the end of the season, but this is also highly variable. Over the past 20 years, rainfall has declined slightly (both annual rainfall, depicted in Fig. 12.12, and seasonal rainfall). The decline in annual rainfall from over 600 mm to below 550 mm is not particularly significant, but the average for the last 10 years is closer to 500 mm than the traditional 600 mm mark! Careful rainfed cropping with water conservation practices should still be possible.

Like Engcobo, Maclear is a high rainfall area, as shown in Fig. 12.13, largely at the edge of the Eastern Cape Maloti Drakensberg Mountains. Although there has been a declining annual rainfall, from over 1000 mm to around 900 mm, rainfall is not the limiting factor in crop production, rather topography and cold winters are! Most of the rain falls between November and March (over 100 mm/month, though these months are highly variable). The spring months up to and including October, often experience less than 50 mm of rain. Where there is arable land, there is usually irrigation available from the many mountain streams, and many farmers have developed ways of coping with the extremes of climate.

Figure 12.14 shows that Patensie seems to buck the trend of declining rainfall, with annual rainfall having increased from around 400 mm at the turn of the millennium to closer to 500 mm over the past 10 years. Rainfall is very well spread out through the year, but it is rare for any month to receive more than 50 mm of rain; this means that the 7-month seasonal rain average is under 300 mm, and this area is high risk for rainfed cropping.

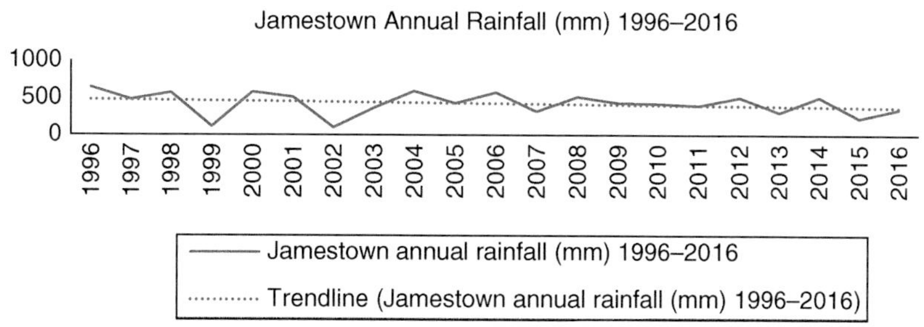

Fig. 12.11. Jamestown annual rainfall (mm) 1996–2016 (MAR 427 mm).

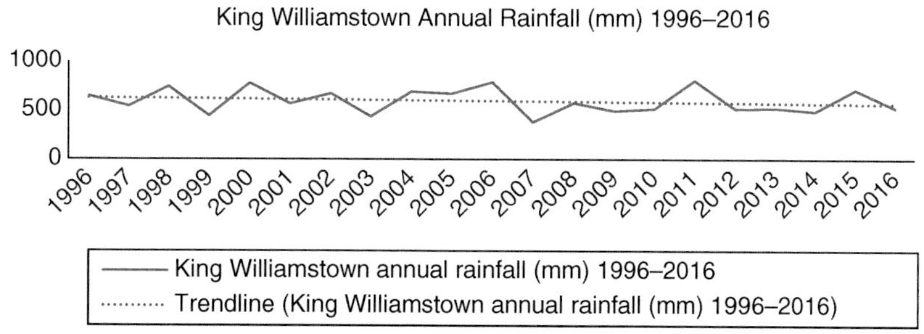

Fig. 12.12. King Williamstown annual rainfall (mm) 1996–2016 (MAR 598 mm).

Queenstown receives most of its seasonal rain during November, December, January, February and March and rainfall in these months is highly variable; there is often rain at the end of the season, but this is also highly variable. Over the past 20 years, rainfall has declined slightly (both annual rainfall, depicted in Fig. 12.15, and seasonal rainfall). The decline in annual rainfall from over 550 mm to about 500 mm is not particularly significant, but over the last 10 years rainfall has often been closer to 500 mm than 550 mm. Careful rainfed cropping with water conservation practices should still be possible.

For Somerset East, although the MAR at 347 mm, and the 7-month seasonal rainfall at 265 mm are too low for rainfed cropping, Fig. 12.16 indicates that the rainfall is less variable than many other areas. Nevertheless, growing rainfed crops without irrigation is not recommended for this area.

Data was not available for Springs for the years after 2010, but rainfall for the 7-month cropping season is only 306 mm, which is too

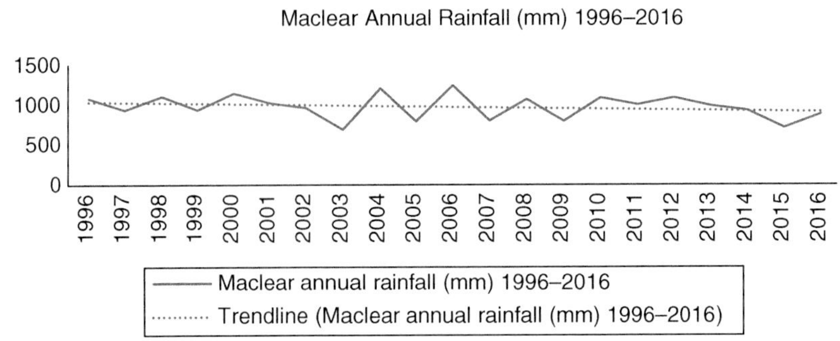

Fig. 12.13. Maclear annual rainfall (mm) 1996–2016 (MAR 975 mm).

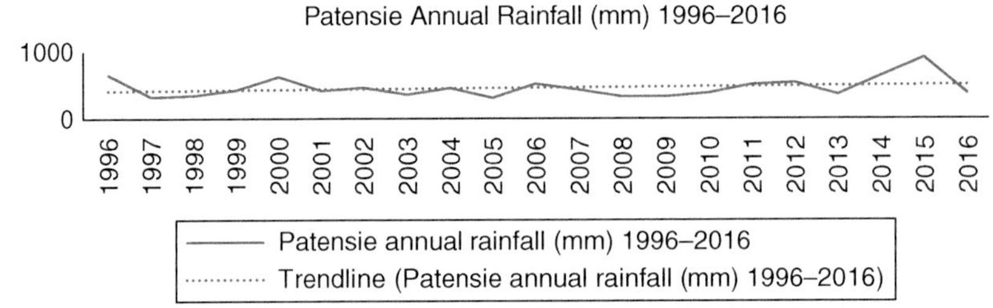

Fig. 12.14. Patensie annual rainfall (mm) 1996–2016 (MAR 461 mm).

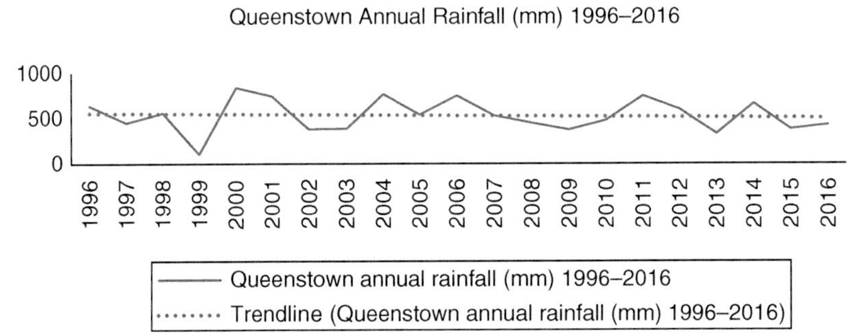

Fig. 12.15. Queenstown annual rainfall (mm) 1996–2016 (MAR 528 mm).

low for rainfed crops, and rainfall appeared to be declining (Fig. 12.17).

Again Stutterheim, like a few other Eastern Cape towns, seems to be bucking the general drying trend in rainfall, with averages having moved from around the 400 mm mark around 2000 to well above the 600 mm mark over the past 7 years (see Fig. 12.18). This is quite a significant shift, but time will tell whether the trend continues. Monthly rainfall is highly variable over the entire season, but with the 7-month seasonal rainfall average at 470 mm, rainfed cropping is a reasonable prospect. Due to the variability, however, water conserving and rainwater harvesting practices should be implemented.

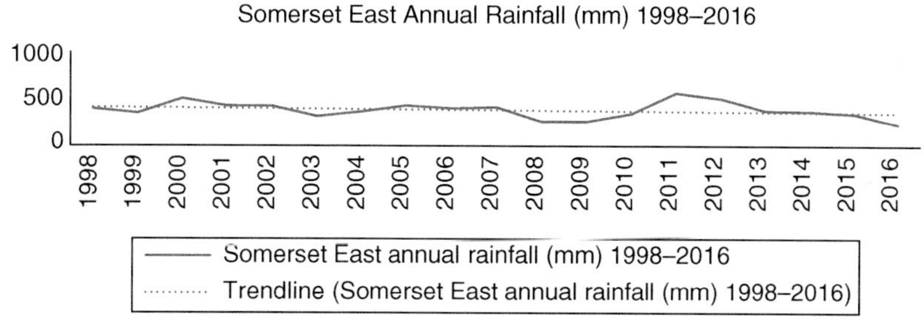

Fig. 12.16. Somerset East annual rainfall (mm) 1998–2016 (MAR 347 mm).

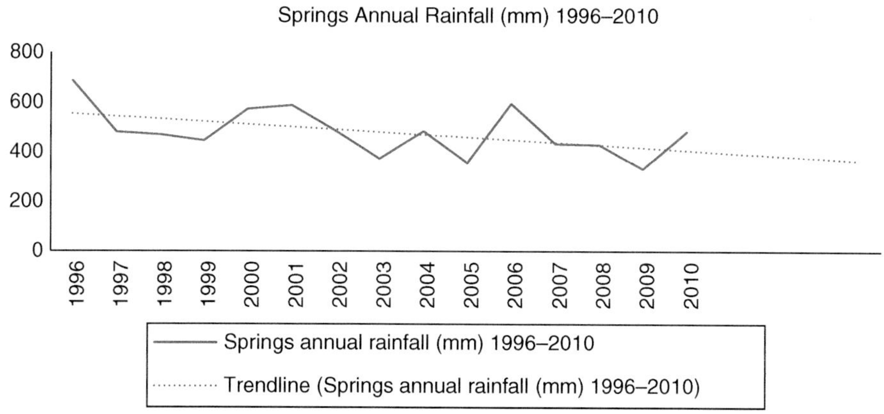

Fig. 12.17. Springs annual rainfall (mm) 1996–2010 (MAR 346 mm).

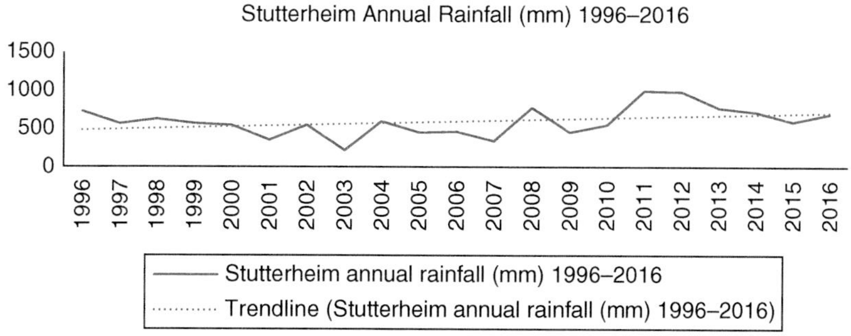

Fig. 12.18. Stutterheim annual rainfall (mm) 1996–2016 (MAR 596 mm).

With MAR of 442 mm, and a 7-month seasonal rainfall of 275 mm, rainfed cropping cannot be recommended for Uitenhage (see Fig. 12.19). The downward trend in annual rainfall (from above 500 mm to below 400 mm) confirms the riskiness of planting crops without irrigation.

Figure 12.20 shows that Umtata, like Barkley East, has declined in annual rainfall from a fairly safe seasonal rainfall to an increasingly marginal situation. Like Engcobo, climate change may well mean that this area may become marginal in the second half of this century. Although the MAR over the past 20 years is not as high as Engcobo (648 mm MAR, 522 mm seasonal rainfall), the trendline shows a decrease from > 700 mm to < 600 mm. Variability for the growing season is not as high as Engcobo, however, meaning that rainfed cropping is likely to remain viable and not as high risk as many other Eastern Cape towns. An area like this lends itself to water storage in dams and to rainwater harvesting and keyline ploughing to conserve moisture in whatever way is possible. Organic farming and the use of mulches would be very important in conditions of such high variability.

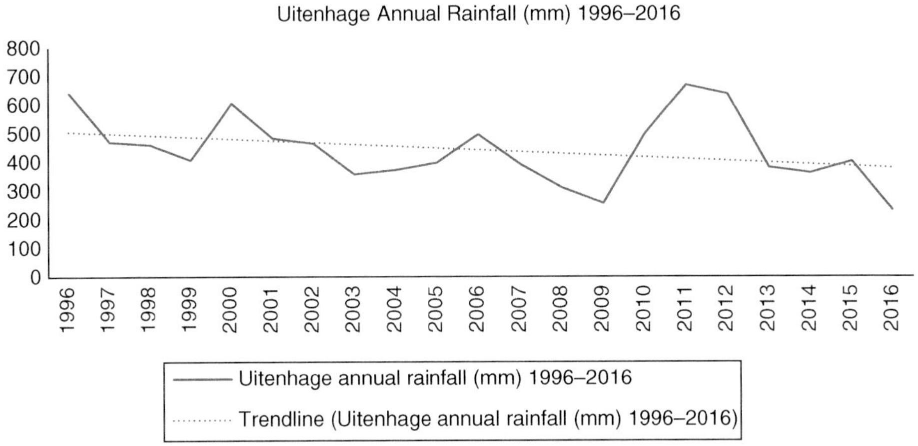

Fig. 12.19. Uitenhage annual rainfall (mm) 1996–2016 (MAR 442 mm).

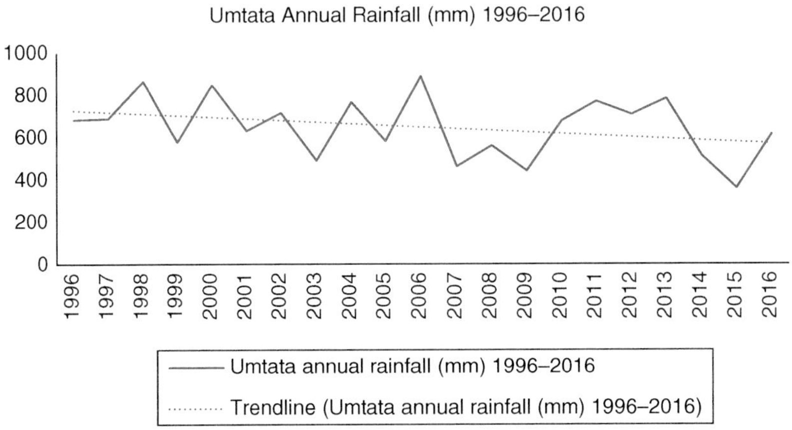

Fig. 12.20. Umtata annual rainfall (mm) 1996–2016 (MAR 648 mm).

What Can We Say About General Trends in Seasonal Rainfall?

The above individual data trend analyses show that the situation is complex, and that averages can be misleading. Each town has its own rainfall pattern, and the large variability makes prediction an uncertain process. Taking all those towns for which complete records exist from 1996 to 2016, a definite downward trend can be seen from the 1996 season to the 2016 season. Although no statistical inferences are justified from the data, Table 12.4 shows a consistent trend of decline in seasonal rainfall for all the towns with complete data sets, simply taking the first and last years in the series.

What Should Be Done Now?

Building resilience into smallholder farming will depend on a set of technical strategies and a set of mentorship strategies, each of which will need to be varied for the three recommendation domains.

Table 12.4. Change over time: difference in seasonal rainfall (total in millimetres for months September–March) between 1996 and 2016.

Location	Seasonal rainfall (mm) 1996	2016	Difference (mm)[a]
Barkley East	714	482	232
Bisho	656	344	312
Butterworth	599	423	176
Cradock	382	259	123
Engcobo	1183	717	466
Fort Beaufort	377	286	91
Graaff-Reinet	358	165	193
Grahamstown	438	240	198
Jamestown	567	215	352
King Williamstown	600	388	212
Maclear	1001	765	236
Patensie	592	299	293
Queenstown	604	316	288
Stutterheim	669	564	105
Uitenhage	598	133	465
Umtata	646	482	164
Average	624	380	244

[a]These values have been rounded to the nearest whole number.

The **technical strategies** will include improving WUE and on harvesting all available moisture carefully. Some techniques (e.g. conservation contours) are well known locally; others (e.g. swales, wetland construction, microclimate development) are relatively unknown. Important agronomic strategies will include organic soil management (increasing colloidal humus using compost and crop rotation, which will improve nutrient- and water-holding capacity) and the use of mulches (to cover the soil and reduce evaporation and soil erosion) (see Chapter 19, this volume). Important livestock management strategies will include holistic resource management approaches and other rotational grazing systems, and may also include 'cut and carry' systems. Local institution building is vital for the economic and knowledge economy development. It will be important to identify *at risk* communities, partly by mapping the relative risk of the various parts of the two districts.

The **mentorship strategies** will include an incentive scheme to encourage commercial farmers to come forward as paid mentors; they should receive adult education training and each be allocated no more than three emerging farmers to mentor on business planning, crop/animal production and marketing.

Emerging farmers should be selected from the existing farmers who are already producing some agricultural products, and who wish to become commercial farmers.

An in-service training programme should be developed, based on the establishment of local producer cooperatives working with the same crops or animals, and forming a study group.

Conclusion

Rainfall is declining in many parts of the Eastern Cape (as can be seen in Figs 12.3–12.20, this applies to ten out of the 18 towns for which data were available), and this will make rainfed agriculture more difficult and more risky in these areas. Of the remaining eight areas, six had stable rainfall patterns and two had seen rainfall increases over the past 20 years.

Strategies for assisting vulnerable households include technical innovations (moisture conservation, SOM management and mulching), as well as a mentorship programme by competent commercial farmers who have been trained in adult eduction.

Attending to capacity building and rural infrastructure development will also be important, as well as agricultural education (especially of young women from the rural areas).

References

AMCEN (African Ministerial Conference on the Environment) (2015) Factsheets 1 to 4: Climate Change in Africa – What Is At Stake? From IPCC, 2007. Available at: https://africacheck.org/factsheets/factsheet-why-do-many-scientists-think-africa-will-be-hardest-hit-by-climate-change/ (accessed 20 May 2017).

Auerbach, R. (1994) *Sustainable Development: Developing What To Sustain Whom?* New Ground, Johannesburg, South Africa.

Auerbach, R. (1995) People, farming and research: how can South African science contribute to sustainable agricultural development? *South African Journal of Science* 91, 3–6.

Auerbach, R. (1999) Design for participation in ecologically sound management of South Africa's Mlazi River catchment. PhD thesis, Agricultural University, Wageningen, the Netherlands.

Auerbach, R. (2013) Transforming African agriculture. In: Auerbach, R., Rundgren, G. and Scialabba, N. (eds) *Organic Agriculture: African Experiences in Resilience and Sustainability*. Food and Agricultural Organization of the United Nations (FAO), Rome, pp. 16–37. Available at: www.fao.org/docrep/018/i3294e/i3294e.pdf (accessed 30 October 2018).

Auerbach, R. (2017a) *Innovations in Smallholder Agriculture. Global Resilience to Climate Change*. ISASAR- IFAD (International Fund for Agricultural Development) Project, Nelson Mandela University, Port Elizabeth, South Africa.

Auerbach, R. (2017b) *IFAD – Climate Change Predictions for the Eastern Cape – Scenario Analysis*. ISASAR-IFAD (International Fund for Agricultural Development) Project, Nelson Mandela University, Port Elizabeth, South Africa.

Auerbach, R. (2017c) *Innovations in Smallholder Agriculture. Approaches and Measures to Building Climate Change Resilience*. ISASAR-IFAD (International Fund for Agricultural Development) Project, Nelson Mandela University, Port Elizabeth, South Africa.

Conway, G.R. (1994) Sustainability in agricultural development: trade-offs between productivity, stability and equitability. *Journal for Farming Systems Research-Extension* 4(2), 1–14.

Cox, L. (2018) Leaked UN draft report warns of urgent need to cut global warming. *The Guardian* 15 June. Available at: www.theguardian.com/environment/2018/jun/15/leaked-un-draft-report-warns-of-urgent-need-to-cut-global-warming (accessed 17 December 2018).

DAFF (Department of Agriculture, Forestry and Fisheries) (2013) *Economic Review of South African Agriculture*. DAFF, Pretoria, South Africa.

DAFF (Department of Agriculture, Forestry and Fisheries) (2017) Abstract of Agricultural Statistics. Available at: https://www.daff.gov.za/Daffweb3/Portals/0/Statistics%20and%20Economic%20Analysis/Statistical%20Information/Abstract%202018.pdf (accessed 1 September 2019).

De Wit, M.J. and Stankiewicz, J. (2006) Changes in surface water supply across Africa with predicted climate change. *Science* 311, 1917–1921. https://doi.org/10.1126/science.1119929

Elbehri, A. (ed.) (2015) *Climate Change and Food Systems: Global Assessments and Implications for Food Security and Trade*. Food and Agricultural Organization of the United Nations (FAO), Rome.

EPOPA (Export Promotion of Organic Products from Africa) (2008) *Organic Exports: a Way to a Better Life?* EPOPA, Swedish International Development Agency, Stockholm.

Folke, C., Carpenter, S., Walker, B., Scheffer, M., Elmqvist, T., Gunderson, L. and Holling, C.S. (2004) Regime shifts, resilience, and biodiversity in ecosystem management. *Annual Review of Ecology, Evolution, and Systematics* 35, 557–581.

Friends of the Earth Africa (2017) *Who will feed Africans? Small-scale farmers and agro-ecology, not corporations!* Friends of the Earth Africa. Available at: www.acbio.org.za/wp-content/uploads/2017/01/FoE-Africa-report.pdf (accessed 7 August 2018).

Gunderson, L.H. (2000) Ecological resilience: in theory and application. *Annual Review of Ecological Systems* 31, 425–439.

Gunderson, L.H. and Holling, C.S. (eds) (2002) *Panarchy: Understanding Transformations in Human and Natural Systems*. Island Press, Washington, DC.
Gunderson, L.H., Holling, C.S. and Light, S.S. (eds) (1995) *Barriers and Bridges to the Renewal of Ecosystems and Institutions*. Columbia University Press, New York.
Holling, C.S. (1973) Resilience and stability of ecological systems. *Annual Review of Ecological Systems* 4, 1–23.
Holling, C.S. (1995) What barriers? What bridges? In: Gunderson, L.H., Holling, C.S. and Light, S.S. (eds) *Barriers and Bridges to the Renewal of Ecosystems and Institutions*. Columbia University Press, New York, pp. 1–24.
IPCC (Intergovernmental Panel on Climate Change) (2014) AR5 Synthesis Report: Climate Change 2014. IPCC. Available at: www.ipcc.ch/report/ar5/syr/ (accessed 7 August 2018).
Mashele, N.J. and Auerbach, R.M.B. (2016) Evaluating crop yields, crop quality and soil fertility from organic and conventional farming systems in South Africa's southern Cape. *South African Journal of Geology* 119(1), 25–32.
Morton, J. (2007) The impact of climate change on smallholder and subsistence agriculture. *Proceedings of the National Academy of Sciences of the United States of America* 104(50), 19680–19685.
Nziguheba, G., Palm, C.A., Berhe, T., Denning, G., Dicko, A., Diouf, O., Diru, W., Flor, R., Frimpong, F., Harawa, R., Kaya, B., Manumbu, E., McArthur, J., Mutuo, P., Ndiaye, M., Niang, A., Nkhoma, P., Nyadzi, G., Sachs, J., Sullivan, C., Teklu, G., Tobe, L. and Sanchez, P.A. (2010) The African Green Revolution: results from the Millennium Villages Project. *Advances in Agronomy* 109, 76–115.
Pimentel, D. (2006) *Impacts of Organic Farming on the Efficiency of Energy Use in Agriculture*. A State of Science Review. The Organic Center, Washington, DC.
Ramaila, M., Mahlangu, S. and du Toit, D. (2011) *Agricultural Productivity in Africa: a Literature Review*. Department of Agriculture, Forestry and Fisheries, Pretoria.
StatsSA (Statistics South Africa) (2017) Mid-year Population Estimates: 2017 (Statistical Release P0302). Available at: https://www.statssa.gov.za/publications/P0302/P03022017.pdf (accessed 29 May 2019).
The Guardian (2018) Water shortages could affect 5bn people by 2015, UN report warns. *The Guardian* 19 March. Available at: www.theguardian.com/environment/2018/mar/19/water-shortages-could-affect-5bn-people-by-2050-un-report-warns (accessed 17 December 2018).
WWF (World Wide Fund for Nature) (2008) *Agriculture: Facts and Trends South Africa*. WWF South Africa. Available at: http://awsassets.wwf.org.za/downloads/facts_brochure_mockup_04_b.pdf (accessed 16 May 2019).

13 Water Efficiency, Energy Efficiency and Suburban Vegetable Production

Raymond Auerbach* and Anastasia Caude
Nelson Mandela University, George, South Africa

Abstract

Rainwater harvesting, solar water heating and organic vegetable production were used to improve efficiency in a suburban home in George, in South Africa's Southern Cape. In a year, treated municipal water consumption decreased from 17 kL/month to 8 kL/month, and 60 kL of rainwater was used for bathing, flushing toilets and watering the garden during this 12-month period. Electricity consumption was slightly decreased using LED lighting, solar water heating, a geyser blanket and energy-efficient appliances. Less power was used in winter for heating of water, but slightly more power was used in summer, as water had to be pumped. Food production from a 7 m^2 vegetable garden made use of recycled bathwater and provided most of the fresh produce for two people. Capital and installation costs for the entire system should be recouped in 7 years.

Introduction

This research project is a sub-project of the ecosystem-based solutions for resilient urban agriculture (ECOSOLA) research project, a joint initiative (funded mainly by the German government) of three partner universities: (i) Oldenburg University (Germany); (ii) University of Dar es Salaam (Tanzania); and (iii) South Africa's (SA) Nelson Mandela University. The Tanzanians (Professor Pius Yanda) are looking at urban development and urban gardens around Dar es Salaam; the Germans (Professors Bernd Siebenhuener and Michael Kleyer) are looking at the impact of climate change on urban food security, and the South Africans (my research team) are working with water-efficient organic peri-urban food production. A sub-project (funded privately) examines efficiency in water, energy and food production for a middle-class home in suburban George (Western Cape, SA). It looked at typical water and energy use for a middle-class home (Figs 13.1 and 13.2), and designed a simple rainwater harvesting and energy-saving system, and will continue to monitor savings in water and electricity, as well as the efficiency of the new organic food production system.

We (Raymond and Christina Auerbach) arrived in the Southern Cape in October 2010, after a period of severe drought, where the Garden Route Dam was almost empty, and draconian water restrictions had been implemented. At the time, I warned the George Educational Summit (2010) about the likelihood of future droughts, as the 7-year drought cycle is fairly predictable, but people are predictable too, and

*raymond.auerbach@mandela.ac.za

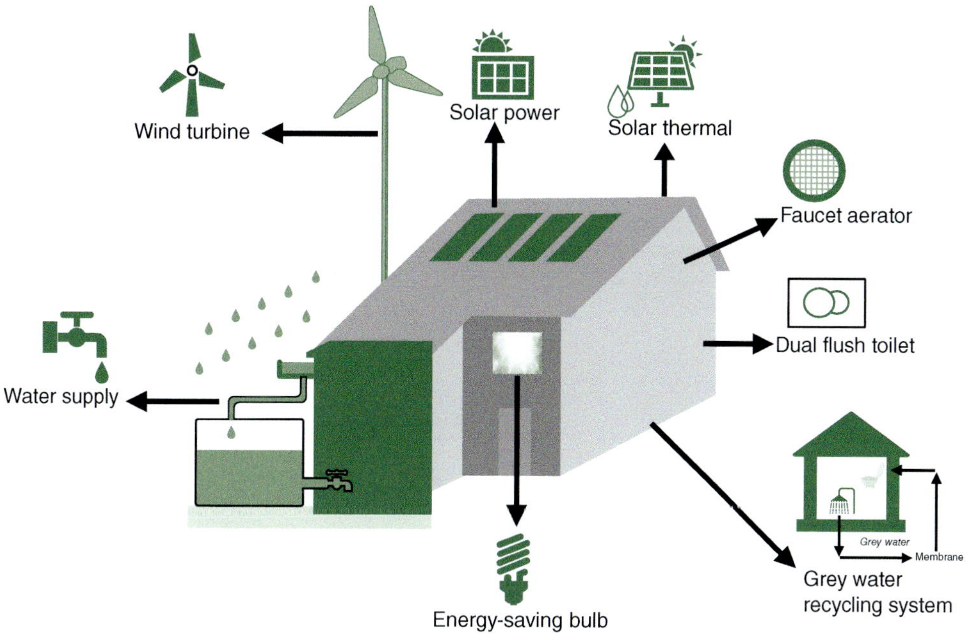

Fig. 13.1. Examples of domestic water and energy use. (From Caude, 2017.)

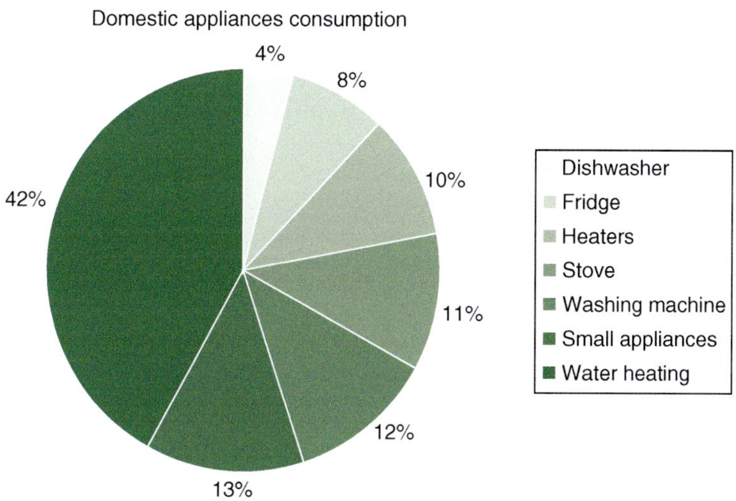

Fig. 13.2. Domestic appliances consumption. (From Caude, 2017.)

forget about drought in the wetter years. I warned that we should not be surprised when we find ourselves faced by a drought in 2017, and each cycle is likely to be a little worse than the one before, given climate change and the increased demands on our limited water resources, which make SA's urban and rural water management vulnerable.

When we bought the house in 2015, we studied efficient domestic energy and water use with the help of Anastasia Caude, a visiting French student (2017 and Table 13.1), and then developed a model to examine water, energy and food as part of the ECOSOLA sustainable urban development project. I developed a rainwater-harvesting system, made minor improvements

Table 13.1. Energy and water-saving measures at the Auerbach home.

Year/month	Implementation	Effect
2014 – pre-Auerbach	Two 2.5 kL rainwater tanks (T3, T4)	Catch rainwater – back yard
2014 – October	Induction stove plate	Low energy cooking
2014 – December	Repair small electric pump (P3)	Water pot plants
2015 – October	Large electric pump (0.5 Kw, P2)	Capacity to redirect water
2016 – January	Two tanks (T3 and T4) inter-connected	Rainwater to top up pool
2016 – January	Geyser blanket and LED lights outside	Save energy
2016 – February	15 LED bulbs	Save energy
2016 – June	Ten LED bulbs	Save energy
2016 – December	Ten LED bulbs	Save energy
2017 – May	Ten LED bulbs	Save energy
2017 – May	One 2500 L rainwater tank (T2)	Catch rainwater – front yard
2017 – June	Sprinklers	Irrigating with bathwater
2017 – August	Solar water heater (G2) and solar-recharged pump (P1)	Rainwater and solar bath
2017 – September	Ten LED bulbs	Save energy
2017 – September	Vegetable garden set up with irrigation	Produce healthy organic food
2017 – October	Tank T1 (500 L) front garden	Catch water off front roof

on energy use efficiency, and then invested in domestic energy efficiency with a solar water heater, as heating water is one of the major uses of electricity in a house (Caude, 2017 and Fig. 13.2). Moreover, in order to contribute to our household food security as urban gardeners, we designed a water-efficient urban food garden to be irrigated from the rainwater harvesting system.

Designing tomorrow's sustainable homestead is important given the future climate and population challenges. To do so, reducing water consumption and protecting water quality are key objectives in sustainable urban design. I had worked for many years on rainwater harvesting (Auerbach, 1999, 2003). The International Water Management Institution defines rainwater harvesting as:

> the collection and/or concentration of runoff water for productive purposes. It includes all methods of concentrating, diverting, collecting, storing, utilising and managing runoff for productive uses. Water can be collected from natural drainage lines, ground surfaces, roofs for domestic uses, stock and crop watering.
>
> (IWMI, 2003)

It facilitates reduction in the use of municipal water, and thus reduces potable water demand, but the rainwater is usually used for toilet flushing, baths/showers or garden irrigation. With a modest financial investment, it is possible to reduce water and energy use substantially, while also producing significant amounts of healthy organic food; this also provides enjoyable exercise without the cost of joining a gym!

Let us take a look at the history of implementation of the domestic model at the Auerbach's house in George, Western Cape (see Table 13.1 and Fig. 13.3).

The installation took place gradually, and was cost-efficient, waiting for special offers on equipment, and using opportunistically available labour.

Evolution of the Water, Energy and Food System

The water, energy and food system evolved over time as follows (see Fig. 13.3):

- The Auerbachs first inherited two tanks (T3 and T4) and a pump with the property in June 2014, but only in December 2014 did they repair the small electric pump (P3) to water the garden, and connected the two tanks so that they could top up the swimming pool with tank water.
- Then, in October 2015, they installed another pump (P2) and in May 2017, a third tank (T2), which was connected to the other two, giving a storage capacity of 7.5 kL.

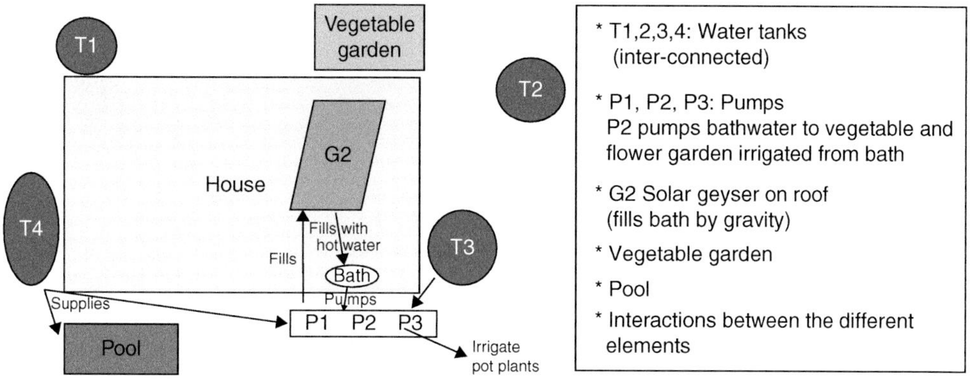

Fig. 13.3. Schematic layout of the water, electricity and food system at the Auerbach house.

The second pump was used to irrigate the garden by recycling bathwater; there was an electric geyser in the roof (G1, not shown in Fig. 13.3).

- In August 2017, a Kwikot 110 L solar water heater (G2) was installed on the roof above the bath, using rainwater from the rainwater tank (T4), and filled by a small battery-operated solar-recharged pump (P1); this system now supplies the bath, the toilet in the main bathroom (by means of buckets) and the entire garden with water from rainwater tank T4, the lowest tank in the series of four. So water for five baths a week (5 × 200 L) is saved, making 4 kL/month less treated municipal water used for bathing, and this rainwater is also recycled to irrigate both the vegetable garden and the ornamental flower garden, which is mainly planted with indigenous plants.
- The new vegetable garden was set up on the front lawn in September 2017, with 7 m^2 of raised beds, irrigated by a micro-jet system using the recycled bathwater, which is also saved in buckets for flushing the toilet.
- The vegetable garden is protected from birds by a removable net, and is fertilized from recycled garden waste, composted in two compost bins placed in corners of the garden, a small bucket worm farm, and two bokashi buckets which process cooked food (using this in the compost bins would attract rats; those that come anyway are recycled by a very efficient non-toxic cat).
- The introduction of LED bulbs has led to a slightly lower electricity consumption during spring, summer and autumn, as has the installation of the geyser blanket on the electric (150 L) geyser (G1).
- Filling the pool using rainwater should save 0.5 kL/month.
- The introduction of garden irrigation using bathwater should reduce water use by 1 kL/month; some energy is used for pumping the water.
- The use of rainwater for bathing should save a further 4 kL/month, and heating this water with a solar water heater should save about 20% of electricity use (half of the consumption shown in Fig. 13.2) per month.

What Change Can We See in Electricity Consumption?

Electricity is purchased for the pre-paid meter at R1000 per purchase; initially, this represented 641 kWh, but from July 2015, R1000 purchased only 546 kWh, and from July 2017, only 537 kWh. As the electric geyser still functions, in order to supply the other bathrooms and the kitchen, savings from the solar rainwater heating system are not as great as they would be if the solar geyser had replaced the electric geyser. Nevertheless, the downward trend in electricity consumption is seen in Fig. 13.4.

Compared year by year, changes in average monthly electricity consumption can be seen in

Fig. 13.5; from household use of 988 kWh/month in the year 2014/2015, the household dropped to an average of 856 kWh/month for the next 2 years, due to the more efficient lighting and the geyser blanket, and in spite of some electricity being used for the water pumps. After installation of the solar geyser in August 2017, and the development of the vegetable garden the next month, consumption dropped further to 760 kWh.

Figure 13.6 shows the seasonal electricity consumption pattern of the Auerbach household in 2014/2015, showing that more energy is used in winter (more heating and cooking), and less in summer. Potential for electricity saving is higher during winter. As a result, the system helps saving energy as far as the weather permits it. This is why people should think of insulating their house in a better way so that in winter electricity consumption does not raise so high.

On the other hand, in summer, electricity consumption is much lower which almost offsets winter's high consumption: the average of winter and summer is 34 kWh/day, similar to spring and autumn averages. The pattern of electricity consumption changed significantly after installation of the solar geyser and irrigation system,

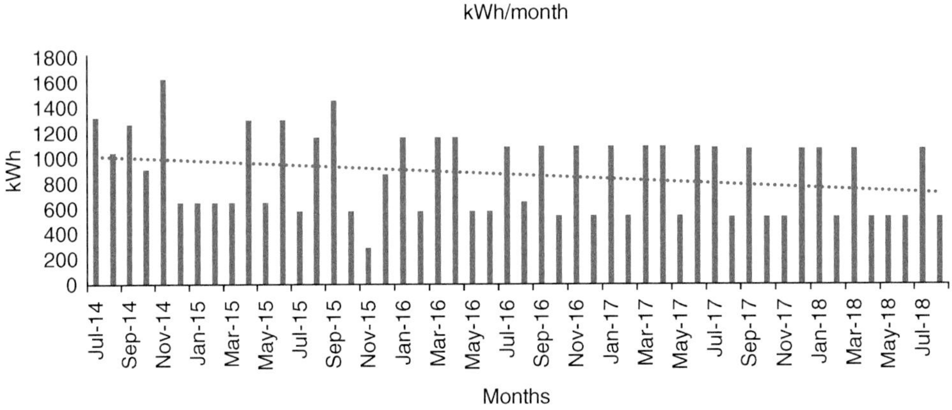

Fig. 13.4. Electricity usage per month – 2014–2018.

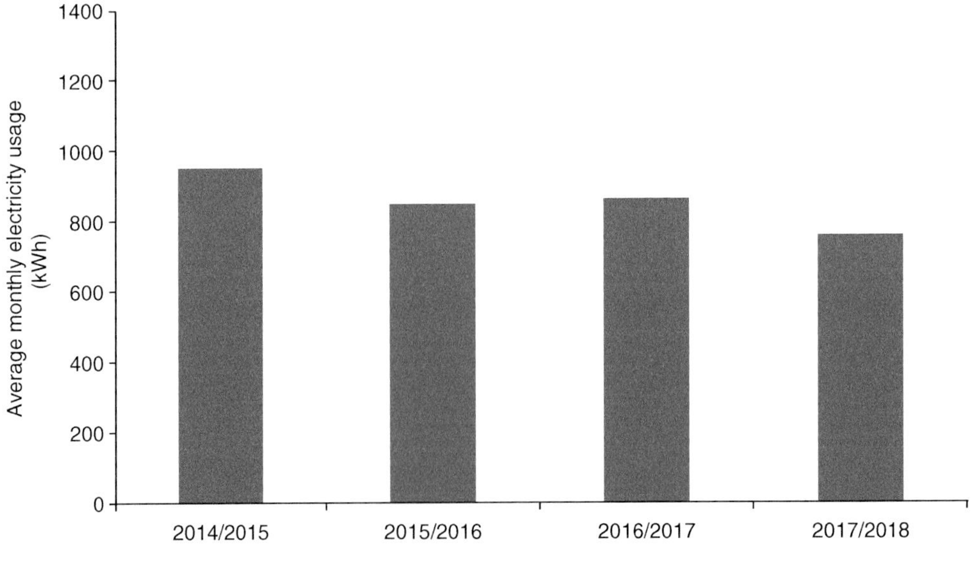

Fig. 13.5. Average monthly electricity usage – 2014–2018.

as can be seen in Fig. 13.7. Comparing average seasonal consumption in 2014/2015 with 2017/2018, we see two minor and two major changes:

- The daily consumption for spring 2017/2018 dropped from 35 kWh to 24 kWh.
- Autumn daily consumption dropped from 38 kWh to 24 kWh.
- Summer consumption rose from 24 kWh to 30 kWh, probably due to more water pumping.
- Winter consumption, on the other hand, fell from 44 kWh to 24 kWh, with the lower water heating requirements. This more even pattern of consumption might be welcomed by power utilities, as they are often hard-pressed to meet the higher winter demand for power.

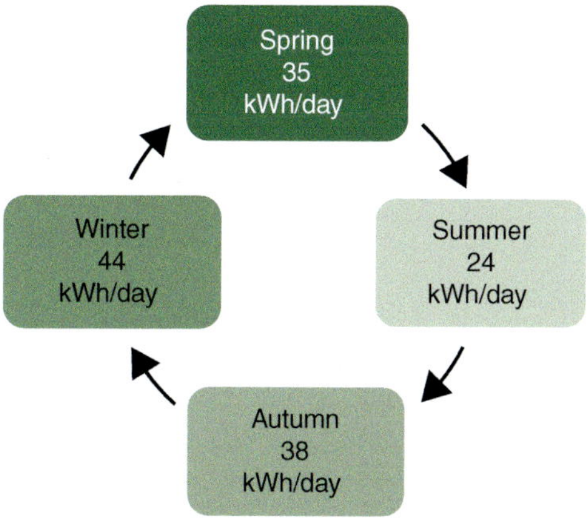

Fig. 13.6. Average electricity consumption each season (kWh/day) in 2014/2015.

Figure 13.2 shows the proportion of energy used by appliances in the household; total appliance use represents about 88% of total household energy use, with the other 12% being used by lighting in a conventional home. Given an average monthly electricity use of 900 kWh, this translates into about 108 kWh being used for electricity monthly. It is estimated that LED bulbs use about 25% of the power used by incandescent bulbs. If all the bulbs had been replaced, this would have saved about 80 kWh/month, but as only about 60 LED bulbs were replaced in the house out of a total of about 120, only half of this saving has been effected, so about 40 kWh/month due to more efficient lighting. The drop from 988 kWh to 856 kWh from the first year of measurement (2014/2015) compared with the next 2 years was thus probably about 40 kWh of savings for efficient lighting, and about 20 kWh saved in water heating due to the geyser blanket. The efficient Bosch dishwasher probably saved about 20 kWh and 240 L/month (using a third of normal electricity, and less than a third of the water). The balance of 52 kWh was probably due to the induction cooker, which uses only about 0.12 kWh to boil a litre of water, compared with an electric stove which uses about 0.17 kWh, and the more efficient refrigerator.

The further drop to 760 kWh/month for the year 2017/2018, can be attributed almost entirely to the solar water heating facility. At R1.86/kWh, this represents a saving of nearly R200/month in electricity costs. The capital cost of R8500 for installing the solar water heater is thus paid off with electricity savings in 46 months.

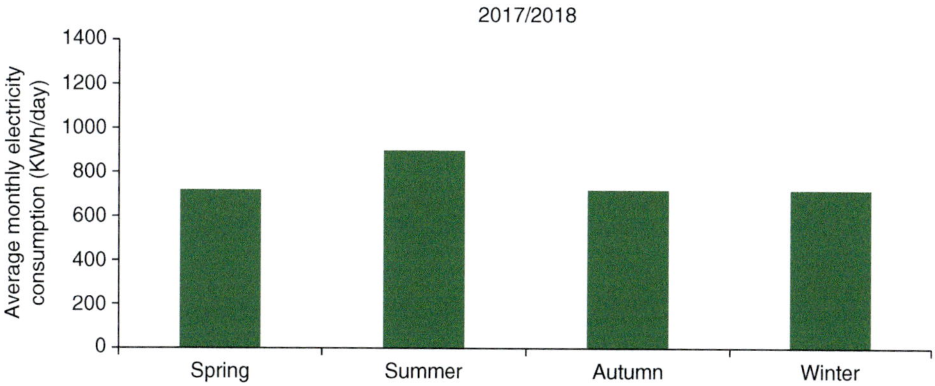

Fig. 13.7. Average monthly seasonal electricity consumption (kWh/day) for 2017/2018.

What About Water Consumption and Water Savings?

As seen in Fig. 13.8, treated water consumption had already decreased slightly since 2015, particularly in 2016. The average monthly consumption for 2014/2015 was 17.5 kL, for 2015/2016 was 16.8 kL and for 2016/2017 was 17 kL/month.

The Auerbach household also does the washing for a guesthouse which we run, and I like to bath when I come home from work; this means that unless there are dramatic lifestyle changes (doing the washing at the guesthouse, having fewer baths, or shallower baths, at least when water is short), total water consumption will remain fairly high. The seasonal water use before water harvesting is shown in Fig. 13.9. The rainwater harvesting system helps to make this lifestyle a little more responsible. Bathing in solar-heated rainwater reduced monthly water use from 17 kL/month to 8 kL/month; this saving was in spite of the fact that the new vegetable garden was watered three times a week, as recycled bathwater was used for this purpose.

The planning was based on the following calculations. Potentially, given an initial rainwater harvesting roof area of about 60 m^2, each 10 mm rainfall event should generate 60 $m^2 \times 0.01$ m = 0.6 kL. The total roof area is about 300 m^2 so there was great potential for harvesting more rainwater. By installing extra tanks and realigning gutters, we now harvest about 200 m^2 of roof area, so that a rainfall event of 10 mm generates 2000 L of rainwater to store. Since our total storage capacity is 8000 L, assuming tanks are empty, 40 mm of rain is enough to fill all tanks.

Luckily, George receives rainfall all year round; the 100-year average for George is 863 mm/year. Monthly rainfall varies from 50 mm during the winter months to 100 mm rain during summer months, so we are safe in assuming that with the 8 kL of water storage in the tanks, and lower use of water during winter, the system will have available on average 5 kL/month during winter and 6 kL/month during summer, even if rainfall in a drought year is zero for 3 months and 50% for another 3 months over any 6-month period. Of this, assuming 20 rainwater baths per month, and assuming that in winter shallow baths of 75 L use 1.5 kL/month and in summer fuller baths of 200 L use 4 kL/month, there should be a surplus of water for other uses of about 2 kL/month throughout the year.

Given the water restrictions imposed due to the drought, consumption had to decrease to below 15 kL/month. The system was installed at the end of August 2017. During the winter months of 2017, consumption averaged 13 kL/month, but the guesthouse was fairly quiet out of season. The busy season for the guesthouse

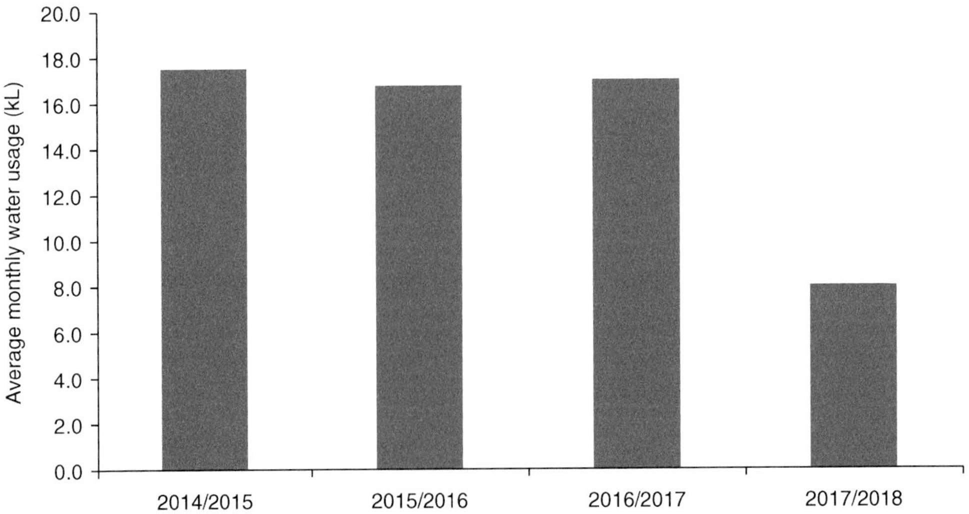

Fig. 13.8. Average monthly water usage – 2014–2018.

runs from September to April, and during those 8 months, an average of about 40 loads of washing must be processed per month, requiring about 25 L per load (1 kL) for the water and energy efficient Bosch Avantix 8 washing machine.

The actual quantity of rainwater used in the year from 1 September 2017 to 31 August 2018 was 44,331 L pumped up to the solar geyser, and 15,500 L used to top up the pool with rainwater in this time period. The total amount of rainwater used in the year was thus 60 kL, reducing the strain on the overstressed Garden Route Dam. If this saving was multiplied by several thousand potential rainwater harvesters, the reduction in water demand during dry years would be considerable.

How Much Water and Time Does a Food Garden Need?

Conventionally grown vegetables need 20–100 mm of water/week, depending on what crops are planted, and on the season (combining evaporation and transpiration). For a food garden, measuring 4 m × 2.5 m (or 10 m^2), depending on what is grown, between 200 L/week and 1000 L/week of water is needed, so 1 kL/month in winter, and 4 kL/month in summer. A well-developed organic soil with high levels of organic matter and mulch to reduce evaporation, uses about half the water of the above conventional example. This means that in the Auerbach's house it is possible to irrigate vegetables from the rainwater harvested off the roof if 0.4 kL is available during the winter and 2 kL is available over the summer months. Growing vegetables is a great resource of food for people and reusing grey water (in period of drought or not) can save a lot of water and money.

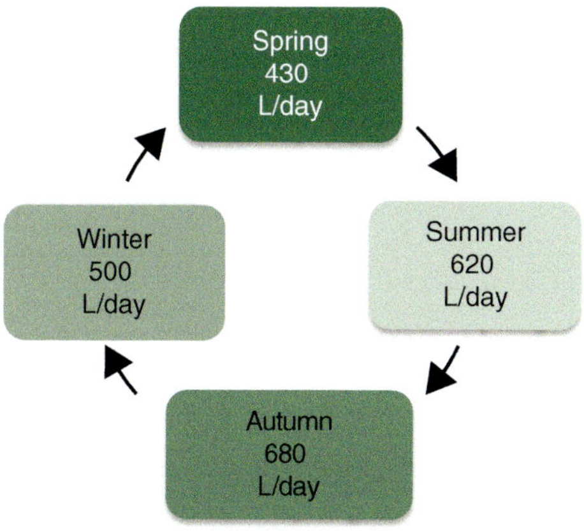

Fig. 13.9. Average water consumption each season (L/day) from 2014 to 2017.

To stay on the organic side, you can also make your own compost easily by using fruit scraps, vegetable scraps, eggshells and coffee grounds using a simple compost bin.

Although gardening is a good form of exercise, it is time-consuming. Growing food is not only knowing how to grow vegetables, but also being organized, a good manager and able to access agricultural inputs cost-effectively. Growing crops makes you grow as a person, and become more responsible and aware of the environment and the world that you will leave to the future generations. The personal health benefits of producing and consuming fresh, healthy food, make the effort worthwhile for us.

Discussion

The major advantage of this system is that a great deal of electricity is used to heat water, and a lot of water is used in bathing. With the solar water heating, much of the electricity used domestically is saved, and the water used to bath is rainwater, so there is no use of expensive, scarce, treated municipal water for this purpose, or for watering the garden. After using the rainwater for a bath, if you do not use soap you can use it to irrigate your food and/or flower garden. The result of this is that you conserve important resources, improve your health and you even save money. In winter, heating the water requires at least 3 h of sunshine, and the water is often not as hot, meaning that only hot water direct from the solar geyser is used to fill the bath, whereas in summer, cold rainwater is pumped through the system to cool and fill the bath (the Kwikot geyser heats the water to above 80°C).

A management requirement is that somebody who understands the system needs to spend 10 min each morning checking the watering of the plants and refilling the solar geyser; this is not onerous, given the semi-automatic watering system, and the proximity of the

Table 13.2. Installation costs of the water and energy efficiency system used in the Auerbach house.

Elements of the system	Cost (R)
Tanks (2500 L × 3 plus a 500 L tank)	10,500
Small water pump	1,800
Main water pump	2,500
Direct current pump	2,000
Pipes and fittings	2,200
Solar water heater	8,500
Constructing vegetable garden and irrigation	4,500
Realigning gutters, fitting solar geyser	5,000
Total	37,000

pumps to the main bedroom. Design of such a system should take into account the convenience of operation in placing elements of the system, and the availability of a practically minded person for maintaining the system. The process of filling the bath is slower than normal, prolonging the time from about 10 min to slightly more than 20 min to draw a full (200 L) bath.

The costs of the system (tanks, pumps and domestic solar-power water heater) require an investment; approximate 2018 costs in SA Rands (R14 = US$1) are as shown in Table 13.2.

As mentioned, the solar geyser is the most cost-effective element of the system, and pays for itself within 4 years. Total savings are estimated at R600/month, so that given a modest maintenance cost of R100/month, and depending on water and vegetable costs, the entire system requires about 7–10 years to pay for itself.

Given the wide distribution of monthly rainfall in the Southern Cape, this area is eminently suited to a system such as described in this chapter. An area with more pronounced seasonal dry periods would require more rainwater storage capacity to achieve similar efficiency.

References

Auerbach, R. (1999) Design for participation in ecologically sound management of South Africa's Mlazi River catchment. PhD thesis, Agricultural University, Wageningen, the Netherlands.

Auerbach, R. (2003) *Rainwater Harvesting – a Literature Survey*. Report for the South African Water Research Commission. Rainman Landcare Foundation, Durban, South Africa.

Caude, A. (2017) George Water and Energy Use. Project report for Institut Polytechnique LaSalle Beauvais ESITPA, Rouen, France.

IWMI (International Water Management Institute) (2003) Multiple Use Water Systems (MUS). Action research project 'Models for implementing multiple-use water supply systems for enhanced land and water productivity, rural livelihoods and gender equity'. Information Worksheet, IWMI, Colombo.

14 Experiential Training of Farmers and University Diploma Students in KwaZulu-Natal and the Southern Cape

Raymond Auerbach*
Nelson Mandela University, George, South Africa

Abstract
This practitioner reflection reports on 30 years of training agroecological farmers in various African contexts. It argues that good theory should be drawn out of good practice, to help farmers to adapt to climate change while producing nourishing food sustainably. Currently, training is often provided by agribusinesses with an interest in promoting the use of their inputs, rather than concentrating on empowering farmers to develop environmentally sound farming systems using locally available resources. Three organic training systems are examined. On-farm systems exposed trainees to farm management and marketing, but not in their familiar context. Training on-site at community gardens was effective only when there was good mentorship and project support. In a university diploma context, 18 months of theory with regular practical activities prepared students for a year of on-farm practical learning. This was supported by guided reflection, and followed by 6 months back in the classroom, integrating theory and practice. In all three systems, learners were challenged with practical activities, after which theory was developed. Organic systems were found to help learners to use locally available resources, especially water, more efficiently. Exposure to good practice, and guided reflection on this, helped learner farmers to understand and integrate good theory into their practice, while practical challenges helped learners to understand what theory means, and how it should be adapted to the local context.

Introduction

> Trainee: I have attended an organic farming course.
> Farmer: So, can you feed yourself?
> Policy maker: If not, why not, and can it feed the world?
> Economist: But do you have a market? Is it viable?
> Politician: This is an expensive investment; what spin will it give me?

How can the interests of the multiple stakeholders above be met in farmer support programmes, given climate change, food insecurity, poor infrastructure and declining food quality? This chapter will argue that farmer training should assist farmers to understand agroecological principles through guided practical experiences. By drawing good theory out of good practice, farmers adapt to the challenges of climate change while producing nourishing food sustainably. This chapter is a practitioner reflection, which applies the above process by showing how three training approaches each yielded insights which informed conceptual shifts strengthening participation and community development. The gap between the promise and the reality of

*raymond.auerbach@mandela.ac.za

participation and community development has been well documented (Carpenter *et al.*, 2016); closing this gap requires structured experiential learning processes.

Addressing food insecurity in times of climate change involves multiple stakeholders. The technical goal of learning about sustainable soil management is at the heart of farmer training, along with plant and animal production skills. These are the tools of the trade of productive farmers the world over. Policy makers are looking for interventions which are 'game changers', where there will be maximum impact on regional food security. Economists tend to be sceptical about unproven interventions unless there is evidence of a growing market, and of profits to be made from products for which there is a demand. Unless political support is forthcoming, local initiatives are unlikely to receive government funding, and may instead attract opposition (UN, 2015). In particular Sustainable Development Goals 2 (zero hunger), 12 (responsible consumption and production), 13 (climate action) and 15 (life on land) require agroecological approaches (UN, 2015). Farmer training therefore has to equip learners to understand the valid concerns of all of these stakeholders.

Climate change imposes a number of new constraints on agricultural production, in terms of adaptation strategies (Elbehri, 2015). Already economically and/or socially vulnerable communities have a new set of challenges to deal with: (i) flooding; (ii) higher temperatures; and (iii) drier crop production conditions (Ayers and Forsyth, 2014). The role of carbon in soil and the atmosphere, decreasing availability of water, dangers of environmental pollution by agrochemicals and fertilizers, the need for conservation of non-solar energy, difficulties of producing food in a warmer environment, and with less stable rainfall patterns, political uncertainties brought about by increasing numbers of climate refugees and the social instability associated with the rising price of food, are all issues which modern farmers will have to deal with (IPCC, 2014; Chapter 7, this volume).

Farmer training takes many forms. Short courses provide detailed technical information about innovations. Government extension services provide some guidance to small-scale farmers. In South Africa (SA), extension study groups often involve large-scale commercial farmers in comparative learning with peer groups of similar progressive farmers; this allows comparative economic analyses so farmers can assess their own performance against that of comparable nearby farms. For small-scale farmers, combinations of short learning programmes, farmers' days, study groups and longer in-depth courses have been found effective, in combination with problem-solving research.

Much of the training on offer is currently provided by agribusinesses, as they have an interest in promoting the use of agricultural inputs such as fertilizers, agrochemicals, seeds, tractors and other farm inputs. Often, these sectoral interests promote environmentally dangerous short-term technologies. Policy makers should be promoting sustainable approaches which do not destroy the resource base from which they emerge. In the context of climate change, carbon needs to be sequestered in the soil as soil organic matter (SOM), rather than contributing to global warming in the form of methane (CH_4) or carbon dioxide (CO_2). Profitable organic farming systems which use water efficiently, recycle plant nutrients, sequester carbon in the soil, use energy efficiently and promote biodiversity while being good for the environment do not often use high levels of agricultural inputs, and therefore lack investor support. However, if it can be shown that these organic systems promote food security, there is political gain in promoting techniques which are environmentally sound, which relieve poverty and which promote health through reduction in poison use and improved nutrition.

Finding effective ways to train farmers so that they gain practical skills, find viable markets, are able to scale up their impact and which are popular with the electorate will be a major part of bringing about sustainable development interventions in the next decade. This requirement then begs the question: How can aspiring farmers gain farming and marketing experience, so that they draw out of this experience ecologically sound production approaches which attract young farmers and farm employees, and which are economically profitable?

The aim of this chapter is to report on practical approaches gleaned from the past 30 years, during which I have developed learning materials for agroecology in various contexts.

As Research Co-ordinator of the Farmer Support Group at the then University of Natal, I was a pioneer of participatory rural appraisal (PRA) in 1991, compiling the first draft of our early PRA manual (Participants, 1993). My work on integrated catchment management led to publication of my doctoral research as a book (Auerbach, 1999). As founder director of the Rainman Landcare Foundation, I developed farmer training curricula in the first decade of this millennium. For 7 years, Rainman trained farmers, sometimes on our farm near Durban, and sometimes *in situ* on various community farms, mostly in KwaZulu-Natal (KZN). We then shared our research and training materials with international organic trainers (see www.ifoam.bio and www.fibl.org). On joining the Nelson Mandela University in 2010, I took on the management of the Experiential Learning programme at the George Campus.

Considering the various approaches available, what has been learned is that good theory should be drawn out of good practice. Successful farmers are right by virtue of their practical success, whether theory agrees or not. Usually, the best way to help learners to explore best practice is to give a few exploratory guidelines, to set up practical learning situations where learners have to find ways of solving problems, and then to assist them with structured group reflection, where some of the learnings from the activities can be recognized and explored (Röling, 1988). This can be followed with the extraction of theoretical principles from practice, and comparisons with the experiences of various case studies. Professor Richard Bawden was able to contribute to the transformation of Hawkesbury Agricultural College in Sydney, Australia in the 1980s, by introducing a business project at the start of first year, compulsory for all students (Bawden, 1992). This allowed students to explore business and production alongside their more formal agricultural studies.

This chapter reports on three approaches to integrated learning for trainee farmers and university diploma students. The various approaches to combining theory and practice in farmer training are evaluated, and the experiences interpreted in the light of adult learning and agricultural extension theory, and throw light on the challenges which climate change will present to farmers.

Theoretical Perspective

Lessons gleaned from experiences in a variety of contexts can contribute to more effective farmer training for agroecology in Africa. Adult learning theory and various theories of extension will be discussed before the introduction of the three case studies. During a sabbatical with us, Bawden helped the Farmer Support Group to incorporate the idea of '**praxis**' into our work with small-scale farmers and showed how experiential learning can link the process of finding out to taking action: what has been found out affects what will be done, and what has been done will affect future ways of finding out (Bawden, 1992; Auerbach, 1994). Praxis is based on Aristotle's idea of the three realms of human activity: *theoria*, *poiesis* and *praxis* (Arendt, 1998); these correspond to theory, production and action, respectively. While Hannah Arendt felt that philosophers should be more involved in practical life, Paulo Freire (1970) defined praxis as reflection and action aimed at the structures to be transformed. David Kolb (1984) argues that praxis is a recurring passage through a cyclical process of experiential learning.

To understand Bawden's conception of praxis, I will start with examining Dr David Kolb's ideas about the adult learning process. In Fig. 14.1, my adaptation of his experiential learning cycle is shown, giving four stages: (i) experience; (ii) reflection; (iii) conceptualisation; and (iv) experimentation. After an experience (which is an actual, concrete, subjectively perceived experience), one may think back on this experience a little later in time. This process of reflection is still subjective (one thinks about a personal experience), but it has now been taken into the abstract world of thought. Having considered carefully what actually happened, one may try to understand the importance of this experience: what does it mean for the future? What can one learn from it? This process of conceptualisation can lead to the formation of an idea about the significance of the experience, which in turn, can lead one to try out ideas, giving rise to further experiences. The conceptualisation is still abstract, but it is no longer subjective. Concepts fit into a wider world of theoretical speculation. A scientific concept will often lead to the formulation of a working hypothesis, leading to a formal experiment.

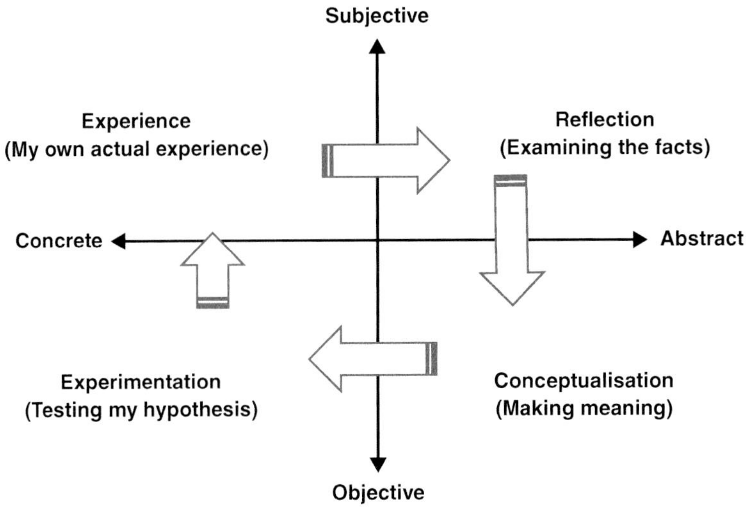

Fig. 14.1. The experiential learning cycle. (Adapted from Kolb, 1984.)

An example is the apocryphal story of Sir Isaac Newton who sat in an orchard in the 17th century, and after seeing several apples falling (experience), reflected on the fact that the apples always fall towards the ground, and then developed the concept of gravity, which he formulated into the Law of Universal Gravitational Attraction according to a book review in *New Scientist*:

> [The Royal Society has re-published] *Memoirs of Sir Isaac Newton's Life* written by William Stukeley, an archaeologist and one of Newton's first biographers, and published in 1752. Newton told the apple story to Stukeley, who relayed it as such: '*After dinner, the weather being warm, we went into the garden and drank tea, under the shade of some apple trees ... he told me, he was just in the same situation, as when formerly, the notion of gravitation came into his mind. It was occasion'd by the fall of an apple, as he sat in contemplative mood. Why should that apple always descend perpendicularly to the ground, thought he to himself?*' After his famous experiments (dropping feathers and iron balls from a tower and measuring how long they took to reach the ground), the working hypothesis was confirmed.
>
> (*New Scientist*, 2010)

If this is indeed how adults learn, teachers and trainers should be putting considerable effort into a learning structure which encourages guided experience, careful reflection on the actual experience, followed by conceptualisation (helping learners to examine what an experience means to them in practice), and perhaps a willingness to unlearn some of the prejudices of the past (MacWilliam, 2013 cited in Erasmus and Albertyn, 2014). Erasmus and Albertyn (2014) continue by pointing out that Kolb's experiential learning theory suggests that learning combines grasping and transforming experiences, implying that change is an integral aspect of the learning process. Adapting experiences to a new context can give rise to innovative solutions.

In my experience, many adults do not spend much time examining the facts and reflecting on them, but often jump to a conclusion based on previous experiences and beliefs. This is how racial, gender and religious stereotypes persist in spite of numerous examples of individuals who do not conform to the stereotype (so, for example: 'Yes, but he's a good Jew/Muslim/black person'; or 'That is pretty good for a woman'). Thus my experience may lead straight to my conceptualisation, without any reflection, skipping Kolb's second stage shown in Fig. 14.1. If higher education is to engage in changing prejudiced attitudes, a part of our pedagogy must engage with helping learners to observe afresh what actually happens in challenging practical situations. Often, this is best done through group reflection. When time is given to examining what happens in a group setting with other learners, new knowledge and beliefs can emerge. This process of joint reflection led to Professor

Niels Röling's theory of 'Platform building for resource use negotiation' (Röling and Wagemakers, 1998; Auerbach, 1999). Röling argues that only after a diverse group of stakeholders has been brought together to consider the legitimate needs and activities of other stakeholders who share resources, does a 'platform' of shared concepts and perspectives begin to emerge. Then it may be possible to negotiate reasonably about use of shared resources.

Let us examine Röling's ideas in the context of classical extension theory. The World Bank, in the 1960s and 1970s, adopted the Training and Visit (T&V) system of extension management, based on the theory of Benor and Harrison (1977). The T&V approach is summarized in Fig. 14.2. Head office develops an extension message, based on knowledge generated from research results. Every month, extension officers are called in and given specific training on what the farmers should do with a specific crop that month. The extension officers then visit client farmer groups and deliver the extension message. The assumptions here are that head office is the owner of the knowledge, and that simply by conveying a series of messages to farmer groups through the extension officer, all problems will be solved, and production will increase.

The T&V system was applied for many years in World Bank projects until it was concluded that T&V did not solve farmer's problems because it was not context sensitive, and was too expensive (Anderson *et al.*, 2006). In his work on extension theory, Röling (1988) calls T&V 'Doing it *for* the farmer'. My own early farming systems research and extension (FSR/E) work (Auerbach, 1994), showed how improving maize production in southern KZN had more to do with easing constraints on production than on supplying technical knowledge. The contribution of this work to FSR/E theory was described in Auerbach (1995), and is summarized in Chapter 1 of this volume.

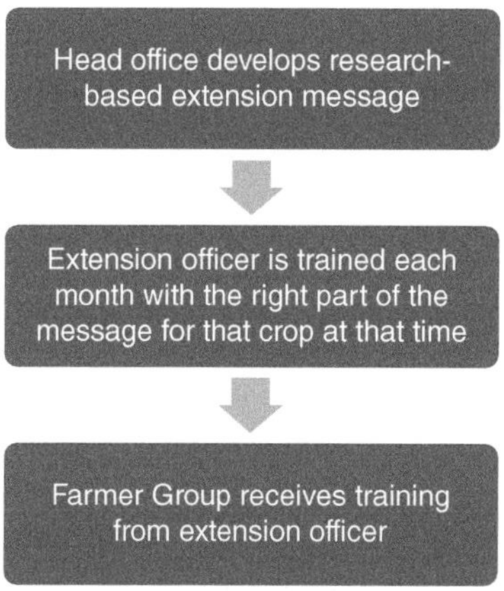

Fig. 14.2. Training and Visit (T&V) system of extension management. (Adapted from Benor and Harrison, 1977.)

In the 1980s, the diffusion of technology conceptual framework became popular (see Fig. 14.3), where the work of progressive lead farmers or master farmers was assisted by research results, and the 'early adopters' would take up the work of these 'innovators' (Rogers, 1962). Through diffusion of technology, most nearby farmers would then adopt the new practices, forming the 'early majority', followed by the 'late majority' adopting the innovation. Eventually even the last group, 'the laggards', would cotton on. Röling (1988) calls this 'Doing it *to* the farmer'. In practice, although some new technologies do follow the diffusion pattern, it was found that there is considerable adaptation and local modification of new practices to suit different conditions. Very often, diffusion is impeded by lack of access to finance, timeous land preparation, agricultural inputs and knowledge (Auerbach, 1994).

Röling (1988) explains the shortcomings of these two theories, and later introduces a number of participatory approaches where there is a process of co-learning between communities, extension agents and researchers (Bawden, 1992; Röling and Wagemakers, 1998). He calls this 'Doing it *with* the farmer', as the emphasis is on helping farmers to innovate in their particular situation. In summary, Röling has explained three approaches to extension: (i) doing for; (ii) doing to; and (iii) doing with the farmers. So much for extension theory; does it explain how farmers learn? Is the process in fact similar to Kolb's experiential learning cycle, as shown in Fig. 14.1?

In his work at Hawkesbury College, Bawden brings together extension theory and adult learning (Bawden, 1992). Just as Röling argues for the empowerment of farmers through a process of co-learning (Röling, 1988), so Bawden argues that agricultural students will benefit from a process of practical learning based on the concept of praxis. Assuming that head office or

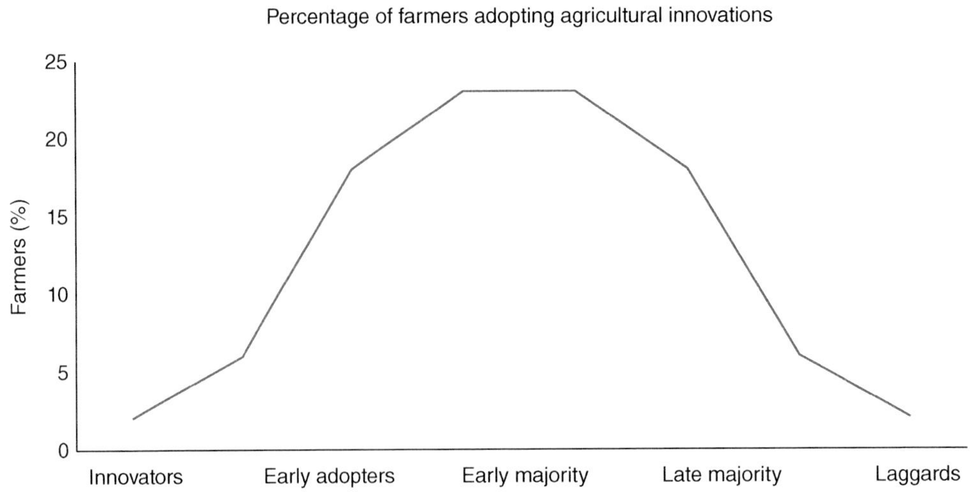

Fig. 14.3. Diffusion of technology extension theory. (Adapted from Rogers, 1962.)

the scientists hold the important knowledge disempowers farmers and inhibits innovation. The introduction of Green Revolution technology by the World Bank led to many problems in Asia and South America (Anderson *et al.*, 2006). It also led to massive waste of resources in Africa, where yields did not increase significantly and farmers could not access the whole range of inputs required (Auerbach *et al.*, 2013). Technology and innovation may thus have unintended side effects, and may even serve to disempower farmers, making them dependent on expensive technology or causing severe environmental and human health damage. Bawden (1992) addresses these unintended negative consequences by using farming experience in farmer training. He argues that praxis draws the theory out of experience.

I conclude from the above that workplace learning is essential in giving a practical balance to theoretical inputs. The actual challenges of a job are very different to the theoretical 'knowing of the issues involved'. Farming from a textbook is a hazardous activity. Both theory and practice are essential, as good theory allows one to generalize from a particular context to other (very different) contexts. Good practice is specific for a given time, place and context; without it the concepts remain concepts. A theoretical cow is very different from a real cow, which may kick you, show affection, demand attention, produce milk, calves (and copious amounts of dung and urine at just the wrong time), get sick, and even die. In addition, I will show through the case studies that experiential learning is more effective if it precedes theoretical instruction. This can stimulate innovation and promote a more profound and grounded understanding of conventional (theoretical) wisdom, and encourage a healthy and independent spirit of inquiry.

The approach outlined in Chapter 4 of this volume by Konrad Hauptfleisch emphasizes the 'Theory-Action-Reflection' triangle, which, he emphasizes, is neither sequential nor hierarchical, and also allows for adaptation to the local context. Both Kolb (1984) and Hauptfleisch emphasize that adult learning requires the teacher to combine subject matter knowledge with the ability to manage a process of exploration and reflection, and that learning involves exploring, preferably in practical situations, and changing what is experienced in one context, to make it more suitable for one's own context.

Methods

Case study research is widely used in organizational studies, as well as in the social sciences (Robson, 2002; Hartley, 2004). In this chapter three case studies using variations of experiential learning are applied, using the insights of Kolb (1984) to guide the process. His approach to praxis is adopted, also informed by Bawden

(1992), and by Röling and Wagemakers (1998), as outlined above. Each of the three case studies reflected in the three models that developed over time are described and discussed. Each time, the actual experience is outlined, followed by a reflection on that experience, and then some more general conceptualisations are presented; hopefully, this will lead the readers to experiment in their own context!

Results and Discussion

I now present the three case studies, using the experiential learning cycle (Fig. 14.1) as a framework for examining the experience, then looking at what was learned in reflecting on the experience, and at what concepts were extracted, leading to experimental design and a new experience. I describe how Model 2 changed in the light of the experience of Model 1, and how these two experiences of the Rainman Landcare Foundation informed subsequent work at Nelson Mandela University, taking the existing model which was found there, and giving rise to Model 3. The conclusion will look at the conceptual insights arising from all three experiences.

Model 1: Rainman training-farm-based learning

Experience

In the years from 2000 until 2005, we tested the training-farm-based model. Students were in class from 8 to 10 a.m., then after breakfast, they spent 2 h in their individual student garden. These were rainfed garden plots, in size only 3 m × 5 m, where three crops were planted, based on the market survey they had done while at their homes. Before attending the programme, they found out what they could sell to their neighbours and for what price, and this informed their choice of crops. In addition, the crops had to be selected as follows: (i) one heavy feeder crop; (ii) one light feeder crop; and (iii) one legume crop. (Heavy feeders need a lot of plant nutrients, light feeders require less, and legumes supply nitrogen to the soil.) This allowed for a good crop rotation, which would prevent most disease and pest problems from reaching epidemic proportions. In the afternoons, students worked on the Rainman organic farm, learning about compost making, irrigation, pest and disease control, weeding, harvesting and preparing crops for marketing through our local community supported agriculture (CSA) scheme, which was part of the Ntshongweni Participatory Guarantee System (PGS) (Katto-Andrighetto and Auerbach, 2009).

Reflection

Although Model 1 was found to be effective, it had two drawbacks. First, students had to find their way to the farm by 8 a.m. each morning as we had no dormitories. Second, little farming took place in the students' home environment during the training period. Although there was a market survey exercise at the start, and a business plan aiming to implement a real project at home at the end of training, many trainees drifted away from agriculture.

We found funds to follow up and establish six local organic farming cooperatives; each of these set up their own PGS to ensure that the produce was organically grown (Katto-Andrighetto and Auerbach, 2009). A secondary (Ntshongweni Organic) cooperative was established to handle marketing and distribution. Four of the six primary (producer) cooperatives functioned quite well and continue to produce organic vegetables. The secondary cooperative was not able to continue without support, and ceased to exist once support was withdrawn. Farmers kept farming organically, as the approach helped them to use soil moisture more efficiently.

Conceptualisation

Rainman Landcare Foundation decided that we had met the local need, and that we should take the training out to the farmers and gardeners in other areas. In partnership with a municipality in KZN, we developed a community-garden-based Model 2, where training was scaled up (100 gardeners in four groups, supported by 20 mentors working on-site north of Durban). Similar training was later also done at two locations in Zululand.

Model 2: Rainman training and on-farm based learning

Experience

The project was set up in partnership with a permaculture training centre (located near the community to be assisted). Rainman set up three processes. First, the new community garden was fenced (eventually) and rudimentary irrigation systems were developed. The garden was ploughed, and plots were allocated to a number of people who were already involved with gardening activities.

Second, the gardeners were offered a 7-month learnership, which included a small stipend, and which required them to attend 1 week of training per month at the Permaculture Centre. This was followed by 3 weeks of mentored activity in the garden, after which the process was repeated for each of the 7 months.

The third process was the selection and training of the gardening mentors, who would complete three 2-week modules, and would support the (mostly elderly) gardeners at the community garden site. This process was problematic, as we had no say in the selection of the mentors, and the project manager and local councillor were both involved in politics. All mentors selected were politically active, and were young (aged 20–30). None of the mentors had significant knowledge of gardening, and few had much interest in gardening. All were keen to receive the monthly payment as mentors. They came to the Rainman farm for 2 weeks of training before the commencement of gardener training. After the first week, I was despairing of the capacity of the group to assist the gardeners.

In the course of this module, I reflected on the activities relating to self-empowerment of the mentors, and ways of treating adult learners with respect and avoiding preaching and lecturing. The role-play activities had failed to elicit a response from the mentors that the old people who were gardening, while uneducated, had many skills which should be respected. Respect, in the view of the mentors, was only due to them because they were older people. I recognized that many of the mentors lacked listening skills, and also lacked empathy. It had also become clear that the informal settlement where our research group lived was a place where many victims of the struggle against apartheid, and also many older impoverished rural people, had fled in order to escape either political persecution or rural poverty. Members of the first group had little respect for members of the impoverished group.

I developed a talking and listening activity in order to allow the mentors to practise listening to each other. I asked them to pair off using a random pairing process, and instructed them to spend 5 min listening to the life story of the other person, and then to tell their own life story in the next 5 min. They would then be given 2 min each to report back to the group on the other person's life experience. I was not ready for the buzz which developed, where each pair of mentors was locked in intimate exchange. There was no way I could bring this to a close after 10 min. Something was obviously working, and so I allowed a little over an hour for the listening process, only reminding participants regularly that one person should be talking, and the other person should be listening. Eventually, we had our feedback session, and story after story emerged of murder of parents and family members, burning down of houses, flight from home, persecution by security police, divided communities where one faction betrayed the other faction, insidious payment for information by government to *impimpis* (paid informants).

The report-back lasted for over 2 h, after which we all sat back, totally shell shocked. There was silence for about 10 min, interspersed with sobbing. When I had some control of my own emotions, I sent the group home for the day, asking them to remember that all of the gardeners had their own difficult story to tell, and that all were worthy of respect, and had experiences which could teach the mentors a great deal. I asked participants to go home and reflect on what had happened. I promised that the next day would start with a group reflection, followed by an attempt to understand the significance of the day's experience.

The next day people behaved very differently. The reflection session was subdued but profound, with powerful insights emerging. We began to develop a vision for empowerment through food security. This was intensely political. Only after the feedback and our breakfast break, did I respond telling a story of an arrogant young American Peace Corps volunteer, who had tried to tell a black community what to do, without himself having any relevant life experience or skills. I told them how this had led me to promise myself that I would not try to teach people skills that I had not practiced myself. This resulted in 20 years of farming before I started training farmers. Many of the mentors accepted that they needed to work on their horticultural skills.

Reflection

Although about half of the mentors were much better after this experience, many did not change the way they related to the gardeners, and the lack of gardening skills meant that they had little to offer the gardeners, who knew more about gardening than the mentors did. However, the experience certainly taught me about the importance of understanding the socio-political context of learners, and how important it is to draw good theory out of good practice, both with adult mentoring and with actual vegetable production. The project had many difficulties to deal with, related to the municipality's lack of responsiveness, but in the end the project did receive an award for the best learnership in the province. Gardeners were able to produce high quality vegetables organically with little additional water and few pest and disease problems.

Model 2 worked well for the gardeners, as they received training which they were able to put into practice in setting up their new plots at the new community garden. Although resources (especially toilets and the fence) were very slow in coming, there was some support, and the gardeners were initially very appreciative. Soon, however, the gardeners started to demand more and more from the municipality. As our role was simply that of trainers, and not project managers, we were no longer formally involved in the project at this point.

Conceptualisation

The two models developed by the Rainman Landcare Foundation were aimed at semi-literate farmers, and were based on the National Qualifications Framework (NQF)2 National Certificate in Mixed Farming Systems; NQF2 level qualifications are designed for mid-secondary school level learners, with only basic numeracy and functional literacy in their mother tongue. Instruction was in isiZulu, and only minimal written competence was required. The mentors completed NQF5 training as Organic/Landcare Facilitators at NQF level 5 (post-secondary school). Training was offered in English, and higher levels of reporting competence were required. Six business management modules were included. Various other mentorship training with trainees selected on the basis of appropriate skill criteria resulted in good skilling and motivation, and many of those Landcare Facilitators are still training farmers. In separate projects, some were trained for the Limpopo and Mpumalanga government extension services and others for smaller civic organizations. Mentorship appears to be very important to the success of trainee farmers.

Model 3: Nelson Mandela University at George (Diploma in Agricultural Management); classroom followed by experiential learning and then integration

Experience

The Nelson Mandela University at George offers an agroecology-oriented agricultural management training approach as part of the School for Natural Resource Management (SNRM), and has a 1-year on-farm practical workplace experience component offered in semesters four and five (middle of the second year to middle of the third year). This is followed by the final theoretical semester (end of the third year) which completes the diploma. Top students may then elect to complete an Advanced Diploma in their fourth year. The workplace learning starts in the South African winter (July), and thus includes this planning period and the spring planting period, as well as the whole summer growing season. By December, many South African farms are ready to harvest; this first semester has a technical focus. The three first-semester assignments are: (i) environmental scan of the farm; (ii) technical report (on any two enterprises, preferably one animal- and one plant-based production sector); and (iii) a personnel project (which looks at compliance with labour law, management of staff and a self-evaluation of the student's management strengths and weaknesses).

The second semester of the workplace learning (January–June) focuses on management, and the three assignments are marketing, profitability/cash flow, and finally an integrated business study, drawing together all five earlier assignments. This is followed by an oral presentation to the first-year students, evaluated by the fourth-year students and staff. By this stage, the second-year students are already off campus having started their workplace experience.

Reflection

The structure of Model 3 is thus based on presenting 18 months of theory followed by 12 months of practical on-farm work, with a final 6 months to integrate theory and practice. The work integrated learning has evolved into an effective period in which the learners are exposed to many experiences on commercial farms across the country (some even choose to go overseas). The six assignments give the opportunity for learners to reflect on technical, social and economic aspects of commercial agriculture, culminating in an oral presentation of their integrated business plan based on all six assignments completed during the year. Students are instructed to look at environmental sustainability issues, but many only pay lip service to this requirement. Many farmers also claim to embrace sustainability, but this is difficult to measure.

Although the Mandela University programme is progressive and has evolved from years of experience on the George Campus, the underlying assumption is that good practice comes out of good theory. The first 18 months in my subjects (Soil Science and Plant Production) had been totally theoretical. My colleague in animal production had already introduced a number of practical activities as she felt that the theory alone was inadequate. I was convinced, and started establishing practical facilities.

In the first year, we set up a weather station, and made compost. In the second year, we planned a permaculture garden. In the third year we found the funds and set the garden up, where students now make compost, design a crop rotation, prepare seedbeds, and plant a vegetable garden which they look after and harvest at the end of the semester. They do a small garden project as part of Plant Production I (with the emphasis on learning how plants grow), and a major project in Soil Science II in the second semester. They are thrown into the first project with very little theory, and are encouraged to experiment. They are helped to reflect on this experience and to use the learning to inform the second project.

They are also exposed to the long-term comparative farming systems research trials, especially in their fourth (BTech/Advanced Diploma) year, where each student completes a crops project. We have observed that students ask more questions and are also prepared to innovate more readily now in their projects. Understanding of basic concepts such as soil acidity, cation exchange capacity and available plant nutrients has improved. They learn in Plant Production classes how to supply nutrients organically, how to set up crop rotations and how to monitor and control pests and diseases. In Soil Science classes, they learn how to improve soil water holding capacity through managing SOM, and how to address soil acidity, making essential plant nutrients available to crops.

Conceptualisation

It was found that a block of theory followed by a block of practice was not an efficient teaching system. Since the establishment of practical facilities, students have been experimenting with soil fertility and plant growth from the start of their diploma. Their attitude to soil science and plant production has improved. They recognize the importance of soil analysis and of integrated pest management, having experienced the problems of plant production first hand. It has been possible to start with some theory reinforced with experimental practice for 18 months. Next, they spend 12 months practising farming in the real world, reinforced with theory through their six assignments. Finally, they have 6 months in which to integrate theory and practice. If they stay for the fourth year, they have the opportunity to design their own experiential learning process.

Conclusion

Climate change will see increases in temperature, decreases in rainfall in many areas, and erratic climate events; organic farming systems show promise in improving water use efficiency

(WUE), and in assisting farmers to use and make available to crops the nutrients which are locally available.

Given the experiences with on-farm training from Model 1, and with the mentors from Model 2, and the other experiences leading to Model 3, the following principles for effective work integrated learning in agroecology are proposed:

- Good practice should *precede* good theory, and should ideally be extracted from learner experiences where possible.
- As quoted earlier from Erasmus and Albertyn (2014), learning only happens if the experience is grasped, examined and transformed; this requires an experiential learning challenge followed by a process of guided reflection.
- Kolb's learning theory (Fig. 14.1) is thus enriched by Röling's 'Platform building for resource use negotiation' and Bawden's combination of adult learning approaches which build on good extension practice. Practical experiences for learner farmers should initially introduce the nature of practical challenges in food production. Once the learners have become familiar with the difficulties of practical soil and plant management, they are more receptive to theories of soil fertility and plant nutrition, as well as integrated pest management. Only then are they in a position to be exposed to the real world of commercial farming.
- Good mentorship, provided by well-trained mentors with experience in agroecological food production, should be available to help learners reflect on their experience, draw out the conceptual implications and integrate theory and practice through this process of praxis.
- Such a structure can help learners to understand what is needed for African food security and food sovereignty in these times of climate change.

Authentic learning and authentic assessment can build in experiences which encourage joint reflection and re-conceptualisation. In this way, learning about farming through guided experience can construct new knowledge, rather than simply reinforcing old habits. This only happens with proactive experienced mentorship, and many training institutions are reluctant to provide adequate resources for this process.

While the Mandela University George Experiential Learning process has adapted to circumstances and resource constraints, it is recommended that this programme should now be evaluated, to determine its effectiveness and the optimal balance between resource use and learner competence. If financial constraints at the university are used to argue for fewer experiential learning resources, the quality of the experience offered to students will decline significantly.

Experiential learning can show farmers how to experiment with local resources. Using organic farming systems can help them to deal with climate change by introducing crop rotations and improving SOM levels. This increases WUE, promotes biodiversity and reduces pest and disease problems. Drawing good theory out of good practice allows learners to adapt to a challenging environment, while giving them the confidence born from the practical achievement of growing nourishing food.

References

Anderson, J., Feder, G. and Ganguly, S. (2006) The rise and fall of training and visit extension: an Asian mini-drama with an African epilogue. Paper WPS 3928, Agriculture and Rural Development Department. World Bank, Washington, DC.

Arendt, H. (1998) *The Human Condition*, 2nd edn. Monoskop. Available at: http://monoskop.org/images/e/e2/Arendt_Hannah_The_Human_Condition_2nd_1998.pdf (accessed 20 December 2018).

Auerbach R. (1994) A farming systems research evaluation of maize production practices in southern KwaZulu. MSc thesis, University of Natal, Pietermaritzburg, South Africa.

Auerbach, R. (1995) People, farming and research: how can South African science contribute to sustainable agricultural development? *South African Journal of Science* 91, 3–6.

Auerbach, R. (1999) Design for participation in ecologically sound management of South Africa's Mlazi River catchment. PhD thesis, Wageningen Agricultural University, Wageningen, the Netherlands.

Auerbach, R., Rundgren, G. and Scialabba, N. (eds) (2013) *Organic Agriculture: African Experiences in Resilience and Sustainability*. Food and Agricultural Organization of the United Nations (FAO), Rome.

Ayers, J. and Forsyth, T. (2014) Community based adaptation to climate change. *Environment Science and Policy for Sustainable Development* 51(4), 22–31.

Bawden, R.J. (1992) Systems approach to agricultural development: the Hawkesbury experience. *Agricultural Systems* 40, 153–176.

Benor, D. and Harrison, J. (1977) *Agricultural Extension: the Training and Visit System*. World Bank, Washington, DC.

Carpenter, M., Emejulu, A. and Taylor, M. (2016) What's new and old in community development? Reflecting on 50 years of CDJ. *Journal of Community Development* 51(1), 1–7.

Elbehri, A. (2015) *Climate Change and Food Systems: Global Assessments and Implications for Food Security and Trade*. Food and Agriculture Organization of the United Nations (FAO), Rome.

Erasmus, M. and Albertyn, R. (2014) *Knowledge as Enablement: Engagement Between Higher Education and the Third Sector in South Africa*. Sun Media, Bloemfontein, South Africa.

Freire, P. (1970) *Pedagogy of the Oppressed*. Herder and Herder, New York.

Hartley, J. (2004) *Essential Guide to Qualitative Methods in Organizational Research: Case Study Research*. SAGE, London.

IPCC (Intergovernmental Panel on Climate Change) (2014) *Climate Change 2014: Synthesis Report*. Contribution of Working Groups I, II and III to the Fifth Assessment Report of the IPCC (Core Writing Team, R.K. Pachauri and L.A. Meyer (eds)). IPCC, Geneva, Switzerland.

Katto-Andrighetto, J. and Auerbach, R.M.B. (2009) Participatory guarantee systems in Africa. *Ecology and Farming* 45, 49–51.

Kolb, D.A. (1984) *Experiential Learning: Experience as the Source of Learning and Development*. Prentice Hall, Englewood Cliffs, New Jersey.

New Scientist (2010) Book Review, Royal Society Manuscript, 18 January 2010. Available at: https://www.newscientist.com/blogs/culturelab/2010/01/newtons-apple-the-real-story.html (accessed 20 December 2018).

Participants (1993) *Participatory Rural Appraisal Manual – Notes from a Training Course held in Bulwer, South Africa*. Midlands Rural Development Network, Pietermaritzburg, South Africa.

Robson, C. (2002) *Real World Research: a Source for Social Scientists and Practitioners-Researchers, Reporting on the Enquiry*. Blackwell, Oxford.

Rogers, E.M. (1962) *Diffusion of Innovations*. Free Press of Glencoe, New York.

Röling, N. (1988) *Extension Science: Information Systems in Agricultural Development*. Cambridge University Press, Cambridge.

Röling, N. and Wagemakers, M.A.E. (1998) *Facilitating Sustainable Agriculture: Participatory Learning and Adaptive Management in Times of Environmental Uncertainty*. Cambridge University Press, Cambridge.

UN (United Nations) (2015) UN Sustainable Development Goals. Available at: https://sustainabledevelopment.un.org/?menu=1300 (accessed 25 January 2016).

Part 3

Supporting Organic Farmers

15 The National Organic Agriculture Movement of Uganda

Jane Nalunga,[1,2*] Raymond Auerbach[2] and Charles Ssekyewa[3]
[1]*National Organic Agriculture Movement of Uganda, Kampala, Uganda;* [2]*Nelson Mandela University, George, South Africa;* [3]*Uganda Martyrs University, Uganda*

Abstract

The growth of the organic agriculture (OA) sector in Uganda started in 1994 as a response to the threat of the Sahara Desert spreading southwards. To enhance the development of the sector, a national OA movement was established in 2001. The National Organic Agriculture Movement of Uganda (NOGAMU) is an umbrella organization that promotes and coordinates OA development in Uganda. Before its inception, organic farming in Uganda developed very slowly over the previous decade. The reasons for establishing NOGAMU are discussed, and its vision, mission, goal, objectives and structure are outlined, as well as NOGAMU's role in the organic sector in relation to activities implemented to support OA development.

NOGAMU'S impact on farmers' livelihoods is elaborated including: (i) increasing adoption of OA by smallholder farmers; (ii) increased food security; (iii) better incomes for smallholder farmers; (iv) access to functional markets; and (v) building social capital. Challenges that affect NOGAMU and the organic sector are also discussed, and we conclude that NOGAMU has played a major role in the development of OA in Uganda.

Introduction

The National Organic Agriculture Movement of Uganda (NOGAMU) is the umbrella organization that unites producers, processors, exporters and other traders, business support organizations and all other stakeholders who are involved, either directly or indirectly, in the organic value chain in Uganda. NOGAMU is a membership-based organization, registered in Uganda as a not-for-profit company limited by guarantee. It was founded in January 2001 with the main goal of uniting and leading the organic sector towards development. Over the years, its direct membership has increased from under 50 individuals and 20 corporate organizations at the inception, to 574 individual category members and 363 corporate category member organizations by the end of 2014, indirectly representing 1,221,000 smallholder farmers across the country (NOGAMU, 2015).

Why Was NOGAMU Formed?

Between the years 1994 and 2000, various non-governmental organizations (NGOs) and community based organizations (CBOs) played a leading role in training farmers on sustainable agriculture with a major focus on organic

*janenalunga@gmail.com

principles and practices, to improve the food security situation after the civil war that ravaged the country between 1980 and 1986 (EPOPA, 2008a, b). By the year 2000 most farmers had attained a basic level of food security, using what is sometimes called 'organic by default'; however, certification and other ways of linking production to the market, were lacking. On the other hand, exporters realized that the demand for organic products on the export market was far higher than what could be supplied, and there was a general lack of coordination between producers and buyers. Stakeholders started exchanging ideas on developing the organic agriculture (OA) sector in Uganda. The first meeting was convened in the year 2000 to discuss forming an umbrella organization with the major role of coordinating and developing OA in Uganda. As a result, NOGAMU was formed and formally launched in 2001 with the following mission, mandate and objectives.

Mission and values

NOGAMU's mission is to coordinate and promote OA development, networking and marketing.

NOGAMU operations are guided by the following values: (i) organic; (ii) integrity; (iii) sharing; (iv) transparency; (v) accountability; (vi) fairness and social justice; (vii) gender equity; (viii) environmentally ethical; (ix) unity; and (x) non-discrimination.

Mandate and scope

NOGAMU's mandate is to coordinate and promote OA development in Uganda. In line with the organization's vision, NOGAMU programmes and activities are oriented towards market access by smallholder farmers and companies. NOGAMU's role is to work towards increased competitiveness and marketing of organic products from Uganda. This is done by facilitating smallholder farmers and other actors in the organic value chain to attain increased competitiveness and access to markets as a foundation for increasing incomes and improving livelihoods in order to alleviate poverty and contribute to attaining Sustainable Development Goals (UNDP, 2015).

NOGAMU's five main focus priority clusters are: (i) marketing and value chain development; (ii) standards and certification; (iii) training, research, extension and education; (iv) advocacy and strategic relations; and (v) institutional development.

NOGAMU's scope is limited to coordinating, facilitating and promoting the entire organic value chain in Uganda by leading, assisting and promoting all actors in the sector, advocating for development of conducive policies and attracting support that would generate momentum required to increase competitiveness of the sector and take advantage of the opportunities presented to Uganda by the rapidly growing global organic sector.

NOGAMU's target group includes smallholder farmers and farmer organizations, processors, exporters and traders, as well as those providing relevant services to the organic value chain.

NOGAMU has members in all the four regions of Uganda (Northern, Eastern, Western and Central), and has organic producers spread through all the agroecological zones of Uganda.

Goal

NOGAMU's overall goal is to contribute to poverty reduction by scaling up incomes of smallholder farmers through increased production and access to organic and fair trade markets.

Specific objectives

The specific objectives of NOGAMU are:

- to promote domestic, regional and export marketing of organic products from Uganda;
- to build capacity and promote training, research, extension and education in OA in Uganda;
- to promote compliance with organic and other complementary standards, and attain certified organic production and improved quality of organic products in Uganda; and
- to advocate for an enabling environment for OA and trade in Uganda.

Membership and Affiliation

NOGAMU is a member of the International Federation of Organic Agriculture Movements (IFOAM), and serves as the IFOAM contact point in Uganda.

NOGAMU is also a member of the African Organic Network (AfrOnet) and East African Organic Network. It is the Country Lead Organization for the African Union (AU) Ecological Organic Agriculture (EOA) Initiative, and works closely with the other East African national organic agricultural movements (NOAMs) including: (i) the Kenya Organic Agriculture Network; (ii) Tanzania Organic Agriculture Movement; (iii) Rwanda Organic Agricultural Movement; and (iv) Burundi Organic Agriculture Movement. They share information and joint implementation of regional programmes and management of the activities related to the East African Organic Products Standards (EAOPS) and the East African Organic Mark, as well as non-organic organizations including the Private Sector Foundation of Uganda, NGO Forum and Uganda Agribusiness Alliance.

Partners

NOGAMU collaborates, networks and partners with a number of organizations at the grass-roots level, as well as at national and international levels in an effort to support and create meaningful impact on target groups. These partners may be implementing, simply collaborating or funding partners.

Implementing partners

These partners participate in undertaking activities along the organic value chain. As NOGAMU is a membership organization, the secretariat facilitates implementation of activities and does not directly carry them out. These activities may be carried out either by grass-roots members of NOGAMU or residents of a target community/region.

Collaborating partners

NOGAMU implements its programmes and projects in collaboration with different member organizations in all the four regions of Uganda. These partner organizations are selected based on their day-to-day activities, technical capacity and geographical coverage. They play a key role in coordination and assist in the implementation of programmes in the respective regions of the country. NOGAMU has four regional partners, one in each of the four geographical regions, to help in coordination and assist in implementation of programmes throughout the country. The regional partners are Lango Organic Farming Promotion for Northern Uganda, Africa 2000 Network Uganda for the Eastern region, Sustainable Agriculture Trainers Network (SATNET) for the Western region and Caritas Kampala for the Central region.

NOGAMU also works with several other member organizations on a case-by-case basis depending on the initiative either started by NOGAMU or the member/partner organization. For example, in 2015 NOGAMU concluded a project with a consortium of seven organic export companies (Biofresh Ltd, Bio Uganda Ltd, Sulma Foods Ltd, Africa 2000 Network, RUCID Ltd, Jali Organic/Be-Organic Ltd and Flona Commodities Ltd), which involved scaling up the production and export of organic dried fruits to regional and international markets through bulking and promotion of a common brand. This project was supported by the Trademark East Africa Challenge Fund.

NOGAMU also partners with other organizations to implement projects that are in line with its mission. In January 2016, for example, NOGAMU signed a contract with PLAN Uganda to implement a project 'Switch Africa Green, Sesame – Green Jobs Uganda'. The overall goal of this project was to increase livelihood opportunities for young sesame farmers through improved eco-agriculture production and green economy inclusion.

NOGAMU also participated in the Productivity and Growth in Organic Value-chains (ProGrOV) project, which was a combined research, development and capacity building project aimed at strengthening research-based knowledge for supporting increased productivity and sustainable growth in OA production and value chains as well as building capacity for future development of OA-based value chains. The project was implemented from 31 December 2010 to 31 December 2016 coordinated by

Makerere University in Uganda in collaboration with the Danish International Centre for Research into Organic Farming Systems (ICROFS), Danish universities as well as universities in East Africa and the NOAMs of East Africa. The project was funded by Danida. The overall objective of ProGrOV was to increase OA productivity and development of agribusiness for economic growth, improved livelihoods and sustainable development in Africa. Previously, NOGAMU worked with the Export Promotion for Organic Products from Africa (EPOPA) project, which will be discussed in more detail later in this chapter (EPOPA, 2008a).

In addition, NOGAMU also works and collaborates with other institutions in both the public and the private sectors to create synergy in the development of a strong and vibrant organic sector in Uganda, the East African region and the rest of Africa. These include, but are not limited to:

- NGOs and CBOs working with farmers at the grass-roots level, which are involved in training and farmer mobilization;
- national trade promotions and business support organizations such as the Uganda Export Promotions Board, Uganda National Bureau of Standards and Private Sector Foundation Uganda;
- government ministries such as: (i) the Ministry of Agriculture, Animal Industries and Fisheries; (ii) the Ministry of Tourism, Wildlife and Antiquities; (iii) the Ministry of Trade, Industry and Cooperatives; and (iv) the Ministry of Water and Environment;
- district local governments; and
- international trade promotion and other organizations such as the AU, the International Trade Centre, United Nations Conference on Trade and Development (UNCTAD), the United Nations Environment Programme (UNEP), the European Union (EU)/Europe-Africa-Caribbean-Pacific Liaison Committee (COLEACP) Pesticide Initiative Programme, and the Centre for Promotion of Imports from Developing Countries in the Netherlands.

NOGAMU is also involved in the implementation of the regional AU-led EOA Initiative (as already mentioned earlier in the chapter) and is part of the Continental Steering Committee chaired by the AU Secretariat. In addition, NOGAMU acts as the Country Lead Organization for Uganda.

Funding partners

Since its inception, NOGAMU has received financial support from various partners including Hivos, Ford Foundation, GTZ (now GIZ), SIDA (the Swedish International Development Cooperation Agency), Swedish Society for Nature Conservation (SSNC), Swedish Development Cooperation (SDC), and Organic Denmark.

NOGAMU'S Role in the Organic Sector

As a national organization, NOGAMU has played various roles in its efforts to develop the organic sector in Uganda. These roles are discussed below.

Coordinating and developing OA in Uganda

Following its mandate NOGAMU has played a major role in coordinating and developing OA in Uganda. Certified OA in Uganda started in 1993 mainly as a response to unfolding market opportunities in Europe and by 1994 some commercial companies were exporting organic products (Taylor, 2006). Increase in certified acreage, however, was slow until 2001, when NOGAMU was started and the fastest development of OA was noted between 2002 and 2007. Increase in certified OA in Uganda may be attributed partly to the efforts of NOGAMU, but also to the various project efforts such as the SIDA-funded EPOPA project (EPOPA, 2008a). Currently, certified organic land in Uganda is 241,150 ha (NOGAMU, 2017; Willer and Lernoud, 2017), which makes Uganda 27th worldwide and the second in Africa after Tanzania in terms of size of area cultivated organically. The number of certified organic producers in Uganda is 190,670 (NOGAMU, 2017; Willer and Lernoud, 2017), making Uganda the fourth worldwide in terms of number of producers after India (585,200), Ethiopia (203,302) and Mexico (200,039) and the second in Africa.

Creating market access for smallholder farmers

Since its inception, NOGAMU has mobilized, trained and linked various farmer groups and producer organizations to local, regional and export markets dealing in different value chains. The biggest organic value chains include Arabica coffee, sesame, cocoa, vanilla, fresh and dried fruits (pineapple, apple banana, mango, papaya, jackfruits), spices and essential oils. Most of the mobilized groups have been supported to attain organic third party certification.

Some of the organic produce sold on the local market is sold through Shop Organic, a retail outlet started by NOGAMU in Kampala. Shop Organic was started in 2002 to serve as an outlet for farmers' organic products, but also as a training medium through which farmers and processors could learn and improve product quality, packaging and product development.

NOGAMU established an organic trade point in 2008 at its office premises to act as a one-stop information centre for all organic sector information in Uganda. The organic trade point has a database of exporters, producers, processors, farmers, buyers, importers, existing markets, prices and volumes of products available.

During the period 2009–2011, NOGAMU coordinated the participation and presentation of Ugandan producers and exporters in the African Pavilion at the BioFach (in collaboration with the Tanzanian Organic Agriculture Movement (TOAM)) and registered a big success. This success translated into a number of requests and orders, especially for dried fruits, from organic importers.

Besides mobilizing producers, NOGAMU sensitizes consumers and creates awareness of organic production and marketing as part of its market development drive. It provides market information to exporters and facilitates their participation in local and international exhibitions. It also links exporters and potential clients to producers.

Developing and reviewing standards

Over the years, NOGAMU has taken an active role in developing and reviewing organic standards. These standards define the procedures for growing, processing, labelling and inspecting organic products with the aim of building trust between farmers, processors and consumers.

With support from EPOPA, NOGAMU started developing the Uganda Organic Standard, in 2001. Between 2002 and 2004 NOGAMU mobilized stakeholders in the organic sector to make various contributions to the standard. Technical support from Grolink Ab and financial support from SIDA through the EPOPA project boosted the process of developing a Ugandan Certification Programme in 2002 (EPOPA, 2008b). The Uganda Certification Programme consisted of developing the Uganda Organic Standard and establishing a local certification company, Uganda Organic Certification Company Ltd (UgoCert), to provide certification services.

The Uganda Organic Standard is a guide to certification based on the Ugandan situation with regard to the international organic principles. It is built on the IFOAM basic standards and is owned by NOGAMU and UgoCert. It serves as a tool for actors in the organic production sector such as advisors, researchers, policy makers and government institutions involved in the development of OA, for ensuring compliance.

After adopting the Uganda Organic Standard in 2004, NOGAMU spearheaded the development of regional standards for the East African region with support from EPOPA. It mobilized fellow NOAMs in the East African member countries, Kenya, Tanzania, Rwanda and Burundi, as well as their national bureaux of standards to become part of the multi-stakeholder process which involved intensive consultations and participation by national governments, the private sector, NGOs, and international institutions. This process was supported by the joint UNEP and the UNCTAD 'Capacity Building Task Force on Trade, Environment and Development' (UNEP-UNCTAD CBTF) initiative.

The EAOPS was adopted in 2007 by the East African Community as the single, official standard for OA production in the region. It was launched in Dar es Salaam on 28 May 2007 (EPOPA, 2008b).

An organic mark (the East African Organic Mark) was registered together with the EAOPS and is owned by the three NOAMs of Kenya, Tanzania and Uganda (Kenyan Organic Agriculture Network (KOAN), TOAM and NOGAMU).

It is used on products that conform to the EAOPS. To use the mark on a product, one has to register with one of the NOAMs in Kenya, Tanzania or Uganda. Any product that bears the East African Organic Mark must comply with the EAOPS or equivalent rules. The mark is currently used on dried fruits, honey, nuts, coffee and tea that are sold as organic, or 'Kilimohai' in East African markets.

In May 2017, NOGAMU signed a memorandum of understanding with the Uganda National Bureau of Statistics to promote the implementation of OA and product-related standards, practices and organic certification in Uganda. This initiative was intended to strengthen inspection, auditing and certification of organic products to meet local and export market requirements.

The EPOPA project

The EPOPA project was important for small farmer development and capacity building in Uganda in the early days. Paraphrased extracts from the Executive Summary of the final report (EPOPA, 2008a) are presented in Box 15.1 for context.

Establishing a local certification body

One of the challenges NOGAMU noted early in the marketing of organic products is the high cost of certification. As an initiative to support organic producers, NOGAMU together with shareholders, established an independent certification body, the Uganda Organic Certification Company Ltd (UgoCert) in 2004, to offer credible certification services against the Uganda Organic Standard and reduce certification fees, initially with accreditation from the certification body Institute of Market-ecology (now amalgamated with Ecocert). Later in 2005 UgoCert attained its independent accreditation to the EU regulation.

Conducting advocacy on policy issues

Since 2004, NOGAMU has played a major role in organizing and facilitating the drafting and development of an organic agriculture policy. In 2003 the Ministry of Agriculture, Animal Industry and Fisheries (MAAIF) instituted the Organic Policy Development Committee to work on the draft policy (Taylor, 2006) and expected it to be ready by the end of 2005. The process however, has been much slower than formerly expected. In October 2016, MAAIF presented to stakeholders the final draft of the National Organic Agriculture Policy, which it plans to present to parliament together with its regulatory impact assessment and the action plan for its implementation.

NOGAMU has also been instrumental in advocating for the inclusion of OA in the Uganda National Development Plan.

With support from EPOPA and SDC, OA has been institutionalized in tertiary institutions such as Uganda Martyrs University where a bachelor's degree in organic agriculture was started in 2005.

Networking

In its role as a coordinating organization, NOGAMU has established various networks with academic institutions (e.g. Makerere University and Uganda Martyrs University at Nkozi), government ministries (e.g. MAAIF and the Ministry of Trade, Industry and Co-operatives) and related organizations (e.g. the National Environment Authority, National Agriculture Research Organization, Uganda Export Promotion Board, Uganda Investment Authority and Uganda National Bureau of Standards) as well as international organizations. These networks have enabled NOGAMU to lobby successfully for inclusion of OA in various programmes, and consideration of organic enterprises, producer organizations and companies for support by private and public institutions.

Impact on Farmers' Livelihoods

There is increasing adoption of OA by smallholder farmers in Uganda. This can be attributed partly to NOGAMU's efforts to sensitize farmers to OA, and supporting them to market organic produce. In general, NOGAMU has

Box 15.1. Paraphrased extracts from the Executive Summary of EPOPA (2008a)

The EPOPA programme was initiated in the mid-1990s by GroLink and Agro-Eco with support from SIDA. In the period 2002–2007 it was considerably scaled up and subsequently phased out in 2008. It operated in Tanzania and Uganda and briefly in Zambia. EPOPA was a 'development through trade' programme with the objective of improving the livelihoods of rural communities through exports of organic products. Exporters were the main partners and the programme worked directly with them to develop exports of organic products. The programme also supported emerging institutions in the organic sector.

Farmers ... sold organic products for approximately US$15 million/year and the total export value is more than double that amount. A total of 110,000 farms have participated, but only 80,000 have actively delivered products to the exporters. ... Some 600,000 people have been beneficiaries of the programme. The cost of the programme for the Swedish taxpayers was one cup of coffee per taxpayer.

In order to set up a successful export project, there was a need to find the right mix of the following:

1. A willing and capable exporter.
2. A production base (i.e. willing farmers with basic knowledge of production in an area with suitable conditions).
3. Market demand.
4. Products that could be competitive in quality and price.

Hardly any funds were made available for investments or other incentives for the participating exporters. The focus of the programme was to create viable businesses, and EPOPA assisted the actors through a wide range of services, from farmer training to marketing and certification.

The participating farmers were smallholders. Most of them ... used almost no agrochemical inputs before participating in the programme, but also did not manage soil fertility effectively. Organic farming itself posed few problems for the participating farmers. Despite the great variety of crops and the large number of farmers, there were no insurmountable problems in the production or with pests. ... [It was hoped that] improved crop rotations, better nutrient recycling, cover crops and green manures and soil conservation would result from project initiatives] but that didn't happen to a very significant extent. Farmers experienced improved food security, largely as a result of increased income, as well as generally improved livelihoods, as demonstrated by improvement in housing, children attending school, and investments in farming.

Government support is needed for the progress of OA: [there is a] lack of supportive policies, but perhaps even more the existence of policies that are harmful to development. Therefore, a programme like EPOPA, despite its private-sector focus, also has to engage in policy dialogue and action. The continued strong demand for organic products and the increased policy support contributed to the success of EPOPA. Other important success factors were:

- clear market focus of the projects and focus on tangible results;
- using commercial actors to link farmers to markets;
- integrating extension work into the commercial chain so that the exporters are responsible for extension work, enhanced by income from the trade; and
- the use of group certification to facilitate the certification process.

Central to the implementation of the projects was the establishment, by the exporters, of a field organization for extension work and for internal control of issues related to certification. All in all, the organization worked, but most of its energy was absorbed by certification issues, and the efficiency of the agronomic advice in many of the projects can be questioned. This was not a main interest of the exporter.

A main challenge to the programme was finding competent and committed exporters. The organic market represented something new for the exporters, and it took quite a while for them to adjust. Project periods were 3 years, but this clearly was too short in most cases; agricultural projects in general need a longer time. Extensions were awarded mainly to improve the sustainability of the venture. Value addition in developing countries is an appealing proposition, but it is not always so easy to achieve. Many of the projects that included value addition experienced major challenges, in particular regarding product design and imported packaging materials and inputs.

Continued

Box 15.1. Continued.

In most of the projects, large groups of farmers were involved, and they did experience a substantial increase in income, expressed as a percentage. However, especially for those producing basic commodities, the increased income was not sufficient to lift them out of poverty. For farmers producing high-value crops, such as cashew, fresh fruits and spices, the increased income is substantial in absolute terms also.

The support to emerging institutions, such as local certification bodies and national organic movements, was successful. There are now organic standards and internationally accredited certification bodies in Tanzania and Uganda and the NOAMs are involved in local market development, advocacy and policy development.

Working with the commercial sector to develop agribusiness involving many smallholders has proved to be successful. Business objectives of commercial actors may not be the same as the objectives of the development cooperation, but with good design, dialogue and pragmatic implementation they can work well together. The organic markets do provide special incentives. The organic production system is well adapted to African smallholders and is sustainable. Apart from the effects on income, organic farming also produces public goods and ecosystem services such as carbon sequestration and biodiversity.

impacted farmers' livelihoods in the four major areas elaborated below.

Increased food and nutrition security

In all areas where NOGAMU has worked, organic farmers are more food and nutrition secure than other farmers in the same locality. This has been reported earlier by Hine *et al.* (2008), in a study where they observed that adoption of OA led to increased sustainability and quantity of food produced per farm. The increase in food production leads to household food security and results in all members of the household having access to enough and nutritious food.

Better incomes

The increase in food production on organic farms enables farmers to sell off bigger surpluses at local markets and hence benefit from higher incomes. Where organic farmers have access to organic markets they benefit from premium prices, which increases their incomes further.

Access to functional markets

Organic farmers linked to buyers by NOGAMU have benefited from access to assured and regular markets. Besides increasing their bargaining power, this has increased their capacity in organic production, diversification and trade, which acts as an insurance for community development.

Building social capital

NOGAMU works with smallholder farmers, who cannot produce enough output consistently to attract a buyer and usually cannot afford to pay certification fees individually to be certified as organic. The strategy used to overcome these two challenges is to mobilize the farmers into groups or producer organizations so that they can bulk their produce and market together. They can also get a certificate as a group, which reduces unit certification costs.

Working in groups helps farmers to interact, increase knowledge and skills transfer, build capacity in solving local challenges and improve social capital resulting in stronger social organizations at local level. One of the approaches used by NOGAMU which seems to build social capital in the local community is the Farmer Family Learning Group approach (Vaarst *et al.*, 2012). When more than one member of a family is supported, the family is more likely to change the way they farm, especially when there is a local learning group.

Farmers participating in OA have improved capacity to participate in self-help projects within communities to provide social services and support others to create a shared vision and functional networks for community development.

Challenges

There are various challenges that affect NOGAMU as an organization as well as those that affect its operations and potential impact.

Lack of a government policy on OA

The process to develop the National Organic Agriculture Policy has been going on since 2004, but it has not been concluded up to now. This means that no funds from government can be expended to support organic activities. Mobilization into viable commercial groups is only supported by the private sector. All activities implemented so far in OA have been supported by the private sector and donors. Lack of government investment in the organic sector limits the impact NOGAMU would have on its target group.

Lack of legislation on organic production standards

Although the Uganda Organic Standard was adopted in 2004, it is not legislated in Uganda as a national standard. This means that organic farmers implement the Uganda Organic Standard and EAOPS as voluntary standards and they are open to impostors and false claims, though the EAOPs is legally in the custody of the East African Community.

Limited research on OA

Research gaps and aspirations in organic farming have been determined, but most ongoing research is focused on validating indigenous practices. There is a lack of comprehensive local research on OA. Organic practitioners rely on experiences and organic research from other geographical areas as well as related conventional agriculture research. Thus far, National Agriculture Research Systems (NARS) have not embraced OA.

Low investment in OA production and marketing

Uganda faces the same agri-financing challenges as other developing countries. Organic companies involved in agriprocessing, farmer cooperatives and small-, micro- and medium-scale enterprises (SMEs) engaged in commodities like coffee, cocoa, chia seed and sesame are interested in expanding their processing capacity, but finance is a major limiting factor. This means that some NOGAMU members are limited in exploiting the full potential of the market opportunities NOGAMU mobilizes for them. Exporters and other traders repeatedly fail to meet the orders and requests for organic products.

Conclusion

OA in Uganda has been promoted by a number of actors for more than 20 years. Its institutionalization under NOGAMU marked a turning point that has seen OA develop faster than during the first 10 years. Promotion and coordination of OA by a single entity has played a major role in development.

Privately led management is not enough to maximize and optimize benefits from existing opportunities. Therefore, getting public sector involvement and support is key to further development of the organic sector. Furthermore, though certification is a marketing tool, sometimes it may limit growth of the organic sector especially when certification costs are not based on realistic market forces of supply and demand, which influence all other monetary factors. Hence the need to adopt other user-friendly and market-related systems such as PGS. In the context of climate smart agriculture and Organic 3.0 (see Introduction and Chapter 2, this volume), the organic sector is likely to become more attractive as a viable alternative system.

References

EPOPA (Export Promotion of Organic Products from Africa) (2008a) *Organic Exports: a Way to a Better Life?* Agro Eco BV and Grolink aB, Hoeje, Sweden.

EPOPA (Export Promotion of Organic Products from Africa) (2008b) *EPOPA Newsletter* No. 9, 2008. Agro Eco BV and Grolink AB, Hoeje, Sweden.

Hine, R., Pretty, J. and Twarog, S. (2008) *Organic Agriculture and Food Security in Africa*. United Nations Organization, New York.

NOGAMU (National Organic Agricultural Movement of Uganda) (2015) *The National Organic Agricultural Movement of Uganda Annual Report 2015*. NOGAMU, Kampala, Uganda.

NOGAMU (National Organic Agricultural Movement of Uganda) (2017) *Annual Report 2017*. NOGAMU, Kampala, Uganda.

Taylor, A. (2006) *Overview of the Current State of Organic Agriculture in Kenya, Uganda and the United Republic of Tanzania and the Opportunities for Regional Harmonization*. United Nations Organization, New York.

UNDP (United Nations Development Programme) (2015) Sustainable Development Goals. Available at: www.undp.org/content/dam/undp/library/corporate/brochure/SDGs_Booklet_Web_En.pdf (accessed 21 December 2018).

Vaarst, M., Nalunga, J., Tibasiima, T., Dissing, A. and Dissing, I. (2012) *The Rwenzori Experience: the Farmer Family Learning Group Approach to Human and Social Capital Building, Environmental Care and Food Sovereignty*. Third World Network, Penang, Malaysia.

Willer, H. and Lernoud, J. (2017) *The World of Organic Agriculture: Statistics and Emerging Trends 2017*. FiBL/IFOAM, Bonn, Germany.

16 Factors Contributing to Adoption or Disadoption of Organic Agriculture in Zambia

Robert Munthali,[1,2]* Raymond Auerbach[2] and Mebelo Mataa[3]
[1]*Klein Karoo Seed Zambia, Lusaka, Zambia;* [2]*Nelson Mandela University, George, South Africa;* [3]*University of Zambia, Lusaka, Zambia*

Abstract

The negative effects of conventional agriculture on the environment in terms of land degradation and pollution have accelerated efforts to develop sustainable agricultural systems. In Zambia the last 20 years or so have seen the promotion of organic farming as a sustainable farming system option, with many farmers adopting the system, but some certified organic farmers later allowed their certification to lapse. This chapter presents a synopsis of the current situation, and examines these developments. Data were collected from secondary sources, focus group discussions and by administering a structured questionnaire. There are approximately 250 farmers (both adopters and disadopters of organic farming); of these the accessible population of adopters and disadopters was 50 farmers selected across identified areas using systematic random sampling methods. The number of organic farmers in the country has been declining, which has affected production. This chapter describes what practices are working effectively in organic production in Zambia, and how they have contributed to adoption.

Factors that have enhanced adoption of organic practices were: (i) farmers know that organic farming encourages biodiversity on the farm such as trees, soil microorganisms, plants and animal life; (ii) food produced through organic farming is perceived to be free from harmful substances such as fertilizers, insecticides and herbicides; (iii) organic farming helps in reducing the effects of global warming and climate change; (iv) manures and composts are much cheaper than synthetic fertilizers where these are available on the farm as local residue materials; (v) in many instances organic farming helps reduce soil erosion on the farm; (vi) it is believed organic foods are comparatively richer in nutritional value when compared with conventional foods; and (vii) organic farming systems do not permit the use of genetically modified organisms.

Some of the reasons contributing to disadoption included: (i) absence of effective organic farming extension services; (ii) it is easier to access inputs for conventional agriculture from government and private companies promoting outgrower schemes; (iii) organic inputs such as manure are bulky and not easy to transport; (iv) making compost from manure in organic farming is seen as both labour intensive and expensive when compared to using synthetic fertilizers; (v) consumers are unwilling to pay premium prices for organic products; and (vi) the local population at present do not appreciate organic foods hence they cannot differentiate them from conventional foods.

It is anticipated that these findings will contribute to developing interventions to improve production and productivity of organic farming systems. Although organic farming is promoted as a plausible production system for sustainable agriculture, it will require comparatively more structural support including specialized extension, certification systems and input providers if its full potential is to be realized and the noted decline in numbers of growers is to be halted.

*munthalirobert@yahoo.co.uk

Introduction

Accelerated agricultural growth is imperative for alleviating poverty (Lyne *et al.*, 2009), and has been identified as the vehicle for economic development and addressing Millennium Development Goals (now Sustainable Development Goals) in Africa (NEPAD, 2003; Chapter 15, this volume). From the time when modern agriculture started there has been a significant upsurge in agricultural production through increased output per unit of land planted, and this is particularly true among high-input precision farmers. However, productivity among the majority of small-scale farmers still remains low (UNCTAD, 2008). Even though modern farming systems such as industrial agriculture have to a large extent resulted in remarkable increases in output for some crops, it has been observed that the benefits of increased food production are short lived. The development has come at a great cost leading to ecological problems such as global warming and pollution resulting from excessive use of chemical fertilizers and pesticides (UNCTAD, 2008; Introduction and Chapter 12, this volume). If agriculture is to continue benefiting from ecosystem services there is a need to balance output with enhancement of environmental conservation, otherwise continued use of injudicious methods of agricultural production will result in impaired ecosystem function, leading to severe consequences for both humankind and ecosystems. It is therefore necessary to develop alternative approaches. In an attempt to mitigate some of the anticipated challenges, in the early 1990s Zambia adopted organic farming as an alternative farming system. This is practised side by side with conventional farming. Unfortunately, despite a good beginning, organic farming in Zambia has had its own serious challenges.

As at the year 2018, total area cultivated under various crops annually in Zambia is estimated at approximately 6 million ha (Braimoh *et al.*, 2018). Of this area only a very small portion is under certified organic cultivation (Willer and Lernoud, 2018). Recent data indicate that total land area under organic management in Zambia is about 0.03% of the total agricultural area in the country farmed by 10,061 certified organic farmers (Willer and Lernoud, 2018), though attempts to identify these farmers proved problematic. The development of the organic sector in Zambia started in 1990 when an idea was conceived to establish an organic association affiliated to the Zambia National Farmers Union (OPPAZ, 2008). The idea culminated in the formation of the Environmental Conservation Association of Zambia. During the same period three agricultural and natural-based enterprises (Agriflora Zambia Limited, Mpongwe Development Corporation Limited and North-Western Beekeeper Project) started exporting organic products to the UK. In 1999 a group of farmers and interested institutions and individuals came together and formed the Organic Producers and Processors Association of Zambia (OPPAZ), which subsequently was registered as a non-governmental organization in 2000, to spearhead the organic agriculture (OA) movement in the country. From that time organic farming in Zambia started growing at an accelerated rate for several years. However, after a few years, production began to decline as demonstrated by the small number of farmers actively involved in organic production at present.

In view of these events it is important to determine critical factors in adoption or disadoption of organic farming systems in the country (FAO, 2013). This chapter describes a research study that examined these factors in Zambia.

Method

The first step in the process was to identify areas to be explored. This information was obtained from the Ministry of Agriculture (MoA) and OPPAZ. It was observed that organic production in Zambia was more predominant in six provinces out of ten, with only ten districts out of a total of 106 in the country actively involved with organic farming. Based on data provided, four sites were selected in areas with high levels of organic farming activities. The four districts recommended by the MoA and OPPAZ were Mazabuka, Kafue, Lusaka, Chongwe and Chibombo.

The next step in the process was drawing samples of organic farmers for the administration of questionnaires. Samples were drawn using a systematic random sampling method. It was important to make a distinction between the generalized populations of organic farmers, and the randomly selected/accessible population of organic farmers (Trochim, 2000). The former is termed as the theoretical population and the latter the accessible population. The total population of certified organic farmers which could be identified comprised approximately 250 farmers both adopters and disadopters; out of this the accessible population of adopters and disadopters was 50 farmers selected across identified areas. Using a systematic random sampling method, the population of organic farmers was listed in a random order. In this respect the sampling fraction used was $f = 50/250 = 20\%$. The interval size, k, was equal to $N/n = 250/50 = 5$. Based on the calculations the random integer selected was from one to five implying a number was selected at random every five numbers, giving a research sample of 50 farmers. That is the number of organic farmers given questionnaires to complete. Within this figure 15 questionnaires were disadopters and the remaining 35 questionnaires the adopters.

Findings

According to information collected from the Chief Executive Officer of OPPAZ (M. Chitalu, Lusaka, 15 October 2016, personal communication by e-mail to Raymond Auerbach), the current number of organic farmers certified in arable crops in Zambia is about 4000; all of them are on group certification schemes of nine small-farmer groups throughout the country. Regrettably, five of these groups withdrew from certification because of perceived intangible benefits of organic certification in the country where the market does not segregate the pricing. According to Willer and Lernoud (2015), in the year 2014 there were approximately 10,055 certified organic farmers operating in the country. At various points it was claimed that the country had between 40,000 and 60,000 certified organic farmers operating; however, these figures included certified bee forage farmers. The discrepancies in the number of organic farmers in the country cast doubt on the reliability of available data. This state of affairs is a clear reflection of the magnitude of the problem facing the organic sector in the country. In order to understand the organic sector it was imperative to look at what practices are operating effectively in organic production in Zambia, and what factors have been influencing adoption and disadoption of organic farming systems, and last but not least, organic certification.

Definitions of OA abound; Wren (2007) defined OA as an integrated farming system that is based on ecological principles, where farmers cycle nutrients on the farm and work with the soil to improve fertility. Rather than using synthetic fertilizers, organic farmers promote biodiversity on their farms, and pests and weeds are controlled through rotation, mechanical means and by supporting diverse populations of plants, insects, and other organisms. The United States Department of Agriculture (USDA) defines organic farming as a production system that avoids the use of synthetic fertilizers, pesticides, growth regulators and livestock feed additives (Lampkin, 1999). Organic farming systems largely depend on crop rotation, crop residues, animal manures, legumes, green manures, off-farm organic wastes, and aspects of biological pest control to maintain soil productivity and tilth, to supply plant nutrients, and to control insect pests and weeds (Lampkin, 1999). The International Federation of Organic Agriculture Movements (IFOAM) definition of OA was given in Chapters 2 and 4 of this volume.

According to the IFOAM (2015) there are four principles of OA, namely:

1. **The principle of health:** OA should sustain and enhance the health of soil, plant, animal and human as one and indivisible.
2. **The principle of care:** OA should be managed in a precautionary and responsible manner to protect the health and well-being of current and future generations and the environment.
3. **The principle of ecology:** OA should be based on living ecological systems and cycles, work with them, emulate them and help sustain them.
4. **The principle of fairness:** OA should be built upon relationships that ensure fairness with regard to the common environment and life opportunities.

Long-term studies have shown that a crop produced by an organic farming system usually yields less than if it was grown in a conventional farming system (Chitja *et al.*, 2009). However, work at the Rodale Institute (LaSalle *et al.*, 2011) indicates that once organic systems are optimized, yields can come close to those of conventional farming, even exceeding conventional yields in dry years. Most of the above long-term studies show that once soil biology has improved, organic treatments take 2–4 years to stabilize at these higher yield levels (LaSalle *et al.*, 2011). Long-term research at Nelson Mandela University has shown that it is possible to eliminate the yield gap between conventional and organic farming systems after 3 years (Auerbach, 2017; Chapters 18–22, this volume).

So far information gathered from small-scale farmers in Zambia shows that maize (*Zea mays*) yields from organic farming systems tend to be less than those of conventional farming systems. However, according to organic farmers this is to a large extent dependant on the type of seed planted (whether an open-pollinated variety or hybrid). It is a known fact that hybrid maize seed cultivars currently on the market are high yielding with yield potential ranging from 6 t/ha to 14 t/ha depending on how the farmers manage their crops. In contrast, organic farmers use an indigenous maize seed variety locally known as Gankata which they save and replant. This is possibly due to its tolerance of low nutrient supply. Using this local variety some organic farmers are able to obtain yields of 4–5 t/ha competing favourably with most improved open-pollinated varieties of maize selling on the market at present. While hybrid seeds and improved open-pollinated varieties are produced by seed companies using outgrowers, the local variety is on-farm saved seed, where a farmer will select cobs from the previous harvest that are not diseased or contaminated and of good germination capacity.

Other than the type of seed used, there are several other factors that contribute to realizing higher yields; good crop management is one such factor. As previously mentioned the majority of organic farmers in the country are small scale and they generally use appropriate technologies. According to data collected during the survey the most common land tillage method is now minimum tillage. Land preparation is done either by hand hoeing or using an ox-drawn plough. Sometimes this is determined by the farmer's level of operation coupled with resource availability, as the majority of small-scale farmers lack resources. Seed is sown by hand after ripping planting holes within the row using a hoe or ripper and rope to make sure rows are straight.

In terms of organic fertilizer application, many tend to apply a split application of manure. The first dosage of manure is usually applied before sowing the seed. The second dosage when the plant is about 30 cm high or about 4 weeks after sowing time. On average about 2 t/ha of compost or animal manure is applied. This was below the recommended minimum of 5 t/ha (Munyinda *et al.*, 2015). In situations where composted manure is used, composting is done on the farm and this generally takes about 8 weeks to make. During crop development stages, the field is at most weeded twice. Many small-scale farmers prefer mechanical weeding, using a hand hoe or ox-drawn cultivator, because currently there is no other method available to farmers, as in organic farming use of herbicides is not permitted.

Similarly, with conventional crops, insect pests and disease occurrences are also common in organic crops. If remedial measures are not carried out in a timely manner there is a likelihood the entire crop could be wiped out. The feedback from farmers so far suggests the most common insect pests in organic crops, in particular legumes, are aphids (*Aphis* spp. or members of the superfamily Aphidoidea). In maize conspicuous pests are maize stalkborers (*Busseola fusca*, *Chilo partellus* and *Sesamia calamistis*) and cutworm (*Agrotis* spp.), while diseases comprise mainly grey leaf spot (*Cercospora zeae-maydis*) and rust (*Puccinia sorghi*). Organic farmers use natural remedies to control insect pests and diseases. For control of specific insect pests and diseases most of the farmers recommended use of ash and chilli powder, claiming it provides the most effective control. In stored grains the most problematic pests are weevils, and these are controlled using neem (*Azadirachta indica* A.) and tephrosia (*Tephrosia purpurea*) powder. Despite the fact that many farmers are content with these control measures, others have complained of sluggish efficacy of natural remedies. Possibly the problem could be attributed to the ratios used in mixing the ingredients by the different farmers because in some instances farmers do not have written instructions.

Adoption and disadoption of OA in Zambia

It is acknowledged that organic farming is a knowledge-intensive farming system (Chitja and Hendriks, 2008) for which there is currently insufficient appropriate location-specific information related to production, pest and disease control, marketing and certification (Chitja *et al.*, 2009). Although in organic farming certain practices have proved to be working effectively, challenges have continued to daunt the system. This is witnessed by the indecisiveness depicted by some organic farmers. In Zambia a good example is in the early stages of development when organic farming was introduced in the country and several farmers including those farming on a small scale converted to organic production. Within a few years, many of the farmers began to abandon organic farming systems, and instead reverted to conventional farming. This is confirmed by data currently available where the number of farmers has reduced by thousands within a few decades, after strong early growth.

In response to questions asked in the questionnaire, respondents brought forward a number of reasons why they decided to abandon organic production systems. Our study showed that some critical reasons leading to increased numbers of disadopters include the following:

- Farmers have a perception that compared with organic farming, conventional farming extension support is readily available from both private companies and government extension workers.
- It has been easier to access inputs for conventional agriculture from government and private companies promoting outgrower schemes. An example is the Farmer Input Support Programme through which the government supports nearly a million small-scale farmers annually with inputs (specifically seed and fertilizer).
- In conventional farming, inputs such as fertilizers are applied with precision and in less bulky volumes that are easy to transport, unlike the bulky manure. Additionally, making compost from manure in organic farming is both labour intensive and expensive when compared with using synthetic fertilizers.
- Consumers are unwilling to pay premium prices for organic products. Apparently, the local population at present does not appreciate organic foods hence they do not differentiate them from conventional foods. The determining factor for most consumers is the price, and unfortunately suppliers of organic food crops have complained that the prices do not segregate.
- There is a perception that conventional farming systems are less labour intensive than organic farming.
- Because of the conversion period required when converting to conventional farming, farmers cannot commence production straight away.
- Agrochemicals are more rapidly effective in controlling insect pests and diseases, and weeds. Weeds are easily controlled using herbicides. This is much cheaper and easier than mechanical weeding.
- In organic farming, stakeholders only perceive a benefit when there is a donor-funded project supporting organic production. Immediately the project comes to an end, farmers are forced to look for other support and resources.
- Non-organic products have a ready market and are in most instances cheap because they do not attract a premium price.
- It is perceived that conventional yields are higher than in organic farming, therefore translating into higher profits per unit area planted.
- Organic farming takes a long time to build soil fertility, and to realize the benefits; farmers have to persevere for some time without immediate returns.

In contrast the proponents of organic farming systems equally have different views and perceptions as to why they have kept supporting this type of farming system. Estimates of the number of certified organic farmers in Zambia vary widely, and this is disturbing, as the actual numbers appear to be much lower than claimed, and even those making the higher claims are not able to provide details to substantiate these claims. It seems that the situation is exacerbated by the lack of coordination between the participants in the organic farming system, partly due to the inconsistencies in information provided by numerous stakeholders. Not even the MoA is able to provide such data. The only organization in the country that was in a position to provide any information is OPPAZ, and even they are not operating at full capacity.

Some of the reasons certified organic farmers or adopters advanced for continuing supporting the system include:

- Organic farming systems are environmentally responsive, excessive use of chemical fertilizers usually exhaust the soils in the long term.
- Organic farming is regenerative farming, meaning that the present generation is able to derive their livelihood from resources currently available and future generations will also be able to sustain livelihoods using the same resources if they are sustainably managed.
- Organic farming encourages biodiversity on the farm such as trees, soil organisms, plants and animal life.
- Food produced through organic farming is perceived to be free from harmful substances such as traces of fertilizers, insecticides and herbicides.
- In organic farming, burning of plant residue materials is not allowed; instead the materials are converted into organic fertilizer by composting. This is helping a lot in reducing the effects of global warming and climate change.
- Organic farmers believe manures and composts are much cheaper than synthetic fertilizers because farmers can make these on the farm using local residue materials.
- In many instances organic farming helps reduce soil erosion on the farm.
- Use of manure in organic farming helps build healthy soils that are rich in organic matter more effectively than in conventional farming. In organic farming the saying goes that 'you feed the soil and not the plant'.
- Pesticides and synthetic fertilizers are not permitted in organic farming. Farmers rather use manure and compost as fertilizers. Insect pests and disease control are achieved using natural remedies.
- Organic farmers also believe both organic and conventional farming systems are labour intensive because most of the practices are similar. Apparently, conventional farmers always single out organic farming as labour intensive.
- In organic farming, soil fertility is replenished using animal manure, green manures and composted manure, which on a comparative basis are cheaper than synthetic fertilizers therefore significantly reducing the cost of production on the farm.
- It is commonly believed that organic foods are comparatively richer in nutritional value when compared with conventional foods.
- Organic farming systems do not permit the use of genetically modified organisms.

Even if there are several benefits farmers derive from organic farming, it must be appreciated this system of farming is not without its challenges. According to information gathered from organic farmers there are equally many problems which organic farmers are faced with. Common problems synonymous with organic farming systems include:

- *Weeds*: This has been found to be a serious problem in fields where farmers use animal and poultry manure. In explanation farmers suggested the problem was more common when using manure collected from animals that are left to feed under free-range or uncontrolled grazing systems.
- *High cost of labour*: Among other factors, dependent on the area that the farmer intends to plant and quantity of manure to be applied per hectare, the cost of making composted manure may be exceedingly high.
- *Labour intensive*: Most of the work is done manually, from land preparation to harvesting. However, this is not too different from labour costs incurred in conventional farming because the practices are almost similar.
- *Lack of resources*: The majority of organic farmers in the country are small scale, just like their counterparts practising conventional farming. The main complaint brought forward when conducting the survey was lack of resources. This has negatively affected the sector, which currently is in a state of decline.
- *Traditional land*: Farmers cannot access financing for investment from financial institutions because the majority of farmers are on traditional land without title deeds. This

has made the situation difficult because without title deeds land cannot be used as collateral to secure support from banks.

- *Insect pests and diseases*: As much as they can be controlled using natural remedies in certain situations, it is quite difficult to achieve complete control. Effectiveness depends mostly on the potency and efficacy of the product used.

It was not easy to determine the actual number of certified organic farmers in the country. Previously, all the organizations engaged in organic production in the country were affiliated to OPPAZ. This is no longer the case. Instead there are a number of organic organizations operating autonomously, mostly in the traditional organic areas. Previously, OPPAZ facilitated certification for all its members. Unfortunately, today organic certification remains a problem. The individual certification category is still a barrier to most of the small-scale farmers because of the high fees charged. As an alternative, the majority of the farmers are now certified under group certification, where each farmer contributes a certain amount of cash to meet the total required, and a quality manager is supposed to manage an Internal Quality Management System. Many of the small-scale farmers are only certified to supply the local markets through PGS.

Conclusion

This chapter highlighted important aspects of organic farming systems in Zambia. It described what practices are working effectively in organic production with particular attention to land preparation, fertilization, and insect pest and disease control. Two other important aspects are factors leading to adoption and disadoption of organic farming systems in the country. At inception the country had a very vibrant organic farming sector that was even able to export organic products to countries within and outside Africa, and as far away as Europe. However, the latest information suggests that organic production has slowly been declining. This is evidenced by the decreasing number of organic farmers in the country. Currently it is not feasible to know the exact number of farmers because organic activities in the country are no longer centrally coordinated as before. Organic certification is another important aspect of organic production. When funds were available, OPPAZ provided organic farmers with the necessary knowledge required in the certification process. However, the cost of obtaining certification is now quite exorbitant especially for the majority of small-scale farmers. Many of them cannot afford individual certification, hence they are encouraged to set up group certification, which is cheaper and affordable because each individual farmer makes a contribution.

Although organic farming is increasingly viewed and promoted as a plausible production system for sustainable agriculture, it is a specialized form of agriculture and requires information beyond the capacity of conventional extension services. It will require comparatively more structural support including specialized extension, certification systems and input providers if its full potential is to be realized and the noted decline in numbers of growers is to be halted.

References

Auerbach, R. (2017) A comparison between organic and conventional farming systems in South Africa's Southern Cape. Paper presented at the 23rd Annual ISDRS (International Sustainable Development Research Society) Conference in Bogota, Colombia, 14–16 June.

Braimoh, A., Mwanakasale, A., Chapoto, A., Rubaiza, R., Chisanga, B., Mubanga, N., Samboko, P., Giertz, A. and Obuya, G. (2018) *Increasing Agricultural Resilience through Better Risk Management in Zambia*. World Bank, Washington, DC.

Chitja, T.J. and Hendriks, L.S. (2008) Emerging issues in smallholder organic production and marketing in South Africa. *Development Southern Africa* 25(3), 135–150.

Chitja, J.M., Hendricks, S.L., Nsahlai, V.I. and Copeland, P. (2009) Decision making for organic crop production choices at Embo, Centocow and Muden. In: Hendricks, S.L. and Lyne, M.C. (eds) *Does Food Security Improve When Smallholders Access a Niche Market? Lessons from Embo Community in South Africa*. African Centre for Food Security, University of KwaZulu–Natal, Pietermaritzburg, South Africa, pp. 29–41.

FAO (Food and Agriculture Organization of the United Nations) (2013) Sustainable Assessment of Food and Agricultural Systems Guidelines. Available at: www.fao.org/nr/sustainability/sustainability-assessments-safa (accessed 31 July 2018).

IFOAM (International Federation of Organic Agriculture Movements) (2015) *IFOAM Approved Principles of Organic Agriculture*. IFOAM, Bonn, Germany.

Lampkin, N. (1999) *Organic Agriculture*. Farming Press, Ipswich, UK.

LaSalle, T., Hepperly, P. and Diop, A. (eds) (2011) *The Organic Green Revolution*. Rodale Institute, Kutztown, Pennsylvania.

Lyne, M.C., Hendricks, S.L. and Chitja, J.M. (2009) Agricultural growth and food security. In: Hendricks, S.L. and Lyne, M.C. (eds) *Does Food Security Improve When Smallholders Access a Niche Market? Lessons from Embo Community in South Africa*. African Centre for Food Security, University of KwaZulu–Natal, Pietermaritzburg, South Africa, pp. 1–10.

Munyinda, K., Mataa, M., Kalinda, T. and Walubita, K. (2015) *Nationally Appropriate Mitigation Action (NAMA) Sustainable Agriculture through Integrated Crop and Livestock Farming*. United Nations Development Programme, United Nations Organization, New York.

NEPAD (New Partnership for Africa's Development) (2003) *Comprehensive Africa Development Programme*. NEPAD, Pretoria.

OPPAZ (Organic Producers and Processors Association of Zambia) (2006) *Organic Sector Development in Zambia*. OPPAZ, Lusaka, Zambia.

OPPAZ (Organic Producers and Processors Association of Zambia) (2008) The status of organic farming in Africa: a case presentation for Zambia. Presentation by OPPAZ at BioFach, Nuremburg, Germany, 23 February.

Trochim, W. (2000) *The Research Methods Knowledge Base*, 2nd edn. Atomic Dog Publishing, Cincinnati, Ohio.

UNCTAD (United Nations Conference on Trade and Development) (2008) *Organic Agriculture and Food Security in Africa*. United Nations Environment Programme (UNEP)-UNCTAD Capacity-building Task Force on Trade, Environment and Development. United Nations, New York.

Willer, H. and Lernoud, J. (2015) *The World of Organic Agriculture. Statistics and Emerging Trends 2015*. Research Institute of Organic Agriculture (FiBL), Frick, Switzerland and IFOAM-Organics International, Bonn, Germany.

Willer, H. and Lernoud, J. (2018) *The World of Organic Agriculture. Statistics and Emerging Trends 2018*. Research Institute of Organic Agriculture (FiBL), Frick, Switzerland and IFOAM-Organics International, Bonn, Germany.

Wren, S. (2007) *African Organic Farmers. Field Crop Manual*. Organic Producers and Processors Association of Zambia (OPPAZ). Pagoda Tree Press, Bath, UK.

17 The Rapid Incineration Field Test as an Accurate, Cost-effective and Practical Tool for Estimating Soil Carbon in Africa

Albert Ackhurst,* Raymond Auerbach and Josua Louw
Nelson Mandela University, George, South Africa

Abstract

Determining soil carbon levels is not an exact science; variability due to sampling technique and method of determination has been documented over the years. Current methods for soil carbon determination are expensive, energy intensive, time-consuming and use potentially hazardous chemicals. To this end a novel but precise rapid incineration field test (RIFT) was developed for determining soil organic carbon (SOC), incorporating principles found in dry combustion and loss-on-ignition but without the constraints of high energy usage or hazardous chemicals.

In order to test effectiveness and accuracy, the RIFT was correlated with a reference method (i.e. dry combustion with a Leco device) while results were also compared with the commonly used Walkley-Black wet chemical oxidation method. Samples from 11 diverse soil forms of the Southern Cape region were analysed according to the three methods. Results indicated that the RIFT correlated well with the Leco technique ($r = 0.9434$; $p < 0.01$) while also being more accurate than Walkley-Black and more consistent than both the other techniques. It is therefore recommended that the use of RIFT for routine SOC analyses be further investigated and validated for accuracy and consistency.

Introduction

Soil organic matter (SOM) is well known to exert a great influence on the physical, chemical and biological properties of soils, and in the process plays an important role in plant production, soil resilience and environmental quality. Adequate levels of SOM are therefore a precondition for sustainable land utilization. SOM levels and dynamics are determined by factors such as precipitation, temperature regime, vegetation and soil type, and reach equilibrium values associated with specific ecological conditions (Scholes and Walker, 1993).

However, soil tillage and disturbance can result in a degradation of soil structure, a loss of SOM and a reduction of microbial activities and diversity (Conant *et al.*, 2011; Chapter 21, this volume). In fact, SOM decline is one of the most relevant land degradation processes, as is evident from the agendas of contemporary international symposia on soil science and environmental quality. The intensity and scale of agricultural practices can influence SOM storage and turnover, which ultimately affects soil security. Therefore, an in-depth scientific understanding of SOM levels, dynamics and cycling is required, especially in a country such as South Africa (SA), where

*albert.ackhurst@westerncape.gov.za

SOM levels are generally low due to prevailing warm and dry conditions (Du Preez *et al.*, 2011).

Traditional and current methods of soil organic carbon (SOC) determination mostly entail destructive *ex-situ* and non-destructive *in-situ* techniques. The former either involve the use of heat to combust dried samples, or subjecting the sample to wet chemical oxidation treatment using the Walkley-Black (WB) method. Levels of carbon are then determined on the basis either of weight loss resulting from the treatment (Kimble *et al.*, 2002; Konen *et al.*, 2002; Rowell and Coetzee, 2003; Gehl and Rice, 2006; McCarty *et al.*, 2010), or of back-titration to determine, from the level of reagents, the original levels of carbon. Dry combustion using a Leco device (which is generally considered the reference method for SOC determination) and the WB method have been most widely used for the production of datasets of SOC around the world (Howard and Howard, 1989; Konen *et al.*, 2002; McCarty *et al.*, 2002, 2010; Blaisdell *et al.*, 2003; Mikhailova *et al.*, 2003; Rowell and Coetzee, 2003; Gehl and Rice, 2006; Chatterjee *et al.*, 2009; Konare *et al.*, 2010). *In-situ* techniques are based on remote sensing imagery, inelastic neutron scattering and spectroscopic measurements like near and mid-infrared and laser-induced breakdown spectroscopy in the field (Gehl and Rice, 2006). A further distinction is that dry combustion is considered a direct elemental analysis, whereas methods such as dichromate oxidation (e.g. WB) and spectroscopy are considered indirect methods (McCarty *et al.*, 2010). Schumacher (2002) states that as the complexity of these methods increases, the required level of operator skill and the accuracy of the techniques escalate.

The cost and time required for these traditional methods of SOC determination led to a call for new techniques to quantify organic soil carbon in units suitable for carbon trading (Chatterjee *et al.*, 2009; Walcott *et al.*, 2009). Conant *et al.* (2011) added that these should be rapid, accurate and inexpensive in order to detect and quantify change in the ecosystem dynamics of carbon. New approaches to SOC determination are essential and should be cheap, fast, simple, accurate and safe. This study investigates the potential of such a method by means of a rapid incineration device developed by the lead author as an affordable prototype for routine soil carbon determinations. The rapid incineration field test (RIFT) for determining SOC is proposed, incorporating principles found in dry combustion and detection of carbon through automated analysis as well as loss-on-ignition, and quantifying carbon content through gravimetric analysis. Rapid incineration requires the direct application of intense heat, above 1000°C, from a physical blue flame under pressure typically delivered through a butane gas torch. The main objective of the study was to explore the RIFT as an alternative method for the determination of SOC by statistically testing the RIFT alongside accepted methods like dry combustion and wet chemical oxidation.

The hypothesis for this study is that the RIFT technique for determining SOC levels is of sufficient accuracy and reliability (precision) to be used for routine soil analysis. In order to test this hypothesis the following key research questions were investigated:

- To what extent do the results of the RIFT and WB methods correlate with dry combustion (the reference method)?
- How accurate is RIFT in terms of SOC assessment compared with WB?
- How variable are the results of the three methods compared with one other?

Material and Methods

Description of the study area

The study was conducted in the coastal area of the Southern Cape region between the towns of George and Sedgefield. In terms of general ecology and geomorphology, the area is complex, with variable soil-forming conditions and associated soil forms (see Table 17.1 for specific details regarding sample plots). From a soil formation point of view, rocks belonging to the Cape Super Group (mostly Table Mountain sandstone and Bokkeveld shale) and the Pre-Cape (granite, phyllite, schist) and Quarternary (aeolian sand) geological time periods dominate (Schafer, 1991). East–west orientated mountains, consisting of Table Mountain sandstone, form the most conspicuous physiographic features in the area, with a profound influence on localized climatic patterns. South of the mountains, a coastal platform with a gentle seaward dip with an altitude range of 180–250 m dominates, followed by a series of coastal lakes in close association with a dune belt of aeolian origin.

Table 17.1. General description of sample plots. Soils are classified according to the Soil Classification Working Group (SCWG, 1991).

Soil group	Location	Coordinates	Habitat	Soil form	Plot no.
Organic	Banks of the Langvlei	33°50′2″ 22°40′34″	Freshwater wetland and dune fynbos	Champagne	S3
Humic	Groenkop Forest	33°56′31″ 22°31′16″	Afro-temperate forest	Kranskop Magwa	S9 S10
	Groenkop Forest	33°57′57″ 22°33′23″	Afro-temperate forest	Nomanci	S5
Orthic on duplex soils	George Campus of the Nelson Mandela University	33°57′44″ 22°32′08″	Transformed areas	Estcourt	S7
	Plantation area at the Garden Route Dam	33°57′31″ 22°30′53″	Transformed areas	Klapmuts	S11
Orthic on podzolic	Floodplain area between freshwater lakes at Rondevlei	33°59′41″ 22°41′53″	Dune fynbos	Lamotte	S1
Orthic on plinthic	Plantation area on George Campus of Nelson Mandela University	33°56′56″ 22°31′19″	Shale fynbos transformed by plantation forestry	Westleigh	S8
Orthic on oxidic	Mid-slope on dune system between two freshwater lakes at Rondevlei	33° 59′45″ 22°41′52″	Dune fynbos	Constantia	S2
Orthic on cumulic	Edge of Langvlei wetland area	33°59′00″ 22°40′36″	Dune fynbos	Tukulu	S4
	George Campus of the Nelson Mandela University	33°58′09″ 22°32′05″	Granitic fynbos transformed by plantation	Oakleaf	S6

The Köppen classification system describes the climate of the study area as warm-temperate (Schafer, 1991). Orographic rainfall patterns are evidenced by an all-year distribution with maxima in autumn and spring, with the mean annual precipitation in the study area ranging from 650 mm to 900 mm, as influenced by altitude and proximity to the mountains (Schafer, 1991).

The Southern Cape is represented by a complex suite of soils ranging from deep sandy profiles developed in coastal aeolianites to shallow residual soils and peaty lithosols in the Outeniqua Mountains (Schafer, 1991). These soils vary in nature due to parent material, climate variation and exposure to soil-forming factors since the mid-Tertiary period, giving rise to lithosols, podzols, duplex and gradational soils; gleysols and often paleosols developed in deep colluvium or tertiary sediments, superimposed with modern soils. Soils representative of the region, with maximum variability in terms of conditions of formation, morphology and classification were selected. The 11 soil forms sampled include the Organic soil group, the Humic soil group, as well as Orthic on podzolized, duplex, plinthic, oxidic and cumulic soil groups (SCWG, 1991). Specific detail regarding each sample plot is provided in Table 17.1.

The natural terrestrial vegetation in the study area includes Southern Afro-temperate Forest, Garden Route Shale and Granitic Fynbos, Cape Lowland Freshwater Wetlands, and Southern Cape Dune Fynbos (Mucina and Rutherford, 2006).

Soil sample collection and preparation

A standard handheld Edelmann auger was used for soil sample collection. Samples were collected in the zone 5–10 cm below the soil surface. Five samples were bulked and homogenized for each site, taken within a radius of 2 m from a central point in the sample plot. Samples were air-dried on standard brown blotting paper until dry to the touch. Samples were then crushed in a mortar and pestle, homogenized again, and sieved with a 2 mm aperture sieve drum to separate and remove macro fractions like stone, woody elements and large organic and root detritus. The process was repeated until all larger fractions were removed, weighed and recorded.

Homogenized soil for each of the 11 soil forms was split into sample sets for testing with the RIFT methodology and for testing by an independent analytical laboratory. The sets included five identical samples of about 5 g each for testing with dry combustion with a Leco device by the analytical laboratory, five identical samples of about 5 g each for testing with wet chemical oxidation (WB method) by the analytical laboratory and five identical samples of about 10 g each for testing with the RIFT method. The RIFT method oxidises SOM, and a ratio of 50% carbon by weight, as proposed by Brady and Weil (2008) was used to convert the SOM values for RIFT to SOC. The following equation (Eqn 17.1) was used for this purpose:

$$\text{SOC}(\%)\text{ from RIFT} = [(\text{dry mass} - \text{incinerated mass})/\text{dry mass}]\times 100/2 \quad \text{(Eqn 17.1)}$$

A further sample of about 50 g of each of the 11 soil forms was used to conduct analyses of a wider range of soil characteristics by the analytical laboratory. This was done to investigate possible correlations between other soil variables and soil carbon characteristics. These variables included: (i) soil textural class; (ii) percentage clay, silt and sand; (iii) pH (KCl); (iv) resistance (Ohm); (v) H^+ (cmol/kg); (vi) available P and K (mg/kg); (vii) exchangeable cations ($cmol_c$/kg) – Na^+, K^+, Ca^{2+}, Mg^{2+}; (viii) cation exchange capacity (CEC) (pH 7) – ($cmol_c$/kg); (ix) base saturation (%) of Na, K, Ca, Mg; and (x) T-value (cmol/kg) (i.e. the sum of all the exchangeable cations in the soil).

Standardization of the RIFT method and prototype device

The RIFT process can be seen as a combination of elements from both dry combustion and weight loss on ignition. This involves the direct application of intense heat (in excess of 1000°C) from a physical flame under pressure on to a preweighed sample for a specific period of time. Weight loss after incineration is then determined gravimetrically in order to derive the oxidised SOM. The flame is typically delivered through a butane gas torch (Fig. 17.1).

Consistent with the literature (Heiri *et al.*, 2001), it was evident from this study that results

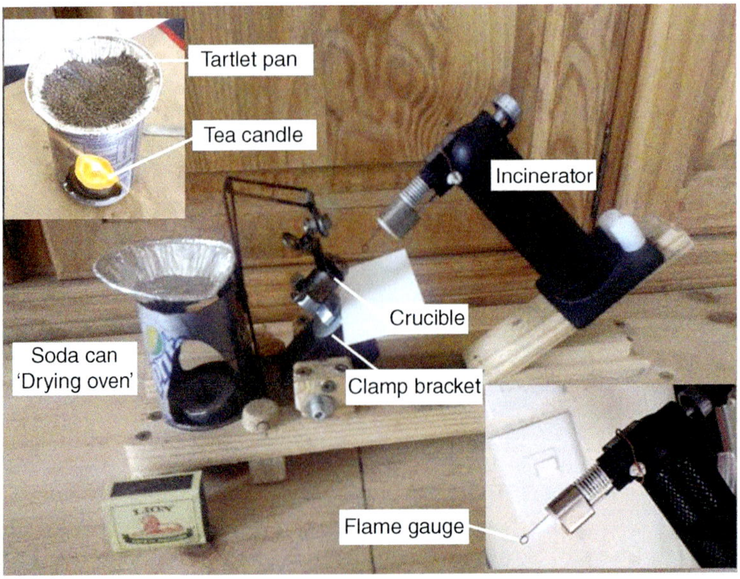

Fig. 17.1. The development of the prototype rapid incineration field test (RIFT) device.

for RIFT, as with other loss-on-ignition techniques, are influenced by sampling methods, sample size, period of exposure, heat intensity and laboratory handling. The nature of the RIFT requires a small sample size suited for rapid incineration (thimble sized, less than 1.5 g) and careful mixing in between consecutive incinerations.

A consistent level of weight loss determination was acquired by measuring the incremental weight loss after each incineration until no further loss within a threshold of 0.005 g was registered. It can then be assumed that the loss that occurred was the result of the consumption of the organic fractions, the bound (crystalline) water and residual moisture. This assumption is also based on the fact that organic matter and bound water starts reducing and evaporating at relatively low temperatures from as low 105°C to 250°C (Heiri *et al.*, 2001; Chatterjee *et al.*, 2009). All consumables appeared fully oxidised or evaporated within two to six incinerations.

It must further be ascertained whether it is necessary to correct the observed weight loss for known quantities of clays. Because of the known influence of clay on the SOC levels and the potential for over or under estimation based on structural water loss (Ball, 1964; Howard and Howard, 1989; Schumacher, 2002; Santisteban *et al.*, 2004; de Vos *et al.*, 2005; Sun *et al.*, 2009) it was thought prudent also to include the percentage of clay in the dataset. It is important to note that the overestimation of organic matter content is likely to occur due to the high temperature oxidation and destruction of fractions of structural water, carbonates and elemental carbon. In general, such fractions are relatively small in natural soils, and if present, it can be corrected for (Schumacher, 2002; de Vos *et al.*, 2005). The soil samples used in this study had a mean pH value of 4.85 (maximum = 6.5) (in KCl), implying that the samples should be free of carbonates. This was expected, since the relatively high rainfall of the area would prevent the accumulation of $CaCO_3$ in natural soils (except for soils close to the coastline).

A prototype RIFT handheld device was put together by combining general utensils and the design was refined to compensate for mechanical and operator inconsistencies related to flame distance and flame intensity (see Fig. 17.1). An aluminium crucible (thimble) with a loading capacity of around 1–1.5 g of dried soil is mounted on an articulated wooden block that can be set at the desired angle (in this case 45°). The incinerator can also be adjusted to be perpendicular to the crucible at a distance of, for example, 4 cm and can be flipped-up and dropped instantly in order to apply (or disengage) the heat on to the sample. A thin copper wire with a small loop is attached to the nozzle and serves as a gauge for flame length.

The device also incorporates a rudimentary 'drying oven' in order to establish consistency related to moisture and soil structure. A simple soda can was cut on the sides and top and fitted with a thick aluminium-foil 8-cm tartlet pan which is heated by a tealight candle in order to dry the sample.

The RIFT process starts by pre-weighing the crucible on a scientific field balance (in this instance an Adam Portable Precision Balance 150 g, 0.005 g) and recording the weight before zeroing the balance. Place about 1.2–1.5 g of the soil sample in the crucible and weigh it again. The crucible is placed in its holding bracket and the butane torch lighter or 'incinerator' is prepared. The steps for a single sample are as follows:

1. The flame is ignited and set to lean with a fully oxygenated blue flame at 2 cm in length.
2. The flexible arm is then swiftly lowered on to the sample at the same moment that a timer or stopwatch is engaged.
3. Samples are incinerated for 60 s at a time after which the heat source is immediately removed by disengaging the incinerator and switching off the flame.
4. The crucible is left in place to cool down for another 60 s.
5. The crucible is then picked up with large forceps and placed on the field balance. The weight is recorded.
6. The next step involves stirring of the sample with a stainless steel microprobe while holding the crucible with the forceps. Stir ten times slowly in a circular motion being careful not to excite the soil particles and lose material out of the crucible. Weigh the sample again and record the starting weight for the next incineration.
7. Repeat steps 1–6 on the same sample in the crucible until no further weight loss is observed (point of depletion). Calculate the total weight loss by adding the individual recorded weight losses together.
8. Clear the crucible and clean it out with a felt rag to remove all previous residues from the inner surface.

Statistical analysis

An analysis of least-squares regression relationships was used for comparing the results of different laboratory techniques in this study. This is a well-used technique for studies of this kind (Howard and Howard, 1989; Miyazawa *et al.*, 2000; de Vos *et al.*, 2005; Ghimire *et al.*, 2007; Szava-Kovats, 2009; Konare *et al.*, 2010; Wang *et al.*, 2011). Further analyses using simple graphical statistics such as the use of scatterplots relative to a line of equality (unity line) where $y = x$, as well as plotting the difference between the methods against their mean in order to measure agreement, was used, following the guidelines of Bland and Altman (1999). It was also important to assess the statistically significant probability (p value) in order to test the validity of the correlations. Box plots were utilized to illustrate the variability between the three methods in terms of distribution and range for each soil form based on their key features. This also contributes to illustrating accuracy and precision within methods. The software used for the statistical analysis was statistica 12 (Statsoft, 2013).

Results

The results of soil analysis for a range of variables are included in Table 17.2. The baseline analysis for SOC by wet chemical oxidation (WB) and dry combustion (Leco), as well as data for the RIFT method are presented in Table 17.3. Data from Table 17.3, in terms of the mean values for each method, for all soil forms, are illustrated in Fig. 17.2 in histogram format. Box plots were used (Statsoft, 2013) in Fig. 17.3 to illustrate the range and variability of data within each soil form for each of the three analytical methods used.

The data provided allows analysis for two aspects of accuracy: (i) how close to 'correct' (using Leco as reference); and (ii) how variable the observations are. This can be illustrated for the 11 soil forms and the three testing methods by using values for accuracy and variability (see Table 17.4). The assumption is that the Leco method is the most 'accurate', due to it being the commonly used reference method (Howard and Howard, 1989; Konen *et al.*, 2002; McCarty *et al.*, 2002, 2010; Blaisdell *et al.*, 2003; Mikhailova *et al.*, 2003; Rowell and Coetzee, 2003; Gehl and Rice, 2006; Chatterjee *et al.*, 2009; Konare *et al.*, 2010). Therefore it will receive a score of 1 throughout for accuracy. The second most and least accurate methods will receive 2 and 3 points, respectively, based on the level of deviation from the reference method. Scores of 1–3 are also allocated for the least (1) to the most (3) variable datasets.

From the accuracy and variability scoring presented in Table 17.4, it is evident that the RIFT method was more accurate than WB (in eight of the 11 samples) and had significantly less variability (seven of the 11 samples) than the other two methods for the five values recorded. The RIFT method seemed to be slightly less effective with the Oakleaf, Klapmuts and Estcourt soil forms in terms of accuracy, but with the Klapmuts and Estcourt soil forms this technique had the greatest precision in terms of variability.

The data for regression analysis were captured into two datasets: (i) per soil form for which the mean of the five observations per soil form is calculated and displayed as 11 data points per testing method; and (ii) as unique or individual linked observations per sample regardless of soil form displayed as 55 data points per testing method. Regression analyses, using scatterplots of Leco versus RIFT and Leco versus WB were employed to compare the techniques consistent with previous work by Howard and Howard (1989), Miyazawa *et al.* (2000), Ghimire *et al.* (2007), Szava-Kovats (2009), Konare *et al.* (2010) and Wang *et al.* (2011) with reference to the 1:1 relationship or lines of equality as described by Bland and Altman (1999), and presented in Figs 17.4–17.7.

Results from a regression analysis with dry combustion (Leco) as reference variable and RIFT and WB as predictor variables illustrates the interrelationships between the outcomes of the three methods by means of a correlation matrix (see Table 17.5). Determination of SOC by means of RIFT is closer related to the reference technique than using WB, based on the regression coefficients presented. Furthermore, from the scatterplots presented in Figs 17.4 and 17.5 (11 soil forms) and Figs 17.6 and 17.7 (55 separate samples), it is visually obvious that RIFT provides a more accurate indication of SOC than the WB technique, as indicated by both the

Table 17.2. Soil analysis report for the 11 sample plots.

Plot	Soil texture	pH (KCl)	Resist. (Ohm)[a]	H^+ (cmol/kg)	Available P (mg/kg)	Available K (mg/kg)	Exchangeable cations (cmol$_c$/kg) Na^+	K^+	Ca^{2+}	Mg^{2+}	CEC (pH 7) (cmol$_c$/kg)[b]	Na^+ (%)	K^+ (%)	Ca^{2+} (%)	Mg^{2+} (%)	T-value (cmol/kg)[c]	Clay (%)	Silt (%)	Sand (%)
1	Sand	5.2	4220	0.30	7	20	0.11	0.05	0.83	0.39	2.89	6.72	3.09	49.37	23.08	1.69	5	2	93
2	Sand	6.5	4430		31	8	0.05	0.02	1.47	0.32	2.58	2.48	1.06	79.24	17.22	1.85	7	4	89
3	Loam	5.1		1.71	7	202	7.00	0.52	7.38	6.20	14.03	30.68	2.27	32.39	27.17	22.80	30	32	38
4	Loam	5.6	140	0.35	14	122	4.90	0.31	11.19	9.50	16.78	18.67	1.19	42.62	36.19	26.26	29	20	51
5	Loam	5.1	1450	1.26	16	249	0.65	0.64	6.97	3.91	12.92	4.83	4.74	51.95	29.10	13.42	22	32	46
6	Loam	4.6	2580	1.01	1	29	0.27	0.08	1.82	1.16	4.54	6.18	1.74	42.04	26.75	4.34	25	26	49
7	Loam	4.9	2110	0.80	4	62	0.32	0.16	3.60	1.58	7.74	4.91	2.43	55.75	24.52	6.46	21	14	65
8	Loam	4.7	1710	0.86	3	42	0.31	0.11	2.65	1.52	6.58	5.66	1.96	48.63	27.96	5.45	19	28	53
9	Loam	3.7	1260	1.06	4	114	0.21	0.29	0.73	0.30	4.25	8.18	11.27	28.08	11.48	2.59	16	35	49
10	Loam	3.9	4430	3.47	3	39	0.18	0.10	0.47	0.45	7.83	3.77	2.16	10.05	9.56	4.66	21	26	53
11	Clay	4.0	3220	1.56	1	53	0.27	0.14	1.44	1.32	5.13	5.63	2.87	30.49	27.99	4.72	25	32	43

[a]Resist., Resistance.
[b]CEC, Cation exchange capacity.
[c]T-value is the sum of all the exchangeable cations in the soil.

Table 17.3. Organic carbon levels (%) for five observations per soil form using dry combustion with a Leco device and wet chemical oxidation (Walkley-Black, WB) conducted by an independent analytical laboratory. The rapid incineration field test (RIFT) methodology follows reflecting the five observations per sample plot. The mean and standard deviation (SD) for each of the three methods are included.

	Leco							WB							RIFT						
Soil form	1	2	3	4	5	Mean	SD	1	2	3	4	5	Mean	SD	1	2	3	4	5	Mean	SD
S2-Constantia	0.14	0.17	0.17	0.20	0.23	0.18	0.03	0.52	2.10	2.17	2.17	2.50	1.89	0.78	0.00	0.21	0.41	0.41	0.62	0.33	0.24
S1-Lamotte	0.21	0.24	0.24	0.27	0.29	0.25	0.03	0.37	2.02	2.25	2.94	3.24	2.16	1.12	0.99	1.00	1.16	1.32	1.32	1.16	0.16
S11-Klapmuts	1.40	1.41	1.47	1.48	1.48	1.45	0.04	1.54	2.00	2.02	2.19	2.48	2.05	0.34	1.84	1.84	2.05	2.07	2.67	2.09	0.34
S6-Oakleaf	0.93	1.31	1.48	1.84	2.12	1.54	0.46	2.58	2.71	2.86	3.15	3.45	2.95	0.35	2.66	3.04	3.19	3.25	3.88	3.20	0.44
S8-Westleigh	1.16	1.47	1.63	1.67	1.90	1.57	0.27	1.57	2.22	2.31	2.52	2.52	2.23	0.39	2.22	2.42	2.43	2.44	2.65	2.43	0.15
S7-Estcourt	1.57	2.17	2.32	2.64	2.56	2.25	0.42	1.55	2.01	2.18	2.18	3.74	2.33	0.83	1.86	2.04	2.07	2.25	2.66	2.18	0.31
S10-Magwa	1.57	2.61	2.62	2.67	2.76	2.45	0.49	1.48	2.15	2.20	3.35	5.13	2.86	1.44	2.26	2.49	2.70	2.71	2.88	2.61	0.24
S9-Kranskop	3.05	3.97	4.16	4.47	6.19	4.37	1.15	1.94	2.43	2.86	3.86	5.81	3.38	1.53	3.56	4.20	4.37	4.42	4.74	4.26	0.44
S5-Nomanci	4.31	6.38	6.46	7.40	7.63	6.44	1.31	3.84	4.63	5.81	6.22	7.01	5.50	1.27	6.11	6.20	6.50	7.23	7.45	6.70	0.61
S3-Champagne	5.19	6.72	6.79	7.34	7.74	6.76	0.97	2.87	3.05	4.27	5.30	6.06	4.31	1.39	4.71	4.94	5.17	5.31	5.71	5.17	0.38
S4-Tukulu	5.82	6.56	7.68	7.90	7.92	7.18	0.94	2.87	3.46	4.09	5.39	6.92	4.55	1.62	5.14	5.77	6.14	6.28	6.61	5.99	0.56

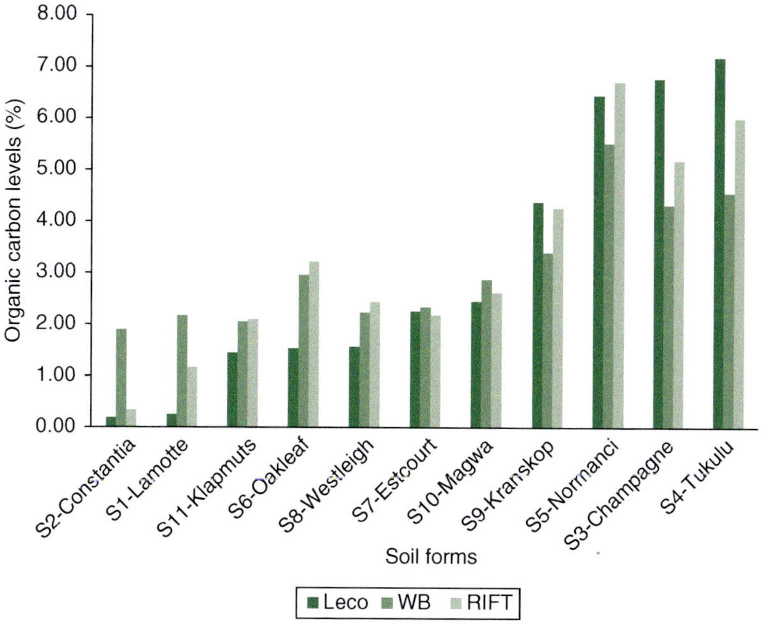

Fig. 17.2. Mean organic carbon levels (%) (n = 5) for the three analytical methods.

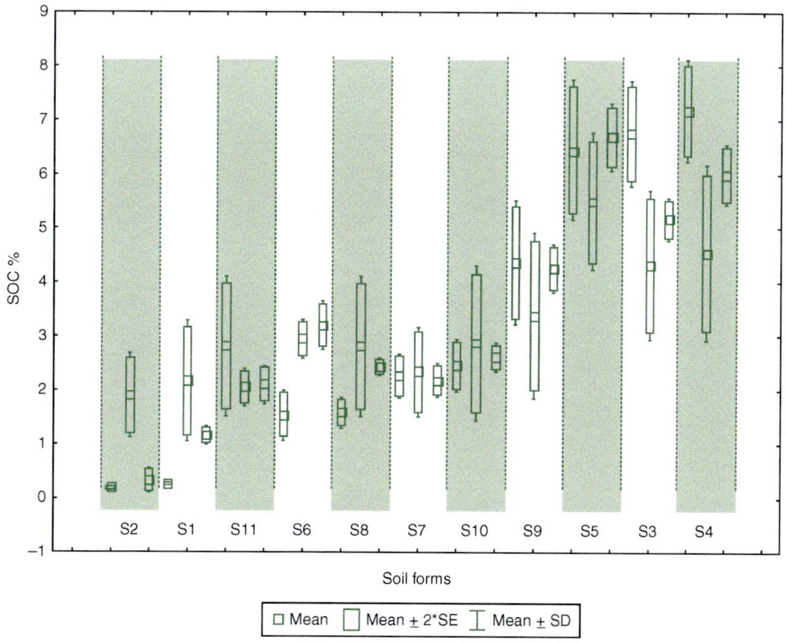

Fig. 17.3. Box plots of soil organic carbon (SOC) (%) for each soil form (n = 5) with the three treatments in the following order: Leco, Walkley-Black (WB) and rapid incineration field test (RIFT). SD, Standard deviation; SE, standard error.

Table 17.4. An analysis of the accuracy for the rapid incineration field test (RIFT) and Walkley-Black (WB) methods, as well as variability between the three methods, for soil organic carbon levels of each soil form. The final score reflects the number of incidences that a method illustrated best accuracy and with the least variability per soil form.

	Accuracy between WB and RIFT			Variability between Leco, WB and RIFT (n = 5)		
Soil form	Leco	WB	RIFT	Leco	WB	RIFT
S2-Constantia	1	3	2	1	3	2
S1-Lamotte	1	3	2	1	3	2
S11-Klapmuts	1	2	3	3	2	1
S6-Oakleaf	1	2	3	2	1	3
S8-Westleigh	1	3	2	1	3	2
S7-Estcourt	1	2	3	2	3	1
S10-Magwa	1	3	2	2	3	1
S9-Kranskop	1	3	2	2	3	1
S5-Nomanci	1	3	2	2	3	1
S3-Champagne	1	3	2	2	3	1
S4-Tukulu	1	3	2	2	3	1
No. of incidences of highest accuracy and least variability	–	3	8	3	1	7

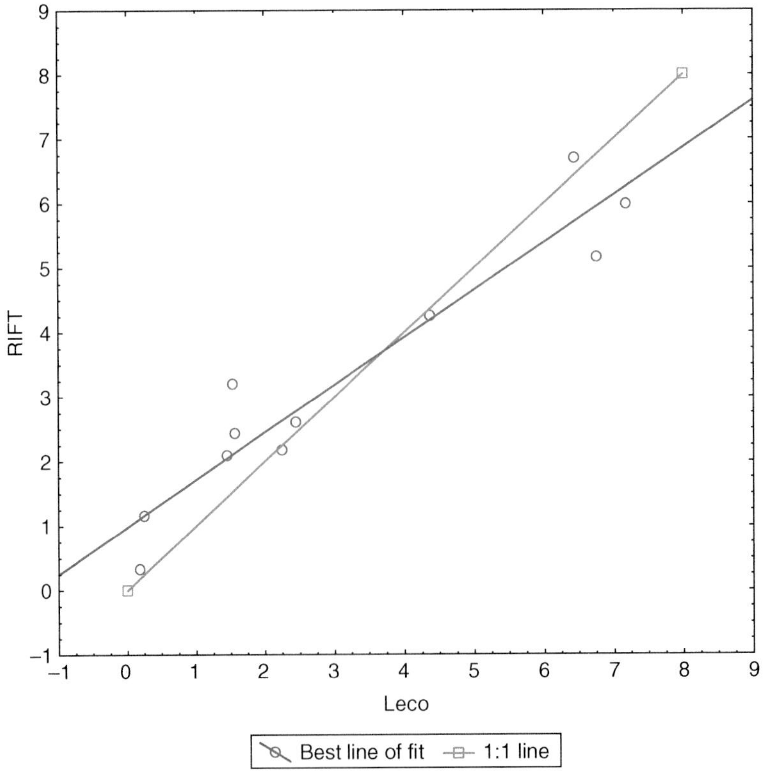

Fig. 17.4. Scatterplot of RIFT versus Leco, together with the line of equality, using the mean values of the 11 sample plots. Regression equation $y = 0.9838 + 0.7349*x$; correlation coefficient $r = 0.9520$, $p < 0.00001$; $R^2 = 0.9063$.

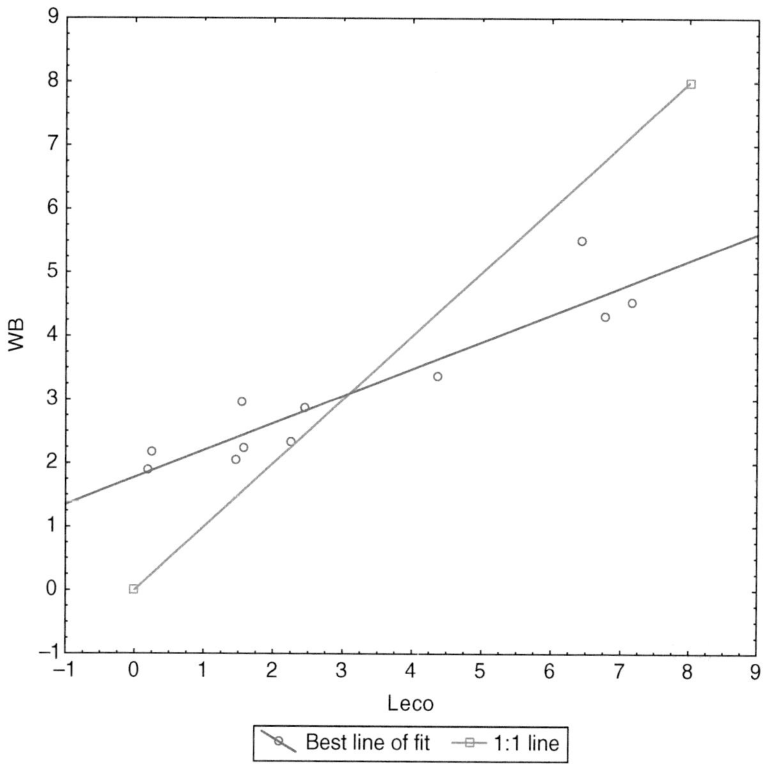

Fig. 17.5. Scatterplot of WB versus Leco, together with the line of equality, using the mean values of the 11 sample plots. Regression equation $y = 1.7765 + 0.4263*x$; correlation coefficient $r = 0.9304$, $p < 0.00003$; $R^2 = 0.8657$.

grouping of data points as well as the deviation from the line of equality. Both the RIFT and WB techniques over-predict and under-predict at low and high SOC levels, respectively, relative to Leco, but the magnitude of this deviation is more prominent with the WB method. The statistical relationships between results from RIFT and WB, relative to the reference Leco technique, are provided with each figure.

The deviations in data obtained through RIFT and WB, relative to Leco, are also illustrated in Table 17.6. The differences in the means were squared in order to render the figures positive, consistent with McCarty *et al.* (2002), so that the accumulated values could be compared (where $\Sigma\Delta^2$ = the sum of the differences from the means squared). From the figures presented it is evident that RIFT ($\Sigma\Delta^2 = 8.841$) provides a significantly more accurate indication of SOC than WB ($\Sigma\Delta^2 = 24.311$) for the range of soils sampled.

Regression analyses between the residual values listed in Table 17.6 and the other soil characteristics tested in this study did not reveal any significant relationships. The correlation between the percentage of clay and the residual value between RIFT and Leco proved insignificant ($r = 0.4943$; $p = 0.1222$; $R^2 = 0.2443$). The correction and refinement of RIFT values to acknowledge potential weight loss due to crystalline water associated with clay is therefore not possible from this study.

Discussion

The selected soils were representative of the spectrum of soil-forming conditions and soil forms found in the study area. The sample of 11 soil forms included sufficient variation in general diagnostic features, topsoil texture, as well as SOC. For the purpose of an exploratory study,

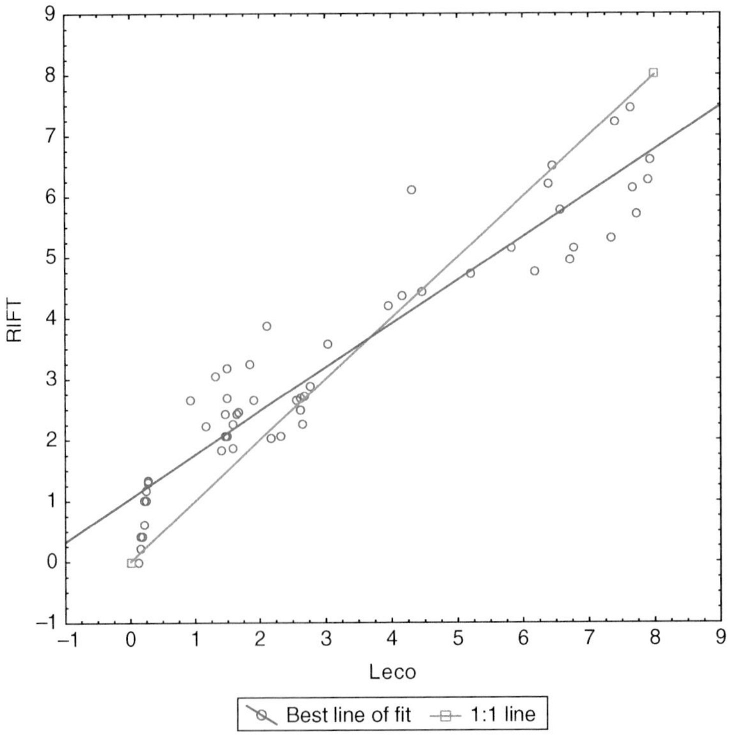

Fig. 17.6. Scatterplot of RIFT versus Leco, together with the line of equality, using data from the 55 individual observations. Regression equation $y = 1.0415 + 0.7164*x$; correlation coefficient $r = 0.9434$, $p < 0.01$; $R^2 = 0.8899$.

this selection of soils therefore lends itself to a comparative study dealing with analytical techniques for SOC determination.

Significant differences in the variability of data for the three techniques were found. In many cases, the proposed RIFT technique showed lower levels of variability around mean values than both the Leco and the WB methods, indicating higher levels of reliability and consistency. This variability could be attributed to operator (in-)consistency and error but it is most likely that structural or chemical properties of the particular soils caused more or less variability in the readings due to the particular chemical or thermodynamic effect of the different analytical techniques.

The RIFT technique showed a higher correlation with the reference Leco method than the WB test, as indicated by the correlation coefficients and deviations from the 1:1 lines of equality. An analysis of residual values between RIFT and WB, relative to Leco, also points to higher levels of accuracy for the RIFT in the majority of cases. Both the RIFT and the WB techniques showed over-prediction at low levels of SOC, and under-prediction at high levels, but the magnitude of the deviation is more pronounced with the WB technique. Explanations for these variable deviations need to be sought in the nature of the particular techniques (i.e. consumption of organic matter through incineration and wet chemical oxidation for RIFT and WB, respectively). It appears that a more complete consumption of organic material results from the RIFT method.

Furthermore, it was noted that lower levels of SOC coincided with sandy soil textures, in which case structural water would be less adsorbed or pervasive leading to more complete consumption during the incineration process. In the case of high SOC soils (which also contained at least 25% clay in this study), both water and organic products are better protected and will offer more resistance to complete consumption.

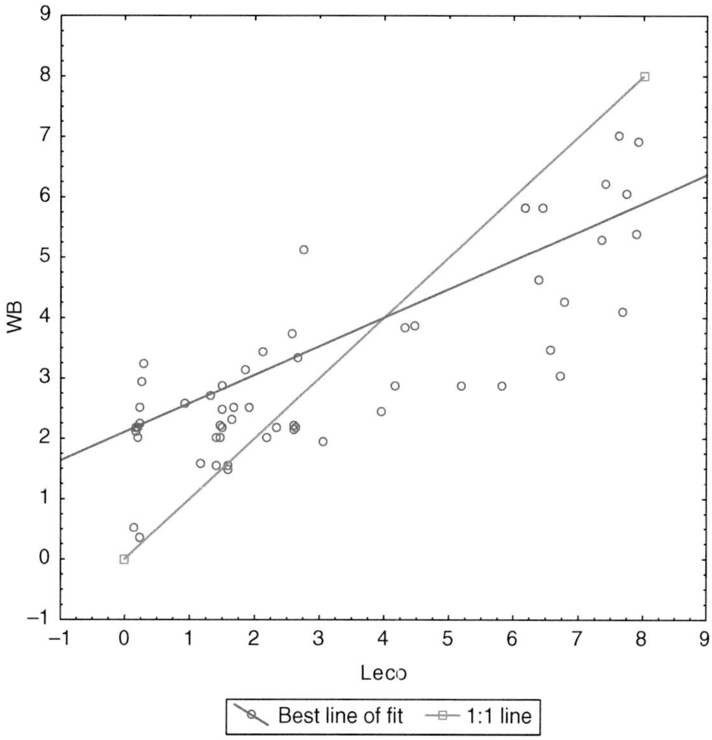

Fig. 17.7. Scatterplot of WB versus Leco, together with the line of equality, using data from the 55 individual observations. Regression equation $y = 1.629 + 0.4734{*}x$; correlation coefficient $r = 0.8047$, $p < 0.01$; $R^2 = 0.6476$.

Table 17.5. Pearson correlation statistics for the relationship between Leco, Walkley-Black (WB) and rapid incineration field test (RIFT) ($n = 11$).[a]

Variable	Leco	WB	RIFT
Leco	1.000000		
WB	0.930433*	1.000000	
RIFT	0.951999*	0.966076*	1.000000

[a]Marked correlation coefficients (*) indicate $p < 0.001$.

This explanation is supported by the fact that in the instances where both SOC and the percentage of clay were high, it took many incinerations (from four to six) before the consumables were deemed depleted. Conversely, where SOC and the percentage of clay were very low, it took only two to three incinerations before weight loss stabilized. This could be remedied by more intense incineration or an extended incineration regime in future.

Soil samples from different regions may have unique RIFT–SOC relationships, depending on particular climatic conditions, vegetative cover, topographical position and soil texture. It is likely that different soil textural classes will behave differently with the application of intense heat on account of unique balances of different soil fractions. It is known that loss-on-ignition overestimates the organic matter content (Gehl and Rice, 2006) and may require unique correction equations to compensate for structural water from clays, oxide colloids (sesquioxides), carbonates, and elemental carbon like charcoal (de Vos *et al.*, 2005). This study did not succeed in using additional soil variables for the refinement or calibration of results obtained through the RIFT procedure. None of the soil variables tested showed any correlation with the levels or residual values of SOC provided by this technique. This study was limited by the small number of samples, and should be regarded as of an

Table 17.6. Deviations in values for soil organic carbon obtained through rapid incineration field test (RIFT) and Walkley-Black (WB), relative to Leco, quantified as sum of differences from the means squared ($\Sigma\Delta^2$).[a, b]

Soil form	Leco	WB	RIFT	RIFT Δ^2	WB Δ^2
S2-Constantia	0.182	1.892	0.331*	0.022	2.924
S1-Lamotte	0.250	2.164	1.159*	0.827	3.663
S11-Klapmuts	1.448	2.046	2.093	0.416	0.358
S6-Oakleaf	1.536	2.950	3.203	2.780	1.999
S8-Westleigh	1.566	2.228	2.432	0.749	0.438
S7-Estcourt	2.252	2.332	2.177*	0.006	0.006
S10-Magwa	2.446	2.862	2.608*	0.026	0.173
S9-Kranskop	4.368	3.380	4.257*	0.012	0.976
S5-Nomanci	6.436	5.502	6.697*	0.068	0.872
S3-Champagne	6.756	4.310	5.167*	2.524	5.983
S4-Tukulu	7.176	4.546	5.989*	1.410	6.917
				$\Sigma\Delta^2 = 8.841$	$\Sigma\Delta^2 = 24.311$

[a]Instances where RIFT values were more accurate than WB values are marked with an asterisk (*).
[b]Under the RIFT Δ^2 and WB Δ^2columns, $\Sigma\Delta^2$ indicates the sum of differences between the two methods.

exploratory nature, and future research needs to consider significantly larger sample sizes from pedologically more diverse areas.

The RIFT technique proved to be a very rudimentary procedure requiring minimal equipment. Apart from a gravimetric device and an incinerator, the only other expenditures are basic items such as aluminium thimbles, butane gas and basic laboratory equipment like spatulas, mortar and pestle, etc. No expensive and potentially harmful chemicals are used, no electricity (except for minimal charging of devices) and no sophisticated laboratory equipment (e.g. drying ovens, desiccators, muffle furnaces, combustion chambers or chemical analysers) are required. Furthermore, drying ovens and muffle furnaces used in alternative techniques are usually used for between 8 h and 24 h at temperatures varying from 105°C (for oven drying overnight) to an average of 450°C (for loss on ignition incineration) implying comparatively high basic electrical costs. The RIFT device is mobile, light and can be deployed in any protected workspace in close proximity to the field.

Conclusion

The increased intensity and scale of agricultural practices, together with concerns about soil degradation and sustained soil quality require an improved understanding of SOM levels, dynamics and cycling. The RIFT showed considerable potential as an analytical technique for the routine analysis of SOC since it correlates well with dry combustion with a Leco device and is more accurate than WB for the selected soils. The RIFT also illustrated more consistency than the other two techniques in the sense that it showed lower variability around mean values. Data from this study indicate that the hypothesis, that the RIFT technique for determining SOC levels is of sufficient accuracy and reliability for routine analysis, can be accepted. Some questions remain with regard to the over-prediction and under-prediction at low and high SOC levels, respectively, relative to the Leco technique but initial observations indicate that it could be a function of the prevalence and extent of crystalline water. Therefore, better understanding is required and the technique needs to be tested under a wider range of ecological conditions, including different climate zones, vegetation types, soil textural classes and parent material types. A larger sample of soil types might also improve the significance of correlations found.

The RIFT technique proved that it is possible to introduce efficient and less time-and energy-consuming methodologies for soil analysis with the added benefit of low maintenance and operational costs and promising levels of accuracy warranting further investigation and consideration of this technique as a valid test for SOC.

References

Ball, D. (1964) Loss-on-ignition as an estimate of organic matter and organic carbon in non-calcareous soils. *Journal of Soil Science* 15(1), 84–92.

Blaisdell, R., Conant, R., Dick, W., Doberman, A., Izaurralde, C., Ransom, M., Rice, C., Robertson, P. and Stuth, J. (2003) *Recommended Procedures for Collecting, Processing, and Analyzing Soil Samples in CASMGS Research Plots*. Consortium for Agricultural Soils Mitigation of Greenhouse Gases 4.1 Working Group: 5. Throckmorton Plant Science Center, Department of Agronomy, Kansas State University, Manhattan, Kansas.

Bland, J. and Altman, D. (1999) Measuring agreement in method comparison studies. *Statistical Methods in Medical Research* 8(2), 135–160.

Brady, N.C. and Weil, R.R. (2008) *The Nature and Properties of Soils*, 14th edn. Pearson Education, New Jersey.

Chatterjee, A., Lal, R., Wielopolski, L., Martin, M.Z. and Ebinger, M.H. (2009) Evaluation of different soil carbon determination methods. *Critical Reviews in Plant Sciences* 28(3), 164–178.

Conant, R.T., Ogle, S.M., Paul, E. and Paustian, K. (2011) Measuring and monitoring soil organic carbon stocks in agricultural lands for climate mitigation. *Frontiers in Ecology and the Environment* 9(3), 169–173.

De Vos, B., Vandecasteele, B., Deckers, J. and Muys, B. (2005) Capability of loss-on-ignition as a predictor of total organic carbon in non-calcareous forest soils. *Communications in Soil Science and Plant Analysis* 36(19–20), 2899–2921.

Du Preez, C.C., van Huyssteen, C.W. and Mnkeni, P.N.S. (2011) Land use and soil organic matter in South Africa 1: a review on spatial variability and the influence of rangeland stock production. *South African Journal of Science* 107(5–6), 27–34.

Gehl, R.J. and Rice, C.W. (2006) Emerging technologies for *in situ* measurement of soil carbon. *Climatic Change* 80(1–2), 43–54.

Ghimire, R., Shah, S.C., Dahal, K.R., Shrestha, A.K., Adhikari, C., Lauren, J.G. and Duxbury, J.M. (2007) Equation for predicting soil organic carbon using loss on ignition for Chitwan valley soils. *Journal of the Institute of Agriculture and Animal Science (IAAS)* 1, 229–232.

Heiri, O., Lotter, A.F. and Lemcke, G. (2001) Loss on ignition as a method for estimating organic and carbonate content in sediments: reproducibility and comparability of results. *Journal of Paleolimnology* 25(1), 101–110.

Howard, P.J.A. and Howard, D.M. (1989) Use of organic carbon and loss-on-ignition to estimate soil organic matter in different soil types and horizons. *Biology and Fertility of Soils* 9(4), 306–310.

Kimble, J.M., Lal, R. and Follet, R.F. (2002) *Agricultural Practices and Policies for Carbon Sequestration in Soil*. Taylor & Francis, London.

Konare, H., Yost, R.S., Doumbia, M., McCarty, G.W., Jarju, A. and Kablan, R. (2010) Loss on ignition: measuring soil organic carbon in soils of the Sahel, West Africa. *African Journal of Agricultural Research* 5(22), 3088–3095.

Konen, M.E., Jacobs, P.M., Burras, C.L., Talaga, B.J. and Mason, J.A. (2002) Equations for predicting soil organic carbon using loss-on-ignition for North Central U.S. soils. *Soil Science Society of America Journal* 66, 1878–1881.

McCarty, G.W., Reeves, J.B., Reeves, V.B., Follet, R.F. and Kimble, J.M. (2002) Mid-infrared and near-infrared diffuse reflectance spectroscopy for soil carbon measurement. *Screening* 66(18), 640–646.

McCarty, G.W., Reeves, J.B., Yost, R., Doraiswamy, P.C. and Doumbia, M. (2010) Evaluation of methods for measuring soil organic carbon in West African soils. *Journal of Agricultural Research* 5(16), 2169–2177.

Mikhailova, E.A., Noble, R.R.P. and Post, C.J. (2003) Comparison of soil organic carbon recovery by Walkley-Black and dry combustion methods in the Russian chernozem. *Communications in Soil Science and Plant Analysis* 34(3–14), 1853–1860.

Miyazawa, M., Pavan, M.A., de Oliveura, E.L., Ionashiro, M. and Silva, A.K. (2000) Gravimetric determination of soil organic matter. *Brazilian Archives of Biology and Technology* 43(5), 475–478.

Mucina, L. and Rutherford, M. (2006) The vegetation of South Africa, Lesotho and Swaziland. *Strelizia* 19, 749–790.

Rowell, M.J. and Coetzee, M.E. (2003) The measurement of low organic matter contents in soils. *South African Journal of Plant and Soil* 20(2), 49–53.

Santisteban, J.I., Mediavilla, R., López-Pamo, E., Dabrio, C.J., Blanca Ruiz Zapata, M., José Gil García, M., Castaño, S. and Martínez-Alfaro, P.E. (2004) Loss on ignition: a qualitative or quantitative method for organic matter and carbonate mineral content in sediments. *Journal of Paleolimnology* 32(3), 287–299.

Schafer, G.N. (1991) *Forest Land Types of the Southern Cape*. Council for Scientific and Industrial Research (CSIR): Division of Forest Science and Technology, Pretoria.

Scholes, R.J. and Walker, B.H. (1993) *An African Savanna. Synthesis of the Nylsvley Study*. Cambridge University Press, Cambridge.

Schumacher, B.A. (2002) Methods for the determination of total organic carbon (TOC) in soils and sediments. *Carbon* April, 25.

SCWG (Soil Classification Working Group) (1991) *Soil Classification – a Taxonomic System for South Africa. Memoirs on the Agricultural Natural Resources of South Africa No. 15*. Department of Agricultural Development, Pretoria.

Statsoft (2013) statistica 12 for Windows. Statsoft Inc., Tulsa, Oklahoma.

Sun, H., Nelson, M., Chen, F. and Husch, J. (2009) Soil mineral structural water loss during loss on ignition analyses. *Canadian Journal of Soil Science* 89, 603–610.

Szava-Kovats, R. (2009) Re-analysis of the relationship between organic carbon and loss-on-ignition in soil. *Communications in Soil Science and Plant Analysis* 40(17–18), 2712–2724.

Walcott, J., Bruce, S. and Sims, J. (2009) *Soil Carbon for Carbon Sequestration and Trading: a Review of Issues for Agriculture and Forestry*. Bureau of Rural Sciences, Department of Agriculture, Fisheries & Forestry, Canberra.

Wang, Q., Li, Y. and Wang, Y. (2011) Optimizing the weight loss-on-ignition methodology to quantify organic and carbonate carbon of sediments from diverse sources. *Environmental Monitoring and Assessment* 174(1–4), 241–257.

18 The Nelson Mandela Long-term Comparative Organic Farming Systems Research Trials: Baseline Study and Trial Design

Raymond Auerbach,* N'wa-Jama Mashele and Catherine Eckert
Nelson Mandela University, George, South Africa

Abstract

The Nelson Mandela long-term comparative organic farming systems research trials (the Mandela Trials) were planned and established in 2014 with peer review from the three sites where organic comparative research has been ongoing for the past 20–30 years (Switzerland, Denmark and Pennsylvania in the USA). Organic and conventional farming systems, as well as monocrop and rotational vegetable production, were compared; crops grown were cabbage (monocrop and rotational), with sweet potatoes and cowpeas as the rotated crops. Owing to variation of soil fertility on the research trial site, an indicator crop (Caliente mustard (*Brassica juncea*)) was planted on the long-term research trial site before any treatments commenced, in order to quantify variation in soil fertility and plant growth. Variation between replications accounted for much of the variability, showing a fertility gradient across the site. Correlations were found in soils between soil pH and available soil P, and between available soil P and leaf P content in the crop. Replication 4 was found to have lower yields and soil fertility than the other three replications, especially plots 37, 38, 39 and 40, which were also found to have some building rubble below the 15 cm of topsoil. Soils were generally acidic and low in available soil P, due largely to high levels of exchangeable Al. It was decided that all plots would receive 1 t/ha of dolomitic lime each year for the first 3 years of the research trial, and that although all four replications would be measured and analysed, Replication 4 might produce atypical results, in which case results from this replication might have to be regarded as outliers in future analysis.

Introduction

Climate change predictions threaten the already volatile food production system; research is thus needed to ensure that informed decisions are made in terms of choosing agricultural systems. A site was identified, fenced and prepared and long-term trials were established. The Nelson Mandela long-term comparative organic farming systems research trials (the Mandela Trials) compare two farming systems (organic and conventional) and investigate the effects of these systems on several production parameters. Water use efficiency (WUE) results are reported in Chapter 19, pest and disease control in Chapter 20, soil microbiology in Chapter 21, and soil fertility and yield in Chapter 22 of this volume. Current food production systems and food use patterns are unsustainable, relying heavily on the use of fossil fuels, chemicals and other

*raymond.auerbach@mandela.ac.za

external inputs (Chapters 1, 5 and 7, this volume). The Mandela Trials aim to contribute towards finding sustainable agricultural solutions.

One of the issues facing South African (SA) citizens is not food availability but rather food accessibility (Drimie and McLachlan, 2013; Frayne *et al.*, 2014; Chapter 1, this volume). A large number of households are faced with inadequate food supply and poor nutrient intake (Drimie and McLachlan, 2013; McLachlan and Landman, 2013). The volatility of food prices forces a majority of those living below minimum standards to spend a large proportion of their income on acquiring energy-dense food which has low nutritional value (May and Rogerson, 1995; Zezza and Tasciotti, 2010). These foods include highly refined carbohydrates, fats and sugars, and these are bought mainly because they are cheaper than healthier alternatives (Frayne *et al.*, 2013; McLachlan and Landman, 2013). The nutritional imbalance of such a diet contributes towards nutrition-related diseases (McLachlan and Landman, 2013; Chapter 7, this volume).

Food security is linked to and influenced by a number of rapidly changing and complex factors such as climate change, environmental degradation, and urbanization and population growth. Thus any hopes of ensuring food security for all persons should follow an interdisciplinary approach to offer feasible solutions (Drimie and McLachlan, 2013; Chapters 8, 9, 10 and 11, this volume). Our study looks at the impacts of two existing practices and their feasibility and sustainability.

Similar Established Research

There are a number of established and long-running research trials comparing organic and conventional farming systems in Europe and America.

- The DOK (D-biodynamic, O-organic and K-conventional) trials in Therwil, Switzerland compare three farming systems on soil biology and chemistry, above-ground fauna and the yield components of winter wheat. The trial was started in 1978, with the primary objective of scientifically assessing the feasibility of long-term organic agriculture (OA) (Mäder *et al.*, 2006; Professor A. Gattinger, FiBL Germany, 2014, personal communication). The results have shown that organic systems utilize natural resources better than the conventional systems, there is less erosion and greater soil aggregate stability which creates conducive conditions for diverse soil microorganisms (Mäder *et al.*, 2006; Fließbach *et al.*, 2007). The yields from organic farming systems were found to be at 80% of conventional systems yields, while production or input costs were much less (nitrogen input was lowered by 65–70%, synthetic fertilizer input was lowered by 35–40%) in the organic systems (Mäder *et al.*, 2006).
- Rodale Institute farming systems trials were established in 1981. The trials compare the long-term effects of three farming systems: (i) manure-based organic farming; (ii) legume-based organic farming; and (iii) agrochemical-input-based farming, on soybean and maize crops. In recent years genetically modified crops as well as no-tillage systems were included to better represent American farming. Soil health, stability and water holding ability increased in the organic systems while the conventional systems soil remained unchanged (Rodale Institute, 2011). The organic system yields were found to be nearly identical to the conventional system yields. However, in times of drought the organic system yields were significantly higher than yields of the conventional system (Lotter *et al.*, 2003; Rodale Institute, 2011). External input costs were marginally lower in the organic systems and even without the premium pricing of organic products, the organic system was still competitive with the conventional system.
- The International Centre for Research into Organic Farming Systems (ICROFS) is based in Denmark (Aarhus University Foulum Campus) and was established in 1996. It coordinates and analyses the effects of organic research and development on the organic sector (ICROFS, 2013). A long-term OA arable crop rotation trial was set up in 1997 at the university. The trial compares: (i) organic farming systems with green manure crop (with/without a catch crop and manure); (ii) organic farming systems without a green manure crop (with/without a

green manure crop); and (iii) conventional systems with and without catch crops and manure. The trial investigates a number of parameters such as yield, plant nitrogen uptake, nitrate leaching and total carbon development in the soil (L.S. Sorensen, Aarhus University, Foulum, Denmark, 2014, personal communication).

Chapter 3 of this book examined the impacts of these long-term comparative trials on policy, farming and consumption in Switzerland, the USA and especially Denmark.

Establishing a Baseline

The above-mentioned research and that of Dr Bo Petterson in Sweden (described in Chapters 1 and 6, this volume) provided guidance in establishing similar long-term research in SA. The reasoning behind the baseline study was to quantify site variability and estimate its impact, in order to document accurately the changes that will be observed over the course of the research. Details of the research treatments are given in the Mandela Trials research protocol (Auerbach and Mashele, 2014).

Soil properties can vary greatly spatially within very small areas; this has implications for the nutrient concentrations, soil pH, soil organic matter (SOM) content and even soil texture (Robertson *et al.*, 1988; Qiu *et al.*, 2019). The spatial variability of nutrients plays a role in the structure or abundance of a plant population on a given area, and also impacts yield potential (Robertson *et al.*, 1988). Plants have the ability to tune the rate at which their roots absorb nutrients in the soil according to the requirements set by their growth rate (Smith and Loneragan, 1997). This trait is limited by the fact that most of the macronutrients found in the soil are found in fixed forms and thus may not be readily accessible to the plants (Qu *et al.*, 2014). There is a distinction between the total content and the available content of the macronutrients in the soil, as the plants have access only to the available nutrients, and only these water-soluble nutrients can be regarded as potentially effective for plant nutrition (Qu *et al.*, 2014).

The texture of a soil influences the cation exchange capacity (CEC) of a given soil as this complex determines the total amount of exchangeable cations (Nathan, 1999) the soil is able to adsorb, from which plant roots can extract the required nutrients.

Non-friable soil structure limits root penetration, as the medium is not supportive enough for plants to develop vigorous/extensive root systems to enable them to scavenge deeper and further for nutrients. These soils are often prone to waterlogging, and excess water reduces O_2 and increases CO_2 to toxic levels. Very often it appears as though the factors act independently and are the sole reason for a particular result, however literature cited above indicates that plant yield is the result of many interacting factors, including soil organic carbon (SOC), soil acidity and plant available levels of N, P and K.

Materials and Methods

Study area

The Mandela Trials are located on the George Campus of the Nelson Mandela University in the Western Cape (the Main Campus is in Port Elizabeth in the Eastern Cape). The location of George in SA is shown in Fig. 18.1; the research trial site location at the George Campus (5 km north-east of George) is shown in Fig. 18.2. The hexagonal permaculture garden mentioned in Chapter 14 (this volume) can be seen to the north of the trial site, just south of the loop road.

The soil is grey in colour, acidic and predominantly sandy loam, with a soil carbon content before ploughing of about 4.3%; according to the South African Binomial System of soil classification, it is the Estcourt soil form (SCWG, 2009). The soil profile consists of an orthic A horizon on an E horizon at a depth of 70 cm, which becomes waterlogged during certain periods due to an underlying impervious gravelly clay layer. This results in the lateral movement of water and nutrients, and the formation of a highly leached gley horizon. The climate is temperate, exhibiting an all-year-round rainfall pattern with a mean annual precipitation of 863 mm (ICFR, 2013).

The site chosen for this research is located on the south-eastern part of the George Campus

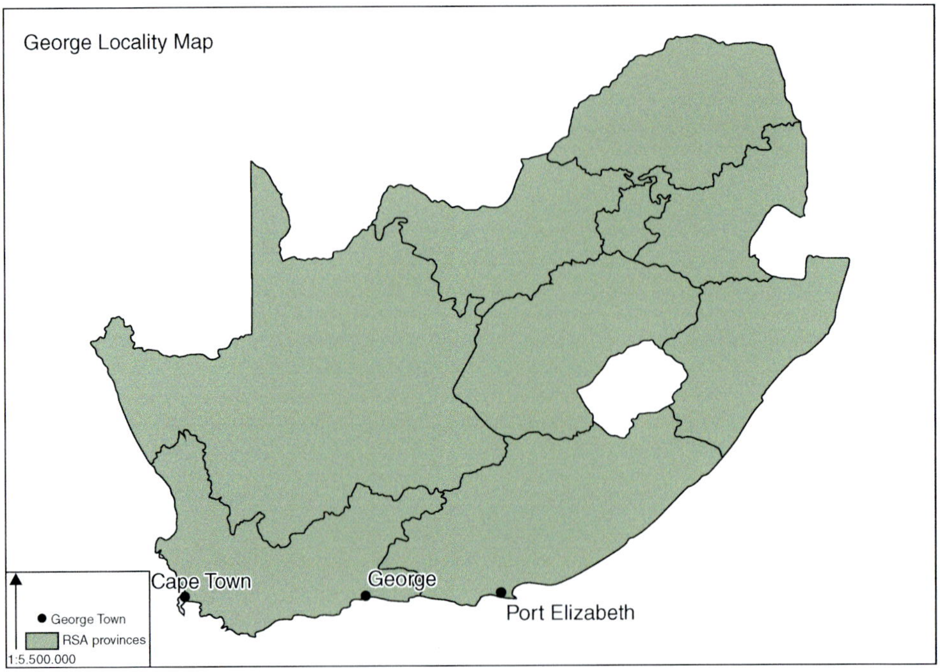

Fig. 18.1. Map of South Africa showing George's location in the Western Cape. RSA, Republic of South Africa.

(22°32′6.546″ E, 33°57′49.289″ S) on ground that had been undisturbed for about 20 years and was dominated by Kikuyu grass (*Pennisetum clandestinum*). The site was fenced with a 2.4-m high electric fence to keep the baboons and bush pigs out. Several trees within the site were felled to minimize differential shading effects.

The initial indicator crop was planted before any treatments were applied to the plots. Crops planted later in a rotational sequence from November 2015 onwards are: (i) cabbage (*Brassica oleracea*) – heavy feeder; (ii) sweet potato (*Ipomoea batatas*) – light feeder; and (iii) cowpea (*Vignia unguiculata*) – legume. The experimental area is 1500 m², and is divided into four replications. Each replication has ten plots of 5 m × 6 m each (gross) and a net area of 4 m × 5 m. The experimental design is a complete randomized block design, with four plots organic, four plots conventional and two control plots in each replication. Each organic and each conventional block is split into a monocrop plot and three rotational plots within each replication, making 40 plots in all.

The schematic plot layout is shown in Fig. 18.3. An undisturbed plot is located outside the fence to the east of the trial site, and is still under Kikuyu grass (plot 41), near the middle shed shown on the right in Fig. 18.2. The pine plantation to the south of the plot was felled after the harvesting of the indicator crop (bottom of Fig. 18.2). This would be to the right of Replication 1 on the schematic diagram in Fig. 18.3. Due to high P levels, two plots were discarded and used as extra controls; Replication 3 included two adjacent plots (plots 29 and 30), while Replication 4 included four adjacent plots (37, 38, 39 and 40), as can be seen in Fig. 18.3.

The plots were subsequently allocated to one of three farming systems: (i) organic (compost, cover cropping, natural pest and disease control); (ii) conventional (synthetic fertilizer, chemical pest and disease control); and (iii) control (no compost or synthetic fertilizer, chemical pest and disease control). At the time of the baseline investigation, no fertilizer treatments had been carried out on any of the plots.

Soil sampling

Using a handheld soil auger, eight samples were collected to a depth of 15 cm each at random

Fig. 18.2. Satellite photograph of a portion of George Campus showing the location of the Mandela trial site and the plot layout within the red rectangle. (Adapted from Google Earth, 2018.)

spots and bulked into a single sample for each of the plots. The samples were then sent off for analysis at Bemlab in Somerset West. They were analysed for pH (KCl), soil carbon (Walkley-Black (WB) and Leco methods), CEC, resistance and bulk density as well as the normal crop nutrients, including available P by Bray 2, which was low (see Table 18.1). The WB soil carbon tests showed initial soil carbon of about 4.1%, but this dropped to about 2.5% in all of the cultivated plots; Leco showed a smaller drop in soil carbon, but exhibited more variability and seemed less reliable (see study on soil carbon determination in Chapter 17, this volume). A subsoil sample was taken from a depth of 45 cm.

Soil biology

In addition, three samples were collected from a depth of 10 cm for biological analysis. The three biological analyses carried out by Sheila Storey of Nemlab (see Chapter 21, this volume) showed there was an average clay content of about 20%, and classified the soil as sandy loam, with:

- total numbers of bacteria between 26,000/g and 58,000/g; and
- moulds at 17–20 cfu/ml and very few nematodes.

It is notable that plots 21, 22 and 23 have rather deep E horizons (Table 18.2), but more

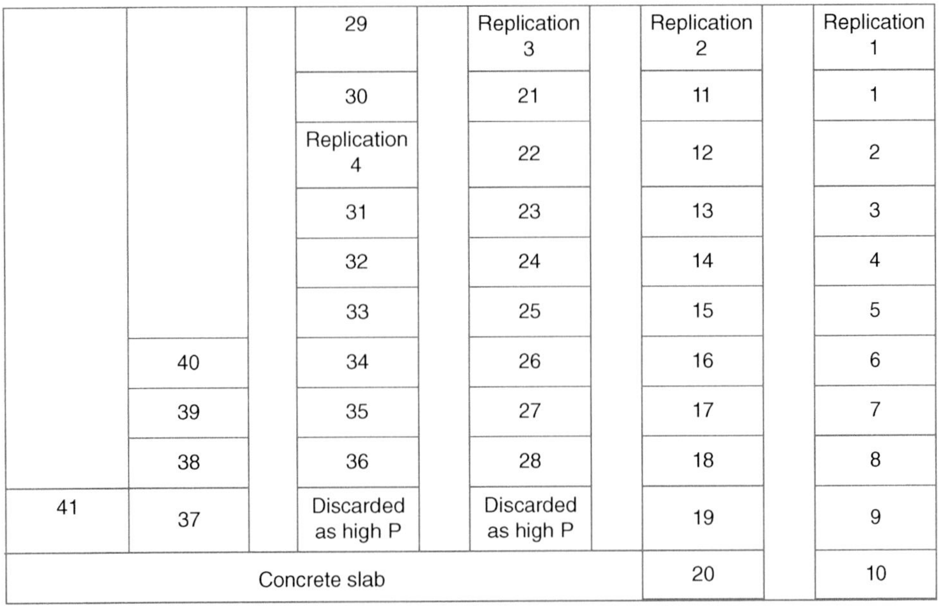

Fig. 18.3. Schematic representation of the plot layout. (The undisturbed plot 41 is outside the fenced trial site.)

important is that plots 37, 38 and 39 (which have probes in them) as well as plot 40 (which has no probe) are all on shallow soil with underlying rubble, and with very deep and thin E horizons.

These plots show signs of some disturbance, possibly due to an old roadway (which can be seen in early photographs of the site), and these four plots showed atypical growth, and should cause the atypical yield results from Replication 4 to be viewed with some suspicion. The mustard yield heights (determined with a disk meter, see Table 18.1) from Replication 4 (plots 31–40) averaged about 10 cm, while the other three replications averaged nearly 13 cm.

Indicator crop

Once the initial plot layout had been agreed upon, and before any treatments had been applied to any of the plots, soil samples were taken from each of the plots to determine the baseline soil fertility. As is evident from Table 18.1, there was considerable variation between plots, and more especially between replications. It was therefore decided that an indicator crop should be planted to quantify the initial variability in terms of biomass yield and actual nutrient uptake. This chapter describes how and why this was done.

Plant growth and nutrition are subject to and influenced by the interaction of various factors. The aim of this study is to investigate how site location and soil nutrient variability affected the growth of Caliente mustard (*Brassica juncea*). Several elements are considered essential for plant growth and development and each plant has its own optimum nutrient element requirements as well as a lower nutrient threshold, below which plant growth will be suboptimal (Silva and Uchida, 2000; Han *et al.*, 2011). Six of the essential elements: nitrogen (N), phosphorus (P), potassium (K), magnesium (Mg), calcium (Ca) and sulfur (S), are required in large amounts by plants in order to complete their life cycle and these have been classified as the macronutrients (Maria, 2004; Brady and Weil, 2008; Fernández and Hoeft, 2009). Other nutrients required in smaller quantities are classified as micronutrients (Tucker, 1999; Nathan, 2009).

Deficiency symptoms occur when one or more elements are deficient. More than one nutrient deficiency can occur in a plant at a given time, however, it is not clear whether the

Table 18.1. Initial soil chemistry (Bemlab) analyses and height of indicator crop.

		Available		Exchangeable cations (mg/kg)						
Plot no.	pH (KCl)	P (mg/kg)	K (mg/kg)	Na^{+}	Ca^{2+}	Mg^{2+}	Leco carbon (%)	WB carbon (%)[a]	CEC ($cmol_c$/kg)[b]	Mustard crop height (cm)[c]
1	5.6	13	177	18	1236	259	3.62	2.49	5.49	20.8
2	5.3	10	148	16	1078	223	3.92	2.61	6.15	21
3	5	9	154	18	950	206	3.18	2.44	7.01	16.4
4	4.9	5	150	18	878	205	4.43	2.47	6.97	6.7
5	4.9	5	173	16	770	181	3.14	2.42	6.45	7.2
6	4.8	6	147	14	676	157	2.53	2.05	6.45	9.5
7	5	6	159	12	654	142	2.69	2.3	6.41	14.1
8	5	11	137	9	798	157	2.81	2.07	5.65	16.4
9	5	11	108	9	644	131	2.83	2.22	5.53	21.5
10	4.90	10	117	39	632	138	2.38	2.18	4.10	6.5
11	5.2	8	163	18	1082	222	4.09	2.52	7.43	13.5
12	5.3	11	123	21	1098	215	4.16	2.59	6.28	11
13	5.1	6	145	14	1044	204	3.36	2.45	7.22	11.6
14	5.1	6	172	16	1052	223	4.31	2.6	8.28	12.9
15	5.1	8	138	18	1100	204	3.61	2.52	7.58	11.7
16	5.1	7	194	18	1104	227	3.71	2.63	7.49	11.2
17	5.2	7	221	14	1060	220	4.14	2.59	6.21	10
18	5.2	6	201	12	982	200	2.94	2.42	7.26	17.4
19	5.1	5	152	9	812	163	2.67	2.25	5.91	15.1
20	5	4	142	16	750	136	2.49	1.92	4.98	6.5
21	5.4	8	221	14	1016	215	2.86	2.35	6.58	17.8
22	5.4	5	176	9	798	169	3.32	2.29	5.94	12.5
23	5.3	5	200	14	964	193	2.75	2.47	6.49	10.4
24	5.3	5	207	16	1050	203	3.08	2.57	6	3.7
25	5.4	5	197	14	1070	194	3.44	2.55	5.98	8.7
26	5.4	6	200	14	1020	176	3.33	2.49	6.31	5.1
27	5.5	6	167	14	1096	160	3.36	2.43	6.54	12.9
28	5.5	10	149	9	1124	154	2.76	2.32	5.78	18.3
29	5.3	6	171	14	1060	208	2.63	2.58	6.35	13
30	5.1	4	143	14	824	175	3.59	2.45	5.94	10.4
31	5.2	7	153	14	894	181	2.91	2.42	6.05	10.1
32	5.3	6	164	14	860	163	3.43	2.57	6.93	7.8
33	5.3	5	204	14	980	192	3.02	2.54	6.01	11.3
34	5.5	5	211	14	1038	182	3.1	2.61	6.06	13
35	5.8	6	146	14	1172	136	2.17	2.15	5.6	6.7
36	6.2	7	128	12	1258	128	1.91	1.95	5.97	9.9
37	5	7	111	18	906	115	2.19	1.74	6.37	5.6
38	5	6	125	21	890	126	2.55	1.99	5.61	8.1
39	5.2	3	148	21	914	132	2.43	1.94	5.55	17
40	5	4	157	21	978	143	2.69	1.8	5.97	13.7
Mean 1–40	5.2	6.8	162.5	15	958	179	3.1	2.3	6.3	11.9

[a]WB, Walkley-Black.
[b]CEC, Cation exchange capacity.
[c]Height of mustard crop was measured with a disk meter.

effects of these simultaneous occurrences are interactive or not (Wallace and Romney, 1980). The most common assumption is that yield is affected by the most limiting nutrient as reflected in Liebig's 'Law of the Minimum': there is only one limiting factor (scarcest resource) and only that factor affects the yield of the plant (Wallace and Romney, 1980); once that deficiency is addressed, another nutrient may become the limiting factor. Of the six macronutrients N and P have been reported as the most limiting elements to plant growth (Maria, 2004; Han *et al.*, 2011; Schröder *et al.*, 2011); they also pose a threat to pollution of water resources when available in excessive amounts in the soil (Robertson *et al.*, 1988). However, determining the limitations of crop yield is often complex, due to unpredictable interactions, especially given the variation in topsoil depth shown in Table 18.2.

Table 18.2. Depth of E horizon (cm) from start of E horizon to end of gley layer, for those plots on which capacitance probes were located.

Plot no.	E horizon depth (cm)	Notes
1	66–110	
3	54–98	Shallow
4	65–110	
6	73–110	
8	70–110	
11	64–90	
12	56–94	Shallow
14	69–101	
16	72–110	
17	63–110	
21	78–95	Deep
22	78–99	Deep
23	80–93	Very deep
26	61–95	
27	70–110	
33	60–90	
34	68–80	
37	87–93	Very deep and thin
38	80–90	Very deep and thin
39	89–94	Very deep and thin

Plant growth- and yield-influencing factors include: (i) light; (ii) moisture; (iii) soil properties; (iv) temperature; (v) pests and diseases; and (vi) nutrient concentrations (Smith and Loneragan, 1997). These factors are not independent of one another and this needs to be taken into account when attempting to diagnose the cause for a given crop yield (Wallace and Romney, 1980; Kosaki *et al.*, 1989). The various factors act differently, some affect the soil's ability to supply nutrients to the plant, and others either limit or facilitate the plant's ability to take up the available nutrients (Corey and Schulte, 1973). In a study by Wallace and Romney (1980) investigating the effects of multiple nutrient deficiencies on plants by omitting iron (Fe) and manganese (Mn) separately and simultaneously, the results showed that interactions between the nutrients greatly affect the results of multiple limiting factors in a plant.

Meteorological data

In the year of establishment of the trials (indicator crop), total rainfall was 1022 mm (see Fig. 18.4),

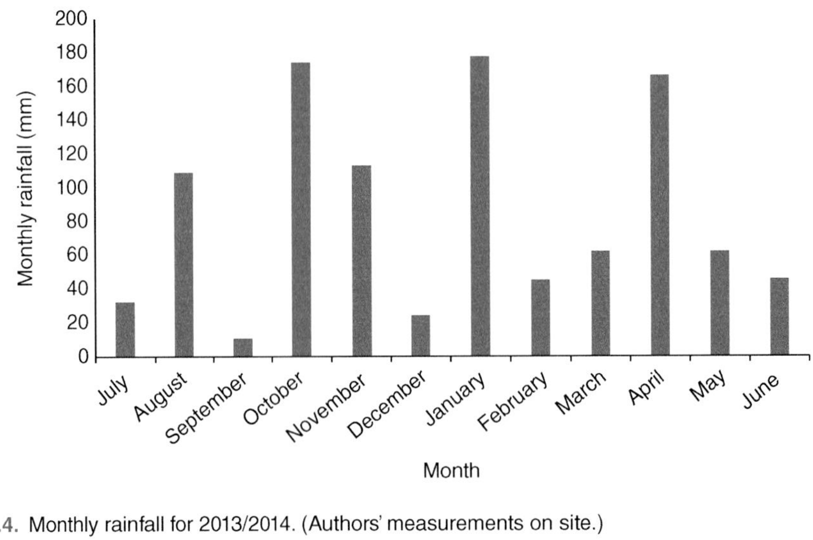

Fig. 18.4. Monthly rainfall for 2013/2014. (Authors' measurements on site.)

of which 250 mm fell while the crop was growing (21 March–25 May 2013), compared with the long-term average annual rainfall for George of 860 mm.

Biofumigant indicator crop choice

The plant chosen as the indicator crop was Caliente mustard (*B. juncea*). This is a hot variety of the mustard plant, which has been found to contain high levels of glucosinolates that suppress weeds, nematodes and diseases (Clark, 2007; Björkman and Shail, 2010; see detailed discussion in Chapter 21, this volume). Under optimal soil nutrient conditions the crop yields high biomass and improves soil quality through its extensive root system which encourages aggregate formation (Clark, 2007). The green plant matures in 4–6 weeks; the leaves can reach a length of 38–46 cm. The plant grows well under cool conditions with full sun, in moist but well-drained soils with a pH of 5.5–8.5 (Clark, 2007; Björkman and Shail, 2010). The site was suboptimal with regards to pH, available P and exchangeable Al, and because of this and the fact that no nitrogen was supplied, low yields were expected.

Site preparation and planting

On the 14 March 2014 the site was ploughed twice using a chisel plough with a weed roller (see Fig. 18.5), and the weeds chopped using a mechanical power tiller. A week later the field was chisel ploughed to eliminate any emerging weed seedlings.

On 21 March 2014 Caliente mustard seeds were broadcast over the prepared soil and lightly

Fig. 18.5. Site preparation using a chisel plough.

raked in. A substantial number of seeds germinated. However, field observation showed low seedling density in some of the plots. Doves and pigeons were observed feeding on the seedlings, especially in Replication 1, adjacent to the pine plantation to the south of the site. Replacement seeds were soaked in water overnight and scattered over areas with poor seedling growth 10 days later. In total just under 1 kg of Caliente mustard seed was planted over the 1500 m^2 experimental area.

Leaf analysis

To test and possibly diagnose the cause of the visible nutrient deficiency symptoms that were noted in some of the mustard plants, leaf samples were collected following guidelines from Bemlab (Dr Pieter Raath, 2014, personal communication). The sampling was done over a period of 2 days, samples from Replications 1 and 2 were collected on the 26 May 2014, and Replications 3 and 4 samples were collected on the 27 May 2014. The sampling varied in each plot according to the dominating plant growth stage and 30 leaves were collected from random plants within each plot.

In plots where more than 50% of the area was covered by stunted discoloured plants, more than 30 leaves from these stunted plants only were collected. In plots where there were more green mature plants than stunted plants, only the leaves of these green plants were collected. The harvested leaves were sealed in labelled bags provided by the laboratory and sent off to the laboratory on the day of collection by courier for analysis.

Yield height measurements

The heights of the mustard plants were taken using a weighted metal disc pasture meter (Bransby and Tainton, 2010).

The measurements were done systematically, starting with plot 1 in Replication 1 and ending with plot 40 in Replication 4. In each plot, five measurements were taken in five different areas within the plot. An average in centimetres was calculated from the readings and the result gave a representative yield height for each plot (Table 18.3).

Results

Soils

As shown in Table 18.1, there is considerable variation in the soil analysis results. Out of the 40 plots, four were classified as sandy soils and the remaining 36 plots were classified as sandy loam soils, but these four samples were close to sandy loam status. The pH was measured in KCl and the site consists of moderately acidic soils ranging from pH 4.8 to pH 6.3, but with only four plots falling below pH 5.0, and two plots above pH 5.6. Available soil P (determined by the Bray 2 method) was found to be very low (average 7 mg/kg). Two adjacent plots, one in Replication 3, the other in Replication 4 were found to have substantially higher amounts of P (24 and 23 mg/kg, respectively) in comparison with the other plots, and were therefore excluded from the subsequent experiment. This was done in order to eliminate any site effects as the high initial P level would confound the yield results of the plants growing in these plots. SOM was determined by the WB and by the Leco method; results from the WB method were more uniform ranging between 1.9% and 2.5% carbon (down from 4.1% in the unploughed sample 41).

Data on soil properties and yield height

A number of soil properties were tested for correlation with the yield heights using descriptive statistics. Positive correlations were found with available soil P and soil pH ($r = 0.416840$, $r = 0.303097$, respectively, and p value < 0.05 for both). The correlation between yield heights in the different replications was negative ($r = -0.249380$, $p < 0.05$), with Replications 1 and 2 having the highest yields and Replication 4 the lowest (see Fig. 18.6). Once again, Fig. 18.6 shows how Replication 4 is significantly lower in height than the other three replications, having a mean height of about 10 cm, while

Table 18.3. Leaf analysis and yield height measurements of Caliente mustard crop.[a]

Plot no.	Leaf analysis N (% DM)	P (% DM)	K (% DM)	Ca (% DM)	Mg (% DM)	Na (mg/ kg)	Mn (mg/ kg)	Fe (mg/ kg)	Cu (mg/ kg)	Zn (mg/ kg)	B (mg/ kg)	Mean yield height (cm)[b]
1	2.93	0.18	3.03	2.87	0.49	322	26	652	8	395	37	20.8
2	3.47	0.17	3.6	3	0.59	350	41	322	7	346	42	20.4
3	3.5	0.17	4.09	2.61	0.53	406	68	279	6	243	37	15.4
4	3.19	0.09	2.86	3.32	0.82	629	61	273	6	260	33	6.7
5	3.96	0.18	4.46	2.52	0.65	682	64	313	5	163	36	7.4
6	3.73	0.14	3.88	2.61	0.56	505	63	318	6	194	32	9.4
7	3.97	0.19	4.75	2.74	0.51	459	63	281	8	160	38	14
8	4.23	0.19	3.87	2.67	0.48	372	64	688	7	165	38	15.5
9	3.65	0.19	4.02	2.88	0.57	259	75	318	6	150	43	9.1
10	3.07	0.14	4.34	3.28	0.67	335	97	345	7	181	42	6.1
11	2.88	0.21	3.09	3.87	0.56	164	39	405	11	599	35	12.9
12	2.54	0.17	2.86	3.55	0.51	208	58	397	6	311	38	10.4
13	2.75	0.14	2.76	3.49	0.61	265	76	393	6	307	44	21.4
14	2.89	0.12	2.79	3.25	0.71	314	63	510	5	366	42	7.2
15	3.7	0.2	3.2	3.12	0.59	174	74	444	7	252	37	11.7
16	3.74	0.17	3.39	3.16	0.6	197	70	362	7	261	40	10.9
17	2.97	0.12	3.08	3.71	0.77	189	76	339	7	340	41	10
18	3.52	0.16	3.73	3.14	0.68	195	74	287	7	261	42	17.4
19	3.58	0.19	4.03	3.19	0.67	232	66	331	7	247	40	15.1
20	2.87	0.12	3.36	2.88	0.58	249	72	411	5	222	39	13.1
21	3.04	0.17	3.16	2.38	0.46	154	57	224	7	157	31	17.8
22	2.87	0.15	3.22	2.75	0.47	204	61	257	7	192	32	12.5
23	3.03	0.11	2.59	3.05	0.57	128	62	179	7	306	28	10.3
24	2.52	0.08	2.31	3	0.72	311	59	238	6	326	27	3.7
25	3.32	0.12	2.55	2.53	0.49	234	44	252	6	216	25	8.7
26	2.84	0.1	2.96	2.92	0.59	263	54	218	6	229	27	5
27	3.04	0.11	2.73	2.93	0.43	140	51	279	6	194	27	12.9
28	2.91	0.13	2.92	2.46	0.33	153	38	216	6	117	28	18.3
29	2.29	0.12	2.82	2.69	0.55	257	62	180	5	161	30	13
30	2.61	0.09	2.29	2.53	0.53	271	69	183	4	162	25	10.4
31	1.89	0.06	1.76	2.74	0.55	345	65	202	3	124	26	10.3
32	2.23	0.07	1.89	2.44	0.6	277	68	168	3	171	26	7.8
33	2.22	0.08	2.16	3.02	0.63	274	68	151	4	181	27	9.4
34	1.61	0.07	2.25	2.55	0.6	172	79	167	4	194	29	12.3
35	1.4	0.08	2.24	3.5	0.45	235	69	200	4	148	35	15.9
36	1.47	0.12	2.51	3.42	0.36	167	50	182	4	90	32	13.1
37	2.53	0.15	2.57	2.88	0.48	182	61	366	5	163	33	6.1
38	1.93	0.1	2.3	2.57	0.56	225	83	160	3	191	29	9.6
39	2.33	0.1	2.48	2.53	0.54	245	85	213	5	189	27	5.4
40	2.36	0.08	2.15	2.45	0.56	229	71	198	4	179	27	7.8

[a]DM, Dry matter.
[b]Values are the mean of five measurements taken from different areas within each plot.

the mean of the other three replications is almost 13 cm. The second lowest yields were from Replication 3, and the trend demonstrates a fertility gradient, with Replications 1 and 2 yielding significantly higher than Replications 3 and 4. But it must be pointed out once more that plots 37, 38, 39 and 40 had abnormally low yields, averaging less than 6 cm against the average of almost 12 cm for the research site.

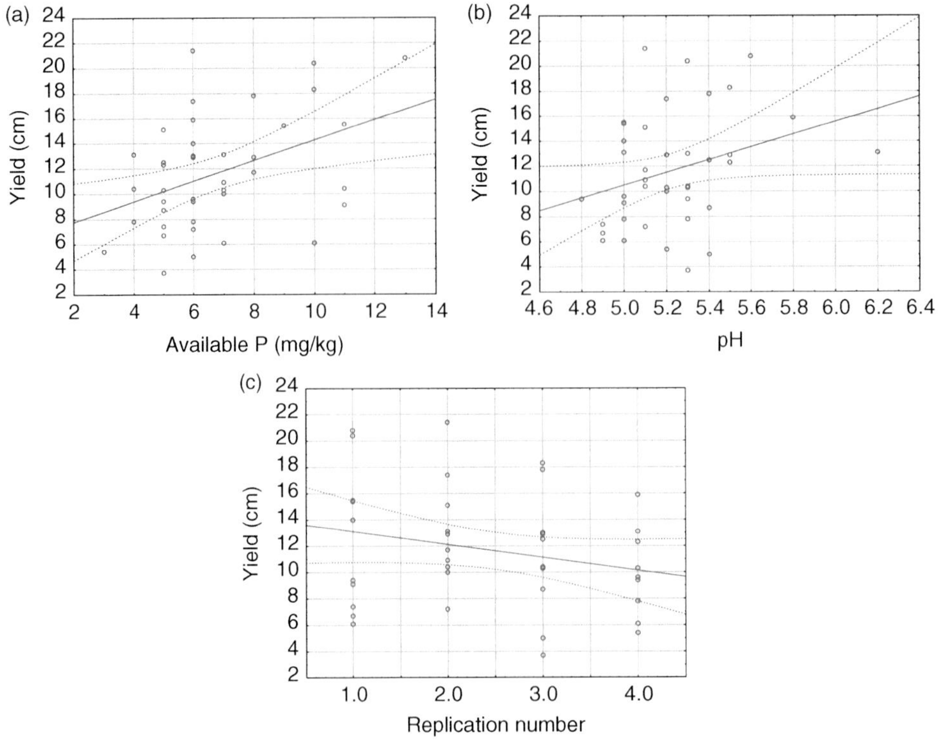

Fig. 18.6. Correlation between yield heights of mustard and available soil phosphorus (P) levels (a). The correlation is significant (r = 0.416840, p < 0.05). Correlation between the yield height and the soil pH (b) is significant (r = 0.303097, p < 0.05). A correlation was found between yield height and the replication number (c). There is a negative but significant correlation (r = –0.249380, p < 0. 05), representing a fertility gradient.

Growth and visible nutrient deficiency symptoms

Although initially vigorous, the growth of the mustard plants was extremely variable between the plots (Fig. 18.7a). Plant growth data were collected for each plot during June. Some plots had healthy-looking tall green plants; other plots had stunted plants with purple leaves (Fig. 18.7b). By July, all plots were showing severe deficiency symptoms (Fig. 18.7c). Replication 1 had a small percentage of stunted purple-yellow plants, and a few plots which had bare patches; however, 60% of the plants growing on this replication had healthy, green, broad leaves and tall plants.

Replication 2 had an even distribution of stunted green plants and tall, broad-leafed, green plants. Replication 3 had a high percentage of stunted yellow-purple and green plants; while 95% of the plants growing in Replication 4 were stunted and had a purple-yellow colour. The root systems of the plants also varied depending on where they were growing: plants in Replication 4 had shallow root systems, while the healthy green plants found in parts of Replication 1 had deep and extensive root systems.

The stunted purple-yellow leafed plants had shallow root systems. In order to understand these plant growth patterns, leaf samples were collected on the 26 and 27 May 2014 and sent to Bemlab for analysis. In each plot 30 leaves were collected randomly, with more leaves being collected in plots that had stunted plants. This was to done to account for the small size of the leaves, to ensure that enough material was available as required for the analysis.

Fig. 18.7. Growth and symptoms observed on the mustard plants. (a) The field covered by mustard plants a month after planting (April); (b) purpling of leaves in plants (June); (c) stunted, thin plants with yellow leaves throughout (July). (Photographs courtesy of N'wa-Jama Mashele.)

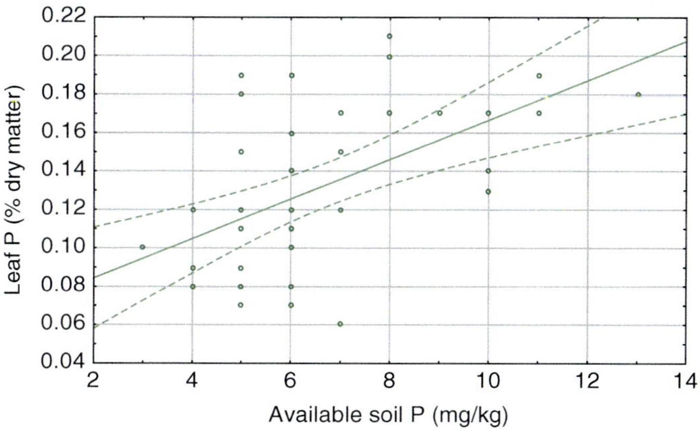

Fig. 18.8. Correlation between leaf P and soil P ($r = 0.558270$, $p < 0.05$).

Pest and disease problems were visible only at a later stage in the mustard plants more especially in the much healthier looking plants in Replication 3.

Leaf analysis

Leaf analysis results give an indication of the amount of nutrients the plant was able to take up (Bornman *et al.*, 1989; Louw and Scholes, 2003). Nutrients in the leaves are subject to a number of variations, depending on the season and physiological state of the plant (Louw and Scholes, 2003). Thus the measurements are dependent on which part of the plant is analysed as well as the age of the plant.

Descriptive statistics were used to correlate soil P and P in the foliar tissue of the mustard plants. Figure 18.8 shows a strong positive correlation between the two factors ($r = 0.558270$, $p < 0.05$).

The leaf analysis confirms that Replication 4 has lower N (mean 2% dry matter (DM)) and lower P (mean 0.09% DM) against the mean for Replications 1–3 (mean 3% DM and 0.15% DM, respectively).

Discussion

Nutrient availability and deficiencies

The growth and subsequent yield of the mustard plants were affected by nutrient availability. Nutrient stress results in the mobilization of nutrients from the older plant parts to the younger developing parts of the plant. This impacts on the expression of deficiency symptoms and the relationship of plant nutrient concentration to growth and yield (Smith and Loneragan, 1997). The appearance of visible symptoms in plants (as in Fig. 18.7) is an indication of metabolic irregularities caused by nutrient deficiencies and toxicities (Grundon *et al.*, cited in Reuter and Robinson, 1997; Han *et al.*, 2011).

The presence of elements in a soil does not guarantee their availability to the plant, due to complex soil characteristics (Schröder *et al.*, 2011). If soils have a poor structure, are too wet or have poor drainage, this inhibits the proper uptake of nutrients by plant roots (Fernández and Hoeft, 2009). The symptoms most common in plants growing in soils with poor drainage are stunted growth and foliage discoloration. These symptoms were observed in most of the mustard plants growing over the experimental area, gradually spreading from Replication 4 towards Replication 1.

In some of the plots there were signs of delayed maturity with plants having small leaves, these were short (± 8 cm) in comparison with plots that had plants with very broad mature leaves and a height of 30–40 cm. This may be indicative of the nutrient imbalances experienced by the plants in the different plots, with the broader, longer leaves being plants that were able to utilize the available nutrients as opposed to those that were hindered by plant physiological processes and/or soil nutrients.

The observed visual symptoms (purple leaves and veins, yellowing of older leaves) are induced by deficiencies in the macronutrients, N and P. Nitrogen forms a part of proteins, enzymes, chlorophyll and growth regulators; it plays a role in all enzymatic reactions (Bromley, n.d.; Tucker, 1999; Silva and Uchida, 2000). The leaf analyses showed that some plants had insufficient supply of nitrogen for growth resulting in a number of physical defects. Phosphorus deficiency often shows in purpling of the leaves, and may result in stunted roots; a deficiency in nitrogen results in stunted plants with minimal aboveground DM, and yellowing and chlorosis in the older leaves as the nutrient is translocated to the younger leaves.

Soil pH

The pH of the soil is important as it determines which elements the plant is able to take up and which it cannot. It affects the microorganism population and activity (Brady and Weil, 2008; Nathan, 2009), by reducing the population of microorganisms needed to transform N, P and S into forms which the plant can use (Fernández and Hoeft, 2009). The site had a pH ranging from pH 4.8 to pH 6.3, and most of the plots have below optimum conditions for nutrient uptake for most of the nutrients as indicated in Fig. 18.9.

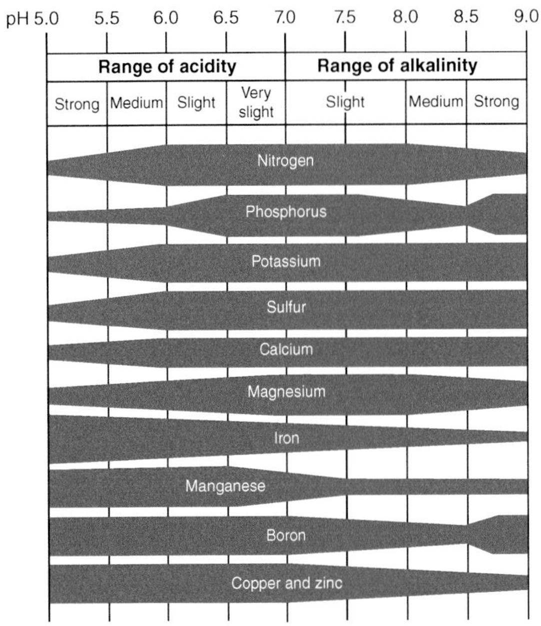

Fig. 18.9. Availability of essential elements to plants at different pH levels in mineral soils. (From Benton Jones, 2012.)

For the plant-growth-limiting elements N and P the pH is severely limiting especially for the uptake of P. At low pH, P is immobile in soils due to a number of chemical reactions (Schröder *et al.*, 2011). In acidic soils, P becomes locked up due to the presence of iron or aluminium (Al) resulting in insoluble phosphorus compounds which the plant cannot use (Silva and Uchida, 2000; Fernández and Hoeft, 2009); given the high soil aluminium levels, the apparently low availability of P is not surprising. With increase in pH, K already becomes available at pH 5.8, while P does not become readily available until soil pH exceeds 6.5.

Phosphorus content in the analysed leaves was low for all the plots sampled. This may be due to the fact that soil phosphorus levels were also found to be below minimum requirements, as there is a positive correlation between the two factors. There was also a significant correlation between soil phosphorus and yield height ($r = 0.42$, Fig. 18.6a) and an even stronger correlation between soil phosphorus and foliar P levels ($r = 0.56$, Fig. 18.8). Phosphorus forms part of the genetic transcript (DNA and RNA), it speeds up plant maturity, stimulates cell division and plays a role in photosynthesis (Bromley, n.d.; Tucker, 1999; Silva and Uchida, 2000). An early indication of P deficiency is slow-maturing plants, a more severe degree of deficiency results in the purpling of both the leaves and the stems (Silva and Uchida, 2000) caused by an accumulation of anthocyanin pigments (as can be seen in Fig. 18.7). Soils with low P levels exhibit limited root growth which then limits plant growth (Schröder *et al.*, 2011), hence the severely stunted plants that were visible on the field.

Yield height varied between replications, showing a fairly weak correlation; however, from the data collected and visual observations, Replications 1 and 2 had a higher amount of taller plants than Replications 3 and 4. The reason for such a growth pattern is difficult to ascertain, as the variations observed (pH and soil nutrient levels) were all distributed within each replication. Proximity to the pine plantation may have been a contributing factor, due either to protection from the wind or greater availability of soil moisture.

The poor growth of the mustard plants could not conclusively be attributed to one particular factor. There were weak but significant correlations of the yield height with: soil phosphorus, soil pH and replication, and other factors that were tested showed trends but yielded no statistically significant results. Though soil analysis results give an idea of the characteristics of the soil and the foliar analysis indicates just how much of those soil nutrients the plant was able to take up, the yield of a given crop is the end product of complex plant processes. Although there was variability in the initial soil fertility levels, and also in mustard growth and development, the generally low levels of soil nutrients make this a suitable site for soil fertility research. It was expected that all treatments which improve soil nutrients and ameliorate soil acidity would show enhanced growth.

Conclusion

The literature reviewed shows that plants are subject to external and internal factors that are not independent of each other. Though these factors act differently, they affect each other and the overexpression or under expression of one factor however minute may hinder the plant from functioning optimally. In the case of this experiment, soil nutrient levels were poor, and the soil was moderately acidic. The experiment served its purpose and we were able to ascertain the initial status of the experimental area. It also highlighted challenges for the next season, thus the next phase of the trial would require some soil amendment strategies along with plant rotations and appropriate planting times. The major conclusion was that Replication 4 is considerably poorer initially than the rest of the site and that, in particular, plots 37, 38, 39 and 40 failure to perform adequately, even after receiving mineral and biological support, might have to be discarded. The overall low fertility status of the soil makes it well suited to a long-term comparative trial, as both conventional and organic treatments are likely to improve as soil fertility improves, due first to amendments with fertilizers, and secondly to improved soil biology. The low available soil P levels were likely to prove challenging for organic farming systems.

References

Auerbach, R. and Mashele, N. (2014) Research protocol for the Nelson Mandela Long-term Comparative Organic Farming Systems Research Trials on the George Campus of the Nelson Mandela Metropolitan University. *Internal Research Paper* 1, Agro-ecology Group, Nelson Mandela University, George, South Africa.

Benton Jones, J. (2012) *Plant Nutrition and Soil Fertility Manual*, 2nd edn. CRC Press, London.

Björkman, T. and Shail, J.W. (2010) *Cornell Cover Crop Guide for Fall Mustard*. Cornell University, Ithaca, New York.

Bornman, J.J., Ranwell, J.F., Venter, G.C.H. and Vosloo, L.B. (1989) *FSSA Fertilizer Handbook*, 3rd edn. Translated from Afrikaans by E.B. Dickson, E. Greenfield and L.B. Roos. Fertilizer Society of South Africa (FSSA), Hennopsmeer, Pretoria.

Brady, N.C. and Weil, R.R. (2008) *The Nature and Properties of Soils*, 14th edn revised. Pearson Prentice Hall, Upper Saddle River, New Jersey.

Bransby, D. and Tainton, N. (2010) The disc pasture meter: possible applications in grazing management. Originally published in 1977 in: *The Proceedings of the Annual Conferences of the Grasslands Society of Southern Africa* 12(1). Available at: http://www.tandfonline.com/doi/abs/10.1080/00725560.1977.9648818 (accessed 30 October 2018).

Bromley, B.J. (nd) Nutrient deficiency symptoms. *Mercer County Horticulture* 10. Available at: https://mgofmc.org/docs/nutrientdeficiency.pdf (accessed 2 September 2019).

Clark, A. (2007) *Managing Cover Crops Profitably*, 3rd edn. Sustainable Agriculture Research and Education (SARE). Available at: https://www.extension.iastate.edu/alternativeag/info/ManagingCoverCropsProfitably_0812.pdf (accessed 2 September 2019).

Corey, R.B. and Schulte, E.E. (1973) Factors affecting the availability of nutrients to plants. In: Walsh, L.M. and Beaton, J.D. (eds) *Soil Testing and Plant Analysis*, revised edn. Soil Science Society of America, Madison, Wisconsin, pp. 603–643.

Drimie, S. and McLachlan, M. (2013) Food security in South Africa a first step to transdisciplinary approach. *Food Security* 5, 217–226.

Fernández, F.G. and Hoeft, R.G. (2009) *Managing Soil pH and Crop Nutrients*. Illinois Agronomy Handbook. Available at: http://extension.cropsciences.illinois.edu/handbook/pdfs/chapter08.pdf (accessed 8 July 2014).

Flieβbach, A., Oberholzer, H.-R., Gunst, L. and Mäder, P. (2007) Soil organic matter and biological soil quality indications after 21 years of organic and conventional farming. *Agriculture, Ecosystems and Environment* 118, 273–284.

Frayne, B., Crush, J. and McLachlan, M. (2013) Urbanisation, nutrition and development in Southern African citizens. *Food Security* 6, 101–112.

Grundon N.J., Robson, A.D., Lambert, M.J. and Snowball, K. (1997) Nutrient deficiency and toxicity symptoms. In: Reuter, D. and Robinson, J.B. (eds) *Plant Analysis, an Interpretation Manual*, 2nd edn. CSIRO (Commonwealth Scientific and Industrial Research Organisation), Canberra, pp. 37–51.

Han, W.X., Fang, J.Y., Reich, P.B., Woodard, I. and Wang, Z.H. (2011) Biogeography and variability of eleven mineral elements in plant leaves across gradients of climate, soil and plant functional type in China. *Ecology Letters* 14(8), 788–796.

ICFR (Institute for Commercial Forestry Research) (2013) *Germuizenhuizen, I. Climatic Profile of the Saasveld Weather Station Site in the Southern Cape*. ICFR, Pietermaritzburg, South Africa.

ICROFS (International Centre for Research in Organic Food Systems) (2013) Available at: www.icrofs.org (accessed 30 November 2018).

Kosaki, T., Wasano, K. and Juo, A.S.R. (1989) Multivariate statistical analysis of yield-determining factors. *Soil Science and Plant Nutrition* 35(4), 597–607.

Lotter, D.W., Seidel, R. and Liebhardt, W. (2003) The performance of organic and conventional systems in an extreme climate year. *American Journal of Alternative Agriculture* 18(3), 146–154.

Louw, J. and Scholes, M. (2003) Foliar nutrient levels as indicators of site quality for *Pinus patula* in the Mpumalanga escarpment area. *Southern African Forestry Journal* 197(1), 21–30. https://doi.org/10.1080/20702620.2003.10431718

Mäder, P., Fließbach, A., Dubois, D., Gunst, L., Jossi, W., Widemer, F., Oberson, A., Frossard, E., Oehl, F., Wiemken, A., Gattinger, A. and Niggli, U. (2006) The DOK experiment (Switzerland). *ISOFAR Science series*, No1: *Long Term Field Experiment in Organic Farming*. Verlag Dr Köster, Berlin, pp. 41–58.

Maria, R.M. (2004) Soil nutrient status, spatial variability of soil chemical properties and fertilization requirements in Cabo Delgado, Nampula and Manica provinces of Mozambique. MSc thesis, University of Hawai'i, Hawaii.

May, J. and Rogerson, C.M. (1995) Poverty and sustainable cities in South Africa. *Habitat International* 9(2), 165–185.

McLachlan, M. and Landman, A.P. (2013) Nutrition-sensitive agriculture – a South African perspective. *Food Security* 5, 857–871.

Nathan, M.V. (2009) *Soils, Plant Nutrition and Nutrient Management*. Missouri Master Gardener Core Manual. MU extension, University of Missouri, pp. 1–12. Available at: http://extension.missouri.edu/p/MG4 (accessed 14 July 2014).

Qiu, W., Curtin, D. and Beare, M. (2019) Spatial variability of available nutrients and soil carbon under arable crop in Canterbury. The New Zealand Institute for Plant and Food Research Limited. Available at: https://www.researchgate.net/publication/308255674_Small-Scale_Spatial_Variability_of_Plant_Nutrients_and_Soil_Organic_Matter_An_Arable_Cropping_Case_Study (accessed 2 September 2019).

Qu, M., Li, W. and Zhang, C. (2014) County-scale spatial variation of macronutrient availability ratios in paddy soils. *Applied and Environmental Soil Science* 2014, Article ID 689482, 10 pp. https://doi.org10.1155/2014/689482

Reuter, D. and Robinson, J.B. (1997) *Plant Analysis, an Interpretation Manual*, 2nd edn. CSIRO, Canberra, Australia.

Robertson, G.P., Huston, M.A., Evans, F.C. and Tiedje, J.M. (1988) Spatial variability in a successional plant community: pattern of nitrogen availability. *Ecology* 69(5), 1517–1524.

Rodale Institute (2011) Rodale Institute Farming Systems Trial: Celebrating 30 Years. Available at: http://rodaleinstitute.org (accessed 23 September 2013).

Schröder, J.J., Smit, A.L., Cordell, D. and Rosemarin, A. (2011) Improved phosphorus use efficiency in agriculture: a key requirement for its sustainable use. *Chemosphere* 84, 822–831.

SCWG (Soil Classification Working Group) (2009) *Soil Classification: a Taxonomic System for South Africa*. Originally Vol. 15 of *Memoirs on the Agricultural Natural Resources of South Africa* (1971). Department of Agricultural Technical Services, Pretoria.

Silva, J.A. and Uchida, R. (2000) Essential nutrients for plants: nutrient functions and deficiency symptoms. In: Silva, J.A. and Uchida, R.S. (eds) *Plant Nutrition Management in Hawaii's Soils, Approaches and Application to Tropical and Subtropical Agriculture*. College of Tropical Agriculture and Human Resources, University of Hawaii at Manoa, Hawaii, pp. 31–56.

Smith, F.W. and Loneragan, J.F. (1997) Interpretation of plant analysis: concepts and principles. In: Reuter, D. and Robinson, J.B. (eds) *Plant Analysis, an Interpretation Manual*, 2nd edn. CSIRO (Commonwealth Scientific and Industrial Research Organisation), Canberra, pp. 3–33.

Tucker, M.R. (1999) *Essential Plant Nutrients: their Presence in North Carolina Soils and Role in Plant Nutrition*. Agronomic Division of North Carolina Department of Agriculture and Consumer Services, Raleigh, North Carolina.

Wallace, A. and Romney, E.M. (1980) Some problems in study of simultaneous multiple nutrient deficiencies in plants. *Journal of Plant Nutrition* 2(1-2), 213–216.

Zezza, A. and Tasciotti, L. (2010) Urban agriculture, poverty, and food security: empirical evidence from a sample of developing countries. *Food Policy* 35, 265–273.

19 Comparative Water Use Efficiency and Water Retention in the Mandela Trials

Catherine Eckert,[1]* Raymond Auerbach[1] and Simon Lorentz[2]
[1]Nelson Mandela University, George, South Africa; [2]University of KwaZulu-Natal, Durban, South Africa

---

Abstract

With the world's growing population and our limited natural resources, there is a need to produce more food using fewer inputs, especially fresh water. Water is a critical resource in agriculture and may be more of a limiting factor than other crop growth requirements. As water availability is impacted by climate change and competition from human consumption and other industries, methods of improving crop water use efficiency (WUE) through conservation of water and the enhancement of crop growth need to be employed to meet our growing demands sustainably.

The research assessed the differences in soil water status between organically farmed crops with a grass mulch and conventionally farmed crops without mulch, to use the water resource more efficiently. This research, conducted at the Mandela long-term organic farming systems research trial site at Nelson Mandela University, George Campus in South Africa's Southern Cape, is part of a larger research project (the Mandela Trials) in which various researchers have covered agronomy, microbiology and pest and disease control.

The organic treatment had a significantly higher soil water content (SWC) than the conventional treatment at all soil depths over the two seasons 2016/17 and 2017/18. Theta probes showed that in both seasons, for the top 6 cm of soil the organic treatment had a higher SWC than the conventional treatment. The minimum and median SWC in the organic treatment were much higher than that of the conventional and control. However, in the second season the control treatment had a higher minimum SWC than the organic, but a lower median SWC (there was almost no plant growth on the control plots due to nutrient deficiencies). The differences in the SWC of the organic and conventional treatment for 2016/17 was significant, however, differences for 2017/18 were not significant. The capacitance probe data for the 10 cm depth show that the minimum and median SWC of the organic treatment were higher than the conventional. The differences in SWC between all three treatments are statistically significant for both seasons. The combined SWC capacitance probe data of the 0–50 cm soil profile showed that the organic treatment had a higher SWC and median than the conventional treatment. However, the control treatment had the highest SWC and median of all the treatments. There was a significant difference in the organic and conventional treatments in both seasons, but no significant difference between the organic and control treatments in the first season.

Soil carbon was significantly higher in the organic treatment, than the conventional. Organic farming methods preserve and promote an increase in soil organic matter (SOM), thus improving the soil structure and increasing the soil's water holding capacity. From this research, it is recommended that organic farming practices can be used to help conserve SWC, keeping it available to crops for longer and helping farmers make more efficient use of this scarce resource. This is especially relevant for low rainfall areas which are affected by water shortages. Improved SWC availability should be coupled with good agronomic practices to increase productive water losses (plant use) and the conversion of water to yields, increasing WUE. Adding organic matter to the soil will improve resilience and help sequester carbon, and thus mitigate climate change. More research is needed to assess what proportion of the increased soil water retention can be attributed to the influence of the mulch and what effect the SOM had.

*eckertcj88@gmail.com

Introduction

The world's limited soil and water resources are under enormous pressure to provide food, fuel and fibre to meet the demands of the world's ever-increasing population, predicted by the United Nations (2007) to reach 9 billion by the year 2050 and require an estimated 60–70% more food. This needs to be done in a sustainable manner which makes more efficient use of our scarce resources and mitigates the impacts of climate change (Lal, 2009; García-Tejero *et al.*, 2011; Aune, 2012; Chartzoulakis and Bertaki, 2015). A major limiting factor in meeting the agricultural production needs of the growing population is fresh water (García-Tejero *et al.*, 2011; Chartzoulakis and Bertaki, 2015). Water is a critical resource in many sectors including agriculture and it has been suggested that water may be more of a limiting factor than other crop growth requirements in some areas (Raza *et al.*, 2012; Chartzoulakis and Bertaki, 2015).

There are sufficient water resources on a global scale, however, distribution is poor and irregular and many water resources are affected by pollution, have been degraded and are overexploited (García-Tejero *et al.*, 2011). Competition for water from industries and urbanization is also having a negative impact on water availability for agricultural use (Chartzoulakis and Bertaki, 2015). According to Bhaduri *et al.* (2014) food and water security will ultimately depend on water use efficiency (WUE) in agriculture as the agricultural sector is the largest and most inefficient consumer of water. Currently an estimated 70% of global water usage can be attributed to agriculture, due to the production needs of the world's population (García-Tejero *et al.*, 2011). Presently, out of the world's total croplands, approximately 280 million ha are under irrigation and produce roughly a third of the world's food supply. The balance of this, some 1250 million ha is rainfed and produces the remaining two-thirds of the world's food supply (García-Tejero *et al.*, 2011; Bhaduri *et al.*, 2014). In developing nations such as sub-Saharan Africa, South Asia and Latin America, rainfed agriculture predominates. As the productivity of rainfed agriculture is improved, it is expected to grow faster than irrigation productivity in countries with shortages of surface water and groundwater (Bhaduri *et al.*, 2014). To meet the demands of the growing population and remain ecologically sustainable, WUE needs to be improved (García-Tejero *et al.*, 2011). As water available for agricultural usage steadily declines, our ability to meet the world's growing demands for food, feed and fibre is challenged (Chartzoulakis and Bertaki, 2015).

Many African nations live in extreme poverty and are reliant on rainfed agriculture and the natural resources of the region to derive a living (IPCC, 2007). Climate change will have a huge impact on the livelihoods of many people on the African continent, exposing millions to water stress, affecting food security and increasing malnutrition (IPCC, 2007). Up to a 50% decline in yields from rainfed agriculture is expected by the year 2020, in addition to a shortened growing season, lower potential yields and decreasing arable land (IPCC, 2007). Rainfed crops grown in warmer tropical and subtropical areas, where they are already close to their maximum temperature threshold, will show a decline in productivity (IPCC, 2002).

Africa is highly susceptible to a changing climate and already experiences frequent floods and droughts. African nations face multiple issues, such as low economic capacity, lack of technology and heavy reliance on rainfed agriculture as well as adverse weather conditions, and this reduces their ability to adapt to these often-sudden changes (IPCC, 2002). In addition, many farmers throughout Africa are facing challenges of poor soil fertility, pests and diseases and poor access to inputs. In Africa, on average, 21% of the gross domestic product is contributed by agriculture and in some countries it is as much as 70% (IPCC, 2002; Parry *et al.*, 2007). The continent is reliant on agriculture. In regions such as the Sahel, East and Southern Africa, the change in climate could have devastating consequences. In the lower rainfall areas of Africa, a change in climate will have adverse results as water-providing rivers would be impacted and decrease, leading to less production and food security overall (IPCC, 2002). Rising sea levels along with droughts and floods and other weather extremes will negatively impact the availability of food and the amount of available water, ultimately halting healthy African development (IPCC, 2002).

As reported earlier in this volume:

> de Wit and Stankiewicz reported on their research findings in the journal *Science* (2006) in an article entitled 'Changes in surface water supply across Africa with predicted climate change', predicting that perennial drainage density (the amount of runoff generating perennial streams) will decrease sharply with climate change. They point out that in higher rainfall regimes (mean annual rainfall (MAR) of about 1000 mm), given a 10% decrease in average precipitation, the decrease in drainage would already be about 17%, but if the same percentage decrease occurs in drier areas (MAR of about 500 mm), the decrease in drainage density would be about 50%.
>
> (Chapter 1, this volume)

In northern, southern and western Africa there will also be a greater rate of decline in the available workable agricultural land due to desertification as a result of less annual rainfall. Overall, the area of arid and semi-arid land in Africa is projected to increase by 60–90 million ha by the 2080s (IPCC, 2007). A decrease in annual rainfall is also expected to affect the availability of water in major rivers throughout Africa. Productive lands will turn into arid and unworkable areas, forcing people to move on to better land capable of producing food and providing a livelihood (IPCC, 2002).

In the context of the world's water crisis we need to make more efficient use of water as a scarce resource. The purpose of the study reported in this chapter was to investigate the influence of organically grown crops with a grass mulch in comparison to conventionally grown crops without mulch, on the soil water status in the topsoil of rainfed agriculture. This allows farmers to make more efficient use of water.

We hypothesize that grass mulch in the organically grown plots improves WUE by: (i) conserving soil water content (SWC) through reduced evaporative losses from the soil surface; (ii) improved water retention through increased infiltration; and (iii) better soil aggregation and retention of fine soil particles through reduced soil erosion. Although improved SWC is related to soil organic matter (SOM), the nature of this relationship was not covered in this research. It is hoped that this will form part of a follow-up project.

Factors Affecting WUE in Crops

Soil water promotes many physical, biological and chemical processes in soil. It is a solute for plant nutrients, aiding in transport and absorption, as well as being a plant nutrient itself. Apart from air and sunlight, water is the most important factor influencing productivity of plants and they need a constant water supply throughout the growing season for efficient production (Chesworth, 2008). Yields increase with increasing plant available water within the root zone up until field capacity, after which there is little effect (Chartzoulakis and Bertaki, 2015). Crop production is often limited by water, but it has been shown that there are many other factors affecting WUE (Passioura and Angus, 2010) including climatic and edaphic factors, crop type and management.

The influence of mulches on SWC and soil properties

Mulches are materials which are applied to the soil surface to help conserve water, suppress weeds, prevent erosion and reduce surface runoff (Singh *et al.*, 2014; Kadera *et al.*, 2016; Kyere *et al.*, 2018). However, the effects of mulches may be influenced by the type of mulch used as well as precipitation and climatic factors (Kadera *et al.*, 2016). Mulches can be of an organic nature such as plant residues and animal wastes or synthetic such as plastic films or synthetic polymers which may be photo- and biodegradable. The selection of a mulching material is dependent on the crop type and management (Kadera *et al.*, 2016).

Mulching conserves soil water by reducing evaporation, inhibiting weed growth, moderating soil temperature, improving soil structure and soil biota (Kadera *et al.*, 2016; Kyere *et al.*, 2018). In addition, organic mulches contribute to an increased soil nutrient status through addition of SOM and recycling of nutrients as mulches break down (Ramakrishna *et al.*, 2005; Sinkevičienė *et al.*, 2009). SOM promotes healthy soils by regulating many chemical, biological and physical processes. Organic matter plays a key role in increasing infiltration of water, regulating soil temperature and reducing erosion. Mulching can not only help to cool the soil by shading, but the increased ability of the soil to hold water will also have a cooling effect which is particularly applicable in areas which are already near the maximum growing temperatures for crops (Stagnari *et al.*, 2009; Aune, 2012; Singh *et al.*, 2014). Crusting of the soil surface is also reduced as large raindrops are intercepted and prevented from compacting the soil surface. This helps to prevent wind and water erosion. SOM acts like a sponge, absorbing water and allowing time for it to soak into the soil below. The organic matter is a reservoir of plant nutrients which are released as it is broken down. Mulches also inhibit the growth of weeds, thereby reducing their competition for water, nutrients and light (Singh *et al.*, 2014; Kyere *et al.*, 2018). Organic mulches consisting of biodegradable materials such as bark chips, straw, crop residues and grass (Stagnari *et al.*, 2009) are cheaper and more environmentally friendly options than plastic mulch (Kadera *et al.*, 2016).

Soil properties and their influence on water holding capacity

SWC can be defined in many ways according to the parameters and the intended use (Romano, 2014). It is the liquid component held in the soil that contains solutes such as organic compounds, gases and salts (Or *et al.*, 2012) and is the volume of water held in a given volume of soil (Romano, 2014). The status of the soil's moisture content can be defined by the volumetric water content (the amount of water in a given volume of soil) and the soil water potential (the force with which the water is held in the soil profile) (Or *et al.*, 2012; Romano, 2014). Soil moisture is a spatially and temporally dynamic property, and due to this is constantly changing (Or *et al.*, 2012).

Field capacity, permanent wilting point and plant available water (PAW) are common terms used to describe soil water status in an agricultural context. Field capacity is the moisture remaining in the soil profile after the free water has drained away, while the permanent wilting point is the soil moisture level at which plants will wilt and die (Or *et al.*, 2012). The moisture in the soil profile between field capacity and the permanent wilting point is available for plants to use and is known as PAW. Soil moisture content above field capacity is also available to plants; however, this is not considered to be PAW since it does not usually remain in this state for long periods of time (Or *et al.*, 2012).

SWC is influenced by climatic, physiographic and edaphic factors, both directly and indirectly. Soil quality is made up of both dynamic and inherent soil properties; the former changes with land use management, while the latter does not. The inherent properties of soils include depth, texture, type of clay, etc., and with the land use management, influence dynamic properties of aggregation, water and nutrient holding capacity, bulk density and organic matter content (Jenny, 1941; Or *et al.*, 2012). These affect the soil hydraulic properties of soil water retention and hydraulic conductivity (Weil and Magdoff, 2004). García-Tejero *et al.* (2011) found that the interaction between climate and the soil hydraulic properties causes most of the variability in soil moisture.

Soil water retention is influenced by the soil structure, texture and organic matter content (Coyne and Thompson, 2006; Chesworth, 2008). Sandy soils with a coarser texture generally retain less water while clay soils, which have a finer texture, retain more water. The amount and type of clay in the soil also influence the ability of the soil to retain water. It has been found that sandy soils often retain less than 10% water by weight at field capacity whereas in clay soils this can be over 40% (Chesworth, 2008).

The combination and arrangement of particles which form the soil structure also influence soil water retention through formation of soil pores. Water retention increases with improved soil structure or tilth (Chesworth, 2008). Soil structure affects many physical, chemical and biological soil processes. It is a complex and dynamic soil property which is constantly changing due to various internal and external influences such as biological activity, changes in soil temperature and moisture (Ghezzehei, 2012). It is also influenced by anthropogenic activities such as tillage methods (Coyne and Thompson, 2006; Ghezzehei, 2012). SOM helps to absorb and retain water in the soil and therefore soils with higher organic matter content retain more water than similar soils with less organic matter;

the amount of water stored in the soil can be controlled to a certain extent through weed control, tillage methods, infiltration, runoff and evaporation from the soil surface (Chesworth, 2008).

Study Area

The research reported here in this chapter took place at the Mandela Trial site, a long-term comparison between conventional and organic farming systems, located at the south-eastern end of Nelson Mandela University, George Campus as previously described in Chapter 18 of this volume, and shown in Fig. 18.2.

The climate of the area is considered temperate with perennial rainfall. Annual rainfall can vary from 500 mm to 1400 mm according to South African National Parks (Ackhurst, 2014) or 600–800 mm according to Schafer (1992, cited in Russell *et al.* 2010). The weather station that has been used for this research recorded an annual rainfall of 924 mm for 2014, the year the trial site was set up. For the years 2015, 2016 and 2017, 962 mm, 703 mm and 640 mm have been recorded, respectively. The monthly rainfall varies from 11 mm to 244 mm (Russell *et al.*, 2010). Average daily minimum and maximum air temperatures experienced are 15–25°C in summer and 7°–19°C in winter. For the greater part of the year winds blow in a south-westerly direction, however, in winter it is common to experience warm north and north-east winds. It is uncommon to experience strong winds in the area and wind speeds are normally under 30 km/h (Russell *et al.*, 2010).

Materials and Methods

Capacitance probes were chosen as the selected method for measuring the SWC. Two brands of capacitance probes were used, namely Aquacheck probes and a Delta T Theta Probe (hereafter referred to as Theta probe). The Aquacheck probes remained in the ground throughout the season, giving continuous SWC readings every half an hour. The handheld Theta probe was used to measure SWC in the top 6 cm every 7–14 days.

For each replication, five Aquacheck probes were installed in the centre of each plot, one in a control plot, two probes in organic plots (one monocrop and one rotated) and two probes in conventional plots (one monocrop and one rotated). Figure 19.1 shows the layout of the plots (as previously described in Chapter 18 and Fig. 18.3, this volume (although note different orientation of plan in Fig. 19.1 compared with Fig. 18.3). The 20 probes are numbered at the bottom right of each plot (where installed). The organic cabbage plots received 5 t/ha of compost

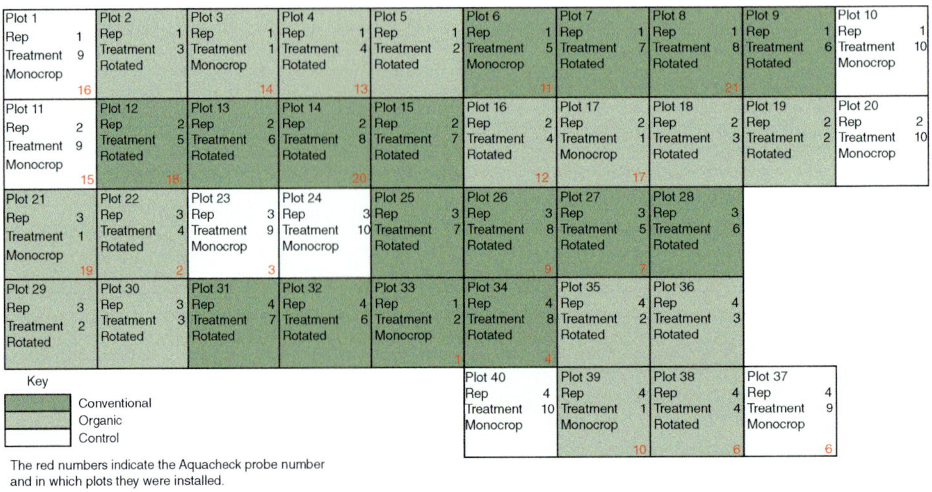

Fig. 19.1. The layout of the experimental plots showing the four replicates, two farming systems and control plots. Rep, Replication.

for both seasons. These were plots 3, 5, 17, 19, 21, 29, 35 and 39 in the 2016/17 season and plots 3, 4, 16, 17, 21, 22, 38 and 39 in the 2017/18 season. Grass clippings cut from the area surrounding the trial site were used as mulch. This was applied evenly over the soil surface at approximately 2 cm thick in the 2016/17 season and 3 cm thick in the 2017/18 season.

The trial site is relatively flat with soils of an Estcourt form, consisting of an orthic A horizon overlying an E horizon and prismacutanic B horizon (SCWG, 1991). An A horizon consists of mineral particles mixed with humus to a greater or lesser degree. The topsoil at the trial site has a soil carbon content of approximately 2.5% (Mashele, 2016). The SA binomial soil classification system (SCWG, 1991) classifies most of South African (SA) topsoils as orthic. 'Orthic' is defined as an ordinary soil which does not fall into the other categories of organic, vertic, humic or melanic topsoils (SCWG, 1991). Because orthic soils occur over a wide range of soil-forming conditions in SA, they vary widely with regards to their soil properties such as mineral composition, organic carbon content, colour and texture, to name a few (SCWG, 1991).

An E horizon is a mineral horizon underlying the A horizon. It is a bleached horizon, having a greyish colour which is paler than the topsoil overlying it. It is formed when there is subsoil which restricts drainage, causing a temporary build-up of water in the soil and a lateral movement of the water, leaching out colloidal matter resulting in this loss of colour (SCWG, 1991).

The B horizon is a mineral horizon subsoil with a blocky, granular or prismatic structure, which has a concentration of organic matter, lime, sesquioxides and silicate clay (SCWG, 1991). It has been found that in prismacutanic B horizons, the lower the clay content, the weaker its structure tends to be (SCWG, 1991).

Data collection and analysis

Rainfall data

The rainfall data were obtained from an on-campus weather station, managed by researcher Tatenda Mapeto, and were already expressed as millimetres of rainfall per day.

Aquacheck probe data

The half-hourly calibrated SWC readings from the Aquacheck probes were averaged for each probe to give a daily SWC reading for the probe at each depth (10 cm, 30 cm, 50 cm). The resulting probe data were averaged into daily readings for each treatment, viz. organic, conventional and control. The organic treatment included data from probes 14, 13, 12, 17, 19 and 2, while the conventional treatment used data from probes 11, 21, 18, 20, 9 and 7. For the control treatment, data from probes 16, 15 and 3 was used. Replication 4 seemed to have some type of rubble in the soil profile. The crops in these plots also did not perform well and therefore it was believed there was some other influence affecting this area apart from the organic and conventional treatments. As a result, Replication 4 was omitted from the analysis.

Theta probe data

The handheld Theta probe data, which represented the SWC in the top 6 cm of soil, were averaged out to a single reading for each of the treatments viz. organic, conventional and control for each sampling date.

The SWC data for the two seasons were compared for two different profile depths, a profile of 0–10 cm and 0–50 cm. The 0–10 cm dataset consists of the 6 cm Theta probe SWC readings and the 10 cm Aquacheck SWC readings for each treatment. The 0–50 cm dataset consists of an average of the average 10 cm, 30 cm and 50 cm daily Aquacheck SWC readings for each treatment to give a single SWC reading for this profile depth for each treatment.

Seasons

The data were gathered over two seasons, namely the 2016/17 season and the 2017/18 season. The 2016/17 season SWC data runs from 22 November 2016 to the 6 April 2017, while the 2017/18 season SWC data runs from 22 November 2017 to the 6 April 2018. The total rainfalls for the 2016/17 season and 2017/18 season were 248 mm and 262 mm, respectively.

Results

2016/17 Season 0–10 cm SWC

The data for the 2016/17 season 0–10 cm are presented as a graph in Fig. 19.2. The Aquacheck data for the three treatments are represented by the line graphs, while the handheld Theta probe data are represented by the scatter plots for the specific sampling date. The rainfall data are represented by the bar chart. When looking at the handheld Theta probe data, which represents the top 6 cm of the soil, the organic treatment with grass mulch always had higher SWC than the conventional treatment with no mulch. The SWC of the control treatment was on one occasion out of the 17 sampling dates slightly higher than the organic plots. The conventional treatment generally had lower SWC than the organic and control treatment, with the SWC of the conventional treatment on two occasions out of the 17 sampling dates being higher than that of the control treatment. When looking at the Aquacheck data, the organic treatment generally had higher SWC than the conventional treatment, while the SWC of the control treatment was more variable. When looking at the rainfall events and their effect on the SWC, in some cases the control plots had a higher peak SWC and in other cases it was the organic or conventional treatment. However, when looking at the drying patterns, the organic treatment generally had a less steep gradient than the conventional treatment and dried out to a lesser degree. The control had much lower plant growth and therefore used less water.

Summarizing the SWC data at a depth of 6 cm for the three treatments using a box and whisker plot showed that the organic treatment had higher maximum, minimum and median SWC values than the conventional or control treatments. The ranges of SWC values are approximately 12–38% for the organic, 6–33% for the conventional and 9–34% for the control. The inter-quartile range (IQR) data in the organic treatment lie between a SWC of approximately 16% and 28%, while for the conventional and control treatments these are approximately 9% and 21%, and 13% and 23%, respectively. The median values are approximately 19% for the organic, 12.5% for the conventional and 14% for the control treatments.

Summarizing the SWC data at a depth of 10 cm for the three treatments using a box and whisker plot gave a range of SWC values of approximately 15–32.5% for the organic, 13.5–33.2% for the conventional and 18–31.6% for the control treatment. The IQR data in the organic treatment lie between a SWC of approximately 18.5% and 25%, while for the conventional and control treatments this is approximately 17.5% and 23%, and 23% and 25%, respectively. The median values were highest for the control treatment (22.5%), followed by the organic (21.5%) and then the conventional (19.8%).

The observed, un-weighted means for the Aquacheck and Theta probe SWC data in the top 0–10 cm for the 2016/2017 season are shown in Table 19.1. From this we can see that the organic treatment had the highest average SWC

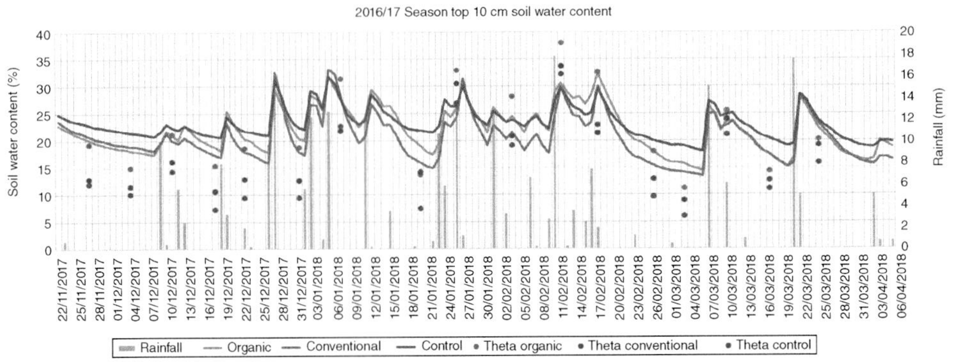

Fig. 19.2. The 0–10 cm soil water content (SWC) data for the three treatments (organic, conventional and control) and rainfall for the 2016/17 season.

Table 19.1. 2016/2017 season SWC means for the three treatments.

Treatment	Theta probe 6 cm SWC (%)	Aquacheck 10 cm SWC (%)
Organic	22	22
Conventional	15	21
Control	17	23

for the season in the 6 cm zone followed by control and then conventional treatments, while in the 10 cm zone, the control treatment had the highest SWC followed by the organic and then conventional treatments.

Using statistica 12, a one-way analysis of variance (ANOVA) was performed on the daily SWC data from the Aquacheck and Theta probes to see if the differences were significant. The ANOVA returned a value of $p < 0.035889$ and $p < 0.0000$ for the Theta probe and Aquacheck data, respectively. This means that there are significant differences between the treatments for the two depths. To find out where the differences lay, a post hoc Tukey HSD (honestly significant difference) test was performed.

The results for the Tukey HSD test were as follows:

- For the Theta probe SWC data at a depth of 6 cm a significant difference between the organic and conventional treatment ($p < 0.030659$) was found. However, no significant differences were found between the organic and control treatment ($p < 0.200687$) or the conventional and the control treatment ($p < 0.652775$).
- For the Aquacheck SWC data at a depth of 10 cm there was a significant difference between the organic and conventional treatment ($p < 0.008042$) and between the organic and the control treatment ($p < 0.017499$). A significant difference was also found between the conventional and the control treatment ($p < 0.000022$).

To analyse drying patterns, gradients of the line from the peak SWC to the driest SWC were calculated for the three treatments using the formula Eqn 19.1:

$$y1 - y2 / x1 - x2 \qquad \text{(Eqn 19.1)}$$

where y is the change in SWC and x is the number of days over which the change occurred. A one-way ANOVA showed no significant difference in the gradients at $p < 0.144987$.

2017/18 Season 0–10 cm SWC

The same analysis was performed on the 0–10 cm SWC data for the 2017/18 season. The data for 0–10 cm are presented as a graph in Fig. 19.3. The Aquacheck data for the three treatments are again represented by the line graphs, while the handheld Theta probe data are represented by the scatter plots for the specific sampling date. The rainfall data are represented by the bar chart. When looking at the handheld Theta probe data, which represents the top 6 cm of the soil, we see a lot of fluctuation in the SWC of the control treatment. The SWC in this treatment is generally higher than the conventional treatment

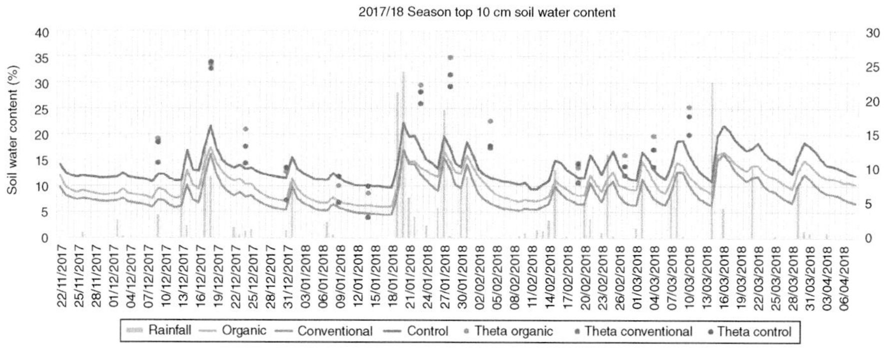

Fig. 19.3. The 0–10 cm SWC data for the three treatments (organic, conventional and control) and rainfall for the 2017/18 season.

except on one sampling occasion out of the 13. When comparing the control and organic treatment we see a lot of variation, with the organic treatments having a higher SWC on eight occasions, versus the control having a higher SWC on five occasions out of the 13 sampling dates. However, when comparing the organic and conventional treatments, we see that the organic treatment had a higher SWC than the conventional treatment on all sampling occasions.

When we review the Aquacheck probe data, the control treatment has the highest SWC throughout the season, followed by the organic and conventional treatments. During rainfall events, the difference in SWC of the organic and conventional treatments became less. However, as seen in the previous season, the organic treatment dried out more slowly and to a lesser extent than the conventional treatment.

A box and whisker plot for the 6 cm SWC data showed that the ranges of data for the organic, conventional and control treatments were 8–35%, 4–32.5% and 10–34%, respectively. The IQRs for the data were 13–25% for the organic, 11–20% for the conventional and 14–24% for the control. The median value for the organic treatment is also much higher at approximately 19%, than conventional (14%) or control (17.5%).

The box and whisker plot for the 10 cm depth showed the ranges of the data to be 6–18% for the organic, 5–17% for the conventional and 10–22% for the control. The IQRs were approximately 8–12% in the organic, 6–10% in the conventional and 14–15% in the control. The median values for the organic, conventional and control were 9.8%, 7.5% and 12.5%, respectively.

The observed, un-weighted means for the Aquacheck and Theta probe SWC data in the top 0–10 cm for the 2017/18 season are shown in Table 19.2. From this we can see that the organic treatment had the highest average SWC for the season in the 6 cm zone followed by control and then conventional treatments, while in the 10 cm zone, the control treatment had the highest SWC followed by the organic and then conventional treatments. This is the same trend which we saw in the previous season.

Table 19.2. 2017/18 season SWC means for the three treatments.

Treatment	Theta probe 6 cm SWC (%)	Aquacheck 10 cm SWC (%)
Organic	20	10
Conventional	16	8
Control	19	14

A one-way ANOVA was performed on the daily SWC data from the Aquacheck and Theta probes to see if the differences in SWC were statistically significant. The ANOVA returned a value of $p < 0.407700$ for the Theta probe data. No significant differences were found for this data. However, when a one-way ANOVA was run on the Aquacheck data, a significant difference was found ($p < 0.00$) between the treatments. A post hoc Tukey HSD test was performed to find out between which treatments the significant difference occurred. The results showed there was a significant difference of $p < 0.000022$ between all three treatments.

The drying patterns were calculated as explained for the previous season and a one-way ANOVA run on the resulting data. There were no significant differences found between the treatments with a value of $p < 0.992113$.

2016/17 Season 0–50 cm SWC

A box and whisker plot for the 0–50 cm SWC data showed the data for the treatments ranged from 16% to 26% for the organic, 13% to 23% for the conventional and 18% to 25% for the control. The IQR data for the organic, conventional and control treatments were 19–22%, 16–19% and 20–22%, respectively. The median values for the organic and control treatments were similar at 20.5% and 20.9%, respectively.

When considering the 0–50 cm soil profile water content data from the Aquacheck probes (Fig. 19.4), there appears to be a much greater difference in the SWC between the organic and conventional treatments, with the organic treatment having a higher SWC than the conventional treatment throughout the season. The observed, un-weighted means were 20%, 18% and 21% for the organic, conventional and control treatments, respectively. When the data were analysed using a one-way ANOVA, it was found that the treatments were significantly different ($p < 0.00$). A post hoc Tukey HSD test further found that there was a significant difference between the organic and conventional treatments

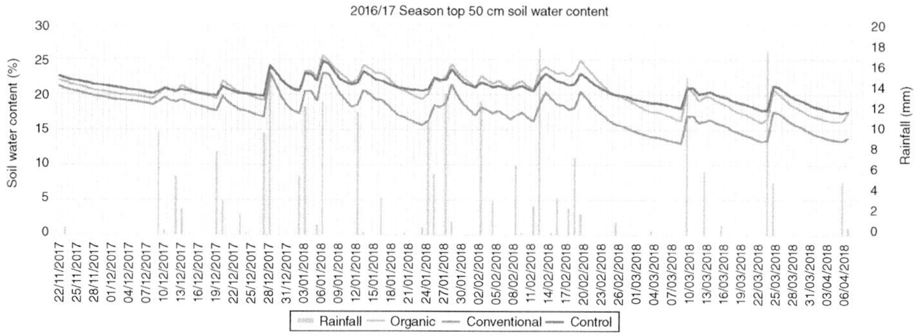

Fig. 19.4. The 0–50 cm SWC data for the three treatments (organic, conventional and control) and rainfall for the 2016/17 season.

and the conventional and control treatments, both with $p < 0.000022$. However, there was no significant difference between the organic and the control treatment ($p < 0.632953$).

2017/18 Season 0–50 cm SWC

A box and whisker plot for the 0–50cm SWC data showed the data for the treatments ranged from 13% to 20% for the organic, 9% to 18% for the conventional and 15% to 21% for the control. The IQR data for the organic, conventional and control treatments were 15–17%, 12–14% and 16–21%, respectively. The median values for the organic and control treatments were similar at 16% and 16.8%, respectively, while the median for the conventional was much lower at 13%.

In the SWC in the soil profile of 0–50 cm (Fig. 19.5), the organic treatment had a higher SWC throughout the season compared with the conventional treatment, while the control treatment had the highest SWC throughout the season of all the treatments. The observed, un-weighted means were 16%, 13% and 17% for the organic, conventional and control treatments, respectively. When analysed using a one-way ANOVA, the treatments were found to be significantly different ($p < 0.00$). A post hoc Tukey HSD test identified significant differences between the organic and conventional treatment ($p < 0.000022$), the organic and control treatment ($p < 0.000026$) and between the conventional and control treatment ($p < 0.000022$).

Soil carbon

SOM improves soil water retention through improvements in soil properties. It also acts as a sponge, absorbing water and retaining it in the soil profile. SOM is roughly 50% carbon by weight (Brady and Weil, 2008) and therefore measuring the soil carbon content gives a good indication of the levels of organic matter in the soil.

Seasonal soil samples taken since the start of the trials in 2014 show that soil carbon levels have been affected by the management practices. This can be seen in Table 19.3. The organic treatment has led to an increase in soil carbon levels from 2.6% to 2.8%, from 2014 to 2018 through the accumulation of soil carbon by the addition of compost, the breakdown of the organic mulch and ploughing in of cover crops. In contrast, during the same period, the soil carbon in the conventional treatment has decreased from 2.7% to 2.5% (Swanepoel, 2018; Chapter 22, this volume).

Discussion

Reviewing the treatments for the two seasons the following can be ascertained:

Soil depth 6 cm

In both seasons, for the top 6 cm the organic treatment clearly had a higher SWC than the

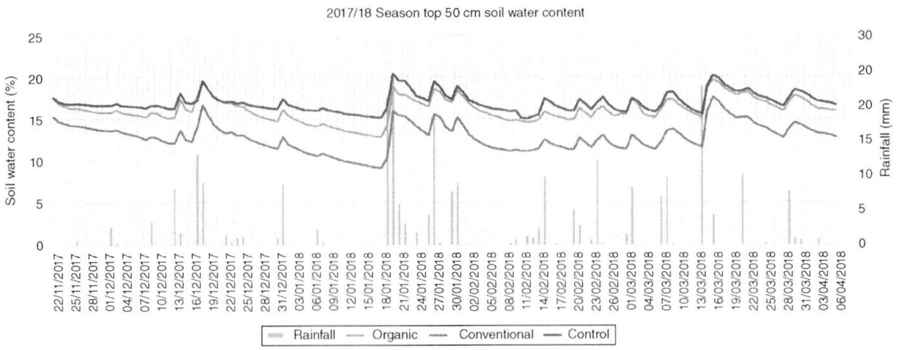

Fig. 19.5. The 0–50 cm SWC data for the three treatments (organic, conventional and control) and rainfall for the 2017/18 season.

Table 19.3. Soil carbon contents from the base season and each successive season.

Treatment	Base (% C)	2014 (% C)	2014/15 Season (% C)	2015/16 Season (% C)	2016/17 Season (% C)	2017/18 Season (% C)
Organic	2.63	2.61	2.66	2.87	2.95	2.82
Conventional	2.76	2.72	2.62	2.52	2.44	2.53
Control	2.44	2.54	2.65	2.43	2.42	2.51

conventional treatment. This is believed to be due mainly to the influence of the grass mulch as it reduces water loss from the soil surface through evaporation. The SWC in the control treatment was more variable and had a higher or lower SWC than the organic treatment throughout the season. The box and whisker plots showed that the minimum and median SWCs in the organic treatment were much higher than those of the conventional and control. However, in the second season the control treatment had a higher minimum SWC than the organic, but a lower median SWC. The control plot had a higher minimum SWC and median than the conventional in season one and two. From these data, we can see that the organic treatment dried out to a lesser extent than the conventional in both seasons and maintained a higher SWC at this profile depth. In the second season, the organic treatment dried out to a greater degree than the control, however, the median SWC in the organic treatment was higher. The differences in the SWC of the organic and conventional treatment for season one was significant, however, differences for season two were not significant.

Soil depth 10 cm

The data for the 10 cm depth show that the minimum and median SWC of the organic treatment were higher than the conventional. However, the control was higher than both the organic and conventional treatments. A reason for this could be that in the control treatment plant growth was extremely poor and therefore likely had lower plant water usage than in the other treatments. In addition, the control plot was only monocrop cabbage, whereas the organic and conventional were a fairer comparison with the same mixture of crop types (cowpea, cabbage and sweet potato). As we have found in the literature review, crop types do affect the crop water usage. When analysing the data further, we can see that the differences in SWC between all three treatments are statistically significant for both seasons (Eckert, 2018).

Soil depth 0–50 cm

The combined SWC data of the 0–50 cm soil profile showed that the organic treatment had a higher SWC and median than the conventional treatment. However, the control treatment had the highest SWC and median of all the treatments. There was a significant difference in the organic and conventional treatments in both seasons, but no significant difference between the organic and control treatments in the first season. Again, as we saw in the top 10 cm, the reason for higher SWC in the control could be down to the poor performance of the control treatment crops which gave virtually no yield. It is thought that the crops in the control plots were also shallower rooted than those in the other treatments, therefore only being able to access the SWC within the top 10 cm, but not in the deeper soil profile. This would explain why the Theta probe data shows a lower SWC for the control treatments on most of the sampling dates as compared to the organic treatment, while the deeper soil profile shows a higher SWC for the control treatment, as opposed to other two treatments.

Conclusion

The purpose of this study was to investigate the influence of organic and conventional farming systems on the SWC in the topsoil of rainfed agriculture. From the findings of the research we can conclude that the organic treatment has a consistently higher SWC than the conventional treatment in both seasons and at all depths of the soil profile analysed. From the literature, we have seen that SOM plays an important role in improving the soil's biological, physical and chemical properties and that organic farming methods preserve and promote an increase in SOM. The addition of compost to the soil as well as the addition of organic matter from the grass mulch as it breaks down over time can contribute to SOM, thus improving the soil structure and increasing the soil's water holding capacity. In the Mandela Trials, organic matter and soil carbon increased significantly in the organic treatments, while decreasing in the conventional treatments.

From this research, it is recommended that organic farming practices can be used to help conserve SWC, keeping it available to crops for longer and helping farmers make more efficient use of this scarce resource by adapting their farming practices. This is especially relevant for low rainfall areas which are affected by water shortages. However, increased SWC availability won't necessarily translate into increased crop yields and therefore the improved SWC availability should be coupled with good agronomic practices to increase productive water losses and the conversion of water to yields, thus increasing WUE. From this research, we can see that the grass mulch and the addition of compost can not only help to increase SWC in the short term, but will also have long-term impacts on soil properties leading to higher soil water retention in the future. Not only will this farming practice increase SWC, but farmers will derive other benefits, as described in the literature, associated with grass mulch and the addition of SOM, reducing their environmental impacts and helping them to become more sustainable. In addition, adding organic matter to the soil will improve resilience of the soil and help sequester carbon and thus help in mitigating climate change.

More research is needed to assess what proportion of the increased soil water retention can be attributed to the influence of the mulch and what effect the SOM had, as preliminary measurements on organic plots with and without mulch in a previous season showed that the mulched plots had a much higher SWC than the unmulched plots.

References

Ackhurst, A. (2014) The use of a rapid incineration field test for determining soil organic carbon in the Southern Cape region. MTech thesis, Nelson Mandela University, George, South Africa.

Aune, J. (2012) Agroecology and strategies for climate change: conventional, organic and conservation agriculture: production and environmental impact. In: Lichtfouse, E. (ed.) *Sustainable Agriculture Reviews* Vol. 8. Springer, New York, pp. 149–165.

Bhaduri, A., Bogardi, J., Leentvaar, J. and Marx, S. (eds) (2014) *The Global Water System in the Anthropocene: Challenges for Science and Governance*. Springer, Cham, Switzerland.

Brady, N. and Weil, R. (2008) *The Nature and Properties of Soils*, 14th edn. Prentice Hall, Upper Saddle River, New Jersey.

Chartzoulakis, K. and Bertaki, M. (2015) Sustainable water management in agriculture under climate change. *Agriculture and Agricultural Science Procedia* 4, 88–98.

Chesworth, W. (ed.) (2008) *Encyclopedia of Soil Science*. Springer, Dordrecht, the Netherlands.

Coyne, M. and Thompson, J. (2006) *Fundamental Soil Science*. Thompson Delmar Learning, New York.

De Wit, M.J. and Stankiewicz, J. (2006) Changes in surface water supply across Africa with predicted climate change. *Science* 311, 1917–1921. https://doi.org/10.1126/science.1119929

Eckert, C. (2018) Top-soil water retention in organic and conventional farming systems in South Africa's Southern Cape. MTech thesis, Nelson Mandela University, George, South Africa.

García-Tejero, I., Durán-Zuazo, V., Muriel-Fernández, J. and Rodríguez-Pleguezuelo, C. (2011) *Water and Sustainable Agriculture*. Springer, Dordrecht, the Netherlands.

Ghezzehei, T. (2012) Soil structure. In: Huang, P.M., Li, Y. and Sumner, M. (eds) *Handbook of Soil Sciences, Properties and Processes*, 2nd edn. CRC Press, Boca Raton, Florida, pp. 2-1–2-14.

IPCC (Intergovernmental Panel on Climate Change) (2002) *Climate Change 2001: Impacts, Adaptation, and Vulnerability*. Cambridge University Press, Cambridge.

IPCC (Intergovernmental Panel on Climate Change) (2007) *Climate Change 2007: Impacts, Adaptation and Vulnerability*. Parry, M.L., Canziani, O.F., Palutikof, J.P., van der Linden, P.J. and Hanson, C.E. (eds) Contribution of Working Group II to the Fourth Assessment Report of the Intergovernmental Panel on Climate Change. Cambridge University Press, Cambridge.

Jenny, H. (1941) *Factors of Soil Formation: a System of Quantitative Pedology*. McGraw-Hill, New York.

Kadera, M., Senge, M., Mojid, M. and Ito, K. (2016) Recent advances in mulching materials and methods for modifying soil environment. *Soil and Tillage Research* 168, 155–166.

Kyere, K., Agyarko, K., Yaw, B. and Asiedu, E. (2018) Influence of grass mulch on soil physical attributes of a luvisol and water requirement of cowpea (*Vigna unguiculata*) in the transition zone of Ghana. *Asian Soil Research Journal* 1(2), 1–13.

Lal, R. (2009) Mother of necessity: the soil. In: Lichtfouse, E. (ed.) *Organic Farming, Pest Control and Remediation of Soil Pollutants*. Vol. 1. *Sustainable Agriculture Reviews*. Springer, Dordrecht, the Netherlands, pp. 5–10.

Mashele, N. (2016) Crop yields from organic and conventional farming systems in South Africa's Southern Cape. MSc thesis, Nelson Mandela University, George, South Africa.

Or, D., Wraith, J., Robinson, D. and Jones, S. (2012) Soil water content and water potential relationships. In: Huang, P.M., Li, Y. and Sumner, M. (eds) *Handbook of Soil Sciences, Properties and Processes*, 2nd edn. CRC Press, Boca Raton, Florida, pp. 4-1–4-24.

Parry, M., Canziani, O., Palutikof, J., van der Linden, P. and Hanson, C. (2007) Technical summary. In: Parry, M.L., Canziani, O.F., Palutikof, J.P., van der Linden, P.J. and Hanson, C.E. (eds) *Climate Change 2007: Impacts, Adaptation and Vulnerability*. Contribution of Working Group II to the Fourth Assessment Report of the Intergovernmental Panel on Climate Change. Cambridge University Press, Cambridge, pp. 23–78.

Passioura, J.B. and Angus, J. (2010) Improving productivity of crops in water-limited environments. *Advances in Agronomy* 106, 37–75.

Ramakrishna, A., Tam, H.M., Wani, S.P. and Long, T.D. (2005) Effect of mulch on soil temperature, moisture, weed infestation and yield of groundnut in Northern Vietnam. *Field Crops Research* 95, 115–125.

Raza, A., Friedel, J.K. and Bodner, G. (2012) Agroecology and strategies for climate change: improving water use efficiency for sustainable agriculture. In: Lichtfouse, E. (ed.) *Sustainable Agriculture Reviews* Vol. 8. Springer, New York, pp. 167–211.

Romano, N. (2014) Soil moisture at local scale: measurements and simulations. *Journal of Hydrology* 516, 6–20.

Russell, I.A., Randall, R.M., Cole, N., Kraaij, T. and Kruger, N. (2010) *Garden Route National Park, Wilderness Coastal Section*, State of Knowledge. South African National Parks, Sedgefield, South Africa.

SCWG (Soil Classification Working Group) (1991) *A Taxonomic System for South Africa*. Department of Agricultural Development, Pretoria.

Singh, L., Beg, M., Akhter, S., Qayoom, S., Lone, B., Singh, P. and Singh, P. (2014) Efficient techniques to increase water use efficiency under rainfed eco-systems. *Journal of AgriSearch* 1(4), 193–200.

Sinkevičienė, A., Jodaugienė, D., Pupalienė, R. and Urbonienė, M. (2009) The influence of organic mulches on soil properties and crop yield. *Agronomy Research* 7(Special issue I), 485–491.

Stagnari, F., Ramazzotti, S. and Pisante, M. (2009) Conservation agriculture: a different approach for crop production through sustainable soil and water management: a review. In: Lichtfouse, E. (ed.) *Organic*

Farming, Pest Control and Remediation of Soil Pollutants. Vol. 1. *Sustainable Agriculture Reviews*. Springer, Dordrecht, pp. 55–84.
Swanepoel, M. (2018) Comparing organic and conventional farming systems in George, South Africa. MSc thesis, Nelson Mandela University, George, South Africa.
United Nations (2007) Press Release POP/952, Department of Public Information. Available at: http://www.un.org/News/Press/docs//2007/pop952.doc.htm (accessed 20 June 2018).
Weil, R.R. and Magdoff, F. (2004) *Soil Organic Matter in Sustainable Agriculture*. CRC Press, Boca Raton, Florida.

20 Biological and Chemical Soil Fumigation and Pest and Disease Management Comparisons in the Western Cape

Abraham van Niekerk,[1]* Raymond Auerbach[1] and Sandra Lamprecht[2]
[1]Nelson Mandela University, George, South Africa; [2]Agricultural Research Council, Stellenbosch, South Africa

Abstract

Brassica spp. are cultivated all over the world; commercial species include cabbage, broccoli, kale, kohlrabi and turnip. In this study the focus was on production of broccoli, *Brassica oleracea*, in the Western Cape province of South Africa and its economically important pests and diseases, diamondback moth (DBM) (*Plutella xylostella*), white blister (*Albugo candida*) and clubroot (*Plasmodiophora brassicae*), and the different methods to control these pests and diseases. The control methods focused on in this study included a commercial chemical control programme, a biological control programme and a holistic approach. There was also a focus on the effectiveness of crop rotation practices versus no rotation crops. The experimental design was a strip-split plot design, with different pest and disease management strategies as the main plot treatment and fumigation and rotation treatment combinations arranged in strips across the main plot treatments. The main plot design was a randomized complete block with four programmes: (i) control; (ii) chemical; (iii) holistic; and (iv) biological. These were replicated four times and laid out in a randomized complete block design. The treatment design of the strip-plot factors was a 2 × 2 factorial with two fumigations (fumigated chemically and fumigated biologically) and two rotations (crop rotated and monoculture) randomly allocated across main plot treatments. Each experimental unit consisted of 40 plants. Plants were evaluated weekly for disease incidence and insect infestation. The data show significant differences between the crop protection strategies. DBM was well controlled by the chemical programme, and although the holistic and biological programmes were better than the control, significantly more moths were present than with the chemical programme. Although biological and holistic programmes took longer to control white blister, by the time of harvest, the holistic programme was more effective, and the biogical programme almost as effective, compared with the chemical programme. Clubroot incidence in the trial was zero.

Introduction

The mustard family of flowering plants (*Brassicaceae*) comprises 338 genera and 3700 species. Many of these are commercially farmed especially the genus *Brassica* which includes cabbage, broccoli, kale, kohlrabi and turnip (Britannica, 2017).

There are many economically significant pests and diseases of this crop; the main focus in this study will be diamondback moth (DBM), *Plutella xylostella*, white blister of *Brassica* spp.

*braamvn@gmail.com

caused by *Albugo candida*, and clubroot of *Brassica* spp. caused by *Plasmodiophora brassicae*. Published information on these pests and diseases, especially white blister and clubroot, is very limited (Nofemela and Kfir, 2005).

DBM, *P. xylostella*

DBM, may have originated from Europe or South Africa (SA), although it has been recorded in 128 countries worldwide. It is found wherever *Brassica* spp. are produced, and is considered the most widely distributed of all Lepidoptera (Dennill and Pretorius, 1995; Saeed *et al.*, 2010). DBM has been found from the cold Himalayan Mountains to the dry Ethiopian region (Marchioro *et al.*, 2017). This is the most important pest of brassicas worldwide (Reddy *et al.*, 2004). It is estimated that DBM causes US$4 billion in losses annually (Zalucki *et al.*, 2012). This worldwide distribution is made possible by the pest's tolerance to high temperatures as well as its high migratory capacity. Thus suitable environmental conditions are exploited (Marchioro *et al.*, 2017).

Understanding pest behaviour, susceptible hosts, reproduction and detection is important so as to manage economically important pests (Sarfaz *et al.*, 2006). Serious damage by DBM occurs in the second and third instar of the larvae, which feeds on the leaves, altering photosynthesis which leads to yield loss, and reducing size and product quality (Correa-Caudros *et al.*, 2016). Farmers have experienced problems with controlling DBM, due to its short life cycle and there has been some recorded resistance to chemical insecticides (Harris and McLean, 1999). There have been situations where growers were forced to plough in all of their standing crop, in spite of applying multiple insecticides, as the pest could not be controlled. This exceptional status of *P. xyslostella* is due to the diversity and abundance of its host plants, lack of natural enemies and its high reproductive rate, with up to 20 generations in 1 year, as well as its insecticide resistance (Saeed *et al.*, 2010). The high incidence of DBM can be explained by the aforementioned ability to reproduce, the lack of natural enemies and its ability to develop resistance to insecticides (Marchioro *et al.*, 2017).

Chemical insecticides are still the preferred method of control for DBM. However, resistance has been recorded with some insecticides worldwide. The reason chemical control is so popular is its practicality, speed and efficiency in population control, but continuous application has contributed to the problem of resistance (Peres *et al.*, 2017). Controlling DBM with pesticides has become more difficult all over the world due the use of single potent toxicants over a long period of time, and resistance to almost all the recommended chemical insecticides has developed (Ghosal *et al.*, 2015). The preferred chemical insecticides that were used were organophosphates, carbamates and pyrethroids, but their continued use has rendered them ineffective in controlling DBM (Correa-Caudros *et al.*, 2016).

The resistance of *P. xylostella* has made it economically impractical to farm with *Brassica* spp. in certain parts of the world. This has forced the industry to start looking at a more holistic approach in controlling this major pest of *Brassica* spp. (Marchioro *et al.*, 2017). The resistance to chemical insecticides has also made room for alternatives to be explored like *Bacillus thuringiensis* (Kfir, 2001). Attempts at biological control as an alternative to reduce populations of *P. xylostella* found that entomopathogenic fungi and nematodes were effective. Using *Beauveria bassiana* for biological control showed promising results, but the mortality is only achieved over a long period of time (9–15 days). Entomopathogenic nematodes, however, can cause a 91% mortality in just 48 h (Correa-Caudros *et al.*, 2016). Various management strategies need to be explored (Sarfraz *et al.*, 2006).

White blister of *Brassica* spp.

Albugo candida is the pathogen that causes white blister or white rust of crucifers. This pathogen is found on almost all *Brassica* spp., including the cultivated vegetable and oil seed brassicas. The fungus can produce two types of infection, local or systemic (Santos and Dias, 2004). There have been 17 different races reported of *A. candida* across the different *Brassica* spp. (Barbetti *et al.*, 2016). White blister was always regarded as a minor disease of *Brassica* spp., but that has changed following severe outbreaks reported this century in the UK, the Netherlands, France, Spain and Portugal and on Brussels sprouts, broccoli, cauliflower and cabbages (Santos *et al.*, 2009). Yield loss of up to 60% has been recorded in some *Brassica* spp. In India, combined infection of *Brassica juncea* leaves and inflorescences caused yield loss of up to 90%, with 63% of this loss through systemic damage (Kaur *et al.*, 2011). This disease has increased in significance in recent years with total crop loss being reported in certain instances (Ploch *et al.*, 2010).

A. candida is an obligate pathogen and is considered to be ancient compared with downy mildew. It is said that white blister was introduced with the cruciferous crops (Kaur *et al.*, 2011). The downy mildew pathogen, *Peronospora parasitica*, commonly co-occurs with white blister and even asymptomatic colonization by *P. parasitica* will speed up the infection by *A. candida* thus increasing disease severity (Barbetti *et al.*, 2016). The localized disease characteristics can be described as the formation of white- to cream-coloured zoosporangial pustules on cotyledons, leaves, stems and inflorescences. It occurs on all plant parts that contain chlorophyll (Kaur *et al.*, 2011). The systemic disease characteristics are caused by oospores in mature stagheads. 'Stagheads' refers to extensive distortion, hypertrophy, hyperplasia and inflorescence sterility. An obligate parasite can only develop on living host tissue, where it produces sexual sporangia or zoospores and thick-walled sexual spores.

The pathogen survives as oospores in crop residues and perennial mycelium in living host tissue (Kaur, 2013). Oospores develop in distorted swellings and galls including the stagheads, and in infected pods and stems. These overwintering spores are quite hardened against drying and extreme temperatures; they are responsible for long-term survival and are liberated when a suitable host is planted (Kaur, 2013). A disease epidemic can be initiated by only a few infected pants that serve as the primary source of infection (Kaur, 2013).

The first symptoms will appear 5–20 days after infection, a new crop of sporangia are released 3–14 days after the first infection to start the second disease cycle, and in cool wet conditions it can complete its cycle every 8–10 days (Kaur *et al.*, 2011). *A. candida* is spread by planting seeds that have been contaminated with oospores, as well as by wind and rain and perennial mycelium in infected live plants (Kaur, 2013).

A. candida has a wide host range which complicates disease control (Choi *et al.*, 2011; Kaur and Sivasithamparam, 2011). Chemical control is quite difficult and only a few products are registered, which are extremely expensive and often farmers cannot afford to apply these. At present there are no alternatives to chemical control against *A. candida*, and this means of control is reported with limited success. Disease-resistant cultivars would be the more environmentally friendly and holistic approach to control, reducing pesticide usage and resistance (Santos *et al.*, 2009). This is still the most efficient and cost-effective means of control of *A. candida* (Barbetti *et al.*, 2016). More research on this disease and its control is needed, since white blister is now an economically important disease which is poorly understood (Ploch *et al.*, 2010).

Clubroot disease of *Brassica* spp.

Clubroot is caused by *Plasmodiophora brassicae*, which is an obligate soilborne plant pathogen of *Brassica* spp., that can cause massive economic loss in production if not controlled (Koike, 2003; Irani *et al.*, 2018). Clubroot is one of the most important diseases of brassicas, and can be found in all brassica production areas worldwide (Labrador Morales *et al.*, 2013). The disease is of global importance and causes yield loss of about 15% although 100% losses have been recorded with severe infection. In the UK, *Brassica* spp. were produced over approximately 15% of all horticultural and arable land in 2014 (McGrann *et al.*, 2017).

The life cycle of *P. brassicae* has two phases. In the primary phase, resting spores in the soil start to germinate as soon as there is a host and soil conditions are optimal. The spores then penetrate a suitable host's root hair, in the form of zoospores. In the second phase, secondary plasmodia form in the cortex of the root, producing galls. These galls prevent the root from functioning normally, and lead to yield loss, as normal functions such as nutrient and water uptake cannot take place (Irani *et al.*, 2018). Clubroot disease is sporadically found in soil with some plants seriously diseased and neighbouring plants having no symptoms at all (Zhao *et al.*, 2017).

P. brassicae in mature secondary plasmodia form resting spores that can survive for a long time in the soil; they are long lived and resistant to severe environmental conditions, making it impossible to prevent the disease with chemical treatment or crop rotation (Irani *et al.*, 2018). The resting spores which can remain viable for over 15 years in the soil in absence of a host make it a very persistent pathogen. The average half-life of the spores is 3.5 years making rotation as a control method not a viable option (McGrann *et al.*, 2017).

Various fungicides, biological controls and soil fumigants have been tested for the control of clubroot disease in brassicas, although their field efficacy has been inconsistent (McGrann *et al.*, 2017). Given the ineffectiveness of traditional chemical control methods, alternative approaches to managing clubroot disease such as biological control have been the most promising. Considerable research has been done in this regard with different bacteria, fungi and crop rotations: *Trichoderma harzianum* is a fungus that has been used as a biofungicide to control various plant pathogens caused by fungi in a wide range of plants worldwide (Labrador Morales *et al.*, 2013; Yu *et al.*, 2015). Although *T. harzianum* shows potential against *P. brassicae* very little recorded research has been done (Yu *et al.*, 2015). Clubroot-resistant varieties have provided effective control against the disease in the production of different *Brassica* spp. (McGrann *et al.*, 2017). However, the evolution of the pathogen has resulted in *P. brassicae* populations that can overcome this method of control as well (McGrann *et al.*, 2017). Although this is a problem disease, resistant cultivars are still the most effective method to control clubroot disease (Irani *et al.*, 2018).

Chemical control of pests and diseases

> Pesticides are widely used to control the growth and proliferation of undesirable organisms that, if left unchecked, would cause significant damage to forests, crops, stored food products, ornamental and landscape plants, and building structures. The use of pesticides in both agricultural and non-agricultural settings provides important benefits to society, contributing to an abundant supply of food and fibre, and to the control of a variety of public health hazards and nuisance pests. Owing to the fact that they are designed to be biologically active, pesticides have the potential to cause undesirable side effects. These include adverse effects on workers, consumers, community health and safety, groundwater, surface water and non-target wildlife organisms. In addition, pesticide use raises concerns about the persistence and accumulation of pesticides in food chains quite distant from the original point of use, and about the role of certain pesticides in causing reproductive failure and endocrine system abnormalities in both wildlife and humans and other species that are not their intended target. It is therefore, important to control the use of pesticides, by carefully weighing the benefits that they confer against any possible adverse effects.
>
> (Government Gazette, 2010)

In spite of this, there are still farmers who believe that pests and diseases can only be controlled with chemical pesticides. Alternatives to chemical pest control solutions that are less harmful to people and the environment, while still effectively controlling pests, are important in the modern world of crop protection (Khan and Damalas, 2015). Chemical pesticides need to be understood strategically and not just applied blindly, as many problems can occur with incorrect use of chemicals (Sarfaz *et al.*, 2006). Chemical pesticides are widely used and very popular because they provide a cheap and effective way for farmers to control various pests and diseases (Mall *et al.*, 2018). Lack of objective information has led farmers to believe that pests and diseases can only be controlled with chemical pesticides (Khan and Damalas, 2015). Human and environmental safety, and resistance of pests and diseases to chemicals are just a few factors playing a role in the use of chemicals (Macharia *et al.*, 2005).

In the USA, a developed country, about 20,000 people are hospitalized every year for pesticide poisoning (Khan and Damalas, 2015). The World Health Organization has reported 3 million acute poisoning events worldwide every year (Khan and Damalas, 2015). Dependence on chemical control has led to pest resistance being reported worldwide (Dennill and Pretorius, 1995). Uneducated and uncontrolled use of chemical pesticides has led to the increasing resistance of pests to pesticides, thus alternatives to exclusive use of chemical control for pests and diseases should be explored (Khan and Damalas, 2015).

Chemical soil fumigation

Historically, soil was chemically fumigated with methyl bromide (MB) for the control of soilborne pathogens. MB, a highly toxic and persistent substance, has been banned worldwide in the last few years, and a gap in the industry has emerged in seeking a suitable replacement such as 1,3-dichloropropene (Shi *et al.*, 2009), for the control of multiple soilborne pests and diseases including weeds, fungi, bacteria and nematodes (Wang *et al.*, 2006). Other replacements for MB are the following: (i) metam sodium (MS) (Sederholm *et al.*, 2017); (ii) chloropicrin; (iii) dimethyl disulfide (Guo *et al.*, 2017); (iv) methyl iodide; (v) propargyl bromide (Wang *et al.*, 2006); and (vi) calcium cyanamide (Shi *et al.*, 2009). These alternatives to MB still need to be studied for their impact on soil ecology, as some of these can be as devastating as MB (Wang *et al.*, 2006). The use of soil fumigants is strictly regulated because of environmental and safety concerns.

The most common soil fumigants used in vegetable production are chloropicrin and MS (Guo *et al.*, 2017). As a soil fumigant MS is the most commonly used, and is the third most used pesticide in the USA (Sederholm *et al.*, 2017). MS salt is hydrolysed, when it comes in contact with water, to methyl isothiocyanate (MITC), a volatile toxic gas which is applied as a broad-spectrum pesticide for its herbicidal, fungicidal and insecticidal qualities. Unfortunately MS can have adverse effects on soil biology, especially on soil microorganisms that are responsible for plant nutrient uptake, nitrogen transformation and pollutant degradation. Recovery of these populations takes time (Sederholm *et al.*, 2017).

As mentioned, calcium cyanamide is also one of the possible replacement products for MB. It is generally used as a fertilizer but it has some fungicidal, herbicidal and insecticidal qualities. Reports state that it is effective in the control of *P. brassicae* (cause of clubroot disease of *Brassica* spp.) (Shi *et al.*, 2009). Calcium cyanamide has some fungicidal properties, and is sold in the European Union (EU) as a fertilizer without national regulations, as it consists of 19% N and > 50% Ca, thus giving the product liming qualities as well as supplying nitrogen. When calcium cyanamide comes into contact with soil moisture, it decomposes to hydrogen cyanamide and hydrated lime. Hydrogen cyanamide has fungicidal properties (Donald *et al.*, 2004), and is a good alternative for liming and a slow-release nitrogen source which has herbicidal and fungicidal properties (Tremblay *et al.*, 2005).

Considering the long-term effect of chemical fumigation on the sustainability of vegetable production, there is a deep need for research into different fumigation and biofumigation alternatives (Guo *et al.*, 2017). The decrease in soil microbial populations with the use of MS has proved to be devastating (Sederholm *et al.*, 2017). Given the need to maintain soil health through management practices, non-chemical alternatives to soil fumigation should be explored (Wang *et al.*, 2006).

Biological control of pests and diseases

Biological control refers to controlling pests and diseases with living organisms. A natural enemy is introduced into the environment of the pest, where it multiplies and becomes effective in reducing or controlling the pest (Britannica, 2018a).

Biological control relies on predation, parasitism, herbivory or other natural mechanisms, but typically also involves an active human management role. It can be an important component of a holistic approach to pest and disease management (FAO, 2018). Using registered biological pesticides should be as effective in controlling pests and disease in brassica production as ordinary pesticides, provided that certain application guidelines are met, as biological products can be more sensitive to apply (Collier and van Steenwyk, 2004). Various Lepidoptera pests can be controlled with *B. thuringiensis*, a soil-living bacterium (Correa-Caudros *et al.*, 2016).

Biological soil fumigation

Soils were commonly fumigated with MB until its ban a few years ago. The product was used to control soilborne pathogens and weeds, and to avoid loss of yield in crops associated with monoculture practices. The negative effect of chemical fumigation products on management of soilborne pathogens, the environment and soil biology has led to the search for environmentally less damaging products as alternatives (Wang *et al.*, 2014). Incorporating cruciferous plant residues (especially mustard varieties) into the soil has been recorded as an alternative to chemical fumigation, furthermore, the incorporation of a legume cover crop into soil can also help to increase soil fertility (Wang *et al.*, 2006). Biofumigation refers to the suppression of soilborne pathogens through toxins released by decomposing organic matter. The volatile chemicals that are released during this process have some fungal, bacterial and nematode control properties (Wang *et al.*, 2014).

Biofumigation is carried out by incorporating *Brassica* spp. plant residues into the soil. These plants have a high glucosinolate (GSL) content that decomposes and is hydrolysed to isothiocyanates (ITC) (Omirou *et al.*, 2011). In hydrolysed brassica organic material, ITC has fumigation properties like MITC in MS (Sederholm *et al.*, 2017). The GSL in brassica plants is biologically inactive, and after tissue disruption and incorporation into the soil it is hydrolysed by myrosinase to a few by-products including ITC which are most toxic to soilborne pathogens (Omirou *et al.*, 2011). As microbes in soil are an important part of the soil ecosystem, biofumigation offers a more ecologically appropriate alternative to toxic chemicals.

Decomposition of organic material, nutrient cycling, pollutant degradation and formation of humic substances are all healthy soil processes (Omirou *et al.*, 2011). Biofumigation is explored as the preferred alternative to fumigation, but the adoption of this practice has been limited due to gaps in the knowledge of mechanisms for disease suppression and control (Wang *et al.*, 2014). The largest group of true fungi is the ascomycetes, including plant pathogens belonging to the genera *Fusarium*, *Pyrenochaeta*, *Sclerotinia* and *Verticillium*. Some research suggests that these can be controlled by biofumigation (Omirou *et al.*, 2011). However, the effect of biofumigation on non-target organisms also needs to be investigated (Wang *et al.*, 2014).

A holistic approach to pest and disease management

A holistic approach to pest and disease management systematically tries to reduce pest and disease numbers on the target plant, and contributes to long-term sustainability by combining judicious use of biological, cultural, physical and chemical control tools in a way that minimizes the risks of pesticides to human health and the environment (Bajwa and Kogan, 2002). Understanding that a holistic approach is a system (rather than a number of haphazard interventions) is extremely important in managing pests and diseases (Way and van Emden, 2000). An adaptable range of pest control methods is explored; these are cost-effective while being environmentally acceptable and sustainable (Way and van Emden, 2000).

Managing diseases and pests with a holistic approach reduces the effects of chemical use (Nga and Kumar, 2008). Farmers need to manage their farms as ecosystems, as chemical use can eliminate the natural enemies as well as pests (Macharia *et al.*, 2005). A holistic approach aims at using the minimum amount of chemical pesticides possible to control a pest, with the incorporation of non-insecticidal control whenever possible (Finch and Collier, 2000). Because of the limited availability of products for biological control and the difficulty of registering these products, a balance should be found between chemical and biological products, which can be used together for resistance management, crop protection and to be economically justifiable.

Methods and Materials

Background

The land for this study was situated on a commercial vegetable farm in Philippi in the heart of the Western Cape vegetable production area, 34°01'33.46" S, 18°32'31.47" E with an elevation of 22 m above sea level, and a long history of brassica production. Similar methods of biological control were incorporated on the long-term comparative organic farming systems trials at the George Campus of Nelson Mandela University (the Mandela Trials).

Trial layout and design

The experimental design was a strip-split plot design, with different pest and disease management strategies as the main plot treatment and fumigation and rotation treatment combinations arranged in strips across the main plot treatments. The main plot design was a randomized complete block with four programmes: (i) control; (ii) chemical; (iii) holistic; and (iv) biological. These were replicated four times and laid out in a randomized complete block design. The treatment design of the strip-plot factors was a 2 × 2 factorial with two fumigations (fumigated chemically and fumigated biologically) and two rotations (crop rotated and monoculture) randomly allocated across main plot treatments (Fig. 20.1). Each experimental unit consisted of 40 plants. The number of plants infested with DBM was assessed weekly. Using assessment keys can be a quick, simple and successful way of assessing the percentage of disease present. It can be used on leaves, individual plants or in small sample areas (James, 1971). White blister severity was measured by assessing the whole block (James, 1971). A disease rating scale of 0–6 was used to determine the percentage infection of the block as follows:

- 0 = 0%;
- 1 = 1–10%;
- 2 = 11–25%;
- 3 = 26–50%;
- 4 = 51–75%;
- 5 = 76–99%; and
- 6 = 100%.

The number of plants infected with clubroot was assessed at the end of the growing season. Whole roots were removed from the soil and washed. Clubroot severity was assessed on a scale from 0 to 3 as per Jordan and Gevens (2011):

- 0 = 0% clubbed;
- 1 = only lateral roots clubbed;
- 2 = < 50% of taproot clubbed; and
- 3 = > 50% of taproot clubbed.

In Fig. 20.1 and Fig. 20.2, the 2 × 2 fumigation and rotation treatment combinations are colour coded. Fumigation (Fum) refers to the semi-permanent bed of soil being treated with chemical fumigation. Fumigation (Bio) refers to the semi-permanent bed of soil being treated with biological fumigation using mustard plants. Rotation (Rot) refers to the crops in the semi-permanent bed cultivated with a crop rotation programme. Rotation (No) refers to the crops in the semi-permanent bed cultivated as a monoculture broccoli crop.

Fig. 20.1. Trial layout. The treatment design of the strip-plot factors was a 2 × 2 factorial with two fumigations ('Fum', fumigated chemically; and 'Bio', fumigated biologically) and two rotations ('Rot', crop rotated; and 'No', monoculture) randomly allocated across four main plot programmes: (i) Biol, biological; (ii) Che, chemical; (iii) Ct, control; (iv) IPM, holistic approach (integrated pest management).

Randomized block design for programme main plot effect

Block	1	2	3	4
	3	4	2	1
	4	1	3	2
	1	2	4	3
	2	3	1	4

where

No.	Programme
1	Control
2	IPM
3	Chemical
4	Biological

Randomized order of fumigation × rotation treatment combinations in strips over programme main plots

3	4	1	2		2	1	4	3		1	3	2	4		4	2	3	1

where

No.	Fumigation	Rotation
1	Bio	Not
2	Bio	Rotated
3	Fumigated	Not
4	Fumigated	Rotated

ANOVA

Source	d.f.
Block	4 – 1 = 3
Programme	4 – 1 = 3
Error (a)	(4 – 1)(4 – 1) = 9
Fumigation	2 – 1 = 1
Rotation	2 – 1 = 1
Fum x Rot	(2 – 1)(2 – 1) = 1
Error (b)	3(4 – 1) = 9
Prog x Fum	(4 – 1)(2 – 1) = 3
Prog x Rot	(4 – 1)(2 – 1) = 3
Prog x Fum x Rot	(4 – 1)(2 – 1)(2 – 1) = 3
Error (c)	9(4 – 1) = 27
Total	4 x 4 x 2 x 2 – 1 = 63

Fig. 20.2. Statistical layout. IPM is the holistic approach programme. d.f., Degrees of freedom.

The four programme sub-blocks in each of the four replications have different pest and disease management regimes. 'IPM', refers to a holistic management programme; 'Che', refers to chemical treatment; 'Biol', refers to biological treatment and 'Ct', refers to control treatment.

The four replications are represented in Fig. 20.1 and Fig. 20.2 as Blocks 1, 2, 3 and 4.

Analysis of variance (ANOVA) was carried out according to the experimental design, using the GLM (General Linear Models) procedure of SAS statistical software version 9.2 (SAS Institute Inc., Cary, North Carolina, USA). A Shapiro-Wilk test was performed to test for normality of variables assessed (Shapiro and Wilk, 1965). The least significant difference (LSD) was calculated at the 5% level to compare treatment means (Ott and Longnecker, 1998). A probability level of 5% was considered significant for all significance tests.

Cultivation practices

Enhancing the quality of the soil is dependent on the physical, chemical and biological properties of the soil. The main focus of agriculture is yield maximization, and as poor agricultural practices and excessive use of nitrogen fertilizer cause a decline in the soil fertility (especially the biological aspects), a beneficial rotation system is critical in brassica production (Ahmad *et al.*, 2014). Rotation crops need to be researched as they can assist in improving soil quality and fertility (Messinga *et al.*, 2015).

Brassicas such as broccoli, cauliflower and cabbage are heavy feeders, which means the plants extract a lot of nutrients from the soil (Venetta, 2011). For this reason the broccoli (*Brassica oleracea*) cultivar 'Star 2204', was planted first in the season. In the second planting of broccoli the cultivar 'Parthenon' was planted as it performs better during cooler months. Broccoli was planted at a density of 30,000 plants/ha.

Radishes (*Raphanus sativus*) although also brassicas, are root vegetables, and part of the mustard family. Radishes are only in the soil for a short period of 4–6 weeks, depending on the weather. Therefore, even though they are of the same family as broccoli, they are a good rotation crop; also they are light feeders (they do not take excessive amounts of nutrient out of the soil; Albert, 2014). The radish cultivar 'Cherry Belle' was planted. The seeding density for radishes was 10 kg seed/ha.

Root vegetables do not thrive with high nitrogen levels, as these cause lush foliage at the expense of the edible root (Growveg.com, 2015).

Monoculture refers to planting the same crop on the same piece of land year after year. It is often not very successful, as non-leguminous crops usually exhaust nitrogen in the soil leading to yield reduction with an increase in pests and diseases (Britannica, 2018b). Practising monoculture can lead to a loss of soil fertility, productivity and higher pest and disease rates (Tshikala *et al.*, 2018). Crop rotation is a system of continuous change of what crop is planted in a location over the course of multiple years or growing seasons. This system can lead to soil improvement and is a vital part in crop protection against pests and diseases (Campbell *et al.*, 1991). In general, crop rotations are known to increase crop productivity and reduce a build-up of soilborne pests and diseases. This is also one of the main reasons for crop rotation as most crops are not hosts to the same pest and disease, and rotation will help with their control (Larkin and Halloran, 2014). Controlling pests and diseases with chemicals is not the only option, crop rotation is a practice that can also be used (Mall *et al.*, 2018).

Soil organic matter (SOM) builds up in the soil when a rotation programme is followed, which leads to increased fertility, better pest and disease management, and helps to control soil erosion (Tshikala *et al.*, 2018). When monoculture of a certain crop is practised, pests and diseases, including soilborne pests that overwinter in plant debris, are likely to increase because of the reliable host that is present (Campbell *et al.*, 1991).

Nitrogen needs to be fixed in the soil in a rotation programme, after the heavy and light feeders. Legumes form nodules on the roots where rhizobia (nitrogen-fixing bacteria) establish themselves, and fix nitrogen (Masson-Biovin *et al.*, 2009). The legume, green bean (*Phaseolus vulgaris*) cultivar 'Douglas' was planted. The seeding density for legumes is 170,000–200,000 seeds/ha. A heavy feeder (broccoli) then followed the nitrogen-fixing legume (Venetta, 2011), starting the second cycle of the rotation.

The following rotation programme was used:

Broccoli → root vegetable radish →
green bean → broccoli

The 'no rotation' programme was broccoli monoculture.

The trial was established on 0.0896 ha (32 m × 28 m = 896 m²). Planting a density of 30,000 plants/ha, there were 2688 *B. oleracea* plants planted for this trial. Each of the treatment programme applications (control, chemical, biological, holistic) had 16 treatments, each with a 5 m² plot with six plants/m² = 480 plants per foliar treatment.

Soil was cultivated before the planting of the broccoli, and a commercial chemical fertilizer programme was applied on all the plots. No organic treatments were included, to avoid confounding variations due to farming system with variations due to pest and disease control, fumigation and rotation alone. Soil cultivation can have the following advantages: (i) it is a form of weed control; (ii) it reduces soilborne pathogens; (iii) it creates structure; and (iv) it helps to retain moisture (McMullen, 2000). All vegetables (rotation crops and broccoli monocrops) for this study were planted in semi-permanent beds, and needed to be cultivated.

Some pathogens can survive in the soil for a very long time. That is why it is important to have a rotation programme (Campbell *et al.*, 1991). Crop rotation will therefore be serving to break the host–pest cycle; these crops can be referred to as disease-suppressive crops (Larkin and Halloran, 2014). One study found that tomatoes in a monoculture programme experienced early blight at a rate of 3% in year one, but increased rapidly to 74% blight in year three (Campbell *et al.*, 1991). Crops in the *Brassicaceae* family used in rotations reduce soilborne diseases, pathogens and nematodes and improve soil health and crop yield (Larkin and Halloran, 2014). It is therefore important that evidence-based information on crop rotation management is made available to farmers, to lead to more sustainable agriculture (Tshikala *et al.*, 2018).

Ploughing was avoided in this study; the beds were only ripped and tilled. Avoiding ploughing saved cultivation time, labour costs, maintenance and fuel costs. The semi-permanent beds in which the vegetables were planted were bedded up. This leads to better water drainage, less waterlogging, less soil compaction, and fertilizer is not lost by being worked in too deeply; organic matter is higher, soil structure is improved which leads to better root development (McMullen, 2000).

Chemical control programme

The chemical control treatments are summarized in Table 20.1.

Calcium cyanamide is a slow-release calcium + nitrogen that has fumigation properties. It was used in conjunction with MS which is registered as a fumigant. Didecyldimethylammonium chloride (DDAC), is used throughout the world as a contact fungicide and as a sanitation product. It was applied with every chemical spray application as DDAC which is an excellent disease-resistance management product. Alpha-cypermethrin is a suspension concentrate insecticide for the control of cutworms (*Agrotis* spp.), American bollworm (*Helicoverpa armigera*) and DBM (*P. xylostella*) and their larvae in brassicas (van Zyl, 2010a, b). Systemic insecticides are used in transplanting to protect young plants against insect pests, until such a time when the plants are big enough to start spraying.

Azoxystrobin + chlorothalonil is a suspension concentrate fungicide with systemic, translaminar and contact properties for the control of white blister (*A. candida*) on brassica. Chlorothalonil is a suspension concentrate contact fungicide for the control of white blister (*A. candida*) on brassica. Tebuconazole is an emulsion in water systemic fungicide for the control of downy mildew, *Peronospora brassicae*, (van Zyl, 2010b). *Hyaloperonospora* is a new genus that accommodates several other *Peronospora* spp., parasitic on brassicas (Constantinescu and Fatehi, 2002).

Chlorfenapyr is a suspension concentrate translaminar insecticide with stomach and contact activity for the control of DBM larvae and large white cabbage moth (LWCM) (*Crocidolomia pavonana*) larvae. Chlorantraniliprole + lambda-cyhalothrin (pyrethroid) is a translaminar encapsulated suspension flowable concentrate with contact and stomach action for the control of DBM, LWCM, cutworm and American bollworm (van Zyl, 2010a).

Biological control programme

The biological control programme is summarized in Table 20.2.

'Biofumigation, as originally defined, is the use, in agriculture, of the toxicity of *Brassica* crop residues to control soilborne plant pathogens' (Motisi *et al.*, 2010). Caliente 199 mustard and Nemat arugula were planted 6 months before the broccoli was planted, and 4 weeks before planting broccoli, it was incorporated into the soil. The decomposing organic matter helped fumigate the soil biologically (Valdes *et al.*, 2012).

Trichoderma harzianum a wettable powder inoculant for the control of root diseases was drenched with *Paecilomyces lilacinus*, a wettable spore concentrate and fungal nematicide. *Trichoderma* spp. can be used to control clubroot (Cheah and Page, 1997). *Trichoderma* spp. also protect the roots and facilitate the absorption of nutrients (Mazhabi *et al.*, 2010). Vegetable juices with *T. harzianum* can be applied as a foliar spray for the control of foliar diseases (van Zyl, 2010b). *Beauveria bassiana* is a biological control agent, registered for the control of DBM. *B. thuringiensis* subspecies *kurstaki* is registered for the control of lepidopterous pests of brassica (van Zyl, 2010a). Azadirachtin isolates from the seeds of the neem tree *Azadirachta indica* has been used to control various pests in vegetables (Darabian and Yarahmadi, 2017).

Holistic control programme

The holistic approach programme in Table 20.3 is a combination of the chemical and biological active ingredients discussed.

Results and Discussion

The effect of different crop protection strategies was evaluated against different pest and diseases at weekly intervals. The following treatments were evaluated:

- control (no pest control programme applied);
- chemical (a chemical control spray programme applied);
- biological (a biological control spray programme applied); and
- holistic (an integrated control spray programme applied).

Evaluation methods were followed, as presented in the 'Methods and Materials' section.

Table 20.1. The chemical control programme used in the trials on *Brassica oleracea*.

Application timing	Problem	Recommendation: Product	Recommendation: Active ingredient	Recommendation: Dosage[ab]	Notes
Fumigation	Nematodes, soilborne diseases	Perlka + herbifume	Calcium cyanamide + metam sodium	500 kg/ha + 900 ml/ha	90 ml Herbifume in 10 L water, drench seedbed with 1L/m^2 (9 ml/m^2/L)
Before planting	Seedling diseases (*Rhizoctonia, Pythium, Fusarium*)	Sporekill drench	Didecyldimethyl-ammonium chloride	50 ml/100 L	
Planting	Diamondback moth, cutworms	Fastac	Alpha-cypermethrin	1 ml/5 ml/ 100 m	Drench over plants. 4 days withholding period
14 days after planting	Downy mildew, white blister	Amistar Opti	Azoxystrobin + chlorothalonil	600 ml/100 L	Only three sprays. 7 day intervals. 7 days withholding period
		Bravo	Chlorothalonil	400 ml/100 L	Only two sprays. 7 day intervals. 7 days withholding period
		TebuCure	Tebuconazole	75 ml/100 L	Only five sprays. 7 day intervals. 7 days withholding period Choose chemicals to repeat and rotate
Every 7–14 days later	Diamondback moth and lepidopterous pests	Hunter	Chlorfenapyr	60 ml/100 L	Only four sprays. 14 day intervals. 7 days withholding period
		Ampligo	Chlorantraniliprole + lambda cyhalothrin	40 ml/500 L	Only four sprays. 14 day intervals. 3 days withholding period
		Fastac	Alpha-cypermethrin	7 ml/100 L	Only two sprays. 14 day intervals. 4 days withholding period
Headforming until harvest	Diamondback moth	Hunter	Chlorfenapyr	60 ml/100 L	Only four sprays. 14 day intervals. 1 days withholding period
		Ampligo	Chlorantraniliprole + lambda cyhalothrin	40 ml/500 L	Only four sprays. 14 day intervals. 3 days withholding period
		Fastac	Alpha-cypermethrin	7 ml/100 L	Only two sprays. 14 day intervals. 4 days withholding period
	Downy mildew, white blister	Amistar Opti	Azoxystrobin + chlorothalonil	600 ml/100 L	Only three sprays. 7 day intervals. 7 days withholding period
		Bravo	Chlorothalonil	400 ml/100 L	Only two sprays. 7 day intervals. 7 days withholding period
		TebuCure	Tebuconazole	75 ml/100 L	Only five sprays. 7 day intervals. 7 days withholding period

[a]Add Nufilm P 30 ml/100 L (3 ml/lL) to spray mixture.
[b]Water volume = 500 L/ha.

Table 20.2. The biological control programme used in the trials on *B. oleracea*.

Application timing	Problem	Recommendation			Notes
		Product	Active ingredient	Dosage[ab]	
Fumigation	Nematodes, soilborne diseases	Caliente mustard + nemat arugula		10 kg/ha	
		Trichoplus	*Trichoderma harzianum*	250 g/ha (0.5 g/L)	Drench every 3–4 weeks
		PL Gold	*Paecilomyces liliacinus*		
Before planting	Seedling diseases (*Rhizoctonia, Pythium, Fusarium*)	Trichoplus	*T. harzianum*	250 g/ha (0.5 g/L)	Drench seedlings every 7 days up to planting
Planting	Soilborne diseases and nematodes	Trichoplus	*T. harzianum*	250 g/ha (0.5 g/L)	Drench every 3–4 weeks
		PL-Gold	*P. liliacinus*	2 kg/ha (4 g/L)	Drench every 3–4 weeks
Planting	Diamondback moth, cutworms	Broadband	*Beauveria bassiana*	1 L/ha (2 ml/l)	Spray 7 day intervals
		Bio-insek	*B. bassiana*	1 L/ha (2 ml/l)	Spray 7 day intervals
		Betapro	*Bacillus thuringiensis*	320 g/ha (0.7g/L)	Spray 7 day intervals
14 days after planting	Downy mildew, white blister	Bio-impilo	Fermented *T. harzianum*	500 ml/100 L	Spray 7 day intervals
		Bio-tricho	*T. harzianum*	500 ml/100 L	Spray 7 day intervals
	Diamondback moth, cutworms	Broadband	*B. bassiana*	1 L/ha (2 ml/L)	Spray 7 day intervals
		Bio-neem	Azadirachtin	500 ml/100 L	Spray 7 day intervals
		Betapro	*B. thuringiensis*	320 g/ha (0.1 g/L)	Spray 7 day intervals
Headforming until harvest	Diamondback moth, cutworms	Broadband	*B. bassiana*	1 L/ha (2 ml/L)	Spray 7 day intervals
		Betapro	*B. thuringiensis*	320 g/ha (0.1 g/L)	Spray 7 day intervals
	Downy mildew, white blister	Bio-impilo	Fermented *T. harzianum*	500 ml/100 L	Spray 7 day intervals

[a]Add Nufilm P 30 ml/100 L (3 ml/1L) to spray mixture.
[b]Water volume = 500 L/ha.

Table 20.3. A holistic control programme used in the trials on *B. oleracea* (referred to as an integrated pest management programme).

Application timing	Problem	Recommendation			Notes
		Product	Active ingredient	Dosage[ab]	
Fumigation	Nematodes	Perlka + herbifume	Calcium cyanamide + metam sodium	500 g/ha + 900 ml/ha	90 ml herbifume/10 L water, drench seedbed with 1 L/m^2 (9 ml/m^2/L)
Before planting	Seedling diseases (*Rhizoctonia, Pythium, Fusarium*)	Trichoplus	*Trichoderma harzianum*	250 g/ha	Every 7 days
Planting	Soilborne diseases and nematodes	Trichoplus	*T. harzianum*	250 g/ha (0.5 g/L)	Drench over plants every 3–4 weeks
		PL-Gold	*Paecilomyces liliacinus*	2 kg/ha (4g/L)	Drench over plants every 3–4 weeks
Planting	Diamondback moth, cutworms	Broadband	*Beauveria bassiana*	1 L/ha (2 ml/L)	Spray with 7 day intervals
		Ampligo	Chlorantraniliprole + lambda cyhalothrin	200 ml/500 L (0.4 ml/L)	Only four applications, 14 day intervals, 3 day withholding period
		Betapro	*Bacillus thuringiensis*	320 g/ha (0.7 g/L)	Spray with 7 day intervals
14 days after planting	Downy mildew, white blister	Copper hydroxide + sulfur	Copper hydroxide + sulfur	500 ml/100 L (5 ml/1L)	Spray with 7 day intervals
		Broadband	*B. bassiana*	1L/ha (2 ml/L)	Spray with 7 day intervals
	Diamondback moth, cutworms	Ampligo	Chlorantraniliprole + lambda cyhalothrin	200 ml/500 L (0.4 ml/L)	Only four applications, 14 day intervals, 3 day withholding period
		Betapro	*B. thuringiensis*	320 g/ha (0.7g/L)	Spray with 7 day intervals
Headforming until harvest	Diamondbackmoth, cutworms	Broadband	*B. bassiana*	1L/ha (2 ml/L)	Spray with 7 day intervals
		Ampligo	Chlorantraniliprole + lambda cyhalothrin	200 ml/500 L (0.4 ml/L)	Only four applications, 14 day intervals, 3 day withholding period
	Downy mildew, white blister	Betapro	*B. thuringiensis*	320 g/ha (0.7g/L)	Spray with 7 day intervals
		Amistar opti	Azoxystrobin + chlorothalonil	600 ml/100 L (6 ml/1L)	Only three applications, 7 day interval, 7 day withholding period
		Bravo	Chlorothalonil	400 ml/100 L (4 ml/1L)	Only two applications, 14 day interval, 14 day withholding period
		Tebucure	Tebuconazole	75 ml/100 L (0.75/ml 1L)	Only 5 applications, 7 day interval, 7 day withholding period

[a]Add Nufilm P 30 ml/100 L (3 ml/1 L) to spray mixture
[b]Water volume = 500 L/ha

The effects of different crop protection strategies on DBM in broccoli production

Results of the number of plants infested by DBM larvae are given in Table 20.4 (means of 40 plants/treatment × four replications). There was a significant ($p < 0.0001$) days after planting × programme interaction for the DBM larvae counts on plants.

Rotation system ($p = 0.5991$) and fumigation ($p = 0.2513$) did not significantly affect DBM incidence. Numbers of infested plants under different management programmes are shown in Fig. 20.3.

White blister infection on broccoli leaves

White blister was noticed and evaluated on the leaves of the broccoli plants. As mentioned in the 'Methods and Materials' section, white blister severity was measured by assessing the whole plot. The disease rating scale of 0–6 was used as explained earlier. The four plots and four replications were scored accordingly to the infection rate. This rating was then converted to a percentage of infection and results for the different management programmes and days after planting are shown in Table 20.5.

Table 20.4. Effect of different management programmes and days after planting on the incidence of diamondback moth (DBM) on broccoli.

	Incidence of DBM (mean no. of infested plants)[a]			
Days after planting	Control	Biological	Holistic	Chemical
35	0.00c	0.00c	0.00c	0.00c
49	2.18b	2.50b	0.31c	0.00c
63	4.37a	2.93b	3.37b	0.00c
74	4.06a	3.00b	2.31b	0.18c
78	4.06a	3.00b	2.31b	0.18c

[a]Means followed by the same letter do not differ significantly at $p = 0.05$ (least significant difference (LSD) = 0.8367).

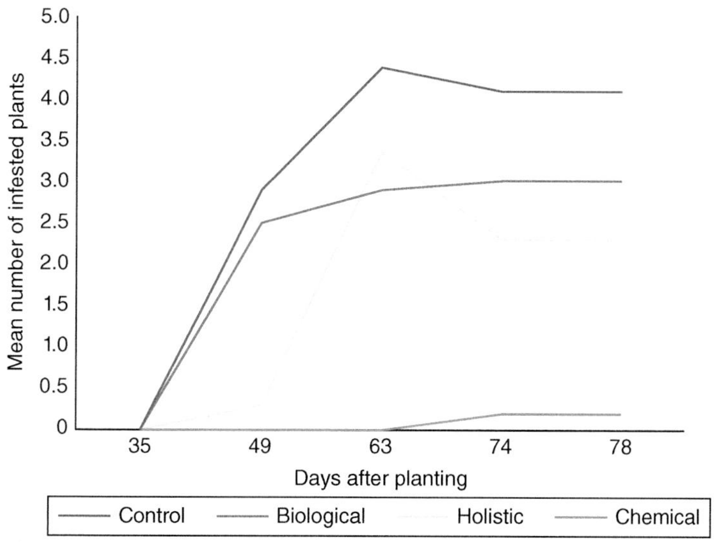

Fig. 20.3. Effects of different management programmes and days after planting on the incidence of diamondback moth (DBM) in broccoli.

There was also a significant ($p < 0.0001$) days after planting × programme interaction for the severity of white blister.

The rotation system ($p = 0.8262$) and fumigation ($p = 0.0946$) did not significantly affect white blister severity.

Severity of infection on broccoli plants grown under different systems are shown in Fig. 20.4.

White blister infection on broccoli heads at harvest

At harvest the infection of white blister on the broccoli heads was evaluated.

If the head is infected the market value goes down, so much so that the crop may not be marketeable, because of the deformation of the heads or stagheads discussed in the introduction. If a head had a blister it was counted as infected, and the mean numbers of infected broccoli heads per ten plants at harvest are shown in Table 20.6.

There were statistically significant differences due to all three treatments compared with the control treatment; the holistic approach gave significantly better control of white blister compared with the biological control programme. The biological programme also produced good control, and was not statistically

Table 20.5. Effect of different management programmes and days after planting on severity of white blister on broccoli.

	Severity of white blister (% infection)[a]			
Days after planting	Control	Biological	Holistic	Chemical
35	12.75hi	7.84ij	1.71j	4.65j
49	36.43bc	29.56b–d	20.18f–h	13.46g–i
63	37.46ab	27.06d–f	28.90c–e	9.40ij
74	32.376b–d	21.43e–g	27.53d–f	8.62ij
78	44.84a	14.78g–i	7.90ij	9.75ij

[a]Means followed by the same letter do not differ significantly at $p = 0.05$ (LSD = 8.0651).

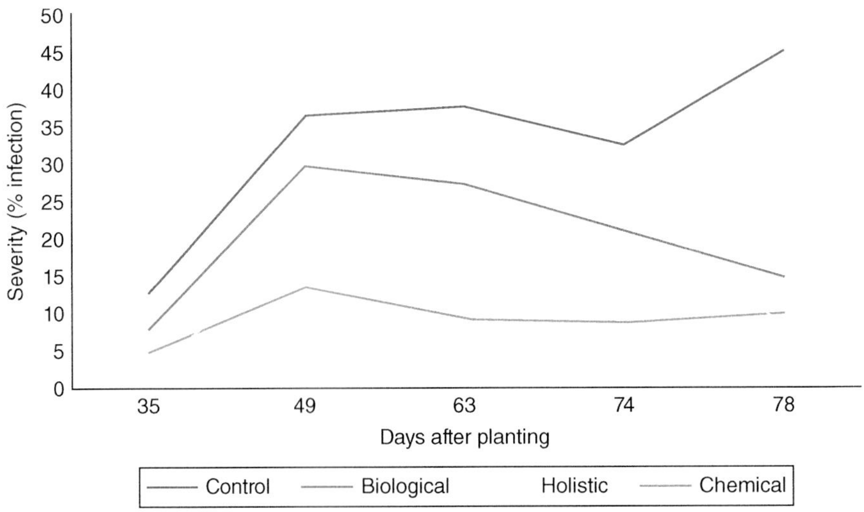

Fig. 20.4. Effect of different management programmes and days after planting on the severity of white blister on broccoli.

significantly different to the chemical treatment programme.

Programmes significantly ($p = 0.0002$) affected the incidence of plants with white blister. As can be seen in Fig. 20.5, the control programme was heavily infested, while the holistic programme had very little infection. Both chemical and biological programmes had moderate levels of infection.

Rotation system ($p = 0.6704$) and fumigation ($p = 0.4018$) did not significantly affect the incidence of white blister on heads.

Research on biological control has increased in the last few years, and with good reason. There are some biological products that can give significant control against pests and diseases. As mentioned in the 'Introduction', farmers still rely largely on chemical methods to control pests and diseases, but with increasing pressure from supermarkets and exporters, and pest and disease resistance to chemicals, different methods to control pests and diseases are now being explored.

In this study the results showed that chemical control was most effective against DBM with 0.18 plants per block infested at harvest. In the 'Introduction' to this chapter, it was mentioned that DBM has the ability to develop resistance to chemical products in a matter of a few seasons, if those chemical products are not used in a rotation and used with care and intelligence. The untreated control had 4.1 plants infested per block. There were no statistically significant differences on DBM control between the biological and holistic programmes at harvest, although the infestation was lower for the holistic approach with an average of 2.3 infested plants, compared with the biological programme for which an average of three infested plants were recorded. This shows that chemicals and biological products can be used in the same programme and against hardened pests for their control; this is a good agricultural practice to use in controlling DBM.

White blister disease on the foliage was controlled effectively with the holistic approach and chemical programme, which had no significant difference and resulted in just 7.9% infection for the holistic approach and 9.8% for the chemical programme, whereas the control had 44% infection. The biological programme had 14.8% infection.

Table 20.6. Effect of different management programmes on the incidence of white blister on *B. oleracea* heads.

Programme	White blister incidence (mean no. of plants/ten plants)[a]
Control	4.18a
Biological	1.56b
Holistic	0.31c
Chemical	1.18bc

[a]Means followed by the same letter do not differ significantly at $p = 0.05$ (LSD = 1.1659).

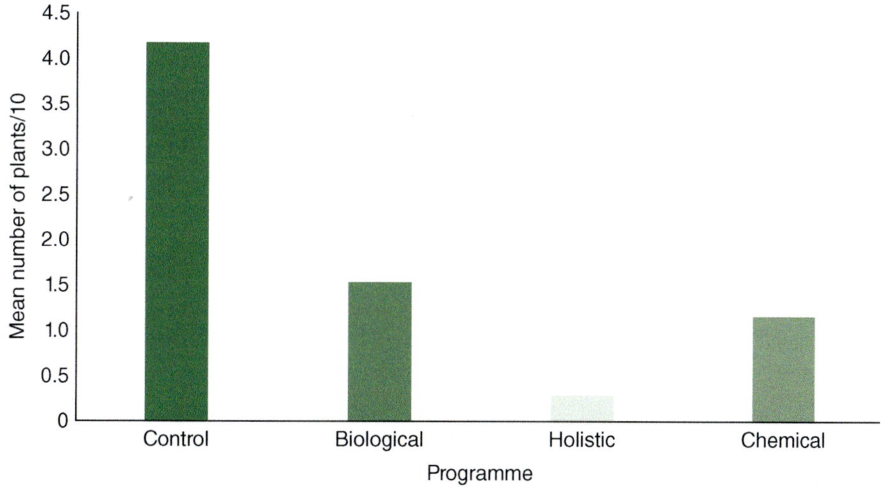

Fig. 20.5. Effect of different management programmes on the incidence of white blister on broccoli heads.

White blister infection at harvest gave interesting results.There were significant differences in all the treatments. In the control, 4.2 plants were infected per ten plants evaluated. In the biological programme 1.6 plants, and in the chemical programme 1.2 plants were infected. The holistic approach programme resulted in significantly better control of white blister than the biological control programme, with only 0.3 plants infected per ten plants analysed. One additional factor to consider is that with the residues that chemical control products leave, there is a withholding period on all chemical products regarding how close to harvest you can apply a product; at this point, application of chemicals ceased. Biological products do not have an application withholding period and can be applied up until harvest.

There was no evidence of clubroot of *Brassica* spp. in any of the treatments.

Conclusion

Data were collected in the study after only one rotation cycle, and unfortunately due to health problems of the farmer, this on-farm trial could not be continued for the planned period of two full rotation cycles. Given a longer research period, more distinct results might have been obtained, but the tendencies noted were already useful.

There was no clubroot infection in the trials for any of the treatments. The data collected relating to DBM and white blister were significant data: it is possible to produce a healthy crop not only with a chemical programme, but also with a more holistic approach, and even with a purely biological approach (which would be acceptable to organically certified farmers).

Unfortunately, in SA there is limited data and published peer-reviewed articles available on the subject of crop protection are of brassicas against pests and diseases. These trials will need to be repeated for a longer period and under a range of conditions to determine the effects of various control methods on clubroot of *Brassica* spp., and to confirm the results obtained for white blister and DBM.

From this study it can be concluded that further research on different crop protection strategies is needed to ensure the best control of pests and diseases of brassicas. The lack of registered biological products is a major problem in building a biological or holistic approach in crop protection programmes. A holistic approach would seem to be the easiest to adopt, for the cautious farmer, the environment and the farm labourers, as it minimizes the amount of chemical residues on edible crops.

The type of registered products that are compatible in a spray tank needs to be researched. As many of the biological products consist of fungi and bacteria, there are few chemical products that can be applied with these biological products, as the chemical products can destroy some of the biological products when they are mixed or applied at different intervals. Research is needed on compatibility of chemical and biological products.

According to Agri-Intel (2018a), SA's database of registered agricultural remedies, there are over 142 chemical pesticides registered for DBM on *Brassica* spp. and only two registered biological products. For white blister there are 12 registered chemical pesticides available and no biological registered products (Agri-Intel, 2018b). For clubroot there are no biological registered products available (Agri-Intel, 2018c).

The prices of the biological products exceed the prices of chemical products, and this is also a major problem in the industry. It is already expensive to farm, with high cost of diesel and implements, labour costs, water costs, fertilizer costs and the cost of crop protection products, and with the relatively low prices that farmers get for their crops at the market. The more biological products are researched and registered, and the more they become common elements of crop protection strategies, the lower the prices of these products will become, assisting farmers to farm with safer, more environmentally friendly products, but still protecting their precious crops.

References

Agri-Intel (2018a) Registered Pesticides on Diamondback Moth in Broccoli. Available at: http://www.agri-intel.com (accessed 7 June 2018).

Agri-Intel (2018b) Registered Pesticides on White Blister on Broccoli. Available at: http://www.agri-intel.com (accessed 7 June 2018).

Agri-Intel (2018c) Registered Pesticides on Clubroot on Broccoli. Available at: http://www.agri-intel.com/agri_intel_crop_protection/search_by_target_post (accessed 7 June 2018).
Ahmad, W., Famanullah, S.Z., Jamal, M. and Shah, K.A. (2014) Recovery of organic fertility in degraded soil through fertilization and crop rotation. *Journal of Saudi Society of Agricultural Sciences* 13, 92–99.
Albert, S. (2014) Growing Radish Problems: Troubleshooting. Available at: http://www.harvesttotable.com/2009/06/radish_growing_problems_troubl/ (accessed 29 May 2015).
Bajwa, W.I. and Kogan, M. (2002) *Compendium of IPM definitions (CID)*. Integrated Plant Protection Centre Oregon State University, Oregon. Available at: http://www.ipmnet.org/ipmdefinitions/index.pdf (accessed 29 May 2015).
Barbetti, M.J., Li, C.X., You, M.P., Singh, D., Agnihotri, A., Banga, S.K., Sandhu, P.S. and Banga, S.S. (2016) Valuable new leaf or inflorescence resistance ensure improved management of white rust (*Albugo candida*) in mustard (*Brassica juncea*) crops. *Journal of Phytopathology* 164, 404–411.
Britannica (2017) *Brassicaceae*. Available at: https://www.britannica.com/plant/Brassicaceae (accessed 5 February 2017).
Britannica (2018a) Biological Control. Available at: https://www.britannica.com/science/biological-control (accessed 22 April 2018).
Britannica (2018b) Monoculture. Available at: https://www.britannica.com/technology/agricultural-technology/Factors-in-cropping#ref558283 (accessed 22 April 2018).
Campbell, C., Biederbeck, V., Zenter, R. and Lafond, G. (1991) Effect of crop rotations and cultural practices on soil organic matter, microbial biomass and respiration in a thin black Chernozem. *Canadian Journal of Soil Science* 71, 363–376.
Cheah, L.H. and Page, B.B.C. (1997) *Trichoderma* spp. for potential biocontrol of clubroot of vegetable brassicas. In: *50th NZ Plant Protection Conference*. New Zealand Plant Protection Society, Auckland, New Zealand, pp. 150–153.
Choi, Y.H., Shin, H.D., Ploch, S. and Thines, M. (2011) Three new polygenic lineages are the closest relatives of the widespread species *Albugo candida*. *Fungal Biology* 115, 598–607.
Collier, T. and van Steenwyk, R. (2004) A critical evaluation of augmentative biological control. *Biological Control* 31, 245–256.
Constantinescu, O. and Fatehi, J. (2002) *Peronospora*-like fungi (Chromista, Peronosporales) parasitic on *Brassicacae* and related hosts. *Nova Hedwigia* 74(3–4), 291–338.
Correa-Caudros, J.P., Suenz-Aponte, A. and Rodriquez-Bocanegra, M.X. (2016) *In vitro* interaction of *Metarhizium anisopliae* Ma 9236 and *Beauvaria bassiana* Bb 9205 with *Heterohabditis bacteriophora* HNI 0100 for the control of *Plutella xylostella*. *SpringerPlus* 5(2068).
Darabian, K. and Yarahmadi, F. (2017) Field efficacy of azadirachrin, chlorfenapyr and *Bacillus thuringiensis* against *Spodoptera exigua* (Lepidoptera: Noctuidea) on sugar beet crop. *Journal of Entomological Research Society* 19(3), 45–52.
Dennill, G.B. and Pretorius, W.L. (1995) The status of diamondback moth, *Plutella xylostella* (Linnaeus) (Lepidoptera: Plutellidea), and its parastitoid on cabbages in South Africa. *African Entomology* 3(1), 65–71.
Donald, E., Lawrence, J. and Porter, I. (2004) Influence of particle size and application method on the efficacy of calcium cyanamide for the control of clubroot of vegetable brassicas. *Crop Protection* 23, 297–303.
FAO (Food and Agriculture Organization of the United Nations) (2018) Biological Pesticides. Available at: http://www.fao.org/tc/exact/sustainable-agriculture-platform-pilot-website/integrated-pests-management/biological-pesticides/en/ (accessed 10 April 2018).
Finch, S. and Collier, R.H. (2000) Integrated pest management in field vegetable crops in northern Europe – with focus on two key pests. *Crop Protection* 19, 817–824.
Ghosal, A., Dolai, A.K. and Chatterjee, M.L. (2015) Bioefficacy of new ready mixed insecticide Novaluron 5.25% + Emamectin 0.9% SC against diamond back moth (*Plutella xylostella*. L.) in cabbage in West Bengal. *Madras Agricultural Journal* 102(1–3), 75–79.
Government Gazette (2010) *Fertilizer, Farm Feeds, Agricultural Remedies and Stock Remedies Act 36 of 1947*. Government Gazette (33899), Notice 1120 of 2010. Government Printers, Pretoria.
Growveg.com (2015) Grow Guides: Crop Rotation for Growing Vegetables. Available at: http://www.growveg.com/growguides/crop-rotation.aspx (accessed 15 May 2015).
Guo, H., Di Gioia, F., Zhoa, X., Ozores-Hampton, M., Swisher, M., Hong, J., Kokalis-Buelle, N., De Long, A. and Rosskopft, E. (2017) Optimizing anaerobic soil disinfestation for fresh market tomato production: nematode and weed control, yield and fruit quality. *Scientia Horticulturae* 218, 105–116.
Harris, B.M. and McLean, B. (1999) Spinosad: control of lepidopterous pests in vegetable brassicas. In: *52nd NZ Plant Protection Conference*. New Zealand Plant Protection Society, Auckland, New Zealand, pp. 65–69.

Irani, S., Trots, B., Waldner, M., Noyidu, N., Tu, J., Kusalik, A.J., Todd, C.D., Wei, Y. and Bonham-Smith, P.C. (2018) Transcriptome analysis of response to *Plasmodiophora brassicae* infection in the *Arabidopsis* shoot and root. *BMC Genomics* 19(23), 1–19.

James, C. (1971) An illustrated series of assessment keys for plant diseases, their preparation and usage. *Canadian Plant Disease Survey* 51(2), 39–65.

Jordan, S.A. and Gevens, A.J. (2011) *Evaluation of the Fumigants Pic-C60 and Pic Plus for Control of Clubroot and Black Rot of Cabbage Grown in Wisconsin, 2011*. University of Wisconsin-Madison, Madison, Wisconsin. Available at: https://wivegdis.wiscweb.wisc.edu/wp-content/uploads/sites/210/2018/08/Gevens-Field-Trials-2011-Cabbage-Black-Rot-Club-Root.pdf (accessed 22 May 2015).

Kaur, P. (2013) Agronomic challenges from novel pathotypes of *Albugo candida* to the emerging *Brassica juncea* industry in Western Australia. *Pytopathologia Mediterranea* 52(3), 418–433.

Kaur, P. and Sivasithamparam, K. (2011) Host range and phylogenetic relationships of *Albugo candida* from cruciferous hosts in Western Australia, with special reference to *Brassica juncea*. *Plant Disease* 95(6), 712–718.

Kaur, P., Sivasithamparam, K. and Barbetti, M.J. (2011) Site of inoculation and stage of plant development determine symptom type and expression in *Brassica juncea* following infection with *Albugo candida*. *Journal of Plant Pathology* 39(2), 383–388.

Kfir, R. (2001) Effect of paristoid elimination on populations of diamondback moth in cabbage. In: Endersby, N.M. and Ridland, P.M. (eds) *The Management of Diamondback Moth and Other Crucifer Pests*, Proceedings of the 4th International Workshop, 26–29 November, Melbourne, Australia. The Regional Institute, Gosford, New South Wales, Australia, pp. 197–205.

Khan, M. and Damalas, C. (2015) Factors preventing the adoption of alternatives to chemical pest control among Pakistani cotton farmers. *International Journal of Pest Management* 61(1), 9–16.

Koike, S.T. (2003) Vegetable disease caused by soilborne pathogens. *Agriculture and Natural Resources (ANR) Publication* 8099, 1–13.

Labrador Morales, M., de Pozo Nunez, E.M. and Garcia Cruz, I. (2013) Effects of *Trichoderma harzianum* over *Plasmodiophora brassicae* Woronin in broccoli, in Escaguey, municipality of Rangel, Merida state. *Centro Agricola* 40(20), 39–44.

Larkin, R.P. and Halloran, J.M. (2014) Management effects of disease-suppressive rotation crops on potato yield and soilborne diseases and their economic implications in potato production. *American Journal of Potato Research* 91, 429–439.

Macharia, I., Lohr, B. and de Groote, H. (2005) Assessing the potential of biological control of *Plutella xylostella* (diamondback moth) in cabbage production in Kenya. *Crop Protection* 24, 981–989.

Mall, D., Larsen, A. and Martin, E. (2018) Investigation the (mis) match between natural pest control knowledge and the intensity of pesticide use. *Insects* 9(1), 1–13.

Marchioro, C.A., Krechemer, F.S., Moraes, C.P. and Foerster, L.A. (2017) A stochastic model for predicting the stage emergence of *Plutella xylostella* under field conditions. *Annals of Applied Biology* 169, 190–199.

Masson-Biovin, C., Giruad, E., Perret, X. and Batut, J. (2009) Establishing nitrogen-fixing symbiosis with legumes: how many *Rhizobium* recipes? *Trends in Microbiology* 17(10), 458–466.

Mazhabi, M., Nemati, H., Rouhami, H., Tehranifar, A., Mahdikhani-Moghadam, E. and Kaveh, H. (2010) Does *Trichoderma harzianum* really increase growth parameters in plants? *Research Journal of Biologial Sciences* 5(11), 739–744.

McMullen, B. (2000) SOILpak for Vegetable Growers. Available at: http://www.dpi.nsw.gov.au/__data/assets/pdf_file/0007/167533/soipak-vegetable-prelim.pdf (accessed 2 September 2019).

McGrann, G.R.D., Yoxall, T., Paterson, L.J., Taylor, J.M.G., Birmpilis, I.G., Walters, D.R. and Havis, N.D. (2017) Control of light leaf spot and clubroot in *Brassica* crops using defence elictors. *European Journal of Plant Pathology* 148, 447–461.

Messinga, A.J., Sharifi, M., Hammermesiter, A., Gallant, K., Fuller, K. and Tango, M. (2015) Soil quality response to cover crops and amendments in a vineyard in Nova Scotia, Canada. *Scientia Horticulturae* 188, 6–14.

Motisi, N., Dore, T., Lucas, P. and Montford, F. (2010) Dealing with the variability in bio-fumigation efficacy through an epidemiological framework. *Soil Biology and Biochemistry* 42, 2044–2057.

Nga, L.T. and Kumar, P. (2008) Contributions of parasitoids and *Bacillus thuringiensis* on the management of diamondback moth in highland crucifer production in Da Lat, Vietnam. *Journal of Asia-Pacific Entomology* 11, 59–64.

Nofemela, R.S. and Kfir, R. (2005) The role of parasitoids in suppressing diamondback moth, *Plutella xylostella* (L.) (Lepidoptera: Plutellidae) populations on unsprayed cabbage in the North West Province of South Africa. *African Entomology* 13, 71–83.

Omirou, M., Rousidou, C., Bekris, F., Papadopoulou, K., Menkissoglou-Spiroudi, U., Ehaliotis, C. and Karpauzas, D. (2011) The impact of biofumigation and chemical fumigation methods on structure and function of the soil microbial community. *Microbial Ecology* 61, 201–213.

Ott, R.L. and Longnecker, M. (1998) *An Introduction to Statistical Methods and Data Analysis*. Duxbury Press, Belmont, California.

Peres, L.L.S., Sobreiro, A.I., Couto, I.F.S., Silva, R.M., Peirera, F.F., Heredria-Vieira, S.C., Cardoso, C.A.L., Manad, M., Scalon, S.P.Q., Verza, S.S. and Mussury, R.M. (2017) Chemical compounds and bioactivity of aquous extracts of *Alibertia* spp. in the control of *Plutella xylostella* L. (Lepidoptera: Plutelladea). *Insects* 8(125), 1–13.

Ploch, S., Choi, Y.J., Rost, C., Shin, H.D., Schilling, E. and Thines, M. (2010) Evolution of diversity in *Albugo* is driven by high host specificity and multiple speciation events on closely related *Brassicaceae*. *Molecular Phylogenetics and Evolution* 57, 812–820.

Reddy, G.V.P., Tabone, E. and Smith, M.T. (2004) Mediation of host selection and oviposition behaviour in the diamondback moth *Plutella xylostella* and its predator *Chrysoperla carnea* by chemical cues from cole crops. *Biological Control* 29, 270–277.

Saeed, R., Sayyed, A.H., Shad, S.A. and Zaka, M. (2010) Effect of different host plants on the fitness of diamond back moth, *Plutella xylostella* (Lepidoptera: Plutelladae). *Crop Protection* 29, 178–182.

Santos, M.R. and Dias, J.S. (2004) Evaluation of a core collection of *Brassica oleraceae* accessions for resistance to white rust of crucifiers (*Albugo candida*) at the cotyledon stage. *Genetic Resources and Crop Evolution* 51, 713–722.

Santos, M.R., Silva, M.J., Ferreira-Pinto, M.M. and Dias, S.J. (2009) Identification of sources and inheritence of chinese *Brassica* vegetable to white blister. *Plant Breeding* 128, 640–644.

Sarfaz, M., Dosdall, L.M. and Keddie, B.A. (2006) Diamondback moth–host plant interactions: implications for pest management. *Crop Protection* 25(7), 625–639.

Sederholm, M.R., Schmitz, B.W., Barbera, A. and Pepper, I.J. (2017) Effects of metam-sodium fumigation on the abundance, activity and diversity of soil bacteria. *Applied Soil Ecology* October, 1–7.

Shapiro, S.S. and Wilk, M.B. (1965) An analysis of variance test for normality (complete samples). *Biometrika* 52(3/4), 591–611.

Shi, K., Wang, L., Zhou, Y. and Yu, J. (2009) Effects of calcium cyanamide on soil microbial communities and *Fusarium oxysporum* f. sp. *cucumberinum*. *Chemosphere* 75, 872–877.

Tremblay, N., Belec, C., Coulombe, J. and Godin, C. (2005) Evaluation of calcium cyanimide and liming for control of clubroot disease in cauliflower. *Crop Protection* 24, 798–803.

Tshikala, S., Fonsah, E., Boyan, G., Little, E. and Gaskin, J. (2018) Crop rotation systems for high-value, cool-season vegetables in the southern United States. *Journal of Food Distribution Research* 49(1), 30–38.

Valdes, Y., Vaiene, N. and Moens, M. (2012) Effects of yellow mustard amendments on soil nematode community in a potato field with focus on *Globodera rostochiensis*. *Applied Soil Ecology* 59, 39–47.

Van Zyl, K. (2010a) *A Guide to Crop Pest Management in South-Africa*. A CropLife South Africa Compendium, 2nd edn. AVCASA, Pretoria.

Van Zyl, K. (2010b) *Chemical Control of Plant Diseases*. A CropLife South Africa Compendium, 2nd edn. AVCASA, Pretoria.

Venetta, D. (2011) Beginner's Guide to Gardening: Crop Rotation. Available at: http://indianapublicmedia.org/eartheats/beginners-guide-gardening-crop-rotation/ (accessed 15 May 2015).

Wang, Q., McSorley, R. and Kokalis-Burelle, N. (2006) Effects of cover cropping, solarisation and soil fumigation on nematode communities. *Plant and Soil* 286, 229–243.

Wang, Q., Ma, Y., Yang, H. and Chang, Z. (2014) Effects of biofumigation and chemical fumigation on soil microbial community structure and control of pepper *Phytophthora* blight. *World Journal of Microbial Biotechnology* 30, 507–518.

Way, M.J. and van Emden, H.F. (2000) Integrated pest management in practice – pathways towards successful application. *Crop Protection* 19, 81–103.

Yu, X., Zhoa, Y., Cheng, J. and Wang, W. (2015) Biocontrol effect of *Trichoderma harzianum* T4 on brassica clubroot and analysis of rhizosphere microbial communities based on T-RFLP. *Biocontrol Science and Technology* 25(12), 1493–1505.

Zalucki, M.P., Shabbir, A., Silva, R., Adamson, S., Shu-Sheng, L. and Furlong, M.J. (2012) Estimating the economic cost of one of the worlds, major pests, *Plutella xylostella* (Lepidoptera: Plutellidae): just how long is a piece of string? *Journal of Economic Entomology* 105(4), 1115–1129.

Zhao, Y., Gao, J., Tian, B., Bi, K., Chen, T., Chen, T., Liu, H., Xie, J., Cheng, J., Fu, Y. and Jiang, D. (2017) Endosphere micropbiome comparison between symptomatic and asymptomatic roots of *Brassica napus* infected with *Plasmodiophora brassicae*. *PLoS One* 12(10), e0185907.

21 Initial Assessment of Selected Biological Soil Health Indicators in Organic Versus Conventional Cropping Systems in Field Trials in South Africa

Mandla Sibiya,[1*] Johan Habig,[2] Sheila Storey[3] and Nico Labuschagne[1]
[1]*University of Pretoria, Pretoria, South Africa;* [2]*Agri Technovation, Wellington, South Africa;* [3]*Nemlab, Klapmuts, South Africa*

Abstract

Agricultural sustainability is built on the premise of promoting soil health. Agricultural practices such as tillage, crop rotation, cover crops and fertilization have been credited with shifts in the soil food web which affects many aspects of soil health. The aim of the current study was to investigate the effects of conventional versus organic cultivation systems on selected biological soil health indicators within the existing long-term comparative farming systems field trial at the George Campus of the Nelson Mandela University in South Africa's Western Cape Province. Our results show that microbial diversity, evenness and enzyme activity was highest in the organic monocrop cabbage treatment. Richness of nematode predators was a function of the abundance of nematode prey that could be improved by organic inputs. The highest nematode diversity was correlated with organic treatments which also indicated the highest fertility (enrichment index). Organic treatments also had the better soil food web structure (structure index) while for the majority of treatments, the decomposition channel was fungal (high channel index). The control and conventional cabbage treatments had more nematodes associated with stressful conditions, possibly because of toxicity from chemicals used in these cropping systems. Nematode populations (functional guilds) were more sensitive to cropping systems than microbial populations (functional diversity) although both biological indicators resulted in similar conclusions. In terms of biological diversity, rotation treatments had an intermediate effect while monocrop treatments performed better. We conclude that the use of organic amendments and cover crops relative to conventional systems has the potential to improve soil quality in the long term, through improved biodiversity and higher organic matter content.

Introduction

The soil food web through its numerous activities is a major determinant of soil physical, chemical and biological properties (Brevik *et al.*, 2015). In addition, it is also responsible for ecological functions such as mineral and nutrient cycling, carbon sequestration, organic matter decomposition and suppression of pest species (de Vries *et al.*, 2013; Henneron *et al.*, 2015; Powell *et al.*, 2015). These functions are determined by the abundance, diversity and interactions of the various trophic groups within the soil (Hohberg, 2003; Minoshima *et al.*, 2007; Zhang *et al.*, 2017). Shifts in the soil food web structure have been a major consequence of

*wikiness@gmail.com

anthropogenic agricultural process (Zhang *et al.*, 2017). In contrast to soil edaphic properties that change very slowly over time, biological properties are quite sensitive to even the smallest changes to their micro-ecosystem (Gryta *et al.*, 2014; Cesarz *et al.*, 2015; Feigl *et al.*, 2017).

The indiscriminate use of synthetic chemicals in conventional systems (Wang *et al.*, 2017) has several implications for the activity and abundance of the underground microfauna (Brevik *et al.*, 2015; Song *et al.*, 2016). Conversely, the exclusion of synthetic chemicals and addition of organic matter in organic farming systems have been found to have a positive effect on the soil microbial activity (Briar *et al.*, 2012). Being acquainted with soil food web dynamics in both time and space forms the basis of responsible soil stewardship (Ferris *et al.*, 2001). Soil biodiversity assessments can either involve structural and/or functional diversity measurements (Cooper and Rao, 2006). Due to the complex and diverse nature of the soil food web, individual community characterization is difficult (Brevik *et al.*, 2015; Tsiafouli *et al.*, 2017). Alternatively, the assessment of the presence and richness of indicator guilds has been proposed (Ferris *et al.*, 2001).

In addition, indices have been developed from functional guilds with high efficiency at various scales over soil ecosystem assessment (Tsiafouli *et al.*, 2017). Nematodes have long been regarded as convenient bio-indicators of a healthy soil (Bongers, 1999; Bongers and Ferris, 1999; Tsiafouli *et al.*, 2017), because of their sensitivity to changes in environmental conditions (Briar *et al.*, 2012), and their assorted feeding behaviours and life strategies (Yeates, 2003; Guan *et al.*, 2018). Scientists have also combined standard microbiological methods with a community approach in order to detect any likely structural and/or functional changes of the soil food web (Feigl *et al.*, 2017). An example of this approach is the Biolog EcoPlate™ method (Gryta *et al.*, 2014, Liu *et al.*, 2015). Soil enzyme activity is one other potential indicator for the analysis of microbial functional diversity as it directly links to the soil biology (Yao *et al.*, 2006).

The influence of long-term cropping systems on the soil food web structure and function remain ambiguous (Naether *et al.*, 2012), and the underlying mechanisms have not been fully explored (Su *et al.*, 2015). Striving to understand biological community dynamics better (Forey *et al.*, 2011) and how they are affected by agricultural management practices should improve our capacity to forecast the impact of management practices on significant soil functions (Fernandez *et al.*, 2016). The aim of the current study was to investigate the effects of conventional versus organic cultivation on soil biological properties within the existing long-term comparative organic farming systems trial initiated by Raymond Auerbach at the George Campus of Nelson Mandela University (the Mandela Trials, Chapters 18–22, this volume).

Materials and Methods

The details of the field trial layout, as well as the various treatments, are comprehensively described in Chapter 18 of this volume. Here we report on the sampling and analyses and will only briefly summarize the treatments. It must be noted that the field trial was initiated during the 2014/15 season, therefore the treatment effects reported here reflect the effect of the treatments over a 4-year period.

Soil sample collection

Two separate sets of samples (one for Biolog EcoPlate™ analysis and one for nematode analysis) were collected from the first three replications of each treatment. Rhizospheric soil (top 15 cm) was sampled randomly within the middle of the three central crop rows within each plot and pooled, rendering a composite sample per treatment replication. Samples were immediately transferred into cooler boxes containing ice packs after sampling to keep them cool until the time of analysis. The following nine treatments were included in the study:

1. Conventional monocrop cabbage – CMCA;
2. Organic monocrop cabbage – OMCA;
3. Conventional rotation cabbage – CRCA;
4. Organic rotation cabbage – ORCA;
5. Conventional rotation cowpea – CRCP;
6. Organic rotation cowpea – ORCP;
7. Conventional rotation sweet potato – CRSP;
8. Organic rotation sweet potato – ORSP; and
9. Control – ConCA.

The first set of samples was immediately transported to the laboratories at the Agricultural

Research Council (ARC) in Pretoria, Gauteng Province for analysis of microbial functional diversity and enzyme activity (Biolog EcoPlate™ analysis). The second set of samples was couriered to Nemlab Diagnostic Laboratory (Klapmuts, Western Cape Province) for nematode analysis.

Determination of functional diversity

Whole-community substrate utilization profiles (CSUP) are assessed when carbon sources are utilized. Soil samples were diluted in sterile distilled water (Buyer and Drinkwater, 1997) and inoculated into Biolog EcoPlates™ (Biolog® Inc., Hayward, USA) containing 31 carbon sources and a control well, in triplicate. The plates were incubated at 28°C. Respiration of carbon sources by microbial populations reduce the tetrazolium dye, causing a colour change which was measured twice daily over a period of 5–10 days at 590 nm to determine average well colour development in each plate (Winding and Hendriksen, 1997). The functional diversity of the soil microbial populations was determined using the amount and equitability of carbon substrates metabolized as indicators of richness and evenness, respectively (Garland and Mills, 1991). The functional diversity of soil microbial communities was quantified with the Shannon-Weaver substrate diversity (H') and evenness (E) indices by calculating the number of different carbon sources utilized by the microbial communities and the degree to which the different substrates were utilized (Magurran, 1988; Lupwayi *et al.*, 2001), respectively. The evenness index was used as a measure of how evenly the different microbial species are distributed within soil microbial populations (Zak *et al.*, 1994).

Determination of soil microbial enzymatic activity

The ability of the soil microbial population to mineralize carbon, phosphorus and nitrogen, was assayed by measuring the ß-glucosidase, alkaline phosphatase, acid phosphatase, and urease activities in the soil. Collected soil samples were air-dried at 40°C for 48 h and sieved (2 mm) before analyses. β-Glucosidase and phosphatase activities were calculated according to Dick *et al.* (1996), by spectrophotometrically determining the release of *p*-nitrophenyl after the incubation of soil with *p*-nitrophenyl glucoside and *p*-nitrophenyl phosphate, respectively, at a wavelength of 410 nm. Urease activity was determined using the method of Kandeler and Gerber (1988), where released ammonia was spectrophotometrically measured after the incubation of soil samples with a urea solution at a wavelength of 690 nm. The results were calculated with reference to the respective enzyme reaction calibration curves.

Statistical analysis of functional diversity and enzyme activity data

Data were subjected to non-parametric statistical analysis using STATISTICA 13 (StatSoft, Inc©). CSUP were statistically analysed by cluster analyses (vertical hierarchical tree plots). A dendrogram was constructed using Ward's clustering algorithm, and the Euclidean distance measured (i.e. the geometric distance between variables in a multidimensional space). Homogenous grouping with Fisher least significant difference (LSD) was calculated at $p < 0.05$ for CSUP as well as soil microbial enzymatic activity. Biodiversity was determined using the Shannon-Weaver diversity index and evenness index, indicating species richness and abundance, respectively.

Nematode dynamics

Nematodes were extracted from the soil by means of the Cobb's decanting and sieving method (Cobb, 1918), and identified to family level. After identification, nematodes were assigned to five different feeding (trophic) groups according to Yeates *et al.* (1993). To each identified family from the various trophic groups such as bacterivore (Ba), fungivore (Fu), carnivore (Ca), omnivore (Om) and herbivore (He) was assigned a colonizer-persister (cp) value (a subscript next to abbreviation of trophic group immediately after each family name) according to Bongers (1990) and Bongers and Ferris (1999). The weighted faunal analysis concept was applied, without plant-feeders, to determine the structure, basal, and enrichment conditions of the soil food web, also the decomposition channel of nutrients was determined (Ferris *et al.*, 2001; Pattison *et al.*, 2008; Kapp, 2013). The soil food web enrichment index (EI), structure index (SI), basal index (BI) and channel index (CI) were calculated as described by Pattison *et al.* (2008) and Kapp (2013).

Analysis of nematode dynamics

Descriptive analysis of nematode functional guild data was done using Microsoft Excel 2016. Multivariate analyses were performed on nematode family data using canonical correspondence analysis (CCA) in xlstat-r 2018 (R version 3.5.0) (Addinsoft, Copyright Company).

Results

Carbon source utilization profiles

The effects of the different treatments on the active bacterial functional diversity are illustrated by means of principal component analysis (PCA) in Fig. 21.1. The cowpea treatments (ORCP, CRCP) clustered mainly to the left, whereas most of the cabbage treatments (ORCA, CMCA, ConCA, CRCA) clustered mainly to the right and the sweet potato treatments (ORSP, CRSP) to the bottom of the graph.

Since the 2-D ordination of the PCA might be unclear in demonstrating the distinction between groups, cluster analysis was performed as an alternative measure to enable a 2-D visualization of the different groups illustrated in Fig. 21.1. A dendrogram was constructed with the aid of

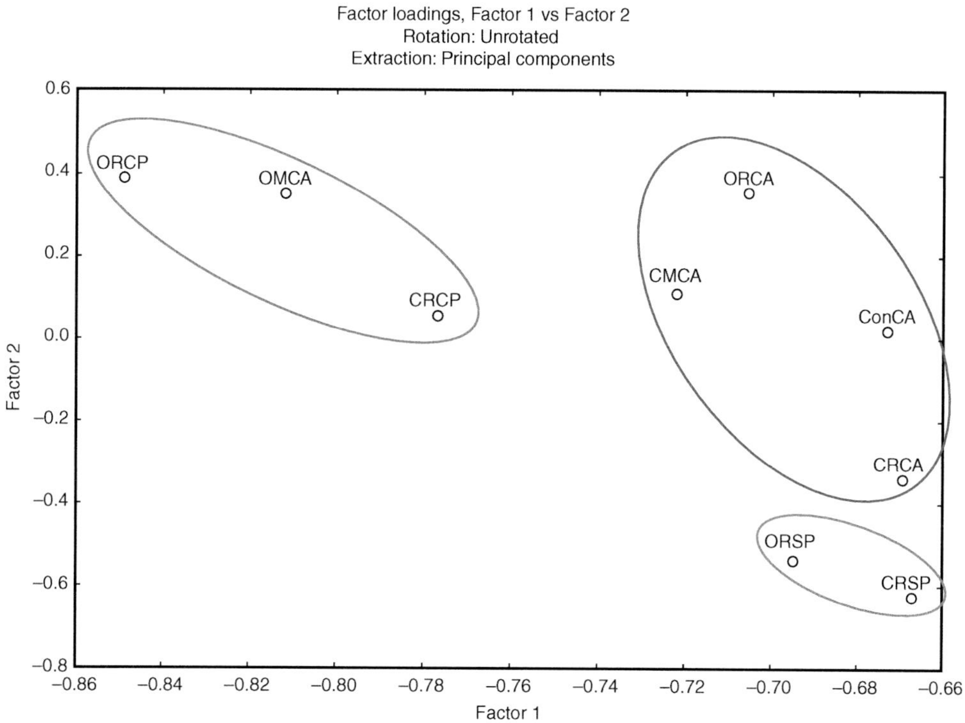

Fig. 21.1. Principal components analysis ordination plot illustrating the differences in the average carbon source utilization profiles in the various conventional and organic cropping system treatments. CMCA, conventional monocrop cabbage; OMCA, organic monocrop cabbage; CRCA, conventional rotation cabbage; ORCA, organic rotation cabbage; CRCP, conventional rotation cowpea; ORCP, organic rotation cowpea; CRSP, conventional rotation sweet potato; ORSP, organic rotation sweet potato; ConCA, control.

cluster analysis to assign treatments into groups, so that treatments in the same cluster are more similar to each other, compared to treatments in other clusters as illustrated in Fig. 21.2. The dendrogram (Fig. 21.2) revealed that the carbon sources utilized in the sweet potato treatments were very similar, thus clustering to the right of the graph. The carbon sources utilized under the various cabbage trials clustered with the cowpea treatments in the middle of the graph. Irrespective of the crop, carbon source utilization was influenced by the fertilizer treatment, with organic and conventional treatments clustering separately from the corresponding crop and rotation practice. It is interesting to note that the carbon sources utilized in two cabbage treatments (ORCA, CMCA) were very similar to each other, but also separate from the rest of the treatments based on fertilizer and rotation practice. No significant differences in CSUPs were observed between the CRCA, CMCA and the control treatments. However, the OMCA treatment differed significantly from the CRCA and the control. No significant differences were observed in the CSUP between the monocropping and rotational treatments, nor between the organic and conventional treatments applied to the same crops.

Microbial diversity indices

Species richness, as well as the abundance (evenness index) of the different kinds of soil bacteria present in the soil community, is shown in Fig. 21.3. The OMCA, CRCP, CRCA and ORCP treatments clustered in the 'healthy soil quadrant', whereas the remaining treatments clustered in the 'transition quadrant'. The closest representation of a healthy soil was demonstrated by the

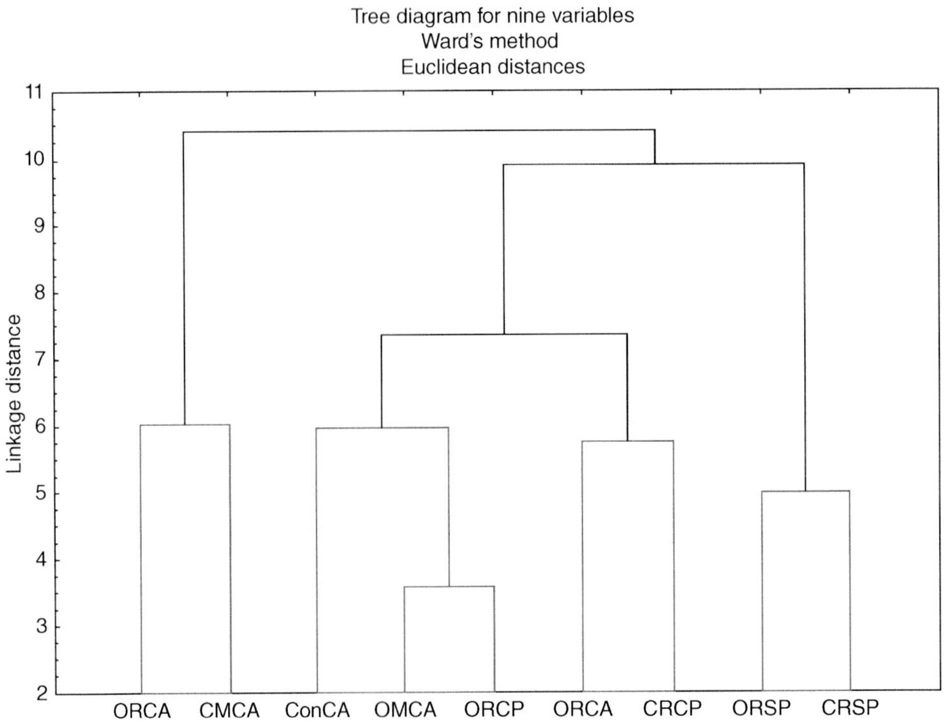

Fig. 21.2. Dendrogram illustrating the carbon source utilization profiles in various conventional and organic cropping system treatments. CMCA, conventional monocrop cabbage; OMCA, organic monocrop cabbage; CRCA, conventional rotation cabbage; ORCA, organic rotation cabbage; CRCP, conventional rotation cowpea; ORCP, organic rotation cowpea; CRSP, conventional rotation sweet potato; ORSP, organic rotation sweet potato; ConCA, control.

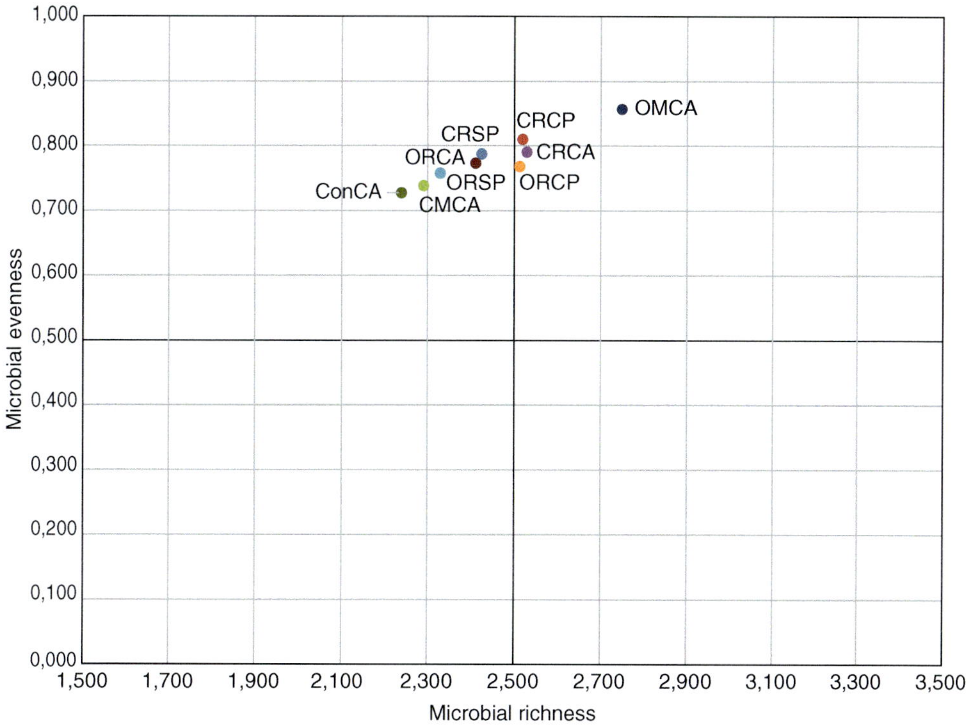

Fig. 21.3. The average soil microbial diversity profile representing the microbial richness and evenness in various conventional and organic cropping system treatments. CMCA, conventional monocrop cabbage; OMCA, organic monocrop cabbage; CRCA, conventional rotation cabbage; ORCA, organic rotation cabbage; CRCP, conventional rotation cowpea; ORCP, organic rotation cowpea; CRSP, conventional rotation sweet potato; ORSP, organic rotation sweet potato; ConCA, control.

soil microbial diversity in the OMCA treatment, whereas the control treatment demonstrated the lowest microbial richness and abundance/evenness of all the treatments.

The Shannon-Weaver substrate diversity index (H') indicated that, depending on the treatments, varying amounts of carbon sources were utilized, with values ranging from 2.24 to 2.75 (Table 21.1). Substrate evenness indices obtained in this analysis ranged between 0.73 and 0.86 (Table 21.1). According to the diversity indices, the species richness and abundance did not differ significantly between treatments, except for the OMCA treatment demonstrating a significantly higher species richness and abundance compared with the CMCA and control treatments. The monocrop cabbage treatment was the only treatment exhibiting a higher soil microbial diversity in the organic treatments compared with the conventional treatments. All the rotational treatments exhibited a higher microbial diversity under the conventional treatments than under the organic treatments. Setting aside the management practices, it would seem that the cowpea treatments exhibited the highest soil microbial diversity, the cabbage treatments exhibited an intermediate diversity, and the sweet potato treatments the lowest microbial diversity.

Soil microbial enzymatic activity

Results of microbial enzyme activity (ß-glucosidase, alkaline phosphatase, acid phosphatase, urease) are presented as a means of determining the potential of a soil to degrade or convert substrates from an organic form into plant-available nutrients (Table 21.2).

Soil microbial communities associated with the various treatments differed in their ability/

potential to mineralize/convert carbon (ß-glucosidase), phosphorus (acid and alkaline phosphatase), and nitrogen (urease). From the results obtained, soil microbial communities in the OMCA treatment exhibited the highest ß-glucosidase, alkaline phosphatase and urease activities, whereas the soil microbial communities in the ConCA treatment exhibited the lowest ß-glucosidase, acid phosphatase and urease activities. No significant differences were observed in phosphatase activity between organic and conventional treatments; urease activity was significantly higher in the ORSP compared with the CRSP treatment; ß-glucosidase activity was significantly higher in the ORCA and OMCA treatments compared with their conventional counterparts, but ß-glucosidase activity was significantly lower in the ORCP treatment compared to its conventional counterpart. On average, all the treatments displayed higher microbial enzyme activity under organic treatments compared with conventional treatments. The exception was the cowpea treatments (CRCP, ORCP) that exhibited higher microbial activity under conventional treatment compared with organic treatment, although the difference was not statistically significant.

Table 21.1. Shannon-Weaver diversity index illustrating soil microbial species richness and the evenness index illustrating soil microbial species abundance in various conventional and organic cropping system treatments.

Treatment[a]	Shannon-Weaver (H')[b]	Evenness (E)[b]
CRSP	2.42[ab]	0.79[ab]
ORSP	2.33[a]	0.76[ab]
CRCP	2.52[ab]	0.81[ab]
ORCP	2.51[ab]	0.77[ab]
CRCA	2.53[ab]	0.79[ab]
ORCA	2.41[ab]	0.78[ab]
CMCA	2.29[a]	0.74[a]
OMCA	2.75[b]	0.86[b]
ConCA	2.24[a]	0.73[a]

[a]CMCA, conventional monocrop cabbage; OMCA, organic monocrop cabbage; CRCA, conventional rotation cabbage; ORCA, organic rotation cabbage; CRCP, conventional rotation cowpea; ORCP, organic rotation cowpea; CRSP, conventional rotation sweet potato; ORSP, organic rotation sweet potato; ConCA, control.
[b]Means followed by different letters differ significantly ($p < 0.05$).

Nematode dynamics

Effect of cropping systems on nematode trophic group and family distribution

A total of 17 nematode families were identified from the field trial. From these families, six (35.29%) were bacterivorous (highest in OMCA treatment), five (29.41%) herbivorous (highest in CRCP treatment), four (23.53%) fungivorous (highest in CRCA treatment), while omnivorous

Table 21.2. Soil microbial enzymatic activities in various conventional and organic cropping system treatments.[a]

Treatments[b]	ß-Glucosidase activity (p-nitrophenol µg/g/h)	Alkaline phosphatase activity (p-nitrophenol µg/g/h)	Acid phosphatase activity (p-nitrophenol µg/g/h)	Urease activity (NH_4-N µg/g/2h)
CRSP	937,739[a]	1493,028[a]	1645,727[ab]	45,843[c]
ORSP	1013,836[abc]	1836,003[ab]	1635,139[ab]	62,431[ab]
CRCP	1114,304[bcd]	1671,787[ab]	1655,180[a]	52,141[ac]
ORCP	943,240[a]	1660,166[ab]	1600,347[ab]	62,281[ab]
CRCA	912,195[a]	1909,414[ab]	1657,552[a]	59,585[abc]
ORCA	1215,664[d]	1867,298[ab]	1628,156[ab]	63,198[ab]
CMCA	966,354[ab]	1802,272[ab]	1659,903[a]	58,717[abc]
OMCA	1144,834[cd]	2149,862[b]	1604,165[ab]	74,644[b]
ConCA	857,984[a]	1581,193[ab]	1582,990[b]	45,503[c]

[a]Means followed by different letters differ significantly ($p < 0.05$).
[b]CMCA, conventional monocrop cabbage; OMCA, organic monocrop cabbage; CRCA, conventional rotation cabbage; ORCA, organic rotation cabbage; CRCP, conventional rotation cowpea; ORCP, organic rotation cowpea; CRSP, conventional rotation sweet potato; ORSP, organic rotation sweet potato; ConCA, control.

(highest in CRCA treatment) and carnivorous (highest in OMCA treatment) at 5.88%, constituted one family each. The higher diversity was observed in the OMCA treatment which had 14 families (82.35% diversity), CRSP treatment had 12 (70.59% diversity), CMCA treatment and CRCP treatment had ten nematode families each (58.92% diversity), while CRCA treatment had nine families (52.94% diversity). All other treatments had a total of 11 nematode families each (64.71% diversity).

The distribution of nematode families in the various treatments is shown in Fig. 21.4. The Aphelenchidae (Fu_2) at 28.90%, Hoplolaimidae (He_3) at 24.04%, Cephalobidae (Ba_2) at 15.61% and Tylenchidae nematode families (He_2) at 12.75%, were found to be the most dominant in the trials. The lowest numbers were the Longidoridae (He_5) at 0.02%, Mononchidae (Ca_4) at 0.05%, Panagrolaimidae (Ba_1) at 0.07% and the Dipterophoridae families (Fu_3) at 0.11%. Only Rhabditidae (Ba_1), Cephalobidae (Ba_2), Aphelenchidae (Fu_2), Tylenchidae (He_2) and Hoplolaimidae (He_3) were found in all treatments. Panagrolaimidae (Ba_1) and Longidoridae (He_5) families were only restricted to the OMCA treatment. On the other hand, CRCP (58.75%) and OMCA (41.25%) were the only treatments to have the Diplogasteridae family (Ba_1). Members of the Monhysteridae (Ba_2) were only found in four of the treatments, namely OMCA (39.59%) and ORCA (38.57%), CRCA (12.97%) and CRSP (8.87%). There were only three treatments with members of the Mononchidae family (Ca_4), namely the OMCA (45.10%) and ORCA (34.31%) and the CRSP (20.59%) treatments. As a consequence of the cropping systems applied, only ORCP (81.55%) and ORSP (18.45%) had members of the Dipterophoridae family (Fu_3). The only treatments to have the Alaimidae family (Ba_4) were ORSP (45.89%), ConCA (30.31%), CMCA (17.85%) and CRSP (5.95%).

From the CCA (Fig. 21.5) eigenvalues from the PCA analysis show that the first two principal components (PC) accounted for 69.61% of the variance of data (PC1: 42.98%, PC2: 26.63%). In the CCA, the ConCA, CMCA and CRCA treatments appeared to relate to each other. Most of

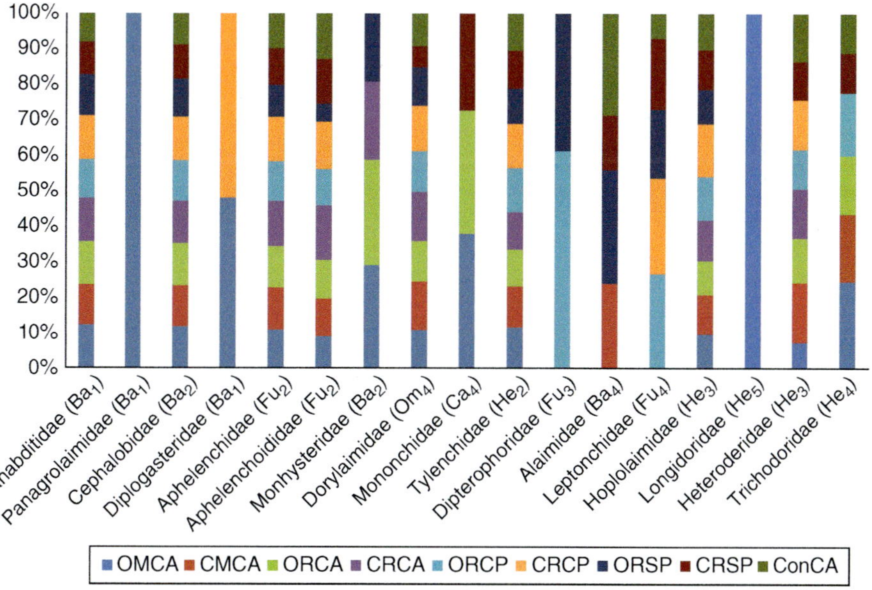

Fig. 21.4. Distribution of nematode families in various conventional and organic cropping system treatments (CMCA, conventional monocrop cabbage; OMCA, organic monocrop cabbage; CRCA, conventional rotation cabbage; ORCA, organic rotation cabbage; CRCP, conventional rotation cowpea; ORCP, organic rotation cowpea; CRSP, conventional rotation sweet potato; ORSP, organic rotation sweet potato; ConCA, control). Ba, bacterivore; Ca, carnivore; Fu, fungivore; He, herbivore; Om, omnivore; numeric value next to nematode family = colonizer-persister (cp) value.

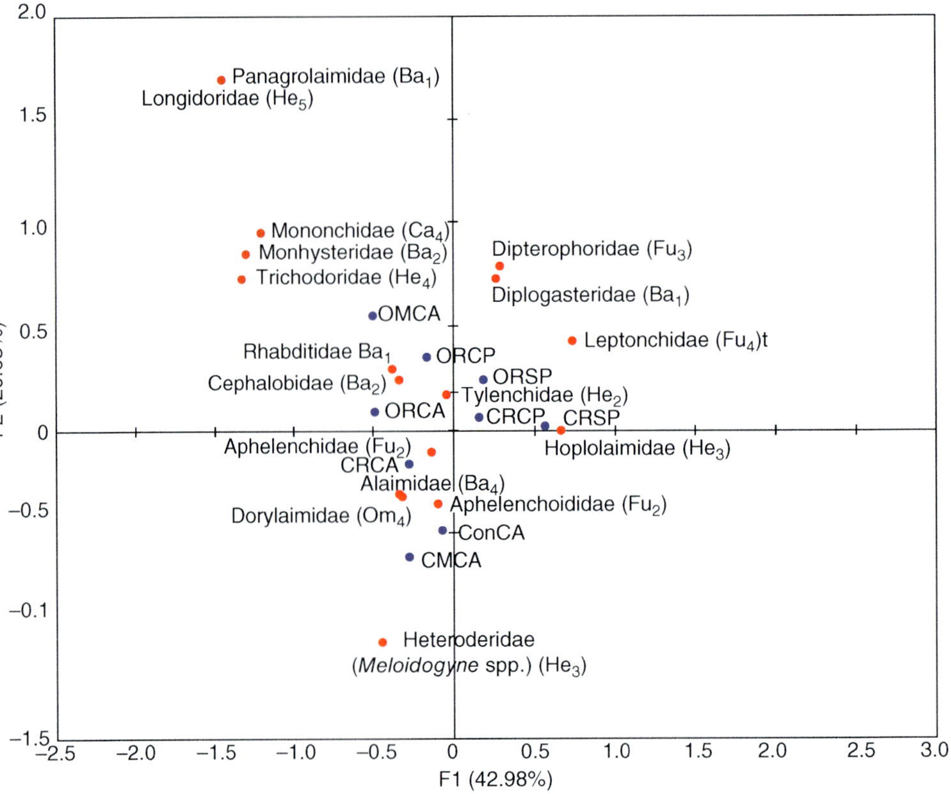

Fig. 21.5. Correspondence analysis of various conventional and organic cropping system treatments associations with nematode families. Cropping system treatments are: CMCA, conventional monocrop cabbage; OMCA, organic monocrop cabbage; CRCA, conventional rotation cabbage; ORCA, organic rotation cabbage; CRCP, conventional rotation cowpea; ORCP, organic rotation cowpea; CRSP, conventional rotation sweet potato; ORSP, organic rotation sweet potato; ConCA, control. Ba, bacterivore; Ca, carnivore; Fu, fungivore; He, herbivore; Om, omnivore; numeric value next to nematode family = cp value.

the organic treatments, OMCA, ORCA and ORSP also seemed closely related. On the other hand, the cowpea treatments ORCP and CRCP closely related to the CRSP treatment and differed from the grouping of the ConCA treatments. An association between the ORCP, CRCP and CRSP group to the nematode families Diplogasteridae (Ba_1), Dipterophoridae (Fu_3), Leptonchidae (Fu_4) and Hoplolaimidae (He_3) seemed to be evident from the CCA. This grouping had the fewest nematode families in association. The ConCA, CMCA and CRCA treatments were strongly associated with the Aphelenchoididae (Fu_2), Aphelenchidae (Fu_2), Heteroderidae (*Meloidogyne* sp.) (He_3), Dorylaimidae (Om_4) and Alaimidae (Ba_4) nematode families. Higher diversity was shown by the OMCA, ORCA and ORSP treatment grouping with an association of eight families, namely the Tylenchidae (He_2), Rhabditidae (Ba_1), Cephalobidae (Ba_2), Monhysteridae (Ba_2), Mononchidae (Ca_4), Trichodoridae (He_4), Longidoridae (He_5) and Panagrolaimidae (Ba_1) families.

Basal, structure, channel and enrichment conditions

All but three treatments were located within Quadrant A of the faunal analysis (Fig. 21.6). The three exceptions which were located within Quadrant D were the ORCA, ORCP and the ConCA treatments. Of the other treatments located in Quadrant A, ORSP seemed to be separated from the other treatments and was the most superior in both the EI and SI (Table 21.3). The other treatments were located closer to each other in a cluster just above the 50% mark of the EI.

The OMCA treatment was next in line in terms of the EI followed by CRCA treatment. The ConCA treatment had the lowest EI of all the treatments closely followed by the ORCP and ORCA treatments in that order. However, in

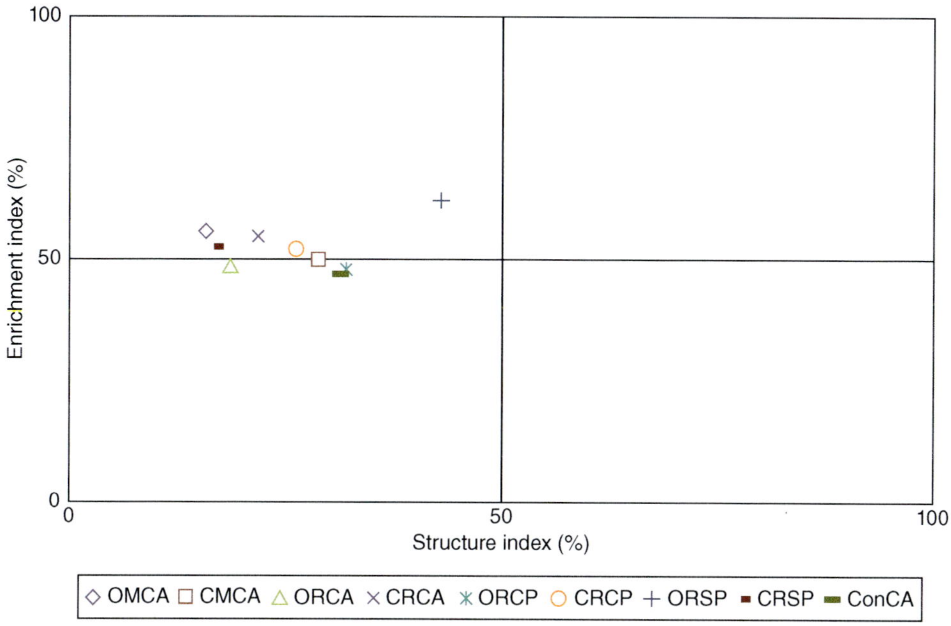

Fig. 21.6. Nematode faunal profile of the nematodes extracted from various conventional and organic cropping system treatments (CMCA, conventional monocrop cabbage; OMCA, organic monocrop cabbage; CRCA, conventional rotation cabbage; ORCA, organic rotation cabbage; CRCP, conventional rotation cowpea; ORCP, organic rotation cowpea; CRSP, conventional rotation sweet potato; ORSP, organic rotation sweet potato; ConCA, control).

Table 21.3. Structure, enrichment, channel and basal nematode indices as derived from the treatments of the Mandela Trials.

Treatments[a]	Structure index (%)	Enrichment index (%)	Channel index (%)	Basal index (%)
CRSP	16.66	52.91	61.11	43.25
ORSP	42.87	62.42	32.75	26.91
CRCP	26.00	52.50	62.99	39.67
ORCP	31.72	48.43	59.74	41.56
CRCA	21.58	55.01	65.01	40.34
ORCA	18.33	49.21	50.40	45.16
CMCA	28.52	50.33	65.56	39.08
OMCA	15.52	56.17	39.23	40.03
ConCA	31.21	47.42	72.97	42.05

[a]CMCA, conventional monocrop cabbage; OMCA, organic monocrop cabbage; CRCA, conventional rotation cabbage; ORCA, organic rotation cabbage; CRCP, conventional rotation cowpea; ORCP, organic rotation cowpea; CRSP, conventional rotation sweet potato; ORSP, organic rotation sweet potato; ConCA, control.

terms of SI, the ORCP treatment was second to the ORSP treatment, while the ConCA and CMCA treatments came in third and fourth, respectively. On the other hand, OMCA, CRSP and ORCA treatments had the lowest SI values. The CI (Table 21.3) for the ConCA treatment was the highest followed by that of CMCA and CRCA treatments. The only CI values below 50% were recorded for the ORSP and the OMCA treatments. In terms of the BI, the highest values were reported from the ORCA, CRSP and ConCA treatments. The lowest values of BI were recorded for the ORSP, CMCA and CRCP treatments.

Discussion

In the present study, selected methodologies (Biolog EcoPlate™, enzyme activity and nematode functional guild analysis) were utilized to assess the effect of organic and conventional cropping systems on selected biological soil health indicators. It must be noted that these effects manifested over a period of 4 years and that more profound shifts could be manifested over a longer period of time (Bardgett and van der Putten, 2014; Tsiafouli *et al.*, 2015).

Effect of cropping systems on carbon source utilization profiles

The Biolog EcoPlate™ method has its fair share of criticism, and work by Yao *et al.* (2000) showed how the pH of the growth media and carbon sources used can affect results as it may not be a representation of the soil from which microorganisms were isolated. None the less, although only restricted to comparative assessment between contaminated or uncontaminated soils (Gryta *et al.*, 2014, Liu *et al.*, 2015), it is still widely used and especially useful when applied in conjunction with other methods (Dong *et al.*, 2008). The mechanism of colour development in Biolog EcoPlates™ is related to differences in carbon source utilization (i.e. food source consumption) which, in turn, appears to relate to the number of viable microorganisms able to utilize the substrates ('food sources') within the wells of the EcoPlate as a sole carbon source (Garland and Mills, 1991).

Results clearly indicated differences in CSUPs between the different sampled treatments. This change also implies a change in microbial functioning between the different cropping systems. The variations in the CSUPs between the two cropping systems may potentially have been influenced by the soil pH, of which some authors consider it to be a 'master variable' (Dick *et al.*, 2000). Tautges *et al.* (2016) and Ellis *et al.* (2001) reported slight increases in pH as a result of organic amendments which may have also had a positive effect on microbial biomass in their studies. As reported in Chapter 22 of this volume, this was also the case in the Mandela Trials. In most cases, however, pH has a greater influence on the bacterial abundance (narrow pH range for optimum growth) than the fungal communities (Bittman *et al.*, 2005; Rousk *et al.*, 2010).

Although in some studies, like those by Berg *et al.* (2002) and Huguet and Rudgers (2010), crop type had an effect on the microbial community, in our studies, the crop type did not influence the microbial biomass, similarly to work by Jangid *et al.* (2011) and Ladygina and Hedlund (2010). Rather, the management system had a greater influence on microbial biomass (Snapp *et al.*, 2010).

Different CSUPs could also be attributed to the stimulation of indigenous soil microbial populations as a result of easily accessible carbon food sources after organic fertilizer application, in the absence of rotational crops. Similar to our study, Tautges *et al.* (2016) and Fließbach *et al.* (2000) also established that organic amendments resulted in a greater CSUP compared with the conventional treatments. These findings and others from Lagomarsino *et al.* (2009), consistent with our findings, subscribe to the hypothesis that a more diverse community is more efficient in resource utilization (Mäder *et al.*, 2002; Shannon *et al.*, 2002; Ros *et al.*, 2006).

Diversity indices

According to the literature, a healthy soil is characterized by the presence of a high number (species richness), as well as the high abundance (evenness index) of the different kinds of soil bacteria present in a community (Sharma *et al.*,

2010) as shown in Fig. 21.3. The Shannon-Weaver substrate diversity index (H') is used to quantify the functional diversity of soil microbial communities based on the number of different carbon sources utilized by soil microbial communities in Biolog EcoPlates™ (i.e. comparable to species richness in the soil). Values of the index typically range between 1.5 and 3.5, but rarely increase above 4.5 for ecological studies.

The high Shannon-Weaver and evenness index suggests that the organic amendments contributed to total soil organic matter (SOM) and potentially increased microbial activity, similar to findings by Gomez *et al.* (2006) and Jangid *et al.* (2008). As a result, this would have a strong impact on the availability of nutrients to the crops growing in association due to higher SOM turnover (Tautges *et al.*, 2016). Also, with higher evenness, there is an overall high resistance to environmental stress (Wittebolle *et al.*, 2009). The choice of fertilizer amendment seemed to have a stronger influence on microbial diversity than the crop rotation systems, similar to results by Li *et al.* (2012). On the other hand, the evenness index is used as an indication of how abundant species are within a soil microbial community (i.e. how close in 'numbers' the different microbial species are in a soil microbial community).

If the abundance of different species in a community is measured, it will invariably be found that some species are rare, whereas others are more abundant/dominant. Substrate evenness assumes a value between 0 and 1, with 0 representing a state where the community comprises one species, while 1 would represent a situation in which all species are equally abundant within a microbial population present in the samples. This means less variation in microbial populations between species, thus, less dominance, and higher diversity. The monocrop cabbage treatment was the only treatment exhibiting a higher soil microbial diversity and evenness in the organic treatments compared to the conventional treatments. This could possibly be attributed to the higher microbial activity as shown by the higher CSUP in this treatment.

Soil microbial enzymatic activity

Enzymatic activity is a relatively sensitive indicator for soil quality/fertility (Swanepoel *et al.*, 2017), hence it is ideal for comparative studies. By implication, the higher the microbial activity (i.e. mineralization rate), the faster the nutrient release from organic substrates that would be made available for plant root uptake. In tandem with the CSUP, the OMCA treatment had the greater enzyme activity, in part due to the greater microbial biomass. In summary, however, enzyme activity was mostly comparable between cropping systems, indicating that management systems did not have a big effect on enzyme activity, similar to findings by García-Ruiz *et al.* (2009). Some studies have, however, accredited higher enzyme activity to seasonal variations during sampling (Schloter *et al.*, 2003).

The production of extracellular enzymes is determined in part by the bacterial and fungal biomass, biological state, structure of the various species and several environmental factors (Allison *et al.*, 2007). Changes in the fungal and bacterial communities in the soil therefore have serious implications on the microbial turnover and nitrogen mineralization (Bittman *et al.*, 2005). Organic amendments have also been reported to increase microbial biomass nitrogen. This may be one other reason why organic treatments had greater enzyme activity in our study compared with conventional treatments. Studies by Bittman *et al.* (2005) and Marinari *et al.* (2010) revealed that elevated levels of nitrogen and carbon fluxes resulted from organic matter application. This increase can directly correlate to the higher urease and ß-glucosidase activity observed in all organic treatments when compared with the conventional treatments in our study. In support of this observation, Lagomarsino *et al.* (2009) reported higher mineralization of nitrogen with organic treatments over the other conventional treatments.

The enzyme ß-glucosidase is derived predominantly from the soil microbial biomass and plays a major role in the degradation of cellulose and carbon cycling (Turner *et al.*, 2002). During this breakdown, labile carbon energy sources are released for the soil food web to utilize (Tautges *et al.*, 2016), hence the higher microbial activity. For that reason, higher ß-glucosidase may be related to the greater SOM from organic systems, which in turn encouraged the greater numbers of fungi and bacteria that produce this very enzyme. Lower enzyme numerical numbers from most of the conventional treatments compared with the organic treatments may have resulted due to direct toxicity of agrochemicals. Incubation trials by Floch *et al.* (2011) showed the sensitivity of enzymes to pesticide dosage applications.

An increase in pH as a result of organic amendments may have also improved the alkaline-phosphatase enzyme production. Acid and alkaline phosphatase are both predominantly found in acid and neutral-to-alkaline soils, respectively, and were used successfully by Dick *et al.* (2000) as pH indicators. From these findings, we can hypothesize that the higher numerical values of the acid phosphatase from the conventional treatments would indicate a situation more towards acid soil conditions. On the other hand, the higher values of acid phosphatase from organic treatments could mean that soil conditions were more or less changing towards neutral. From our results, the cropping systems did not indicate excessive shortages of phosphorus, as phosphatase activity was not significantly higher in any one cropping system, apart from the ConCA treatment with acid phosphatase. Acid phosphatase is an enzyme secreted by plant roots in a situation when there is phosphorus deficiency in the soil (Hayes *et al.*, 1999), in reaction to varying degrees of soil disturbance (Swanepoel *et al.*, 2017). As the available soil P levels were much lower for the ConCA (Chapter 22, this volume), this seems to accord with the actual available soil P situation.

Nematode population dynamics

There were obvious variations in nematode dynamics between the cropping systems in our study. Normally, nematode community shifts are influenced by crop species, crop rotation and rhizosphere condition (Scharroba *et al.*, 2016). Differences in nematode population dynamics between the different cropping system treatments in our study may be attributed to these factors. However, previous studies on comparative farming systems revealed that specific nematode community shifts can exceed crop-related assemblage shifts (Quist *et al.*, 2016). For instance, bacterivore and fungivore nematode communities have been reported to correlate with SOM (Zhang *et al.*, 2017) and cover crops (Ferris *et al.*, 2012). Similar to our findings, most of these families were dominant in organic treatments, alluding to the possibility of fertilizer treatment and cover cropping having a stronger influence than crop type.

Direct toxicity from ammonium-based fertilizers (Hopkins and Shiel, 1996; Tenuta and Ferris, 2004) has been attributed to reduced numbers of the bacterivore (Bulluck *et al.*, 2002) and fungivore (Zhang *et al.*, 2016) nematode families. This may be the reason that bacterivorous nematodes were most prevalent in the organic treatments. Other studies also concurred with our findings (Sánchez-Moreno *et al.*, 2009). Predacious and omnivorous nematodes (including those that rank a high cp value) are generally considered to be quite sensitive to disturbance (Bongers and Ferris, 1999). The only reason some were found in our trial site may be the high availability of prey for carnivorous nematodes and/or the increase in available food sources for omnivorous nematodes (Mills and Adl, 2011), as all systems were highly disturbed through tillage practices.

Zhao and Neher (2013) conducted a study to try to identify nematode guild indicators as a result of agricultural management systems. Normally, a disturbance in the form of organic amendments is usually accompanied by a flush in Ba_1 and some Ba_2 nematode communities (Ferris *et al.*, 2001; Mills and Adl, 2011). Hence, the dominance of the Ba_1 and Ba_2 functional guilds in the organic treatments (Fig. 21.6) indicates a flush in carbon food sources as earlier findings from the microbial diversity assessment proposed. This enrichment improved the overall microbial biomass, making prey available to the nematodes, which caused their numbers to increase (Bongers and Bongers, 1998). Also on a positive note, nutrients immobilized in the prey are later released during the feeding behaviours of nematodes (Irshad *et al.*, 2011), making them available for absorption by plants. As a result, nematodes are also important in nutrient cycling.

High numbers of the *Meloidogyne* spp. in association with legumes have been correlated with reduced nitrogen fixation (Ibewiro *et al.*, 2000). In their study also, there was an inverse relationship between *Meloidogyne* spp. and other plant parasitic nematodes. These findings were in line with our results, where the *Meloidogyne* spp. were inversely related to other He_3 nematodes (Fig. 21.5). The CCA (Fig. 21.5) also indicated a close correlation between the treatments Con-CA, CMCA and CRCA to basal nematodes Aphelenchidae (Fu_2) and Aphelenchoididae (Fu_2). Because of their dominance we assume that these treatments are experiencing stressful conditions (Ferris *et al.*, 2001). This may be a result of the chemical formulations used to control pest and diseases with these trials. In another instance, the organic treatments apart from the ORCP treatment, correlated with the most nematode families (Fig. 21.6), indicating that these treatments encourage diversity, among other things.

Nematode indices

A study found nematode community dynamics to be more affected by pH than heavy metal contamination (Ellis *et al.*, 2001). Korthals *et al.* (1996) reported that the CI was much higher under low soil pH conditions. This observation may explain why the CI is high with most conventional treatments in our study as the pH may have decreased over time. This directly correlates with a higher fungal community and supports our earlier hypothesis that the fungal community has a higher tolerance to fluctuating pH ranges compared with the bacterial community, which is expected to decline under such conditions. In other studies also, long-term exposure to heavy metals had a negative effect on nematode dynamics (Zhao *et al.*, 2011). Conventional systems have long been linked to high heavy metal accumulation in the long term. This may have had a negative influence on the nematode genera in the conventional treatments of our study.

From the faunal analysis in our study (Fig. 21.6), Quadrant A is characterized by nutrient-enriched, highly disturbed soils comprising mostly opportunistic nematodes. Ferris *et al.* (2001) described soil under annual cropping systems to map on the left side of the faunal profile, unlike perennial cropping systems that normally have the opposite effect. This means it was expected that in our study the cropping systems would map on either Quadrant A and/or D (Fig. 21.6). All treatments that mapped in Quadrant A are also characterized by a bacterial decomposition channel, low C:N ratio and a disturbed soil food web. Quadrant D, on the other hand, is commonly associated with stressed, nutrient-depleted, fungal decomposition channel, high C:N ratio and degraded food web conditions (Ferris *et al.*, 2001).

A high SI depicts a system that is more resilient with a buffering that would enable a regulatory effect on herbivorous families (Ferris *et al.*, 2001). This would result in fewer disease incidences resulting from pathogenic nematodes, and a resultant healthy root system for crops growing in such soils. The highest EI recorded in the ORSP and OMCA treatments indicate that these treatments were more fertile (enriched), having greater numbers of enrichment indicators such as bacterivorous and fungivorous nematodes, relative to other nematode families. The most abundant nematodes under stressful conditions (basal conditions) are the cp-2 guilds (Cephalobidae (Ba_2), Aphelenchidae (Fu_2) and Aphelenchoididae (Fu_2)). In our study the treatments with the highest numbers of these nematode families were mostly those associated with low SI, mapping in Quadrant D. Based on this, these treatments were characterized by a disturbed food web, meaning among other factors their basal index was high. This is characteristic of a resource-limited system, under adverse environmental conditions or recently contaminated, as seen by the high numbers of cp-2 nematodes (Ferris *et al.*, 2001). The Fu_2 nematodes are also indicative of a system where the fungal community is supported by organic sources (Ferris *et al.*, 2001).

Conclusion

The results thus far obtained indicate that biological soil health indicators are indeed sensitive to the different cropping systems, that is, fertilizer treatments and to some extent crop type. Soil biological properties are quite sensitive to changes in management systems and far precede detectable changes in edaphic soil properties. This makes them ideal indicators of a healthy soil, within the context of a healthy soil comprising a system that fosters long-term sustainability through ideal biological, chemical and physical soil properties without damage to the environment.

In our study, management practices precipitated different reactions from the soil food web, influencing their content, diversity and abundance within various communities. The combined use of microbial and nematode analyses can become a very useful approach in deciphering the sustainability of a management system. The observation that microbial and nematode data were directly linked was of importance, since the community shifts between the different indicators correlated. In turn, this led to identical conclusions, that the organic treatments supported a greater and more diverse soil food web compared with the conventional treatments. All the measured treatments resulted in a healthier soil profile than the control treatment, which accords with the very low yields of this control treatment (see Chapter 22, this volume). The OMCA seemed to be the stand-alone treatment in terms of superiority, while the ConCA was the poorest treatment in all aspects measured.

Conversely, the rotation treatments of the conventional system seemed to challenge most of the rotation organic treatments although no significant differences were recorded. The toxicity from the agrochemicals used under conventional systems proved to have a negative effect on the biological diversity, this is clearly indicated by the CCA and nematode results from both cropping systems in question. This stresses the importance of nematode data and accompanying nematode indices, as they showed more sensitivity to cropping systems, compared with other indicators used in this study. It was expected that biodiversity would be amplified by the rotation treatments, however, this was not the case, nor was it evident in Chapter 20 of this volume, and the nematode study accompanying this (van Niekerk, 2018). The trial being in its fourth year during sampling, it could be assumed that in the long term, these rotation treatments would surpass the monocrop treatments in terms of biodiversity and resilience.

High diversity contributes to a more resilient soil food web that could potentially be more efficient at performing ecosystem services. Ideally, soil food web dynamics should be monitored over an extended period of time in order to attain a more complete reflection of the impact that different cropping systems might have on microbial diversity and activity as an indicator of soil fertility and health. The current study is part of a larger programme in which the data reported here will be combined with other soil health indicators that can be anticipated to provide a more comprehensive picture of the effects of the cropping systems on soil health (Chapter 22, this volume).

References

Allison, V., Condron, L., Peltzer, D., Richardson, S. and Turner, B. (2007) Changes in enzyme activities and soil microbial community composition along carbon and nutrient gradients at the Franz Josef chronosequence, New Zealand. *Soil Biology and Biochemistry* 39, 1770–1781.

Bardgett, R.D. and van der Putten, W.H. (2014) Belowground biodiversity and ecosystem functioning. *Nature* 515, 505.

Berg, G., Roskot, N., Steidle, A., Eberl, L., Zock, A., *et al.* (2002) Plant-dependent genotypic and phenotypic diversity of antagonistic rhizobacteria isolated from different *Verticillium* host plants. *Applied and Environmental Microbiology* 68, 3328–3338.

Bittman, S., Forge, T. and Kowalenko, C. (2005) Responses of the bacterial and fungal biomass in a grassland soil to multi-year applications of dairy manure slurry and fertilizer. *Soil Biology and Biochemistry* 37, 613–623.

Bongers, T. (1990) The maturity index: an ecological measure of environmental disturbance based on nematode species composition. *Oecologia* 83, 14–19.

Bongers, T. (1999) The maturity index, the evolution of nematode life history traits, adaptive radiation and cp-scaling. *Plant and Soil* 212, 13–22.

Bongers, T. and Bongers, M. (1998) Functional diversity of nematodes. *Applied Soil Ecology* 10, 239–251.

Bongers, T. and Ferris, H. (1999) Nematode community structure as a bioindicator in environmental monitoring. *Trends in Ecology & Evolution* 14, 224–228.

Brevik, E., Cerdà, A., Mataix-Solera, J., Pereg, L., Quinton, J., *et al.* (2015) The interdisciplinary nature of SOIL. *Soil* 1, 117.

Briar, S.S., Barker, C., Tenuta, M. and Entz, M.H. (2012) Soil nematode responses to crop management and conversion to native grasses. *Journal of Nematology* 44, 245.

Bulluck, L., Brosius, M., Evanylo, G. and Ristaino, J. (2002) Organic and synthetic fertility amendments influence soil microbial, physical and chemical properties on organic and conventional farms. *Applied Soil Ecology* 19, 147–160.

Buyer, J.S. and Drinkwater, L.E. (1997) Comparison of substrate utilization assay and fatty acid analysis of soil microbial communities. *Journal of Microbiological Methods* 30, 3–11.

Cesarz, S., Reich, P.B., Scheu, S., Ruess, L., Schaefer, M., *et al.* (2015) Nematode functional guilds, not trophic groups, reflect shifts in soil food webs and processes in response to interacting global change factors. *Pedobiologia* 58, 23–32.

Cobb, N. (1918) Estimating the nematode population of soil. *Agricultural Technical Circular* 1. Bureau of Plant Industry, United States Department of Agriculture, Washington, DC.

Cooper, J.E. and Rao, J.R. (2006) *Molecular Approaches to Soil, Rhizosphere and Plant Microorganism Analysis*. CAB International, Wallingford, UK.

De Vries, F.T., Thébault, E., Liiri, M., Birkhofer, K., Tsiafouli, M.A., *et al.* (2013) Soil food web properties explain ecosystem services across European land use systems. *Proceedings of the National Academy of Sciences of the United States of America* 110, 14296–14301.

Dick, R.P., Breakwell, D.P. and Turco, R.F. (1996) Soil enzyme activities and biodiversity measurements as integrative microbiological indicators. In: Doran, J.W. and Jones, A.J. (eds) *Methods for Assessing Soil Quality*. Soil Science Society of America, Madison, Wisconsin, pp. 247–271.

Dick, W., Cheng, L. and Wang, P. (2000) Soil acid and alkaline phosphatase activity as pH adjustment indicators. *Soil Biology and Biochemistry* 32, 1915–1919.

Dong, X., Huai-Ying, Y., De-Yong, G. and Huang, C.-Y. (2008) Soil microbial community structure in diverse land use systems: a comparative study using Biolog, DGGE, and PLFA analyses. *Pedosphere* 18, 653–663.

Ellis, R.J., Neish, B., Trett, M.W., Best, J.G., Weightman, A.J., *et al.* (2001) Comparison of microbial and meiofaunal community analyses for determining impact of heavy metal contamination. *Journal of Microbiological Methods* 45, 171–185.

Feigl, V., Ujaczki, É., Vaszita, E. and Molnár, M. (2017) Influence of red mud on soil microbial communities: application and comprehensive evaluation of the Biolog EcoPlate approach as a tool in soil microbiological studies. *Science of the Total Environment* 595, 903–911.

Fernandez, A.L., Sheaffer, C.C., Wyse, D.L., Staley, C., Gould, T.J., *et al.* (2016) Associations between soil bacterial community structure and nutrient cycling functions in long-term organic farm soils following cover crop and organic fertilizer amendment. *Science of the Total Environment* 566, 949–959.

Ferris, H., Bongers, T. and De Goede, R. (2001) A framework for soil food web diagnostics: extension of the nematode faunal analysis concept. *Applied Soil Ecology* 18, 13–29.

Ferris, H., Sánchez-Moreno, S. and Brennan, E. (2012) Structure, functions and interguild relationships of the soil nematode assemblage in organic vegetable production. *Applied Soil Ecology* 61, 16–25.

Fließbach, A., Mäder, P. and Niggli, U. (2000) Mineralization and microbial assimilation of 14C-labeled straw in soils of organic and conventional agricultural systems. *Soil Biology and Biochemistry* 32, 1131–1139.

Floch, C., Chevremont, A.-C., Joanico, K., Capowiez, Y. and Criquet, S. (2011) Indicators of pesticide contamination: soil enzyme compared to functional diversity of bacterial communities via Biolog® Ecoplates. *European Journal of Soil Biology* 47, 256–263.

Forey, E., Barot, S., Decaëns, T., Langlois, E., Laossi, K.-R., *et al.* (2011) Importance of earthworm–seed interactions for the composition and structure of plant communities: a review. *Acta Oecologica* 37, 594–603.

García-Ruiz, R., Ochoa, V., Vinegla, B., Hinojosa, M., Pena-Santiago, R., *et al.* (2009) Soil enzymes, nematode community and selected physico-chemical properties as soil quality indicators in organic and conventional olive oil farming: influence of seasonality and site features. *Applied Soil Ecology* 41, 305–314.

Garland, J.L. and Mills, A.L. (1991) Classification and characterization of heterotrophic microbial communities on the basis of patterns of community-level sole-carbon-source utilization. *Applied and Environmental Microbiology* 57, 2351–2359.

Gomez, E., Ferreras, L. and Toresani, S. (2006) Soil bacterial functional diversity as influenced by organic amendment application. *Bioresource Technology* 97, 1484–1489.

Gryta, A., Frąc, M. and Oszust, K. (2014) The application of the Biolog EcoPlate approach in ecotoxicological evaluation of dairy sewage sludge. *Applied Biochemistry and Biotechnology* 174, 1434–1443.

Guan, P., Zhang, X., Yu, J., Cheng, Y., Li, Q., *et al.* (2018) Soil microbial food web channels associated with biological soil crusts in desertification restoration: the carbon flow from microbes to nematodes. *Soil Biology and Biochemistry* 116, 82–90.

Hayes, J.E., Richardson, A.E. and Simpson, R.J. (1999) Phytase and acid phosphatase activities in extracts from roots of temperate pasture grass and legume seedlings. *Functional Plant Biology* 26, 801–809.

Henneron, L., Bernard, L., Hedde, M., Pelosi, C., Villenave, C., *et al.* (2015) Fourteen years of evidence for positive effects of conservation agriculture and organic farming on soil life. *Agronomy for Sustainable Development* 35, 169–181.

Hohberg, K. (2003) Soil nematode fauna of afforested mine sites: genera distribution, trophic structure and functional guilds. *Applied Soil Ecology* 22, 113–126.

Hopkins, D. and Shiel, R. (1996) Size and activity of soil microbial communities in long-term experimental grassland plots treated with manure and inorganic fertilizers. *Biology and Fertility of Soils* 22, 66–70.

Huguet, V. and Rudgers, J.A. (2010) Covariation of soil bacterial composition with plant rarity. *Applied and Environmental Microbiology* 76, 7665–7667.

Ibewiro, B., Sanginga, N., Vanlauwe, B. and Merckx, R. (2000) Influence of phytoparasitic nematodes on symbiotic N_2 fixation in tropical herbaceous legume cover crops. *Biology and Fertility of Soils* 31, 254–260.

Irshad, U., Villenave, C., Brauman, A. and Plassard, C. (2011) Grazing by nematodes on rhizosphere bacteria enhances nitrate and phosphorus availability to *Pinus pinaster* seedlings. *Soil Biology and Biochemistry* 43, 2121–2126.

Jangid, K., Williams, M.A., Franzluebbers, A.J., Sanderlin, J.S., Reeves, J.H., *et al.* (2008) Relative impacts of land-use, management intensity and fertilization upon soil microbial community structure in agricultural systems. *Soil Biology and Biochemistry* 40, 2843–2853.

Jangid, K., Williams, M.A., Franzluebbers, A.J., Schmidt, T.M., Coleman, D.C., *et al.* (2011) Land-use history has a stronger impact on soil microbial community composition than aboveground vegetation and soil properties. *Soil Biology and Biochemistry* 43, 2184–2193.

Kandeler, E. and Gerber, H. (1988) Short-term assay of soil urease activity using colorimetric determination of ammonium. *Biology and Fertility of Soils* 6, 68–72.

Kapp, C. (2013) *Nematode Soil Community Structure and Function as a Bio-indicator of Soil Health in Fynbos and Deciduous Fruit Orchards*. Stellenbosch University, Stellenbosch, South Africa.

Korthals, G.W., Bongers, T., Kammenga, J.E., Alexiev, A.D. and Lexmond, T.M. (1996) Long-term effects of copper and pH on the nematode community in an agroecosystem. *Environmental Toxicology and Chemistry* 15, 979–985.

Ladygina, N. and Hedlund, K. (2010) Plant species influence microbial diversity and carbon allocation in the rhizosphere. *Soil Biology and Biochemistry* 42, 162–168.

Lagomarsino, A., Moscatelli, M., Di Tizio, A., Mancinelli, R., Grego, S., *et al.* (2009) Soil biochemical indicators as a tool to assess the short-term impact of agricultural management on changes in organic C in a Mediterranean environment. *Ecological Indicators* 9, 518–527.

Li, R., Khafipour, E., Krause, D.O., Entz, M.H., De Kievit, T.R., *et al.* (2012) Pyrosequencing reveals the influence of organic and conventional farming systems on bacterial communities. *PLoS One* 7, e51897.

Liu, B., Li, Y., Zhang, X., Wang, J. and Gao, M. (2015) Effects of chlortetracycline on soil microbial communities: comparisons of enzyme activities to the functional diversity via Biolog EcoPlates™. *European Journal of Soil Biology* 68, 69–76.

Lupwayi, N., Arshad, M., Rice, W. and Clayton, G. (2001) Bacterial diversity in water-stable aggregates of soils under conventional and zero tillage management. *Applied Soil Ecology* 16, 251–261.

Mäder, P., Fliessbach, A., Dubois, D., Gunst, L., Fried, P., *et al.* (2002) Soil fertility and biodiversity in organic farming. *Science*, 296, 1694–1697.

Magurran, A.E. (1988) *Ecological Diversity and its Measurement*. Springer, New York.

Marinari, S., Lagomarsino, A., Moscatelli, M., Di Tizio, A. and Campiglia, E. (2010) Soil carbon and nitrogen mineralization kinetics in organic and conventional three-year cropping systems. *Soil and Tillage Research* 109, 161–168.

Mills, A. and Adl, M. (2011) Changes in nematode abundances and body length in response to management intensive grazing in a low-input temperate pasture. *Soil Biology and Biochemistry* 43, 150–158.

Minoshima, H., Jackson, L., Cavagnaro, T., Sánchez-Moreno, S., Ferris, H., *et al.* (2007) Soil food webs and carbon dynamics in response to conservation tillage in California. *Soil Science Society of America Journal* 71, 952–963.

Naether, A., Foesel, B.U., Naegele, V., Wüst, P.K., Weinert, J., *et al.* (2012) Environmental factors affect acidobacterial communities below the subgroup level in grassland and forest soils. *Applied and Environmental Microbiology* 78, 7398–7406.

Pattison, A., Moody, P., Badcock, K., Smith, L., Armour, J., *et al.* (2008) Development of key soil health indicators for the Australian banana industry. *Applied Soil Ecology* 40, 155–164.

Powell, J.R., Welsh, A. and Hallin, S. (2015) Microbial functional diversity enhances predictive models linking environmental parameters to ecosystem properties. *Ecology* 96, 1985–1993.

Quist, C.W., Schrama, M., De Haan, J.J., Smant, G., Bakker, J., *et al.* (2016) Organic farming practices result in compositional shifts in nematode communities that exceed crop-related changes. *Applied Soil Ecology* 98, 254–260.

Ros, M., Klammer, S., Knapp, B., Aichberger, K. and Insam, H. (2006) Long-term effects of compost amendment of soil on functional and structural diversity and microbial activity. *Soil Use and Management* 22, 209–218.

Rousk, J., Bååth, E., Brookes, P.C., Lauber, C.L., Lozupone, C., *et al.* (2010) Soil bacterial and fungal communities across a pH gradient in an arable soil. *The ISME Journal* 4, 1340.

Sánchez-Moreno, S., Nicola, N.L., Ferris, H. and Zalom, F.G. (2009) Effects of agricultural management on nematode–mite assemblages: soil food web indices as predictors of mite community composition. *Applied Soil Ecology* 41, 107–117.

Scharroba, A., Kramer, S., Kandeler, E. and Ruess, L. (2016) Spatial and temporal variation of resource allocation in an arable soil drives community structure and biomass of nematodes and their role in the micro-food web. *Pedobiologia* 59, 111–120.

Schloter, M., Bach, H.-J., Metz, S., Sehy, U. and Munch, J. (2003) Influence of precision farming on the microbial community structure and functions in nitrogen turnover. *Agriculture, Ecosystems & Environment* 98, 295–304.

Shannon, D., Sen, A. and Johnson, D. (2002) A comparative study of the microbiology of soils managed under organic and conventional regimes. *Soil Use and Management* 18, 274–283.

Sharma, S.K., Ramesh, A., Sharma, M.P., Joshi, O.P., Govaerts, B., *et al.* (2010) Microbial community structure and diversity as indicators for evaluating soil quality. In: Lichtfouse, E. (ed.) *Biodiversity, Biofuels, Agroforestry and Conservation Agriculture*. Springer, New York, pp. 317–358.

Snapp, S.S., Gentry, L.E. and Harwood, R. (2010) Management intensity – not biodiversity – the driver of ecosystem services in a long-term row crop experiment. *Agriculture, Ecosystems & Environment* 138, 242–248.

Song, M., Li, X., Jing, S., Lei, L., Wang, J., *et al.* (2016) Responses of soil nematodes to water and nitrogen additions in an old-field grassland. *Applied Soil Ecology* 102, 53–60.

Su, J.Q., Ding, L.J., Xue, K., Yao, H.Y., Quensen, J., *et al.* (2015) Long-term balanced fertilization increases the soil microbial functional diversity in a phosphorus-limited paddy soil. *Molecular Ecology* 24, 136–150.

Swanepoel, P., Habig, J., Preez, C., Snyman, H. and Botha, P. (2017) Tillage effects, soil quality and production potential of kikuyu–ryegrass pastures in South Africa. *Grass and Forage Science* 72, 308–321.

Tautges, N.E., Sullivan, T.S., Reardon, C.L. and Burke, I.C. (2016) Soil microbial diversity and activity linked to crop yield and quality in a dryland organic wheat production system. *Applied Soil Ecology* 108, 258–268.

Tenuta, M. and Ferris, H. (2004) Sensitivity of nematode life-history groups to ions and osmotic tensions of nitrogenous solutions. *Journal of Nematology* 36, 85.

Tsiafouli, M.A., Thébault, E., Sgardelis, S.P., Ruiter, P.C., Putten, W.H., *et al.* (2015) Intensive agriculture reduces soil biodiversity across Europe. *Global Change Biology* 21, 973–985.

Tsiafouli, M.A., Bhusal, D.R. and Sgardelis, S.P. (2017) Nematode community indices for microhabitat type and large scale landscape properties. *Ecological Indicators* 73, 472–479.

Turner, B.L., Hopkins, D.W., Haygarth, P.M. and Ostle, N. (2002) β-Glucosidase activity in pasture soils. *Applied Soil Ecology* 20, 157–162.

Van Niekerk, A. (2018) A comparison of different strategies to control pests and diseases in *Brassica* spp. production in the Western Cape. MSc thesis, Nelson Mandela University, George, South Africa.

Wang, J., Song, Y., Ma, T., Raza, W., Li, J., *et al.* (2017) Impacts of inorganic and organic fertilization treatments on bacterial and fungal communities in a paddy soil. *Applied Soil Ecology* 112, 42–50.

Winding, A. and Hendriksen, N.B. (1997) Biolog substrate utilisation assay for metabolic fingerprints of soil bacteria: incubation effects. In: Insam, H. and Rangger, A. (eds) *Microbial Communities*. Springer, New York, pp.195–205.

Wittebolle, L., Marzorati, M., Clement, L., Balloi, A., Daffonchio, D., *et al.* (2009) Initial community evenness favours functionality under selective stress. *Nature* 458, 623.

Yao, H., He, Z., Wilson, M. and Campbell, C. (2000) Microbial biomass and community structure in a sequence of soils with increasing fertility and changing land use. *Microbial Ecology* 40, 223–237.

Yao, X.H., Min, H., Lü, Z.H. and Yuan, H.P. (2006) Influence of acetamiprid on soil enzymatic activities and respiration. *European Journal of Soil Biology* 42, 120–126.

Yeates, G.W. (2003) Nematodes as soil indicators: functional and biodiversity aspects. *Biology and Fertility of Soils* 37, 199–210.

Yeates, G.W., Bongers, T.D., de Goede, R., Freckman, D. and Georgieva, S. (1993) Feeding habits in soil nematode families and genera—an outline for soil ecologists. *Journal of Nematology* 25, 315.

Zak, J.C., Willig, M.R., Moorhead, D.L. and Wildman, H.G. (1994) Functional diversity of microbial communities: a quantitative approach. *Soil Biology and Biochemistry* 26, 1101–1108.

Zhang, X., Ferris, H., Mitchell, J. and Liang, W. (2017) Ecosystem services of the soil food web after long-term application of agricultural management practices. *Soil Biology and Biochemistry* 111, 36–43.

Zhang, Z., Zhang, X., Xu, M., Zhang, S., Huang, S., *et al.* (2016) Responses of soil micro-food web to long-term fertilization in a wheat–maize rotation system. *Applied Soil Ecology* 98, 56–64.

Zhao, B., Li, X., Liu, H., Wang, B., Zhu, P., *et al.* (2011) Results from long-term fertilizer experiments in China: The risk of groundwater pollution by nitrate. *NJAS-Wageningen Journal of Life Sciences* 58, 177–183.

Zhao, J. and Neher, D.A. (2013) Soil nematode genera that predict specific types of disturbance. *Applied Soil Ecology* 64, 135–141.

22 Soil Fertility Changes and Crop Yields from the First 4 Years of the Mandela Trials

Marike Swanepoel,* Raymond Auerbach and N'wa-Jama Mashele
Nelson Mandela University, George, South Africa

Abstract

Long-term farming systems trials (the Mandela Trials) showed that organic farming systems initially had lower yields than conventional systems in the first 2 years (20% and 31% lower, respectively). In the third (drier) year, the yield gap was closed once soils had improved biologically and available soil phosphate had been supplemented with rock phosphate, and the organic system outyielded conventional by 11%. In the fourth (wetter) year, nitrogen-poor compost caused low yields in the organic system, and conventional yields were 27% higher than organic. Soil became less acid under organic management, and soil organic matter (SOM) levels improved, as did available potassium. The organic farming systems were supplied with less than a quarter of the nutrients supplied to the conventional system, but yields were comparable throughout. From the second season, crop rotation for both organic and conventional systems outyielded cabbage monocropping systems, but this was only statistically significant in the fourth year; the benefit of crop rotation was greater for organic systems, probably because in the fourth year the low nitrogen nutrition in the organic system affected monocropped cabbage severely, while rotated cabbage benefited from the previous cowpea (legume) crop. The rotation was: cabbage (heavy feeder) followed by sweet potato (light feeder) followed by cowpea (legume). Water use efficiency was better for the organic system with mulch helping to reduce evaporation and SOM improving the water and nutrient holding capacity of the soil. Soil microbiology was more diverse for the organic system and biological pest and disease control, although slower and not very effective against cutworm, was able to keep pests and diseases at acceptable levels.

Introduction

Soil forms the foundation of any food system, and for this specific study all the results are influenced by what is happening in the soil. In comparing different farming systems, it is important to classify the soil and to document any nutrient inputs to the soil in the different farming systems. An experimental site was used that had been undisturbed for 20 years before the first trials started in 2014. A baseline soil fertility analysis was carried out to document the soil fertility levels before ameliorating the soil (Chapter 18, this volume). Two farming systems that were expected to have systemically different effects on the soil fertility were applied. The organic farming system made use of compost that is made from natural resources in the surrounding area, and the conventional farming system allows the application of synthetic fertilizers.

*marikeswanepoel17@gmail.com

Compost increases soil organic matter (SOM) which improves the physical, biological and chemical properties of the soil and can lead to an increase in soil pH (buffer acidity). This buffering effect allows P and K to gradually become more readily available. Disadvantages when using compost are the variation in the nutrient content of the compost and the slower nutrient release rate when compared with synthetic fertilizers. Heavy or long-term application of poor quality compost can lead to nutrient, heavy metal or salt accumulation which can adversely affect soil organisms, water quality and plant growth. Synthetic fertilizers on the other hand contain ingredients designed to stimulate plant growth and provide a predictable amount of nutrients and minerals (e.g. N-P-K ratio of 2:3:4 with 30% active ingredients represented as '2:3:4 (30)'). A rapid increase in N, P, K and Zn can be expected when using synthetic fertilizers in the soil, but they can also cause long-term imbalances in soil pH and fertility and are often leached into groundwater or rivers.

Soil fertility can be defined by the ability of the soil to provide the required conditions for plant growth. It is a result of the chemical, biological and physical processes that work together to provide aeration, water, nutrients and stability to the plant as well as the freedom from any growth inhibiting substances (Stockdale *et al.*, 2006). The availability of essential plant growth nutrients is influenced by soil structure and chemical composition and mediated by soil biota that provide essential biological processes (Tamburini *et al.*, 2016). The three primary macronutrients for plant growth are: nitrogen, phosphorus and potassium (NPK) which are required by plants for optimal growth. Larger amounts of these nutrients are required by the plant compared with micronutrients (e.g. molybdenum, boron, zinc, etc.); calcium, magnesium and sulfur are essential secondary nutrients. According to Haghigi (cited in Materechera, 2014), soil properties can be affected by any activity related to land use and agricultural management practices.

In these trials, initially in 2014 for the baseline study, and then for each cycle for the growing seasons 2014/15, 2015/16 and 2016/17, soil samples were collected from each plot after planting near the end of the year, to determine the changes in soil fertility as the trial progressed. The same sampling procedures have been used since the baseline study. Crops were planted in late spring (November) and grew through summer into the next year; crops were harvested in February or March. At harvest time in February 2018, a final set of samples was collected and sent for analysis.

The local climate in the Southern Cape is characterized by rain in most months, with an annual mean of 866 mm. The Southern Cape lies between the winter rainfall area of the Western Cape and the summer rainfall area of the Eastern Cape; natural vegetation is Cape fynbos, characterized by highly leached acidic sandy soils with low available phosphate. Winters are cool, but frost is rarely experienced; summers are warm, but not as hot or dry as the arid Karoo biome situated north of George over the Outeniqua Mountains. The site is shown in Fig. 22.1, with cabbages in the foreground, and sweet potatoes and cowpeas in the background. The detailed location and plot layout were given in Chapters 18 and 19 of this volume.

Methods and Materials

The soil is acidic and low in available phosphate (Mashele and Auerbach, 2016; Chapter 18, this volume, summarized in Table 22.1), and the organic cabbages received less than a quarter of the N, P and K nutrients provided to the conventional crops, yet the yields were comparable (although lower for the first 2 years until the soil biology improved). An initial dressing of 1 t/ha (3 kg/plot) of dolomitic lime was added to all the plots in all treatments to address the acidity and the high exchangeable Al levels in 2014, and was ploughed into the soil.

Initially (season 2014/15), the conventional crops received 400 kg/ha (for the cabbages), and 200 kg/ha for the sweet potatoes and for the cowpeas, of 2.3.4 (30) + 0.5% Zn, and top dressings of 200 kg/ha and 100 kg/ha calcium ammonium nitrate (CAN), respectively. In subsequent seasons, all of these amounts were halved. The CAN contains 27% N + 8% Ca, and this means that it will increase the soil Ca levels and have a slight liming effect, helping to ameliorate soil acidity to some extent in the conventional treatment.

Fig. 22.1. The research trial site showing organic cabbages with mulch in the foreground, and cabbages from the control treatment as well as sweet potatoes and cowpeas in the background.

Table 22.1. Soil fertility changes over 2 years (2014–2016); main results only.

	Treatment			
Parameter (January 2016)	Original	Control	Conventional	Organic
Soil P (Bray II) (mg/kg)	4	10	31	13
pH (KCl)	5.0	5.2	5.2	5.4
Acid saturation (%)	16.6	11.5	11.6	9.8
Exchangable Al (mg/kg)	3.3	2.77	2.98	2.93
Organic C (%) (Leco)	4.4	2.9	3.1	3.5
Soil K (mean) (mg/kg)	160	121	121	156
Cation exchange capacity ($cmol_c$/kg)	7.5	6.9	7.1	7.5

The organic crops received an initial compost dressing of 27 t/ha for all plots in 2014, and subsequent cabbage crops received only 5 t/ha of compost, while the other crops were not composted at all. The organic sweet potatoes and cowpeas thus had to rely on nutrients left over from the cabbage crop, and this was sufficient to yield a reasonable crop. Crop rotation can use nutrients effectively where the heavy feeder crop receives compost, and the subsequent light feeder crop and legume crop find enough residual nourishment; the different root systems and pest and disease susceptibility are also usually enough to confuse most pests and aerobic soil conditions help to keep diseases away.

Together with soil management designed to encourage aerobic soil microorganisms, which thrive on colloidal humus provided by compost, the gradual enlivening of the soil keeps disease organisms and pathogens (which are mostly facultative anaerobes) at bay, and sweetens the soil, raising pH, soil carbon and cation exchange capacity, and lowering exchangeable aluminium and acid saturation.

By mid-2016, the conventional management system had raised available P to acceptable levels; the two dressings of dolomitic lime (2 × 1 t/ha) and the applications of CAN had raised Ca levels, and this resulted in a slight increase in pH, and a decrease in acid saturation. After 2 years, the organic management system had raised SOM levels and decreased acidity slightly, but modest compost applications alone failed to correct soil P deficiencies (see Table 22.1).

All plots then received a third dressing of 1 t/ha of dolomitic lime, making a total of 3 t/ha over the 3 years (in three annual dressings of 1 t/ha), which raised the pH slightly and decreased soil acidity from the original 2014 levels even in the control (unfertilized) treatment (compare columns 'Original' and 'Control' in Table 22.1). Conventional treatments then received the double dose of fertilizer in the first year (2014), and half of this amount in the following 3 years. The organic plots received the heavy initial compost dressing and thereafter, for the next 3 years, only the cabbage plots received 5 t compost/ha/year; the other crops received no fertilizer or compost (for details, see Mashele and Auerbach, 2016; Swanepoel, 2018). The compost used the first year was good-quality purchased compost with a good balance of nutrients, purchased from George Nursery. The compost used in the second and third years was made on site, and had good levels of N and K, but, being made of low P materials, was deficient in P, and failed to raise the available soil P levels adequately.

In the third year, rock phosphate (13.5% P) was added to the organic treatments to correct the P deficiency, which is permitted in organic agriculture (OA), and this raised plant available soil P from 13 mg/kg to 26 mg/kg. Calphos was applied to all the organic plots at 900 kg/ha (2.7 kg/plot) followed by a compost dressing of 5 t/ha (15 kg/plot) of home-made compost, only for the organic rotated and monocrop cabbages. For 2016/17, compost was made on site with natural products (grass and manure). For 2017/18, compost was purchased from Grow Green Organics. The nutrients provided for the 2016/17 season from the home-made compost and the rock phosphate are shown in Table 22.2, and those for 2017/18 in Table 22.3; it can be seen that the N levels of the 2016/17 compost were adequate, while the N levels of the 2017/18 compost were very poor. Compost P levels for both years were very low, the total per ha in 5 t of compost being 1.1 and 1.7 kg P respectively.

Fertilization in season 2016/17

All plots (including the control treatment) received the third dressing of 1 t/ha of dolomitic

Table 22.2. Nutrients applied for different crops in season 2016/17.

Fertilizer[a]	Crops	Amount	Formulation
Conventional compound fertilizer NPK 200/100 kg/ha	Rotated cabbage Monocrop cabbage Cowpea Sweet potato	200 kg/ha = 0.6 kg/plot 200 kg/ha = 0.6 kg/plot 100 kg/ha = 0.3 kg/plot 100 kg/ha = 0.3 kg/plot	2:3:4 (30) + 0.5% Zn; 200 kg contains: • 13.3 kg N • 20 kg P • 27 kg K • 1 kg Zn
Additional top dressing of CAN (conventional plots) 100/50 kg/ha	Rotated cabbage Monocrop cabbage Cowpea Sweet potato	100 kg/ha = 0.3 kg/plot 100 kg/ha = 0.3 kg/plot 50 kg/ha = 0.15 kg/plot 50 kg/ha = 0.15 kg/plot	CAN 27% N; 100 kg contains: • 27 kg N • 8 kg Ca
Compost (organic plots, cabbage crop only; others no compost) 5 t/ha	Rotated cabbage Monocrop cabbage Cowpea Sweet potato	5 t/ha = 15 kg/plot 5 t/ha = 15 kg/plot	5 t compost (dry matter 64.6%) contained: • 24.6 kg N • 1.1 kg P • 2.8 kg K • 26.6 kg Ca • 4.8 kg Mg • pH 6.3 (in KCl)
Rock phosphate fertilizer at 900 kg/ha	All organic plots	900 kg/ha = 2.7 kg/plot	900 kg rock phosphate contains 121.5 kg P

[a]CAN, Calcium ammonium nitrate.

Table 22.3. Nutrients applied for different crops in season 2017/18.

Fertilizer[a]	Crops	Amount	Formulation
Conventional compound fertilizer NPK 2:3:4 (30)	Rotated cabbage Monocrop cabbage Cowpea Sweet potato	200 kg/ha = 0.6 kg/plot 200 kg/ha = 0.6 kg/plot 100 kg/ha = 0.3 kg/plot 100 kg/ha = 0.3 kg/plot	2:3:4 (30) + 0.5% Zn; 200 kg contains: • 13.3 kg N • 20 kg P • 27 kg K • 1 kg Zn
Additional top dressing of CAN	Rotated cabbage Monocrop cabbage Cowpea Sweet potato	100 kg/ha = 0.3 kg/plot 100 kg/ha = 0.3 kg/plot 50 kg/ha = 0.15 kg/plot 50 kg/ha = 0.15 kg/plot	CAN 27% N; 100 kg contains: • 27 kg N • 8 kg Ca
Compost (organic plots, cabbage crop only)	Rotated cabbage Monocrop cabbage Cowpea Sweet potato	5 t/ha = 15 kg/plot 5 t/ha = 15 kg/plot	5 t compost (dry matter 56.6%) contained: • 8.8 kg N • 1.7 kg P • 7.6 kg K • 24.1 kg Ca • 2.8 kg Mg • pH 5.3 (in KCl)

[a]CAN, Calcium ammonium nitrate.

lime, which was ploughed into the soil in August; for the conventional treatments, cabbage crops then received the NPK mixture (200 kg/ha of 2:3:4 (30) + Zn) and CAN (similar to limestone ammonium nitrate, with 27% N and 8% Ca, was applied at 100 kg/ha to all cabbage plots that were conventionally cultivated). The conventional sweet potatoes and cowpeas received 100 kg/ha 2:3:4 (30) and 50 kg/ha CAN top dressing. The organic cabbages received 5 t/ha of home-made compost (the actual nutrients in the compost as applied were calculated after analysis), and the sweet potatoes and cowpeas had no fertilizer or compost (Table 22.2).

Fertilization in the 2017/18 season

In this season no lime or Calphos was added to any of the treatment plots. Compost was applied at 5 t/ha for the organic-rotated and monocrop cabbages. Due to the low P content of compost made on site with natural surrounding plant material that is also highly acidic (fynbos), it was decided to purchase compost made by Grow Green Organic suppliers. The latter compost is made mostly with residues from local pine plantations in this geographical area, and thus also had low levels of P; the Grow Green compost had less than a third of the nitrogen compared with the previous year's home-made compost, but three times the potassium. NPK and CAN was applied to all three crops in the conventional treatments as described above. Control treatments were not fertilized. The nutrients supplied for each treatment are shown in Table 22.3.

Nutrients (NPK only) applied per plot for each treatment in the respective seasons are indicated in Tables 22.4 and 22.5. The fertilizer and compost were weighed individually for each specific plot. Fertilizer and compost were separately placed into each planting hole or furrow for all seasons.

Whereas Table 22.2 shows that the organic cabbage had reasonable levels of nitrogen in 2016/17 (24.6 kg/ha, which is more than half of the conventional cabbage nitrogen application for that season), the much lower level shown in Table 22.3 for 2017/18 (8.8 kg/ha, which is less than a quarter of the conventional nitrogen application) means that the 2017/18 organic cabbage was severely constrained in terms of nitrogen nutrition, even though for the 2017/18 season P supply was slightly better (1.7 kg/ha versus 1.1 kg/ha) and K supply was a lot better (7.6 kg/ha versus 2.8 kg/ha) than for the 2016/17 season.

Results and Discussion

This section is divided into three parts: (i) soil; (ii) pests and diseases; and (iii) crop yields.

Table 22.4. Nutrients (N, P and K) applied for season 2016/17.

Treatment	N (kg/ha)	P (kg/ha)	K (kg/ha)
Organic cabbage	25	1	3
Conventional cabbage	40	20	27
Organic cowpea and sweet potato	0	0	0
Conventional cowpea and sweet potato	20	10	13

Table 22.5. Nutrients (N, P and K) applied for season 2017/18.

Treatment	N (kg/ha)	P (kg/ha)	K (kg/ha)
Organic cabbage	9	2	8
Conventional cabbage	40	20	27
Organic cowpea and sweet potato	0	0	0
Conventional cowpea and sweet potato	20	10	13

Soil Results after Seasons 2016/17 and 2017/18

The yield-limiting implications for organic cabbage in the 2017/18 season of the low nitrogen levels of the compost are fairly dramatic, as will be seen in the presentation of yield data (see 'Yield Results and Discussion' later in the chapter); this illustrates not so much an inherent limitation of organic systems as an object lesson in the importance of properly made compost. Without well-made compost, organic yields will be low, and without adequate available soil phosphate, they will also be constrained. As nitrogen is freely available (80% of the atmosphere is nitrogen), a good compost-making process should contain at least 0.5% of the total weight (about 1% of the dry weight) of N. The volatility of N compounds mean that careful management of the decomposition process is required.

The effects of the higher K levels supplied to the organic treatment in 2017/18 can also be seen in the final soil analysis results shown in Table 22.6. The increase in available P after the application of the rock phosphate to the organic treatment in 2015/16 can also be clearly seen. The decrease in carbon for both the organic and the conventional treatments in the 2017/18 season is difficult to explain, as is the decrease in available P for both organic and conventional in 2017/18, and for conventional in the 2016/17 year.

Table 22.6. Long-term mean soil analysis from 2014 to 2017/18.

Management practice	Season	pH (KCl)	C (Leco) (%)	P (mg/kg)	K (mg/kg)	Ca (mg/kg)
Organic	2017/18	5.6	2.7	16.6	135	859
	2016/17	5.6	3.7	25.4	81	756
	2015/16	5.4	3.5	12.3	124	1076
	2014/15	5.4	2.0	5.5	130	1049
	2014	5.3	3.1	6.2	166	973
Conventional	2017/18	5.4	2.4	10.9	66	841
	2016/17	5.5	2.7	15.6	58	656
	2015/16	5.2	3.1	21.3	121	1042
	2014/15	5.2	2.1	7.6	113	1058
	2014	5.2	3.2	7.2	161	947
Control	2017/18	5.5	2.4	9.0	114	827
	2016/17	5.6	2.7	9.0	88	696
	2015/16	5.3	2.9	10.3	121	1966
	2014/15	5.3	2.0	6.4	109	1002
	2014	5.2	2.9	7.0	159	950

Detailed analysis of variance (ANOVA) was carried out for all soil parameters, and for the changes over time from 2014 to 2018 (Swanepoel, 2018), using Genstat (VSN International, 18th edition); the conclusions of this work are summarized below.

Soil pH

Soil pH is one of the most important soil characteristics for crop production and is a measure of the acidity or alkalinity of the soil (Brady and Weil, 2008). The difference in pH levels determines the availability of different elements to plants for uptake. Soil pH levels range from 0 to 14, with 7 being neutral, above 7 alkaline and below 7 acidic. For most crops the optimal pH level is between 5.5 and 7.0. Soil becomes acid when basic plant nutrients such as K, Ca and Mg in the colloids are replaced by H and Al.

This acidification can be caused by intensive conventional farming methods making use of acid-forming fertilizers, and also by manure, acid rain, the removal of basic cations (K, Ca, Mg) when harvesting crops, or accelerated decomposition of SOM due to tillage practices. The pH decline rate is determined by the buffer capacity of the soil, its physicochemical properties associated with texture, soil minerals, structure and organic matter content (Benton Jones, 2012). Acid soils cause damage to the root zone, which influences the growth pattern of crops and can also cause various deficiencies (Miles and Farina, 2013). The width of the bands in Fig. 22.2 indicates the availability of the nutrients within the indicated pH range. The wider the band the higher the ability of the plant to absorb the nutrient. The narrower the band, the less the plant is able to absorb (Benton Jones, 2012).

In the trials, because of soil acidity and high levels of exchangeable Al, all plots (including the control plots) were treated with1 t dolomitic lime/ha/year for the first 3 years, which was added in 2014, 2015 and 2016, before planting the crops. Lime raises pH and neutralizes exchangeable Al (Brady and Weil, 2008).

Soil pH 2016/17

Soil pH results showed that management practice was a significant factor. Pairwise comparisons

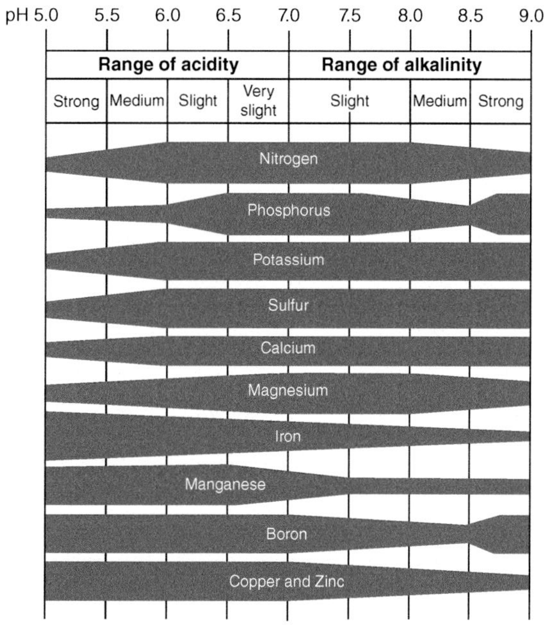

Fig. 22.2. Availability of essential nutrient elements to plants at different pH ranges from soils. (From Benton Jones, 2012.)

showed no significant difference between the pH levels of control and organic practices. However, pH levels for the control (5.6 KCl) and organic practices (5.6 KCl) were significantly higher than the pH levels for the conventional practice (5.4 KCl). The reason for the soil being significantly more acidic could be the use of fertilizer in this treatment since 2014, which has the potential to acidify soil.

Soil pH 2017/18

Soil pH results showed no significant difference between any variances. In management practices, conventional pH levels were 5.4 KCl, organic 5.6 KCl and control 5.5 KCl.

Exchangeable aluminium

High exchangeable aluminium (Al^{3+}) levels in soil are usually associated with low fertility and highly acidic soil, and are known to be one of the most important growth-limiting factors in acidic soils. Soil becomes acid due to the presence of exchangeable Al^{3+} combined with one or more of the following factors: (i) depletion of basic cations such as K^+, Na^+, Ca^{2+} and Mg^{2+} by leaching and crop removal; (ii) application of fertilizers; (iii) atmospheric acidity (e.g. acid rain); and (iv) decomposition of organic residues (Tisdale *et al.*, 1985). Soil P is also fixed by Al and is then not readily available for plant growth.

Exchangeable aluminium levels 2016/17

Results showed that management practice was a significant factor (Table 22.7). Pairwise comparisons showed no significant difference between Al^{3+} levels for conventional and control practices. However, Al^{3+} levels for control and conventional practices were significantly higher than the Al^{3+} levels for the organic practice. The reason for Al^{3+} levels being higher in conventionally managed soils might be that the pH levels were lower in conventional (significantly) and control soils. Levels of Al^{3+} should preferably be less than 2 mg/kg.

Exchangeable aluminium levels 2017/18

Results showed that management practice was a significant factor (Table 22.8). Pairwise comparisons showed no significant difference between Al^{3+} levels for conventional, organic and control practices.

Acid saturation

Acid soil conditions are known to restrict crop growth and with over 30% of all land worldwide available for cultivation being acidic, this is a worldwide problem (Miles and Farina, 2013). Acid saturation percentage can be defined as the total acid cations in chemically equivalent terms expressed as a percentage of the total cations in chemically equivalent terms. An index of the Al activity levels is then provided by this relationship.

Acid saturation levels 2016/17

Results showed that management practice was a significant factor. Pairwise comparisons showed no significant difference between acid saturation levels for control and organic practices. However, acid saturation levels for the conventional practice were significantly higher than acid saturation levels for the control and organic practices (Table 22.9). Conventional practice results could be due to low pH levels and high exchangeable Al^{3+} rates.

Table 22.7. Mean values of exchangeable aluminium (Al^{3+}) in soil for different management practices in 2016/17.

Management practice	Mean (mg/kg)	Significant difference[a]
Conventional	4.98	a
Control	4.98	a
Organic	4.31	b

[a]Different letters indicate significant differences at $p < 0.05$.

Table 22.8. Mean values of exchangeable aluminium (Al^{3+}) in soil for different management practices in 2017/18.

Management practice	Mean (mg/kg)	Significant difference[a]
Conventional	1.29	a
Control	1.53	a
Organic	2.16	a

[a]Different letters indicate significant differences at $p < 0.05$.

Acid saturation levels 2017/18

Results showed no significant difference in any variance.

Soil carbon

SOM is a diverse and complex mixture of organic substances including soil organic carbon (SOC) (Brady and Weil, 2008). SOC comprises about half of the mass of SOM and forms the basis of soil fertility. SOC promotes the structure, physical and biological health of soil, releases nutrients for plant growth and acts as a buffer against harmful substances (Adhikari and Hartemink, 2017). Organic matter in the world's soils contains two to four times as much carbon as is found in all the world's vegetation, and thus it plays a critical role in the global carbon balance (Brady and Weil, 2008). Soil management, climate, land use and soil physical characteristics all influence the inputs of C in soils (Adhikari and Hartemink, 2017). Organic farming relative to conventional farming systems is known for one of its key ecosystem services, namely the accumulation of organic C in the soil (Blanco-Canqui *et al.*, 2017). Main drivers of SOC accumulation under organic farming systems are: (i) crop rotation; (ii) reduced tillage intensity; and (iii) the input of organic amendments (Blanco-Canqui *et al.*, 2017). Compared with conventional farming systems, studies have suggested that organic farming systems can store more SOC (Blanco-Canqui *et al.*, 2017).

Soil carbon levels 2016/17

Results showed that management practice was a significant factor. Pairwise comparisons showed no significant difference between the C levels of conventional and control practices. However, C levels for the organic practice were significantly higher than the C levels for the conventional and control practices (Table 22.10). This result agrees with similar results in studies presented in the literature (e.g. Fließbach *et al.*, 2007; Rodale Institute, 2011) where C levels in organic systems tend to increase over years. The initial amount of soil carbon for these soils was over 4% before the initial cultivation.

Planting methods (monocrop and rotation) proved to be a significant factor (Table 22.11). Pairwise comparisons showed no significant difference between the C levels of control and rotated methods. However, C levels for the monocrop practice were significantly higher than the C levels for the control and rotated practices. The reason for this result is unclear.

Soil carbon levels 2017/18

Results showed that management practice was a significant factor (Table 22.12). Pairwise comparisons

Table 22.9. Mean values of acid saturation in soils for different management practices.

Management practice	Mean (%)	Significant difference[a]
Conventional	12.65	a
Control	9.53	b
Organic	9.13	b

[a]Different letters indicate significant differences at $p < 0.05$.

Table 22.10. Mean values of carbon in soils for different management practices in 2016/17.

Management practice	Mean (%)	Significant difference[a]
Control	2.685	a
Conventional	2.704	a
Organic	3.209	b

[a]Different letters indicate significant differences at $p < 0.05$.

Table 22.11. Mean values of carbon in soils for different planting methods.

Planting method	Mean (%)	Significant difference[a]
Monocrop	3.182	a
Rotated	2.881	b
Control	2.685	b

[a]Different letters indicate significant differences at $p < 0.05$.

Table 22.12. Mean values of carbon in soils for different management practices in 2017/18.

Management practice	Mean (%)	Significant difference[a]
Control	2.433	a
Conventional	2.449	a
Organic	2.737	b

[a]Different letters indicate significant differences at $p < 0.05$.

showed no significant difference between the C levels of conventional and control practices. However, C levels for the organic practice were significantly higher than the C levels for the conventional and control practices.

Phosphorus (P)

P is one of the three main macronutrients required by plants for optimum growth. It functions as a component of certain enzymes and proteins, ATP, which functions as an energy currency that drives most biochemical processes, RNA, DNA and phytin (Brady and Weil, 2008; Benton Jones, 2012). Fundamental processes of nitrogen fixation, photosynthesis, seed production, maturation and flowering are dependent on adequate phosphorus nutrition (Brady and Weil, 2008). A deficiency in P can lead to stunted and slow growth of the crop with the older leaves displaying a purple colour (Benton Jones, 2012).

Phosphorus levels 2016/17

Results showed all the management practices were significantly different from one another, with organic management practices being significantly higher than conventional and control and conventional being significantly higher than control (Table 22.13). These results reflect the addition of 900 kg/ha of rock phosphate in the organic treatments, with organic management pH levels being significantly higher than the conventional pH levels.

Phosphorus levels 2017/18

Results showed that management practice was a significant factor (Table 22.14). Pairwise comparisons showed no significant difference between the P levels of conventional and control practices. However, P levels for the organic practice were significantly higher than the P levels for the conventional and control practice, showing a residual effect from the rock phosphate treatment of the previous year.

Potassium

Potassium is the third macronutrient required for plant growth; concentrations between 1% and 5% are required by most crops (Tisdale *et al.*, 1985). Potassium is absorbed as potassium ion K^+ from soil solutions by crops (Tisdale *et al.*, 1985). K is required in the accumulation and translocation of newly formed carbohydrates and also in maintaining the water status of the plant, opening and closing of stomata and sustaining the turgor pressure (Benton Jones, 2012). A deficiency in K can lead to a burned appearance on the edges of older leaves (scorch), it can also cause plants to easily lodge and be prone to diseases. The quality and production of the fruit or seed of crops will also decrease due to a lack of K (Benton Jones, 2012).

Potassium levels 2016/17

K results showed that management practice was a significant factor (Table 22.15). Pairwise comparisons showed no significant difference between the K levels of control and organic practices. However, K levels for the control and organic practices were significantly higher than the K levels of the conventional practice.

Potassium levels 2017/18

K results showed that management practice was a significant factor (Table 22.16). Pairwise

Table 22.13. Mean values of P in soils for different management practices in 2016/17.

Management practice	Mean (mg/kg)	Significant difference[a]
Conventional	15.6	b
Control	9	c
Organic	25.4	a

[a]Different letters indicate significant differences at $p < 0.05$.

Table 22.14. Mean values of P in soils for different management practices in 2017/18.

Management practice	Mean (mg/kg)	Significant difference[a]
Conventional	10.94	a
Control	10.00	a
Organic	16.56	b

[a]Different letters indicate significant differences at $p < 0.05$.

comparisons showed no significant difference between the K levels of control and organic practices. However, K levels for the control and organic practices were again significantly higher than the K levels of the conventional practice.

Calcium

Calcium is another macronutrient; it is absorbed as the ion Ca^{2+} and is abundant in leaves. The standard concentration ranges from 0.2% to 1.0% (Tisdale *et al.*, 1985). Ca plays a vital role in maintaining cell integrity and membrane permeability and enhances growth and pollen germination. It is important for carbohydrate transfer and protein synthesis and activates a number of enzymes for cell mitosis, division and elongation (Benton Jones, 2012). The presence of Ca may also detoxify the presence of heavy metals in plants (Benton Jones, 2012). A deficiency in Ca can cause the growing tips of leaves and roots to turn brown and die; edges of leaves become ragged as the edges of emerging leaves stick together (Benton Jones, 2012).

Calcium 2016/17

Ca results showed that management practice was not a significant factor (Table 22.17). Pairwise comparisons showed no significant difference between the Ca levels of any of the practices. Although management practice was a factor, post hoc pairwise comparisons did not show any significant difference in Ca levels between the different levels of management practice. Although organic Ca levels were higher than conventional and control practices they were not significantly (Table 22.17). Ca levels should always be about three times the amount of the Mg levels in the soil, and in this season, they are, but in general the desirable level would be 900 mg/kg Ca in the soil.

Calcium 2017/18

Results showed no significant differences between any variance. Results showed no significant differences in Ca levels in soils for different management practices in 2017/18.

Magnesium

Magnesium is a macronutrient and is adsorbed in the form of the ion Mg^{2+}; it is the only mineral constituent of the chlorophyll molecule (Tisdale *et al.*, 1985; Benton Jones, 2012). Mg is thus of vital importance, for without chlorophyll the autotrophic green plant would fail to carry on photosynthesis (Tisdale *et al.*, 1985). Mg serves as a cofactor in most enzymes that activate phosphorylation processes as a bridge between pyrophosphate structures of ADP and ATP and the enzymes molecule (Benton Jones, 2012). It also plays a vital role in stabilizing the ribosome particles in the configuration for protein synthesis (Benton Jones, 2012). Mg deficiencies can lead to the yellowing of older

Table 22.15. Mean values of K in soils for different management practices in 2016/17.

Management practice	Mean (mg/kg)	Significant difference[a]
Conventional	58.3	b
Control	88.1	a
Organic	80.5	a

[a]Different letters indicate significant differences at $p < 0.05$.

Table 22.16. Mean values of K in soils for different management practices in 2017/18.

Management practice	Mean (mg/kg)	Significant difference[a]
Conventional	65.6	b
Control	114.6	a
Organic	148	a

[a]Different letters indicate significant differences at $p < 0.05$.

Table 22.17. Mean values of Ca in soils for different management practices in 2016/17.

Management practice	Mean (mg/kg)	Significant difference[a]
Conventional	656	a
Control	696	a
Organic	756	a

[a]Different letters indicate significant differences at $p < 0.05$.

leaves, with interveinal chlorosis (yellowing between veins) and growth will be slow with plants being very susceptible to diseases (Benton Jones, 2012).

Magnesium 2016/17

Mg results showed that management practice was a significant factor (Table 22.18). Pairwise comparisons showed no significant difference between the Mg levels of control and organic practices. However, Mg levels for the control and organic practices were significantly higher than the Mg levels of the conventional practice.

Low pH soils have lower Mg levels, and in this season the conventional soils had a significantly lower pH (5.46 KCl) than organic and control soils that consequently lead to the Mg levels being significantly lower (213.1 mg/kg) in conventional soils.

Magnesium 2017/18

Mg results showed that management practice was a significant factor (Table 22.19). Pairwise comparisons showed no significant difference between the Mg levels of control and organic and control and conventional practices. However, Mg levels for the organic practice were significantly higher than the Mg levels of the conventional practice.

Table 22.18. Mean values of Mg in soils for different management practices in 2016/17.

Management practice	Mean (mg/kg)	Significant difference[a]
Conventional	213.1	b
Control	252.1	a
Organic	260.1	a

[a]Different letters indicate significant differences at $p < 0.05$.

Table 22.19. Mean values of Mg in soils for different management practices in 2017/18.

Management practice	Mean (mg/kg)	Significant difference[a]
Conventional	153.5	a
Control	170.8	ab
Organic	175.5	b

[a]Different letters indicate significant differences at $p < 0.05$.

Discussion on soil fertility results

The pH in organically managed soil systems of both seasons was higher than the conventionally managed soil pH. In season 2016/17 it had a significantly higher pH of 5.6 KCl compared with the conventionally managed system at pH 5.4 KCl. This affected the level and the availability of various nutrients in the soil by exceeding the optimal pH level of growth for most crops.

Exchangeable Al, K, Na, Ca, Mg, available P, combustible C and acid saturation were measured and their significance in plant nutrition was discussed above. With conventional systems being more acidic, levels of Al^{3+} and acid saturation were significantly higher in the conventional system compared with the organic systems in season 2016/17. In season 2017/18, there were no significant differences between these elements in the organic and conventional systems. Soil C levels were significantly higher in both seasons for organic and rotated systems in general. The ideal amount of C for effective farming practices is around 4%; in these seasons the organic soils contained 3.2% and went down to 2.7% C, which could be due to different ploughing methods applied. Before soils were ploughed in 2014, the soils had 4.4% C.

P and K levels were significantly higher in the organic systems compared with the conventional systems. Although P levels were significantly higher in organic systems they dropped from 25.4 mg/kg in 2016/17 to 16.56 mg/kg in 2017/18. The mean K level in the organic system in 2016/17 was 80.5 mg/kg and this was significantly higher than the conventional system mean level of 58.3 mg/kg. K levels in 2017/18 drastically went up to 148 mg/kg in organic and 65.6 mg/kg in conventional systems. It would be ideal to get the K levels up to 200 mg/kg for both farming systems for optimal crop growth and production. The level of Ca showed no significant difference between treatments in both seasons, with mean levels in 2016/17 of 656 mg/kg in conventional and 756 mg/kg in organic system soils and in 2017/18 it went up to 841 mg/kg in conventional and 859 mg/kg in organic. The desired Ca level for both farming systems would be up to 900 mg/kg.

In both seasons, soil Mg was significantly higher under organic management than under conventional management. In season 2016/17, organic soils had mean levels of 260.1 mg/kg Mg compared with 213.1 mg/kg in conventional soils; and in season 2017/18 the mean levels were 175.5 mg/kg for organic and 153.5 mg/kg for conventional soils. This could be due to lower pH levels in both seasons for conventionally managed soils.

Water use efficiency (WUE) and soil microbiology

WUE has been assessed with help of hydrologists from the University of KwaZulu-Natal (Chapter 19, this volume), and soil microbiology with the University of Pretoria's Department of Microbiology (Chapter 21, this volume). The mulch clearly helped the organic farming system to thrive, particularly in the drier years. The overall positive effect of the organic treatments on soil diversity and richness was documented in Chapter 21:

> The combined use of microbial and nematode analyses can become a very useful approach in deciphering the sustainability of a management system. The observation that microbial and nematode data were directly linked was of importance, since the community shifts between the different indicators correlated. In turn, this led to identical conclusions, that the organic treatments supported a greater and more diverse soil food web compared with the conventional treatments.
>
> (Sibiya *et al.*, Chapter 21, this volume)

On the other hand, there was a slight negative response of the soil biology to crop rotation.

Residual soil fertility as shown by the 2018 mustard indicator crop

An initial indicator crop of Caliente mustard was planted in 2014 and the height of this crop was measured using a disk meter (for results see Table 18.1, this volume), and in October 2018 a final crop of Caliente mustard was planted and measured and after the final crop harvest the leaves from this crop were analysed. While the initial readings for the organic and conventional plots were similar at a low level (crop height 13.1 mm and 12 mm, respectively), both systems had far higher readings 4.5 years later, but with a large difference between farming systems. Mustard plants in the conventional system, having received no fertilizer after the crops were harvested, very rapidly ran out of nutrients and at first appeared stunted and then rapidly changed colour, showing nutrient and moisture stress. The organic plots looked green and healthy, and this was reflected in the disk meter readings (organic mean crop height 120 mm, conventional 89 mm). While leaf analysis showed many small differences, the leaf P levels were similar for both systems. Notable were the N and K levels, where organic averaged 3.2% N and 3.4% K, while conventional averaged 2.8% N and 2.7% K.

Conclusion on soil fertility, WUE and soil microbiology

Overall, organic farming systems are showing very promising results with significantly higher levels in soil pH and important soil nutrients (soil carbon, P, K, Mg and Ca) for optimal plant growth than conventional systems. The conventionally managed system, on the other hand, is becoming more acidic with a lower pH and higher Al and acid saturation, leading to significantly lower nutrient levels that may lead to lower yields. Residual soil fertility after the 4 years of cropping was better in the organic farming system.

Pests and Diseases

Pests and diseases were a factor in both systems, but were controlled reasonably well (organic using biological control methods, conventional using poisons); an experiment measuring pests and disease in organic and conventional vegetable rotations is described in Chapter 20 of this volume, assessed with help from the Soilborne Pathogens Institute of the Agricultural Research Council at Stellenbosch. As pointed out in Chapter 20, there are few remedies registered for organic pest and disease control, and the ongoing problems with cutworms in the organic treatment were never completely solved; the crop rotation helped to minimize pest and disease outbreaks. Cabbage seedlings were first attacked by the common cutworm (*Agrotis longidentifera*) even though cutworm bait was applied to all the conventional and control plots.

Pests and diseases 2016/17

In the 2016/17 season, the organic cabbages were not treated against the common cutworm because there is no certified organic product on the market for it. Most of the cabbage seedlings that were damaged in the organic and conventional plots were replanted but only a few of them grew and reached maturity. The control plots were mainly undamaged by pests. Other main pests causing critical damage to organic and conventional cabbages during the growing season (see Figs 22.3 and 22.4) were the diamondback moth (*Plutella xylostella*), cabbage semi-looper (*Thysanoplusia ni*), green milkweed locust (*Phymateus leprosus*), garden locust (*Acanthacris ruficornis*) and the cabbage aphid (*Brevicoryne brassicae*) (Prinsloo and Vivienne, 2015).

No threatening diseases or deficiencies were found in this season that could affect the cabbage growth, yield or quality. This could be due to the low rainfall season, soil treatments and early treatment against various pests and diseases.

Early damage appeared where baboons bit the small, still-developing heads of 208 cabbages. Most of the damaged cabbages developed several small heads instead of one big head due to this damage.

To be fair to all the treatments, it was decided not to discard the border rows but rather take the weight of the 40 heaviest cabbages per plot after weighing all 70 cabbages separately per plot. This protocol made up for the damage caused by the baboons in all treatments. This data (cabbage weight and damage) was recorded.

Fig. 22.3. Garden locust nymph (*Acanthacris ruficornis*) (a), green milkweed locust nymph (*Phymateus leprosus*) (b), and damage caused by locusts on cabbage (c).

Fig. 22.4. Cabbage aphids (*Brevicoryne brassicae*) (a), cabbage semi-looper (*Thysanoplusia ni*) (b), and damage to cabbage head and leaves (c).

Pests and diseases 2017/18

A few cabbage seedlings were destroyed by cutworms in the organic system where no cutworm bait was applied but there was less damage than the previous season. Almost no cutworm damage was found on conventional and control plots where cutworm bait was applied. Seedlings destroyed by cutworms were replanted. In general, conventional cabbages had more pest occurrence and damage, diseases and cabbage rot symptoms than organic cabbages (Fig. 22.5). Organic cabbages barely had any damage at harvest time and looked much healthier compared with conventional and control cabbages. The most often occurring pests that created the most damage in both systems were the cabbage aphid, the diamondback moth (also called the lesser cabbage moth (*Plutella xylostella*)), the garden locust nymph and the green milkweed locust nymph. Bacterial soft rot caused by the bacterium *Pectobacterium carotovora* (previously *Erwinia carotovora*) was mainly found on conventional cabbage and the crop was exponentially more infected than the previous season. The increase in the incidence of this disease could have been caused by the warm temperatures (25–30°C), wet weather and humid conditions which occurred in this season.

Yield Results and Discussion: Seasons 2016/17 and 2017/18

Introduction

This section will discuss the third and fourth year of the overall trial (seasons 2016/17 and 2017/18) results, with the aim of comparing the yield differences between three farming systems namely, organic and conventional and control. The three main crops planted in these farming systems were cabbage, cowpea and sweet potato, with two varieties of the latter, namely the orange-fleshed sweet potato with a mixture of the white-fleshed sweet potato only in season 2016/17. These crops were selected because cabbages are widely grown by large- and small-scale farmers in South Africa (SA), and sweet potatoes and cowpeas are highly nutritious indigenous African crops commonly grown by small-scale farmers. In terms of soil fertility requirements, the crops can be classified as: (i) heavy feeders (cabbage) because they are heavy nitrogen feeders; (ii) light feeders (sweet potato) that require less nitrogen, potassium and phosphorus; and (iii) legumes (cowpea) that have the capacity to fix nitrogen in the soil. They also have differing root systems, and pest and disease susceptibility, that makes them well suited for a crop rotation. As the results of sweet potato and cowpea yields were inconclusive, these results are not presented here; for most years, these yield results were similar (details in Mashele and Auerbach, 2016; Swanepoel, 2018). The main importance of the rotation crops was to help with pest and disease prevention and to bring N into the soil; in these areas, they seem to have been successful, although yields were not very high. Organic sweet potatoes (R18/kg) were sold at higher prices than conventional (R15/kg) in both seasons.

Due to the limited availability of organic matter, the supply of organic matter to organic systems through high levels of compost is often impractical. Before each season started, cover crops were planted to supply more organic matter for the organic system in the trial. A mixture of oats and serradella was planted as cover crops in all organic treatments over the winter period. These cover

Fig. 22.5. Cabbage aphid infestation (a), caterpillar of diamondback moth (*Plutella xylostella*) (b), and bacterial soft rot damage (c).

crops provide organic matter resulting in a lighter dressing of compost being required, addressing the common problem of SOM availability.

Cabbage, sweet potato and cowpea were each planted in prepared soil as per farming system treatment (organic, conventional and control). In each plot, the growth, pests, diseases and the yield weight were monitored and recorded.

All of this research shows that organic farming brings about certain improvements in soil fertility after about 3 years of organic management (higher SOM, lower soil acidity, better soil biology, aerobic soil conditions, greater WUE). Initially, organic farming systems had lower yields (20–31%), but this yield gap was eliminated in the third year; in this dry year, rainfed crops outyielded equivalent conventional crops by 11% (Table 22.20). The trials are laid out as a complete randomized block experiment with monocrop and rotation as factors and plots split for farming system (organic and conventional), with ten treatments and four replications.

The organic cabbages (averaging 70.6 kg/plot, see 2017 in Table 22.20) outyielding the conventional cabbages (63.7 kg/plot average) in this dry year (604 mm/year of rain against the long-term average of 866 mm/year). The yield gap has disappeared, thanks to adequate available soil P and good quality compost with adequate N.

The following year was wetter (693 mm/year), and the conventional cabbages produced their highest yield ever, with the rotated conventional cabbage yielding 85 kg/plot, as against 76 kg/plot for the monocropped conventional cabbage.

On the other hand, cabbage yield from the organic treatment was lower this season than the previous season, even though the rainfall was heavier. This is ascribed to the poor quality compost not supplying adequate nitrogen for optimal plant growth. However, even with the poor quality compost, the organic treatment rotated cabbages yielded 76 kg/plot, slightly higher than the previous year. With good quality compost, one could have expected similar yields to the conventional rotated cabbage. Both conventional and organic rotated treatments are seeing the benefit of the previous legume crop; in addition, the organic treatment is benefiting from the cover crop (serradella and oats), which has provided both organic matter and some nitrogen from the legume.

Soil was treated with an initial dressing of 1 t/ha of dolomitic lime and in every treatment in season 2016/17 and in season 2017/18 no lime was applied. Calphos was applied to all the organic plots in season 2016/17 (but Calphos was not applied in season 2017/18), followed by a dressing of 5 t/ha of compost/plot for the rotated and monocrop organic cabbages in both seasons. The other organic plots received no compost in both seasons. All the conventional plots in both seasons received NPK and CAN; the CAN contained 27% N and 8% Ca.

Table 22.20. Cabbage harvests and annual rainfall for 2015–2018.

		Year			
		2015	2016	2017	2018
Seasonal rainfall (mm)[a]		754	975	604	693
Management system[b]					
Conventional	monocrop (kg/plot)	48.9	65.5	61.0	75.5
	rotated (kg/plot)	50.5	74.6	66.3	84.6
	mean (kg/plot)[c]	**49.7**	**70.0**	**63.7**	**80.1**
Organic	monocrop (kg/plot)	40.4	41.2	66.0	50.7
	rotated (kg/plot)	40.5	58.3	75.2	75.9
	mean (kg/plot)[c]	**40.5**	**49.7**	**70.6**	**63.3**
Control (monocrop)	mean (kg/plot)	17	6	8	4

[a]Seasonal rainfall values in columns '2015', '2016', '2017' and '2018'are for seasons 2014/15, 2015/15, 2016/17 and 2017/18, respectively (rainfall being measured from 1 May in one year to 30 April in the following year). George mean annual rainfall – 866 mm.

[b]Values are the mass (kg) of 40 cabbages/plot.

[c]Values for conventional mean and organic mean are set in bold so they can be compared easily for each year.

Cabbage results 2016/17

In total, 24 plots were planted with cabbage, eight conventional plots, eight organic plots, and eight control plots. Conventional cabbage grew faster in the beginning but took longer to produce a fully-grown matured head. The organic cabbage grew much slower but formed a fully matured head much earlier compared with the conventional cabbage. The growth of the control cabbage was very poor and produced only a few small heads (Fig. 22.6).

Yield data were captured and recorded when cabbage heads were fully matured. For the variable of cabbage head weight, an ANOVA appropriate for a split plot design with additional control (randomized complete block design at a whole plot level) was used to test for treatment effects using GenStat for Windows 10 (VSN International, 18th edition). Only if the F-value was significant ($p < 0.05$) were treatment differences further investigated using the least significant difference (LSD) statistic. The only farming system factor that was statistically significant was the difference between the control and the other farming system treatments (Table 22.21 and Fig. 22.7).

The treatment control versus no control value was $p < 0.001$, showing that both conventional and organic treatments were significantly higher yielding than the control. There was no statistically significant difference between the average yield of the organic and conventional management practice, supporting the hypothesis that organic yields, initially lower than those of conventional farming systems during the first 2 years, can approach similar yield levels after the third year of organic management. The organic yields were 11% higher than the conventional treatment in this relatively dry year, and quality was superior.

The rotational cropping system (conventional treatment) yielded 11% more than the monocropped cabbage (conventional treatment), but again, this was not statistically significant (Table 22.21 and Fig. 22.8). Organic rotated cabbage yielded 14% higher than organic monocropped cabbage, while the conventional crop rotation system only yielded 9% higher than conventional monocropped cabbage.

Table 22.21. Summary of means ($\bar{x}$), standard deviation (SD) and sample size (n) for various treatments in the trial for 2016/17, to determine the impact of the different treatment factors on cabbage yield.

Treatment	Sample (n)	Mean weight (kg) ($\bar{x}$)	SD
Control	8	0.152	0.237
None	16	1.678	0.367
Conventional	8	1.592	0.358
Organic	8	1.764	0.380
Monocrop	8	1.588	0.221
Rotated	8	1.769	0.471
Conventional monocrop	4	1.526	0.294
Conventional rotated	4	1.658	0.448
Organic monocrop	4	1.649	0.130
Organic rotation	4	1.879	0.534

Fig. 22.6. Conventional cabbage growth (a) compared with control cabbage growth (b) at the same date.

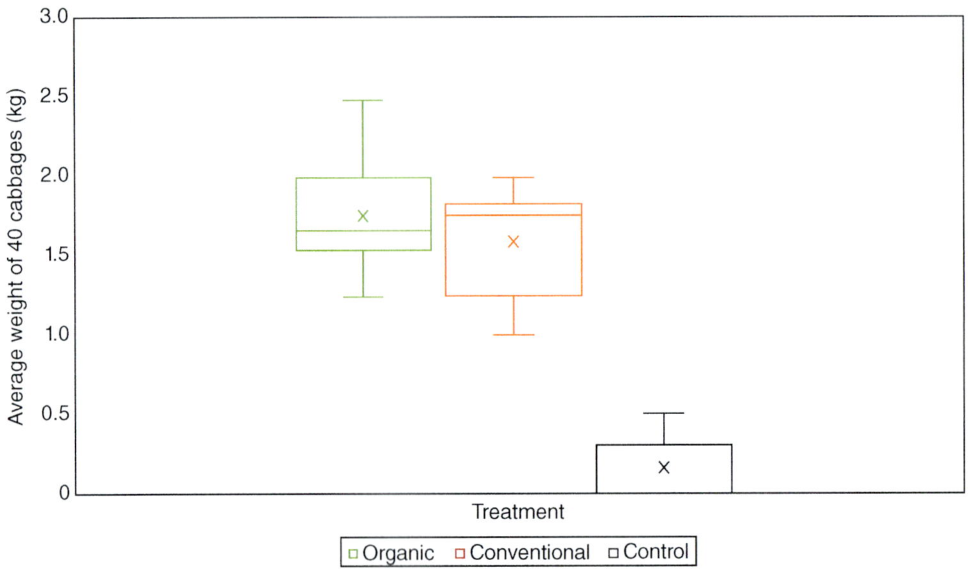

Fig. 22.7. Average cabbage weight differences (kg for 2016/17) of organic, conventional and control management practices showing the interquartile range (IQR), mean (×) and median of each treatment.

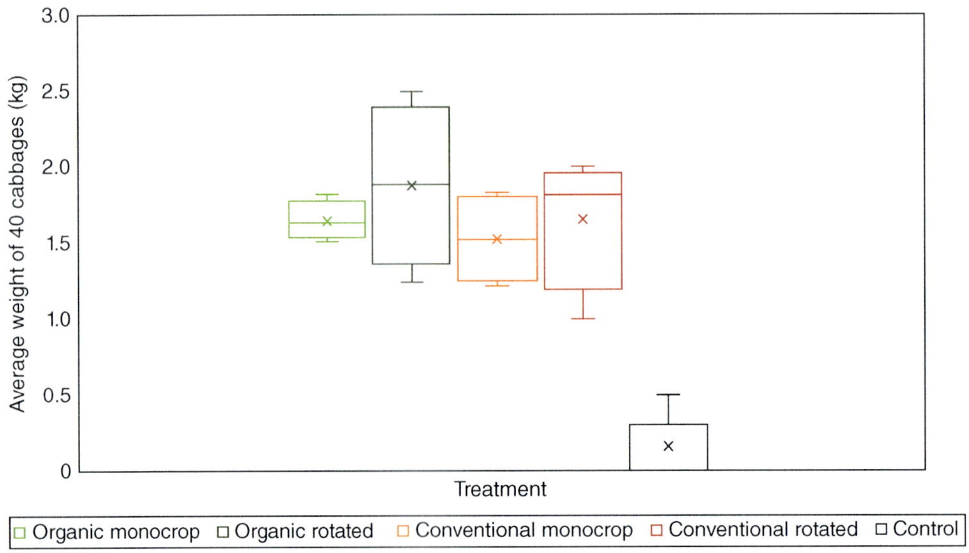

Fig. 22.8. Average cabbage weight differences (2016/17) between organic monocrop, organic rotated, conventional monocrop, conventional organic and control treatments in different management practice.

Cabbage results 2017/18

In total, 24 plots were planted with cabbage, eight conventional plots, eight organic plots, and eight control plots. Growth was slower due to less rain in this season. Conventional and organic cabbages grew at the same pace, but the conventional cabbages were ready to harvest before the organic cabbages. Some of the organic cabbage plots (plots 38 and 39) struggled a lot and never really matured like most of the other organic plots (Fig. 22.9). As mentioned in Chapters 18 and 19 of this volume, plots 37, 38, 39 and 40 consist of old rubble under about 50 cm of top soil; the soil in these plots is very hard unlike the other plots. In general, however, organic cabbage heads were smaller and looked healthier than the conventional heads. Control cabbage growth was very poor and never matured to harvest. There was no external damage from baboons as was experienced in the previous season.

Only if the *F*-value was significant ($p < 0.05$) were treatment differences further investigated using the LSD statistic. Prior to all analyses, the assumptions appropriate for a valid ANOVA were checked. Results showed that planting method and management practice were significant factors (Table 22.22). Yields of organic and conventional cabbages can be seen in Figs 22.10 and 22.11).

Conventional yields were significantly higher than organic yields, but if the fourth replication yields are discarded as outliers, there is no statistically significant difference in yields. However, the effects of the low-nitrogen compost appear to have limited organic yields in this season.

Since the second year of the Mandela Trials, rotated cabbage outyielded monocropped cabbage by a larger margin each year, but in the 2017/18 season for the first time, crop rotation, yielding 27% higher, was statistically significantly better than monocropping. Organic rotated cabbages yielded 50% higher than organic monocropped cabbage, and conventional rotated cabbage yielded 12% higher than conventional monocropped cabbage.

The better performance of the organic rotated cabbage probably reflects access to residual N from the previous cowpea crop, showing the benefits of rotation as a source of N nutrition, even in the absence of adequate fertilization. Although there was a large difference between yields of monocropped conventional and organic cabbage (average cabbage head weight of 1.9 kg versus 1.3 kg, or 49% higher average yield), the difference between rotated conventional and organic cabbage yields was only 11% (2.115 kg versus 1.897 kg per head on average). This can be seen in Tables 22.20 and 22.23, and Figs 22.11 and 22.12).

Table 22.22. Summary of means ($\bar{x}$), standard deviation (SD) and sample size (n) for various treatments in the trial for 2017/18, to determine the impact of the different treatment factors on cabbage yield.

Treatment	Sample (n)	Mean weight (kg) ($\bar{x}$)	SD
Control	8	0.032	0.060
None	16	1.792	0.384
Conventional	8	2.001	0.216
Organic	8	1.582	0.411
Monocrop	8	1.577	0.421
Rotated	8	2.006	0.182
Conventional monocrop	4	1.887	0.194
Conventional rotated	4	2.115	0.192
Organic monocrop	4	1.267	0.346
Organic rotation	4	1.897	0.095

Fig. 22.9. Organic cabbage growth from plots 38 and 39 (a) compared with a randomly selected conventional cabbage plot at the same growth stage (b).

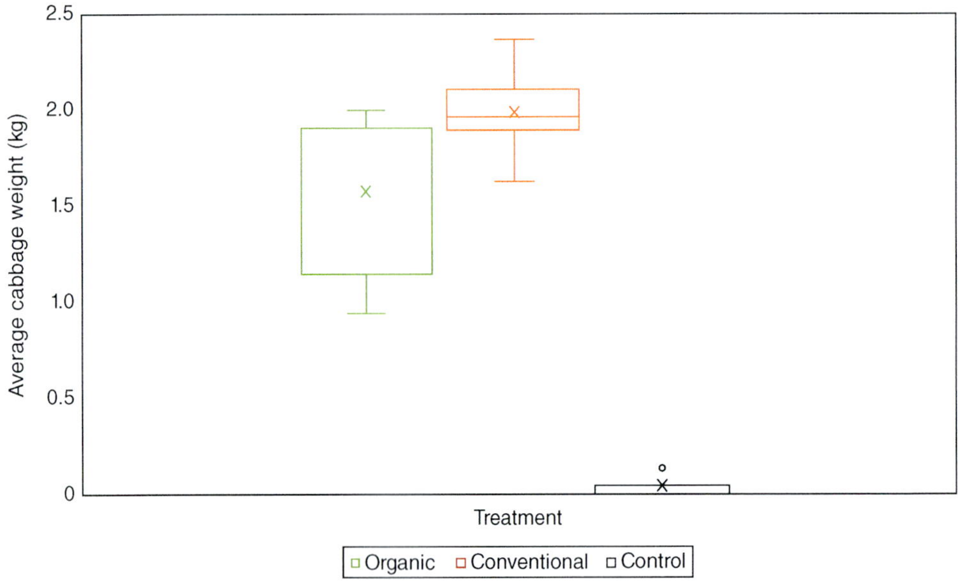

Fig. 22.10. Average cabbage weight differences (in kilograms) of organic, conventional and control management practices showing the interquartile range (IQR), mean (×) and median of each treatment.

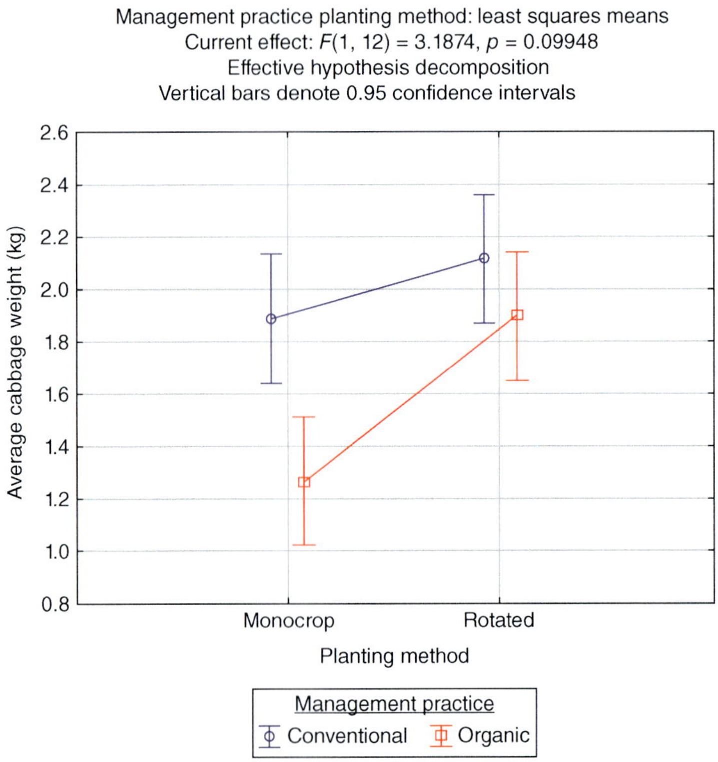

Fig. 22.11. Differences in average organic and conventional cabbage weight when grown under monocropping and crop rotation.

Table 22.23. Average cabbage weight compared in different management practices (conventional and organic) with the effect of two different planting methods (monocrop and rotated).[a]

	Conventional × Monocrop	Conventional × Rotated	Organic × Monocrop	Organic × Rotated	None
Conventional × Monocrop	/				
Conventional × Rotated	0.227*	/			
Organic × Monocrop	0.62*	0.847*	/		
Organic × Rotated	0.01	0.217*	0.63*	/	
None[a]	1.562*	2.082*	1.235*	1.865*	/

[a]Asterisk (*) indicates significance at *F*-probability < 0.2150 (least significant difference).

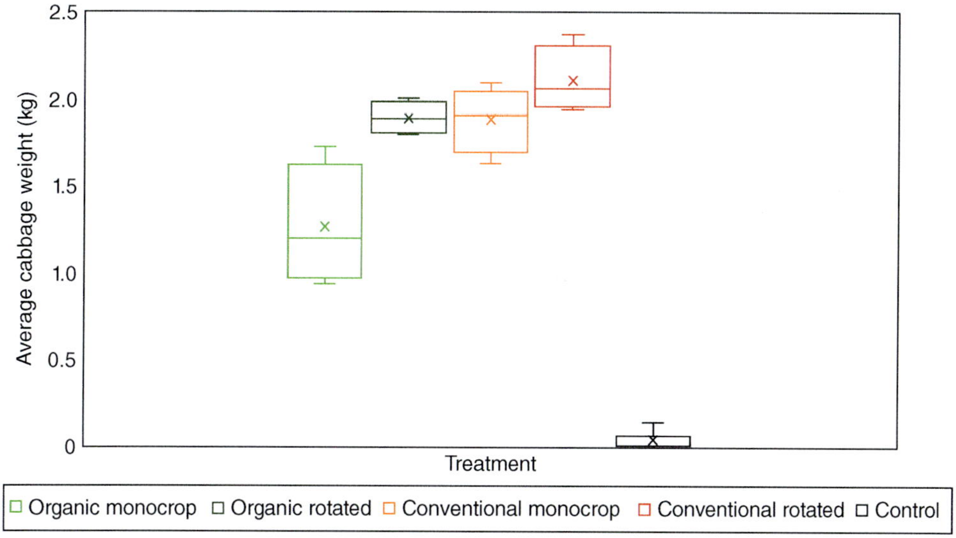

Fig. 22.12. Average cabbage weight difference between organic monocrop, organic rotated, conventional monocrop, conventional organic and control treatments in different management practices.

Conclusion

Organic farming systems improve soil fertility by improving microbial diversity, increasing soil organic carbon, soil pH and availability of phosphorus and potassium, and decreasing soil acid saturation and exchangeable Al. Providing that good quality compost is used, even low rates (5 t/ha, applied to every third crop) are able to produce acceptable levels of crop yield.

Organic farming systems are able to deal with pests and diseases in most cases, through creating aerobic soil conditions which inhibit facultative anaerobes and allow beneficial aerobic soil organisms to dominate. Crop rotations use soil nutrients efficiently and help in pest and disease suppression. Cover crops help to increase SOM, and the legumes bring nitrogen into the soil. Mulch conserves soil moisture and suppresses weeds, allowing organic crops to become more water efficient.

Further research is needed into combatting specific pests such as cutworm, and into long-term effects of organic farming on soil fertility, ecosystem services and nutrient density of the food produced. A range of food quality parameters should be investigated further through continuing research in various locations.

Since prices obtained for organic produce were on average 30% higher than conventionally farmed produce, and input costs were much lower, the economic performance of the organic farming system was competitive. However, detailed research is needed on the profitability of organic farming systems, including a true cost accounting approach, which includes the hidden costs of conventional farming in terms of global warming, fossil-fuel use, impacts of poisons on people and ecosystems and food quality related to human health.

Organic food systems appear to have many benefits over and above the direct short-term profits generated, and these should be quantified where possible. Not all benefits can be quantified economically, and ways need to be developed to describe and evaluate the non-quantifiable benefits.

References

Adhikari, K. and Hartemink, A.E. (2017) Soil organic carbon increases under intensive agriculture in the Central Sands, Wisconsin, USA. *Geoderma Regional* 10, 115–125.

Benton Jones, J. (2012) *Plant Nutrition and Soil Fertility Manual*, 2nd edn. CRC Press, London.

Blanco-Canqui, H., Francis, C.A. and Galusha, T.D. (2017) Does organic farming accumulate carbon in deeper soil profiles in the long term? *Geoderma* 288, 213–221.

Brady, N.C. and Weil, R.R. (2008) *The Nature and Properties of Soils*, 14th edn. Pearson Prentice Hall, Upper Saddle River, New Jersey.

Fließbach, A., Oberholzer, H.R., Gunst, L. and Mäder, P. (2007) Soil organic matter and biological soil quality indicators after 21 years of organic and conventional farming. *Agriculture, Ecosystems & Environment* 118, 273–284.

Mashele, N.J. and Auerbach, R. (2016) Evaluating crop yields, crop quality and soil fertility from organic and conventional farming systems in South Africa's southern Cape. *South African Journal of Geology* 119(1), 25–32.

Materechera, S.A. (2014) Influence of agricultural land use and management practices on selected soil properties of a semi-arid savanna environment in South Africa. *Journal of Arid Environments* 102, 98–103.

Miles, N. and Farina, M. (2013) *Soil Acidity and its Management in Crop Production*. Grain SA. Available at: http://www.grainsa.co.za/soil-acidity-and-its-management-in-crop-production (accessed 23 November 2017).

Prinsloo, G.L. and Vivienne, M.U. (2015) *Insects of Culivated Plants and Natural Pastures in Southern Africa*. Entomological Society of Southern Africa, Hatfield, South Africa.

Rodale Institute (2011) The Farming Systems Trial, Celebrating 30 Years. Available at: www.rodaleinstitute.org (accessed 20 March 2016).

Stockdale, E.A., Shepherd, M.A., Fortune, S. and Cuttle, S.P. (2006) Soil fertility in organic farming systems – fundamentally different? *Soil Use and Management* 18, 301–308.

Swanepoel, M. (2018) Comparing organic and conventional farming systems in George, South Africa. MSc thesis, Nelson Mandela University, George, South Africa.

Tamburini, G., Berti, A., Morari, F. and Marini, L. (2016) Degradation of soil fertility can cancel pollination benefits in sunflower. *Oecologia* 180(2), 581–587.

Tisdale, S., Nelson, W. and Beaton, J. (1985) *Soil Fertility and Fertilizers*, 4th edn. Macmillan, London.

Part 4

Upscaling the Organic Sector in Africa

23 Urban Agriculture: Challenges and Opportunities in Urban Water Management and Planning

Maren Wesselow,[1]* Christina Kifunda,[1] Raymond Auerbach[2] and Bernd Siebenhüner[1]

[1]*Carl von Ossietzky University, Oldenburg, Germany;*
[2]*Nelson Mandela University, George, South Africa*

Abstract

Africa is urbanizing rapidly, and many unemployed or poor people are trying to produce food in urban and peri-urban areas. Many local authorities see this as high-risk food production, mainly because of the risk of use of contaminated water. However, with good planning, urban gardens can supply healthy food and exercise, while providing environmental amenity value and building community solidarity. In Dar es Salaam, many rivers are heavily polluted and gardeners are pumping clean water from springs, where they can afford to do so. Many use shallow wells, and sometimes pumps are later installed, or deep wells are developed for irrigation. Experiences from different African countries show that rainwater harvesting and water conservation can contribute to increasing plant available water, and methods which could be useful include swales, grass mulch, zaï pits, Fanya juu terraces and crescent embankments. Planners should see urban gardens as an opportunity for developing community green spaces rather than a threat to public health and orderly development.

Introduction

Africa is among the fastest urbanizing regions in the world (UN, 2014; UN-Habitat, 2016). Population growth and rising economic opportunities attract ever more people to move from rural areas to cities and urban fringes. As recorded in Chapters 1 and 7 of this volume, climate change is now also a major factor in migration from rural areas to urban centres. Some of the world's fastest growing future megacities are to be found in the region, such as Nairobi, Kinshasa and Dar es Salaam. In low-income countries with relatively slow industrialization and large unskilled workforces (Potts, 2012), urbanization leads to high levels of poverty, unemployment and food insecurity accompanied by ineffective infrastructures and institutions. Unlike in rural areas, the urban poor have to purchase their food and food expenses often require almost all of their disposable income. This makes them especially vulnerable to price fluctuations and food insecurity (Poulsen *et al.*, 2015). What is more, settlement takes place mostly on marginal land, without defined and formal land titles or in highly vulnerable areas that are increasingly threatened by the impacts of land use and climate change such as flooding, water scarcity, erosion or mud slides.

*maren.wesselow@uni-oldenburg.de

Encountering these challenging conditions, inhabitants of urban and peri-urban areas increasingly resort to various forms of urban agriculture and horticulture. In small home gardens and open spaces, they grow their own food, keep livestock and produce other non-food products like wood, charcoal and fibres. Through urban and peri-urban agriculture, urban dwellers can reduce their food expenditure, obtain a more stable food source and income, and minimize the negative impacts of variable wages or food prices (Poulsen *et al.*, 2015).

Urban and peri-urban agriculture also helps to green the city, to maintain buffer and reserve zones, and has positive impacts on the microclimates of cities. It can improve health both through better nutrition and through healthy outdoor activities. In contrast to rural agriculture, urban agriculture tends to be smaller sized, more dispersed, more adaptive and integrated with non-agricultural land uses, activities and services (Mougeot, 1999). Additionally, producers can exploit the proximity of the urban demand, specialty markets and quality-conscious consumers (Fig. 23.1).

However, agriculture in many urban areas of sub-Saharan countries faces various environmental and political challenges. Due to high land pressure in urban areas, urban farmers operate on smaller plots, on marginalized land, often with contaminated soil and with limited access to water, which may also be polluted.

Therefore, this chapter examines the conditions and strategies for urban and peri-urban farmers in African cities to advance organic solutions and to develop more systems-based approaches. These can help to tackle the core problems such as limited water availability, high pollution levels and deteriorating ecological functions in urban areas. This chapter draws on a case study of urban and peri-urban agriculture in Dar es Salaam in Tanzania which is part of the German government-funded ECOSOLA project (ecosystem-based solutions for resilient urban agriculture in Africa), and on a survey of organic agriculture (OA) carried out for the German agency GIZ (Deutsche Gesellschaft für Internationale Zusammenarbeit) (Mashele and Auerbach, 2017).

Fig. 23.1. A peri-urban market near Kampala, Uganda. (Photograph by R. Auerbach 2008.)

Challenges of Urban Agriculture in the Case of Dar-es-Salaam, Tanzania

As urban encroachments degrade natural water source areas, like watersheds, wetlands, springs and groundwater recharge areas, the quality, quantity, affordability and reliability of water constitute a key challenge for urban crop farmers (Fig. 23.2). Dar es Salaam in Tanzania receives on average 1148 mm rainfall/year with seasonal rains in October/November and March/April (referred to as *vuli* and *masika* seasons). Farmers in Dar es Salaam cultivate demand-driven and water-intensive crops such as aubergines, tomatoes, carrots, onions and leafy vegetables. While farmers struggle with water scarcity in dry seasons, they face the risk of flood events during rainy seasons. As rainfall is unpredictable and unstable, farmers rely on irrigated agriculture. However, their access to water is problematic and often informal.

In view of increasingly erratic rainfall patterns, access to water will be a decisive factor for the development of urban agriculture in Southern Africa. The water-holding capacity of the soil will also be an important factor. The research results from Chapters 18–22 of this volume show OA is more water efficient than conventional agriculture, and that mulching can help conserve moisture. While Dar es Salaam is a high-rainfall tropical city and already has problems with erratic rainfall, most of the Southern African urban areas are much drier, and will increasingly experience moisture- and temperature-related production challenges.

Fig. 23.2. Farmers dig wells to access groundwater for irrigation in Dar es Salaam. (Photograph by M. Wesselow.)

Chapters 7, 8 and 9 of this volume showed how water availability and access to urban land are interconnected in Cape Town and peri-urban KwaZulu-Natal (KZN), and how volatile food prices are in times of drought. Chapters 10, 11 and 12 showed the need to build capacity for peri-urban and rural food producers to participate in shortened value chains, and how gardens contribute to building community solidarity. Chapter 13 gave practical examples of urban rainwater harvesting (RWH), while Chapters 14, 15 and 16 showed how to support farmers in accessing these markets.

Against the background of climate change, African food production needs to become more water efficient; this is especially important for urban food production (Thomas *et al.*, 2007; Montmasson-Clair and Zwane, 2016). Water efficient techniques encompass measures to improve the water-holding capacity of soil, to reduce water runoff and plant requirements (e.g. hydroponics or organoponics, drip irrigation, zero tillage, cover crops, zaï pits) and RWH, as well as techniques to recycle and reuse water resources. Other methods such as keeping flood plains free from housing construction help to reduce negative impacts from flooding.

Access to water for Farmers in Dar es Salaam

Farmers in Dar es Salaam state that water is one of the most pressing challenges they face and consequently they have developed creative ways to access water, most of which are informal. In high-income areas people use piped water for irrigation, while in low-income areas farmers use water from open wells and rivers (Nganyanyuka *et al.*, 2014).

The official supplier of water and sewerage services in the city is Dar es Salaam Water and Sewerage Corporation (DAWASCO). DAWASCO is responsible for the operation and management of water supply and sanitation services in Dar es Salaam city, parts of Kibaha and Bagamoyo in the coast region. The area for which DAWASCO is responsible has a total population of 5,781,557 of whom 3,931,459 (68%) are served by the utility through 256,290 domestic connections and 371 additional kiosks (EWURA, 2017).

However, the Ministry of Water and Irrigation admits that due to a lack of maintenance of pipes and valve meters, old infrastructure and illegal water connections, around 50% of water (49% in 2015, 53% in 2016) is lost (Ministry of Water and Irrigation, 2016). The water leaking out from defective pipes and valves is sometimes informally used for irrigation. A farmer from Kinondoni district describes how he depends on irrigation water from a broken water pipe:

> We depend on water that flows in this valley from up there. This water is from a DAWASCO pipe, which burst. When they burst up there they flow down here. So we use that water for our irrigation. When DAWASCO manages to repair and block, we get a water shortage.
>
> (Farmer from Kunduchi ward, 2017, personal communication)

Due to the lack of an efficient distribution network, and high prices, people resort to deep and shallow wells in order to develop groundwater (Figure 23.2). The rapid rise in the use of shallow wells increases the risk of groundwater contamination from pit latrines. Saria and Thomas (2013) found that water from shallow wells in Dar es Salaam city is bacteriologically and chemically contaminated. Kyessi (2005) describes how shallow wells are gradually upgraded in a step-by-step process starting with traditional open shallow wells, a hand pump in a second stage, and deep wells with motor pumps and raised water storage tanks as a third stage. Many farmers have boreholes and use water pumps to access groundwater.

When groundwater is accessed near the sea, salt water intrusion makes water unsuitable for irrigation purposes (Mtoni, 2013). A farmer in the Mapinga ward explains: 'So if I use the drilled water which has some salt, I see the crops do not do well in a long time' (Farmer, 2017, personal communication). The salinization of groundwater in coastal areas is an indicator of the high groundwater-pumping rates. The declining groundwater level in Dar es Salaam makes groundwater sourcing an unsustainable strategy (Ferguson and Gleeson, 2012; Javadi *et al.*, 2015). A major joint study by the Universities of Sokoine and Ghent is examining sustainable management of groundwater resources around Dar es Salaam; part of the study led to the publication of an MSc thesis in geology by de Witte, whose study concludes:

> This means that every year about 70% of the recharge is removed by extraction. This is not a sustainable way of water extraction... Consequences of this overexploitation are a decrease of the groundwater level and salt water intrusion in areas close to the ocean.
>
> (de Witte, 2012)

Local rivers and streams (e.g. Msimbazi and Mpinji rivers) are very feasible sources of water for irrigation purposes. Many farmers cultivate vegetables along the riverbanks (Bahemuka and Mubofu, 1999) and they have agreements for sharing the water resources (Wesselow, 2019). However, water from these rivers is polluted by heavy metals (Bahemuka and Mubofu, 1999). The main sources of this pollution are industrial effluents, and indiscriminate disposal of domestic or sewage drainage directed to the rivers (untreated or partially treated).

Vegetables take up these metals by absorbing them from contaminated soils, as well as from deposits on parts of the vegetables exposed to the air from polluted environments (Zurera-Cosano *et al.*, 1989). A farmer from Mzimuni ward explains that farmers were forced to use water from a spring as the local river is polluted by textile industries:

> The water we were supposed to use for irrigation from Mzimuni River valley is polluted by wastes from textile industries. Now we are using high costs to run the machine to pump water for irrigation from Mzimuni spring to our farms.
>
> (Farmer from Mzimuni ward, 2018, personal communication)

Sewage is often discharged untreated into streams. Biological agents discharged from households and industries were also present in the water including coliform bacteria (Muster, 1997). Pesticides (which are widely applied in urban agriculture in Dar es Salaam) also contribute to the pollution of the rivers (Mwevura *et al.*, 2002).

Another problem for farmers in Dar es Salaam is that rivers and streams are highly seasonal. While they dry up in the dry season, they may flood during the rainy season. Cultivation areas along the riverbanks often serve as flood plains during this time of the year. As rain falls erratically, floods can cause harvest failures. The poor water quality and erratic quantity make rivers an insecure source for irrigation water (Kiangi, 2014).

In some parts of the city, farmers use wastewater to irrigate their crops posing environmental and health threats for farmers and consumers. The heavy metal contamination from wastewater irrigation is of serious concern due to its implications for human health (Kihampa, 2013).

Rainwater Harvesting Techniques as an Opportunity to Balance Erratic Rainfall

As shown in Chapters 19 and 22 of this volume, organic farming methods improve the water- and nutrient-holding capacity of soil, and mulches reduce evapotranspiration. As soil organic matter (SOM) increases, water retention improves, and when mulches and RWH structures are present, water infiltration improves dramatically.

Rainwater harvesting structures may include zaï pits, Fanya juu terraces, half-moons, and these African innovations from different countries are summarized below (from Auerbach, 2003).

Zaï pits

Zaï pits are dug into degraded soils to collect and concentrate water at the plant. They were traditionally small pits, but now are often larger, with compost added. This technique spread from Burkino Faso to Niger (there, they are called 'improved tassa'). They were rapidly adopted in the 1990s when they were clearly seen to help improve yields in dry years. This seems to be a very effective rehabilitation system for degraded lands. Reports are that 12,000–25,000 zaï pits/ha are used; 1 ha of zaï takes about 60 × 5 h workdays to construct. During the dry season, zaï trap litter and fine sand deposited by the wind. They also create a microclimate which protects young plants against wind and runoff. Weeds do not grow on the crusted, barren land between the zaï. Mulching has also been widely adopted, using 3–6 t/ha of grass. This supplies nutrients, conserves water and also attracts termites which open up the compacted and crusted soil.

Fanya Juu terraces

Fanya juu terraces are formed by throwing soil up slope from a trench to make bunds on the contour, which eventually become bench terraces. They are usually protected by a cut-off drain or diversion ditch (Fig. 23.3).

Half-moons

Low lying crescent embankments (about 100 cm high) are used to harvest runoff on areas up to 100 ha. Lateral structures and cross embankments are also constructed to maximize surface runoff from the *wadis*. Organic matter is added, and fields are levelled, breaking up clods. Vegetation is planted on the embankments to stabilize them; stones and plastic sacks may also be used to reinforce weak points.

Vetiver grass on swales

Auerbach (2003) describes his own RWH system on his KZN smallholding of 8 ha:

> vetiver grass is effective when planted on swales. The effect of the swale is to catch water which falls on the area above the swale, and to slow the water down, maximising infiltration. The vegetation on the crest holds the soil of the swale in the event of intense rainfall causing

Fig. 23.3. Trees have grown on this Fanya juu ditch in Uganda and these form a windbreak and trap water. (Photograph by R. Auerbach, 2009.)

> runoff flow to overtop the swale. The swale also creates a moist microclimate in the furrow above the swale wall, which often becomes highly productive, as plant available moisture is much greater here. Swales are different to contour bunds commonly erected in soil conservation programmes. Soil conservation aims to remove water from the field without damage to the soil; swales promote infiltration.
>
> (Auerbach, 2003)

Swales and mulch in combination can effectively halve the amount of water required for irrigation (Fig. 23.4).

Integrating Rainwater Harvesting Structures into the Urban Context

Nuhu Hatibu and Henry Mahoo (2000) describe the history and effectiveness of RWH in semi-arid parts of Tanzania, and summarize some of the most important processes. In many parts of Tanzania, the fact that there is a dry spell in the middle of the growing season makes the aridity even worse, and is compounded by the unpredictable nature of this dry spell – its timing varies from place to place, and from year to year. Clearly in such an environment, RWH strategies can make the difference between a viable crop yield and a crop failure. Responses in the past have often been inappropriate, such as cut-off drains, which increase runoff, and tree planting, which decreases the amount of plant available water. Approaches other than drought-resistant varieties have now been shown to be effective. In Tanzania, approaches such as '*Mashamba ya Mbugani*' are traditional. In this case farmers plant high-water-demand crops lower down the landscape to exploit the concentration of rainwater flowing into the valley bottoms from surrounding high ground. These approaches concentrate on increasing water supply, rather than reducing demand. If rainwater is not managed, it quickly evaporates or flows away as flash floods, often into saline sinks. Microcatchment systems have

Fig. 23.4. Swales (planted with vetiver grass) and natural mulch can halve water use by reducing evaporation. Cliffdale (near Durban), KwaZulu-Natal. (Photograph by R. Auerbach, 2009.)

catchment areas which generate runoff, and cultivated basins where runoff is concentrated.

According to Hatibu and Mahoo (2000), planning of both agricultural and urban areas should give more attention to RWH. Factors influencing runoff include: (i) land surface (slope, length, vegetation, roughness and erosion risk); (ii) soil type; (iii) rainfall characterisics (rainstorm amount, intensity, distribution); and (iv) catena sequences. Types of runoff include rill-flow, gully-flow and ephemeral stream-flow. Assessing RWH potential requires identification and quantification of naturally occurring runoff and its current use, assessment of effects on downstream users if water is harvested, and assessment of alternative water sources. The essence of RWH is to capture water where it falls, in order to meet local needs. Strategies for doing this seek to improve infiltration, minimize root-zone water loss and improve crop water use and productivity. This is done through runoff harvesting (e.g. strip catchment tillage, basin systems, semicircular hoops, conservation bench terraces) and floodwater harvesting (e.g. cultivated reservoirs, stream-bed systems, hillside conduits and ephemeral stream diversion).

In urban areas, where many gardens are relatively small in area, and where much of the surface is hardened (roads and buildings), RWH can contribute significantly to water resources for irrigation. Water may be stored in tanks (see Chapter 13, this volume, where four rainwater tanks with a total storage of 8000 L were found to provide adequate water for a home vegetable garden of 7 m^2 plus an extensive flower garden), but then systems need to be developed to make use of this water efficiently, and to provide adequate storage for periods of drought. This is an expensive solution, but ofen represents a worthwhile investment.

Because urban areas have many hardened surfaces, good urban design can plan to move water from these hard surfaces (mostly roofs and roads) and lead this water to wetlands, dams or water tanks, where the water can be stored for use during dry periods (Fig. 23.5). As for the open

Fig. 23.5. Where water can be stored (a), even in dry Giyanai, South Africa, good quality crops can be produced (b). (Photograph by R. Auerbach, 2008.)

soil, here the objective is to 'slow down, spread and sink' the water, as advocated by many African water strategists, such as Robert Mazibuko (KZN) and Zephaniah Phiri Maseko (Zimbabwe). Maximizing directed runoff from hard surfaces and minimizing runoff from the fields will help to use all rainfall more effectively. Building SOM and making use of mulches will help to hold the moisture where it can be productive. Water quality is often a problem in urban agriculture; sometimes, artificial wetlands can be constructed, and often solutions have to do with formalising sanitation.

Use of mulches can reduce evaporation, as shown in Chapter 29 of this volume; in OA, grass mulches are preferred, as plastic mulches are expensive, and difficult to recycle.

Water use efficiency depends on using good irrigation technology: drip irrigation systems can function at low pressure, and use less water than sprinkler systems (Fig. 23.6).

Water quality is often a problem in urban agriculture (see Fig. 23.7 where polluted water was used for an urban gardening project in Kampala). Sometimes, artificial wetlands can be constructed, and often solutions have to do with formalizing sanitation.

Conclusion

Urban agriculture will play an increasingly important role in feeding Africa's rapidly urbanising population. Urban authorities are challenged to provide urban farmers with adequate infrastructure for irrigation and security. Moreover, the urban design can support farmers by conserving natural wetlands, and use natural ways of rain water collection. City Engineers and Planners should be working together to harvest water of hardened surfaces, lead this water into wetlands for purification, store the water safely and make it available for organised, productive use, especially by poorer communities. With some community organisation and education about water efficient irrigation systems, food can be produced cost-effectively where needed.

Fig. 23.6. Drip irrigation in Dar es Salaam. (Photograph by M. Wesselow.)

Farmers should be helped to join together to develop social groups to address water use for irrigation, and strategies for urban gardening. They could collaborate in the construction of deep wells, with local government assistance, and learn from other communities in various locations, using farmer to farmer approaches, and farm family learning groups. Urban gardens can

Fig. 23.7. Polluted water, urban gardening project, Kampala. (Photograph by R. Auerbach, 2006.)

happen haphazardly, or city authorities can make them a safe, environmentally and socially acceptable feature of their city.

Farmers should be helped to join together to develop social groups to address water use for irrigation, and strategies for urban gardening. They could collaborate in the construction of deep wells, with local government assistance, and learn from other communities in various locations, using farmer-to-farmer approaches, and farm family learning groups. Urban gardens can happen haphazardly, or city authorities can make them a safe, environmentally and socially acceptable feature of their city.

References

Auerbach, R. (2003) *Rainwater Harvesting: a Literature Survey*. Report for the Water Research Commission, Rainman Landcare Foundation, Durban, South Africa.

Bahemuka, T. and Mubofu, E. (1999) Heavy metals in edible green vegetables grown along the sites of the Sinza and Msimbazi rivers in Dar es Salaam, Tanzania. *Food Chemistry* 66, 63–66. https://doi.org/10.1016/S0308-8146(98)00213-1

De Witte, I. (2012) Characterization of the status of Dar es Salaam aquifer in view of salt-water intrusion and nitrate contamination. MSc (Geology) thesis, Ghent University, Ghent, Belgium. Available at: https://lib.ugent.be/fulltxt/RUG01/001/892/507/RUG01-001892507_2012_0001_AC.pdf (accessed 31 December 2018).

EWURA (Energy and Water Authorities Regulatory Authority) (2017) *Water Utilities Performance Review*. Report for the FY 2016/2017, EWURA, Dar es Salaam, Tanzania.

Ferguson, G. and Gleeson, T. (2012) Vulnerability of coastal aquifers to groundwater use and climate change. *Nature Climate Change* 2, 342.

Hatibu, N. and Mahoo, H. (2000) *Rainwater Harvesting for Natural Resources Management: a Planning Guide for Tanzania*. RELMA – SIDA, Nairobi, Kenya.

Javadi, A., Hussain, M., Sherif, M. and Farmani, R. (2015) Multi-objective optimization of different management scenarios to control seawater intrusion in coastal aquifers. *Water Resources Management* 29(6), 1843–1857.

Kiangi, A.G. (2014) Assessment of the quality of domestic water in Kinondoni District, Dar es Salaam, Tanzania. Master's degree thesis, The Open University of Tanzania, Dar es Salaam, Tanzania.

Kihampa, C. (2013) Heavy metal contamination in water and sediment downstream of municipal wastewater treatment plants, Dar es Salaam, Tanzania. *International Journal of Environmental Sciences* 3(5), 1407–1415. Available at: http://www.ipublishing.co.in/ijesarticles/thirteen/articles/volthree/EIJES31142.pdf (accessed 6 January 2019).

Kyessi, A. (2005) Community-based urban water management in fringe neighbourhoods: the case of Dar es Salaam, Tanzania. *Habitat International* 29, 1–25.

Mashele, J. and Auerbach, R. (2017) *Participatory Guarantee Systems in South Africa*. A report compiled for the German government (GIZ). South African Organic Sector Organisation, Johannesburg, South Africa.

Ministry of Water and Irrigation (2016) *Water Sector Development Programme*. Water Sector Status Report 2016. Ministry of Water and Irrigation, United Republic of Tanzania, Dodoma, Tanzania.

Montmasson-Clair, G. and Zwane, M. (2016) *Climate Change Adaptation and Agriculture in South Africa: a Policy Assessment*. Report compiled for South Africa World Wide Fund for Nature (WWF-SA). WWF-SA, Pretoria.

Mougeot, L.J. (1999) Urban agriculture: definition, presence, potentials and risks, and policy challenges. *Cities Feeding People Series Report* 31. Presentation at the International Workshop on Growing Cities Growing Food: Urban Agriculture on the Policy Agenda, 11–15 October 1999, La Habana, Cuba.

Mtoni, Y. (2013) Saltwater intrusion in the coastal strip of Dar es Salaam Quaternary aquifer, Tanzania. PhD thesis, Ghent University, Ghent, Belgium.

Muster, G. (1997) Environmental problems of urban agriculture: a case study of Dar es Salaam, Tanzania. MA thesis, School of Oriental Studies, University of London, London.

Mwevura, H., Othman, C. and LMhehe, G. (2002) Organochlorine pesticide residues in sediments and biota from the coastal area of Dar es Salaam city, Tanzania. *Marine Pollution Bulletin* 45, 262–267.

Nganyanyuka, K., Martinez, J., Wesselink, A., Lungo, J.H. and Georgiadou, J. (2014) Accessing water services in Dar es Salaam: are we counting what counts? *Habitat International* 44, 358–366.

Potts, D. (2012) *Whatever Happened to Africa's Rapid Urbanisation?* Africa Research Institute (ARI) Counterpoint. ARI, London.

Poulsen, M.N., McNab, P.R., Clayton, M.L. and Neff, R.A. (2015) A systematic review of urban agriculture and food security impacts in low-income countries. *Food Policy* 55, 131–146.

Saria, J.A. and Thomas, I.M. (2013) Water quality in selected shallow wells in Dar es Salaam. *Journal of the Open University of Tanzania* 11(1), 13–24.

Thomas, D., Twyman, C., Osbahr, H. and Hewutson, B. (2007) Adaptation to climate change and variability: farmer responses to intra-seasonal precipitation trends in South Africa. *Climatic Change* 83(3), 301–322. https://doi.org/10.1007/s10584-006-9205-4

UN (United Nations) (2014) *World Urbanization Prospects: the 2014 Revision*. Department of Economic and Social Affairs, UN, New York.

United Nations Human Settlements Programme (UN-Habitat) (2016) Urbanization and development: emerging futures. *World Cities Report 2016*. UN-Habitat, Nairobi.

Zurera-Cosano, G., Moreno Rojas, R., Pozo-Lora, R. and Rincon, F. (1989) Mineral elements in canned Spanish liver paté. *Food Chemistry* 32, 217–222. https://doi.org/10.1016/0308-8146(89)90050-2

24 A Future Strategy for Organic Development in Southern Africa

Raymond Auerbach[1,2]* and Matthew Purkis[2,3]
[1]*Nelson Mandela University, George, South Africa;* [2]*South African Organic Sector Organisation, Johannesburg, Gauteng Province, South Africa;* [3]*University of Johannesburg, Johannesburg, South Africa*

Abstract

This volume brought together research on concepts and global perspectives on organic food systems, on methods of building capacity in times of climate change, on supporting organic farmers and on upscaling the organic sector in Africa. This chapter summarizes these findings and concludes that four priorities have been identified: (i) to diversify the farmer base; (ii) to develop climate smart organic agriculture; (iii) to educate young people and consumers about organic food systems; and (iv) to support agricultural education, especially for farm women. The potential for developing organic markets and aligning emerging organic farmers with them is discussed in detail, including restaurants, retail, government feeding schemes, a variety of box schemes, office parks and corporate canteens and direct sales to consumers. Several 'smartphone apps' are explained in this regard, and a 3-year project is outlined to develop the organic sector and appropriate technical and extension support. The need for monitoring and evaluation is outlined, and a system for measuring farm sustainability progress is developed for individual farmers at single farm level. Intervention schedules for 1 year, 5 years and 10 years are then developed for the South African organic sector. A system for using existing farming skills and infrastructure to build vibrant and diverse sustainable farming systems is outlined, using apprenticeships, mentorship and research.

Introduction

The book has presented a global overview and a conceptual framework to describe the situation in which farmers, food processors, canteen planners, school authorities, chefs and consumers find themselves, and it suggests how they can progress towards a more sustainable future. Tools, approaches and research results have been described, each one of which has a role to play in transforming Southern African food systems from the current situation, where 7 million South African (SA) families are food insecure, and levels of obesity and non-communicable diseases such as diabetes, high cholesterol, chronic heart conditions, autism and cancer are rapidly increasing, and where stunting in infants remains unacceptably high (as shown in Chapter 7, this volume). Addressing these problems will require political will and energetic activism, as well as good science. Chapter 23 of this volume made some suggestions about urban food production in Africa; this chapter will look specifically at transforming food systems in SA, with organic agriculture (OA) as the vehicle for this transformation.

*raymond.auerbach@mandela.ac.za

Transforming SA agriculture requires four sets of changes:

- Presently, 10,000 mostly white farmers produce 80% of food sold in formal markets; half of this food is produced by 300 mega-farms – this is a very narrow and racially profiled base, and very risky for future SA food security; *the farmer base needs to expand and diversify, involving more women, more young people and more cultural diversity*.
- With climate change, many of our current food-producing areas will become, at best, marginal for rainfed cropping; research-based strategies for improving water use efficiency (WUE) and sequestering soil carbon will be critically important. *Water efficient and environmentally sustainable organic food production systems are needed*.
- In general, SA food choices are currently unwise, with a great deal of the child support grant monies spent on empty calories; obesity, diabetes, heart disease and stress-related sickness have become major problems; *organic food systems offer a healthy, nourishing and environmentally sound alternative, and public nutrition education is needed*.
- Women make most of the decisions about food purchase, and also about smallholder farming, and yet much of the agricultural training targets men; as the case of Ghana showed, *educating women (rural and urban) about organic food systems will have positive repercussions on household food security*.

These changes can be summarized as:

1. Diversifying the farmer base;
2. Developing evidence-based climate smart OA;
3. Food education for young people and consumers; and
4. Agricultual education, especially for farm women.

In this final chapter, after looking at the need for a paradigm shift and reviewing what the research of the past 10 years has shown us about agricultural development, we will analyse the market opportunities and how they affect the four areas mentioned above. We will then examine an existing project proposal which aims to start by training 50 organic ambassadors to work on the above four sets of changes. After that, we will look at how sustainability indicators can help to build a database for agroecology in SA, and how the organic sector will help build this database, and finally, we will present a summarized programme for the organic sector for 1 year, 5 years and 10 years in table form as an action plan.

Paradigm Change: Moving from Two Conflictual Narratives to Transformation

As Thomas Kuhn (1962) tells us:

> [those who] learned the bases of their field from the same concrete models ... will seldom evoke overt disagreement over fundamentals ...[as their]... research is based on shared paradigms ...[and]... the same rules and standards for scientific practice. That commitment and the apparent consensus it produces are prerequisites for normal science, i.e., for the genesis and continuation of a particular research tradition.
> (Kuhn, 1962)

The old paradigm in agricultural research was: maximize yields per hectare, whatever the environmental cost. This helped us to treble yields per hectare, but also has destroyed the integrity of our food system globally. Health is in decline, the environment is under stress, farmers can only make profits by increasing sizes of herds, intensity of cropping and mechanisation of the process. The result is industrial agriculture, climate change and massive health problems. That it is now time for a paradigm shift is clear to thousands of environmentally responsible farmers, consumers and policy makers!

This is a scientific revolution based on several changes in scientific consensus, and the way we think about agricultural knowledge systems; as outlined in this volume:

- 'Agricultural extension' has become: communication and innovation.
- 'Maximizing industrial food production' became: develop sustainable food systems.
- 'Government by multi-nationals' needs to become: strengthen local government institutions (response, not control), develop food sovereignty.

A geographical paradigm shift is needed, understanding sustainable cities (with inner-city food gardens) within peri-urban intensive agricultural production, nested in farming areas which provide essential ecosystem services, employment, health and beauty. Catherine Macombe (2018) believes that cities will soon have to limit their size to a food footprint, which she defines as the size of the city which can be sustained by the area of land around the city; as transport becomes more expensive, this will constrain the growth of cities, and limit their size to what can be sustained environmentally.

A prerequisite for SA policy makers is an effective database of agroecological farming activities, and in order for these activities to develop and become mainstream, a process of institutional development and capacity building needs to take place. This requires a framework for measuring progress towards more sustainable agriculture, based on an understanding of social, environmental and economic criteria. Monitoring and evaluation of emerging agroecology bright spots is needed, with applied research assisting farmers to develop soil fertility and improve WUE.

Our experience as an organic sector in Southern Africa over the past 50 years leads us to certain conclusions based on broad experience of supporting farmers (large and small) with training, extension, quality management, scientific advice on soils, crops, pests and diseases, economic planning support, market linkages and capacity building. It is clear that a developmental state (as SA prides itself on being) should be putting considerable resources into transformation of agriculture. Yet, Minister Gugile Nkwinti (Department of Rural Development and Land Reform) estimated in 2009 that 90% of SA land reform projects were failing. The reasons for this are well known, but the lack of political will in implementing a workable agricultural transformation programme has been a feature of the ANC government since 1999. The initial Rural Development White Paper of 1996 had many positive features, but in subsequent years, most of these were abandoned. The National Development Plan (NDP) rediscovered some of these features (NDP 2030, 2012), but under the Zuma government the NDP languished in a political wilderness.

'White monopoly capital' was used as a rallying cry to unite those who were suspicious of anything to do with big business, technology or mentorship by experienced business people to bring about a transformation which uplifted all, and this attitude is vilifying one section of the SA population, while plundering the state for selfish ends. It has been shown that the phrase 'white monopoly capitalism' was part of a carefully orchestrated plan by the Gupta family, assisted by the public relations company Bell-Pottinger, to distract SA's attention from systemic plundering described in the Public Protector's *State of Capture* report (Madonsela, 2017).

From the late Peter Mokaba's call 'Kill the farmer, kill the Boer' to allegations of targeted farm murders over the past 20 years, the (white) commercial farming sector has been portrayed on the one hand as racist, capitalist, exclusive and ruthless, and on the other hand as the totally innocent victims of reverse racism. Unfortunately, these two polarized pictures reflect our difficult history, and each perspective appears as 'current unacceptable injustice' to members of particular social groups. There is some truth in allegations of farmer racism, and the experience of too many black SA citizens has been: (i) poor pay for farm labourers; (ii) the institutionalized apartheid farm prison labour system; (iii) continuing poor conditions of many farmworkers; and (iv) the difficulties of accessing good education for children of people working on many SA farms.

There is also good reason for farmers to be concerned about the future, whether they are white or black, large or small scale, organic or conventional; government has not supported agriculture effectively over the past 30 years, and over 40% of skilled labour was lost to SA agriculture over the past 20 years. Crime, in rural as in urban areas, is a major problem for all SA citizens, and even more so for visitors from elsewhere in Africa, who often have to cope with xenophobia as well!

However, Presidents Mandela and Mbeki each gave valuable indicators for the transformation of SA agriculture (as shown in Chapter 1, this volume). President Zuma chose to ignore these answers while diverting funds intended for agricultural transformation to the enrichment of a few corrupt persons. President Mandela advocated the Reconstruction and Development Programme (RDP), aimed at giving disadvantaged SA citizens access to housing, water and energy, and at transforming our educational system. President Mbeki advocated spending at least 10% of gross domestic product (GDP) on rural development, in order to address rural poverty, an example which Ghana has followed with positive results (as explained in Chapter 1, this volume) by doubling spending on agricultural education, targeting farm women, with resulting drastic decreases in poverty and food insecurity.

We will introduce a set of measurable indicators developed with the International Federation of Organic Agriculture Movements (IFOAM) as part of the Sustainable Organic Agriculture Action Network (SOAAN) activities and which were presented to the IFOAM 'Organics 3.0' Conference in Goesan, South Korea in 2015. The conference built on earlier work done for the Bonn Sustainability Indicators Work Camp held in 2012, which came up with the very general 'sustainability flower' idea, and which was then developed into more concrete indicators usable by farmers to assess the sustainability of their own farms (Auerbach, 2015). These indicators will be presented towards the end of this chapter, with ideas of what is needed to grow the SA organic sector in a healthy and well-balanced way, and of how to measure progress. The SA organic sector needs to move forward with some introspection and a process of listening to consumers about what they want, and attention to interacting with policy makers. We need to take careful note of how the Danish organic sector developed dramatically when good research, consumer feedback and interaction with policy makers became efficient and was seen as important (Chapter 3, this volume). First, let us review the main findings of our research thus far.

Review of the Main Findings of this Book

Educate farm women

Chapter 1 showed how the President of the International Fund for Agricultural Development (IFAD) pointed out that 'To farm successfully, women need agricultural resources and inputs, as well as access to rural finance, education, and knowledge. They also need rights to the land they farm and a voice in the decisions that affect their lives' (IFAD and WFP, 2013). Chapter 1 went on to comment:

> Later that week, we presented the President of Ghana with an award, after the ministers of Agriculture and of Education reported to us how Ghana had halved poverty and food insecurity: the key intervention was education of farm women, and this was achieved by doubling of the agricultural education budget in Ghana. The Minister of Education (a qualified social worker) spent time with us, and commented that Thabo Mbeki's insights on rural development [using 10% of GDP for rural development] had inspired them to invest in rural infrastructure and people. FARA [the Forum for Agricultural Research in Africa] formally recognized this achievement during this Agricultural Science Week in 2013 in Accra 'Africa feeding Africa through Science and Technology', with the acknowledgement of progress towards a food secure Ghana. If we understand and respect local institutional dynamics, much can be achieved.
>
> (Chapter 1, this volume)

Build farmer skills progressively and systematically

Chapter 1 of this volume also presented a model (Fig. 1.1) of farmer progression from sub-subsistence, through efficient subsistence farming to semi-commercial and perhaps into commercial farming systems. It was shown that national food self-sufficiency (Fig. 1.2) may be achieved with a few technically efficient mega-farms, but that this is short term and not very sustainable, socially or environmentally. While many approaches to household food security are more equitable, they are not always socially or environmentally sustainable in the long run, unless capacity-building processes accompany them. The significance of the triple bottom line was analysed in Auerbach (2018a), concluding that the paradigm shift from a dual economy structure with a few large commercial, short-term-oriented farmers, and a large number of small-scale, short-term-oriented smallholders, being the legacy of our apartheid history, has to change. We need to build on the whole heritage of SA, including both the science-based technical skills and the indigenous technical knowledge, while broadening the farmer base and insisting on social and environmental sustainability.

Build local institutional capacity

The importance of capacity building and developing local institutions was reinforced by several chapters, showing how: (i) participatory rural appraisal (PRA) approaches are important; (ii) Participatory Guarantee Systems (PGS) may be employed; (iii) value chains may be shortened; (iv) integrated catchment management (ICM) may be employed to create 'platforms for resource use negotiation'; and (v) the impacts of drought and climate change need to be understood and planned for if future disasters and food and water crises are to be mitigated. This takes place in the context of food systems which need to be overhauled, and African food systems which have many sound traditions, but which have been eroded by the advent of fast foods rich in empty calories, salt, fats and sugars. These are produced as demanded by the global industrial food corporations, and only recently have consumer organizations appeared in Southern Africa to raise awareness concerning the health issues associated with these changes, and the accompanying erosion of food sovereignty.

Educate consumers about sustainable food systems and health

At the heart of sustainable food systems is sustainable agriculture, but it is only one part of the food system; the links between producers, processors, food preparers and consumers need to be strengthened. Experiential learning will be a vital part of the processes of training organic farmers and training other players in the food system to take a more holistic and responsible approach, and to keep the value chains as short as possible. Quality management requires conscious consumers, demanding quality, healthy food, and conscious producers, caring for their environments and providing ecosystem services which are valued by local government and resource managers.

Given SA's political past and the alienation of many people from sustainable use of the land, a multi-dimensional revolution is required in approaches to rural development and food systems' restoration. The history of the organic and allied movements was summarized in Chapters 1 and 2, while Chapter 3 of this volume outlined what the United Nations Conference on Trade and Development (UNCTAD) recommended in terms of how governments could support an emerging organic sector. It also showed how research, especially in Denmark, helped the organic sector to support both farmers and consumers. These two strands, the need to provide effective education and training to rural people, especially rural women, and the need for government to support research into various aspects of organic food systems, are fundamental to the emergence of a successful organic sector, and to transformation of SA agriculture socially, environmentally and economically. These findings are summarized in Auerbach (2018b), which concludes that OA research should target both OA production problems, and consumer requirements and perceptions, and work with government to help formulate evidence-based policy.

Chapter 3 of this volume quoted the UNCTAD (2008) Executive Summary, where Point 4 recommended:

> An action plan for the organic sector should be developed based on analysis of the state of the sector, participatory consultations, a needs assessment and proper sequencing of actions. The action plan should state measurable targets for the organic sector to help agencies and stakeholders focus their efforts.
>
> (UNCTAD, 2008)

For this to happen, both the South African Organic Sector Organisation (SAOSO) and PGS-SA need to develop into credible, well-resourced national organizations, with a strong grass-roots presence in the nine provinces, each with a training centre, and with marketing arms (Fig. 24.1). All three SAOSO and PGS-SA consultative workshops in 2018 were told by farmers that a soil and food analysis facility is needed, which can give objective, research-based advice to farmers and consumers on soil, environmental, animal and human health issues.

This proposed structure for the organic sector needs to meet the four areas of transformation identified above:

1. Diversifying the farmer base;
2. Developing climate smart OA;
3. Food education for young people and consumers; and
4. Agricultual education, especially for farm women.

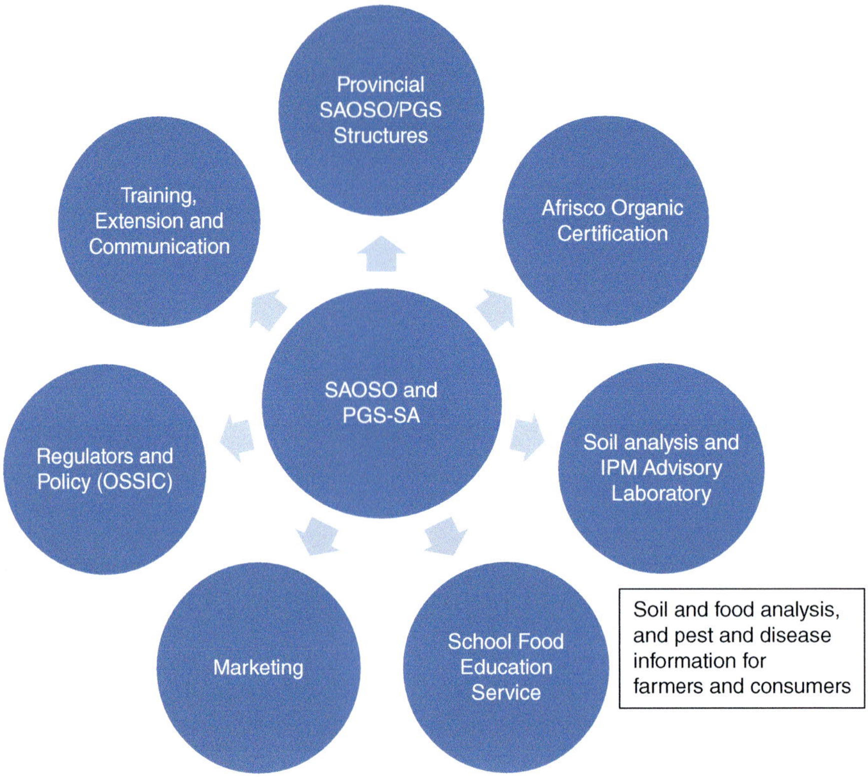

Fig. 24.1. A proposed structure for the organic sector in South Africa (SA). IPM, Integrated catchment management; OSSIC, Organic Sector Strategy Implementation Committee; PGS, Participatory Guarantee System; SAOSO, South African Organic Sector Organisation.

However, they need to be based on a sound market analysis, and ongoing marketing development. The Biological Systems Laboratory has arisen directly out of the research described in this book, and the consultations with the organic sector in 2018.

Chapters 1–3 of this volume sketched the historical and research context; chapters 4 and 14 looked at experiential training and the developing international community of praxis. Chapter 5 defined food systems, and Chapter 6 gave practical examples of how these function in practice; Chapter 7 showed the impacts on food systems and household food security when weather shocks affect food prices.

Chapter 8 showed how PRA can help farmers to plan and conceptualise their food production systems, while Chapters 9 and 11 showed how the value chain can be shortened, allowing money to circulate in local community investment programmes, and how smartphone apps can help farmers to communicate and access the market. The usefulness of PGSs in accessing high-end markets was outlined in Chapter 10, while Chapter 12 used the Eastern Cape province of SA as an example of how rainfall has changed over the past 20 years, and what the implications are for vulnerable small-scale farmers reliant on rainfed agriculture.

Chapters 13 and 19 deal with WUE, first in urban then in rural contexts, while Chapters 15 and 16 look at the organic sectors in Uganda and Zambia, respectively. Uganda has put a support structure in place and has developed training and marketing capacity, and that sector has grown substantially, and although it faces lack of government support and some stagnation at present, there are over 1 million organic farmers in the country who derive significant benefit from OA. Zambia was developing rapidly, but due to institutional decay and lack of accountability, the National Organic Agricultural Movement has declined sharply in its capacity to serve the needs of organic farmers in Zambia.

An accurate and cost-effective technique suitable for farmers to test soil carbon is described in Chapter 17 (the Rapid Incineration Field Test), and this technical innovation could help farmers to quantify their contribution to carbon sequestration.

Chapters 18–22 describe in detail the findings of the Mandela Trials, first the baseline data for the site, then comparative WUE, then the biological pest and disease control approach and finally the soil microbiology. These findings are combined in Chapter 22 with discussion of the soil chemical changes and a comparison of agronomic yield components.

The final section deals with urban food production in Africa and an organic sector plan.

Market Potential

There is much scope to develop market linkages that would allow for the farmer to access a better price on a local market level. Coordination of the supply and demand between the producers and the overall regional market requirements will enable an inclusive value chain to develop over time. The existing market potential is really broad for smallholder farmers, and currently large corporates such a Tsogo Sun, school feeding schemes, restaurants, hospitals, corporate canteens, local markets, independent retail stores and the informal sector are all potential markets.

The advocacy and marketing strategies would be developed on the back of innovation in supply chain management, sourcing local, short supply chain, carbon-effective logistics solutions and supporting local economic development, as the basis of a sound marketing strategy to leverage local support and create awareness around agroecology and PGS in the market.

Market potential can be uncovered through a survey process in the selected region in collaboration with the University of Johannesburg with the use of the #ResearchGo app which will map the economic activities in the focus areas, and will inform the approaches and interventions, in order to connect these regional localized value chains. Ensuring that ordinary people can access a wider food basket and broader nutrient profile is a priority.

The following are potential role players in an inclusive value chain:

- *Wellness Warehouse and Fruits Unlimited* – The current total of 27 retail stores, established in 2018 and 2019, give this new chain national coverage. This business entity has endeavoured to set up a national chain of independent retail stores and have committed to use the SAOSO Organic Standards for Production and Processing, and to source, as far as possible, from local smallholder producers.
- *Jacksons Real Food Market* – This is based in Gauteng with two stores, and is looking to expand coverage into the Western Cape. Jacksons Real Food Market have committed to sourcing from local smallholder producers and advocate for passing 50% and more of the retail value back to the producer.
- *High-end restaurants*, chefs associations and innovative businesses such as Deli Delicious – These have links through Izindaba Zokudla, a regional farmers' forum that is coordinated by the University of Johannesburg, and Khula, an online marketplace looking to develop market linkages for farmers in Gauteng and KwaZulu-Natal (KZN).
- *Local farmers' markets* – Smallholder producers can supply products to these existing farmers' markets and new marketplaces that may become established which aim to supply the local community members in that area, with a primary focus in the townships.
- *Independent retail outlets* – Independent retail outlets such as Jacksons Real Food Market, Wellness Warehouse, Organic Emporium, Ethical Suppliers and other small food-related businesses could be supplied with fresh produce and free-range eggs as a starting point. Processed goods could also potentially be sold in such outlets.
- *Local informal traders* – Street vendors and 'spaza' shops are an extremely relevant market to consider. Having fresh produce that is grown locally will steer the food back into the communities through the informal market exchange. Grading of produce at

regional agrihubs could allow for produce with cosmetic blemishes to be sold to the street traders at a lower price than retail.

- *Government feeding schemes* and community feeding schemes have been identified as a market for smallholder producers by government, as well as farmers' associations and other sector bodies. Linking the smallholder producer to this market will allow for local produce to be supplied where it is needed most: schools and hospitals, for example. The main challenge is the procurement process, which often sees government taking up to 120 days to pay producers for fresh produce or processed goods. This time frame is not financially viable for the smallholder farmer. School feeding schemes currently feed 9 million mouths a day, but often with processed and dried goods, and rarely with local fresh produce.
- *Government-driven programmes*, agrihubs and irrigation schemes, now all aim to supply smallholder produce into these programmes. Processed goods such as fermented sauerkraut and vegetables would be a great addition to the school feeding schemes. A diverse basket of food goods would supplement the diets of the learners, and increase the nutritional value and immune system development specifically for children in Early Childhood Development Centres.
- *Box schemes* – A variety of box schemes and food packages could cater for various living standards measure (LSM) brackets. Existing businesses such as Munching Mongoose could cater for the high bracket LSM. Local box schemes, distributed from agrihubs as potential microbusiness for youth, could serve the lower LSM bracket. Aggregating the produce from micro-producers will unlock such opportunities in the value chain for smallholder farmers.
- There are new developments and *inner city regeneration* programmes (e.g. the Maboneng Precinct) where cost-effective flat rental opportunities and retail spaces could cater for a young upcoming market. Local restaurants and apartment blocks could be supplied by short supply-chain solutions. Hydroponic rooftop farming is being advocated in the cities with the Urban Agriculture Initiative developed by the Chamber of Mines, aiming to establish 100 hydroponic rooftop farms in Johannesburg central business district (CBD). These types of microbusinesses, adapted with agroecological principles, could develop a thriving community of rooftop farmers in the CBD of Johannesburg supplying the local market options directly. Allotment gardens should be developed by local authorities with appropriate water supply and security, and should be integrated with municipal organic-waste composting programmes.
- *Office parks and corporate canteens* – This is a viable market for a short supply chain. Many canteens procure from the central market; the market agents source fresh produce to supply these various corporate canteens. Delivery of fresh produce to the corporate canteens will create a consistent market with large volumes for smallholders to supply.
- *Agro-processors and independent packhouses* – These are some of the low-priced market participants as they often trade fairly close to the central market prices. Offloading excess produce to packhouses will ensure a base whereby volumes of produce could be marketed consistently.
- *Direct to end consumer* – The consumer can come to the local farm and harvest their own basket of food for the week, developing a personal relationship with the farmer, reconnecting the end consumer with the production of fresh produce. Such community supported agriculture (CSA) is a feature of the European and US organic sectors.

Corporate and Government Involvement

Government involvement

Currently the documented support by government to smallholder farmers is very limited. Some cooperatives receive agricultural inputs but often it is sporadic and mostly conventionally orientated. Regional community seed banks have been established in the Eastern Cape and Limpopo to support the saving of seed within the communities. The aim is to ensure the preservation of indigenous varieties. The central gene bank has been established by the Department of Agriculture, Forestry and Fisheries (DAFF) to preserve selected varieties and the genetics of selected seed stock. This government initiative could be supported through the development of peer-to-peer seed certification systems, to ensure the quality of the seed stock being deposited into the bank is stable, uniform and distinct.

Currently there is no specific programme that is geared towards smallholder farmers, but DAFF classifies smallholder farmers as subsistence farmers. As shown in this volume in Chapter 1, Fig. 1.1, and also in Chapters 9, 10 and 11, participation in shortened value chains can benefit smallholder farmers, and can help them to add value to their produce and to access high-value niche markets. The main current area for government support is for commercial market-orientated farmers. Government is stuck in an old paradigm, which only sees large-scale industrial agriculture as 'modern'; government needs to support organic smallholder farmers nationally as part of a movement towards becoming a developmental state. All partners have a duty to work with government to deliver on the policies that have been promulgated, to ensure a food secure and food sovereign SA.

The development of programmes should not rely on government cooperation and funding; innovation generally comes from the entrepreneurs. Access to the city infrastructure such as the agrihubs will enable micro-farmers and the people from the surrounding communities to benefit from fully functional city infrastructure. These agrihubs and agri-parks are crucial for the establishment of regional inclusive value chains and were intended for this purpose, but they lack overall coordination.

Government should outsource the management of these hubs to local independent service providers. Supporting government with the agroecological extension service will be one of the potential avenues of collaboration and would result in more capacity within the government-driven extension service. Working with the various departments to achieve food security is important, as several government departments should coordinate national food security. Breaking down the 'silos' of government is a key task in establishing a functional working relationship with the public sector.

The corporate sector

The corporate sector certainly has a role to play. Much work can be done on linking the benefits of an agroecological value chain to climate change and the Sustainable Development Goals (SDGs). These goals highlight the need for accountability around environmental damage of corporate practice. Corporate social investment funds (if managed transparently) can contribute to getting start-up projects and microbusinesses funded in the short term.

The retail sector has a lead role to play in supporting local sourcing, linking this to local economic development. This could begin the process for developing consumer interest in local and clean food. Consumers will drive the demand for local produce, corporate marketing campaigns should inform the public around conscious consumer choices, with government incentives. The corporate sector could also start many community-driven food gardens by investing in the programme to support and develop new and existing gardens.

A Potential Project Design

The challenges of smallholder farmers have been well understood and documented by independent researchers and organizations aiming to support smallholder farmers within the agricultural sector. There has been limited impact with previous projects due to the independent nature of these programmes that often are not focusing on the systemic challenges that smallholder farmers are presented with. These needs are highlighted in the points below.

- There is a need for a variety of training offerings for farmers, extension officers and the community, ideally aligned to the Sectoral Education and Training Authority (SETA) training accreditation. The training in the provinces needs to be consistent and there needs to be a degree of quality control, and ongoing development through Training of Trainers (ToT) training packs developed by various stakeholders in the agricultural sector such as sector bodies, non-governmental organizations (NGOs) and other service providers in the agricultural sector. Visual training material is often the best way to bridge the language gaps that are evident in the different provinces of SA.

- Extension officer training could be accompanied by a mentorship programme that will guide the new extension officers to open up areas of business in the agricultural sector and supporting the smallholder farmer with sound agroecological advice.
- The existing model of trainers transferring their knowledge and skills in the community on a voluntary or self-funded basis is not sustainable. They need vehicles, laptops and funds to host workshops. The scope to develop this training into sustainable businesses is very evident if these trainers are initially supported and incubated to become viable service providers.
- Farmers have expressed interest in training in a variety of skills sets that allow for the farmer to be fully equipped to develop farms. This should include rainwater harvesting, construction and electrical skills, among others. Farmers also require basic support in mapping, design, project planning and implementation on the land.
- Farmers need basic resources such as tools, clothing and protective gear, which might be accessed through sponsorship. As theft is a big issue, fencing is required to secure the animals and areas of production. Security of gardens will play a role in sustainability and financial viability in the long term.
- Farmers need access to land through the land redistribution programme, land bank or even public spaces owned by government or private stakeholders. Farmers require mentorship and support with business plans and record keeping so that they can access finance.
- In terms of the value chain, the biggest challenge for farmers is getting their food to a centralized distribution point such as the regional agrihubs and access to cold storage facilities that will extend the market potential for the smallholder farmers. Solutions need to be found with potential partners in business and government. Farmers must participate in the creation of upstream business opportunities with the community to enhance the socio-economic impact.
- Community engagement was key to forming trust and to encouraging a 'cultural revolution' and this needs to continue through markets and events. The PGS system is a tool for community participatory processes to occur.
- Research and technology will play an important role. This would include needs analysis within communities, integrating data collection into the role of the extension officers. Solutions to water and energy challenges need to be explored through technologies. The Whole Food Movement advocates food processing which does not denature food; at the same time, food systems are more than just the food production or processing, and peer learning from farmer-to-farmer should play a major role in agricultural extension.

Potential Contribution of Key Stakeholders

The potential contributions from key stakeholders have been identified nationally with many active NGOs and representative bodies that could potentially add value to a programme of this nature. In order to facilitate smooth collaboration between partners a shared vision is extremely important; this needs to be agreed upon by all participating stakeholders. What has been captured in this assessment is that there are many organizations actively contributing to various aspects of food security, but they generally lack the required national impact to see the development of an alternative food system that supports the development of smallholder farmers. It is evident that this type of work happens in pockets all around the country with little co-ordination around a national vision of food security within a spirit of collaboration. The streamlining of project support, with various lead organizations in selected provinces would see a coordinated effort, to ensure effective management of projects and allocated resources. This would ultimately enable documented and monitored impact on the ground with farmers and other target beneficiaries.

Other areas highlighted by the viability assessment include involvement with:

- SAOSO and PGS-SA – in respect of growing awareness, supporting the standards, the development of training, driving value chain opportunities and the research agenda. Their roles would also include lobbying government for policy change and scanning funding opportunities for projects. The development of linkages with regional and international organizations with similar objectives was seen as an important role that needed to be facilitated.

- Organizations with developmental project experience, such as Lima Rural Development Foundation (LIMA) – for access to regional hubs and partnership on programme roll out. An agroecological extension service could add value to LIMA's stretched national extension service.
- Oxfam – as there is potential around supporting a number of their programmes, and they are interested in partnering to achieve a shared vision.
- Gender CC – as they have regional presence, strong links with women in agriculture and relationships with an international donor funder network.
- Solidarity Economy – This NGO has a strong presence in the Eastern Cape with farmers' associations on the ground. Much positive work has been achieved around cooperative development. Organizationally they are very much in alignment with SAOSO's shared vision of a food-sovereign SA.
- Media partners such as Indigo Media – as they can assist with facilitating TV adverts, radio slots, social media and print media that can disseminate a consistent message to all stakeholders of the nutritional benefits of organic food and the ethical value chain that supports the development of smallholder farmers, youth and women in agriculture.
- Research partners (e.g. the Agricultural Research Council, the African Centre of Biodiversity, SA Food Lab, universities, the Water Research Commission, conservation agencies and more recently the Biological Systems Laboratory) – as this would lead to a national agroecological research agenda.
- Other sector bodies (e.g. the Biodynamic Association of Southern Africa and the permaculture community) – as collaboration with them would strengthen the movement nationally.
- Technology partners – as collaboration through ICT solutions and supporting farmers through information exchange and appropriate technology transfer is essential. UJ PEETS is a strong partner within the existing university network as well as Indigo Media with the Aparate farmer database/platform technology.
- Partners with objectives that align with the shared vision – it is important to identify such partners and link with them online to shared resources and events; engaging with farmers' associations, unions and existing agricultural cooperatives.
- Community-based organizations, farmers' forums and NGOs – to work together in a cohesive manner to achieve the collective vision of a food secure SA. SA Food Sovereignty Campaign can drive the newly formed policy and general awareness around systemic change.
- Government at national (e.g. the Organic Sector Strategy Implementation Committee (OSSIC)), provincial and local levels – to advocate for support for infrastructure and value-chain-related activities.
- Input suppliers, seed suppliers and service providers – to be endorsed through the programme for the roll out of training programmes and shared services to smallholder farmers.
- The retail sector – to provide opportunities to smallholder farmers through the buy-in of an inclusive value chain with ethical distribution of profits and to develop strategic market linkages for sustainability in the value chain.

SAOSO Sector Plan Recommendations

SAOSO in partnership with many of the stakeholders mentioned in this chapter should develop a programme that supports: (i) smallholder farmers with training for best practice; (ii) an innovative extension service; (iii) integration of innovation and appropriate technology into the supply chain; (iv) access to local markets, endorsement and certification; and (v) peer-to-peer information exchange systems that see innovation throughout the value chain. The project would catalyse business creation among youth and women in the agricultural sector and establish a local market. That will enable a viable livelihood to be achieved for a large percentage of farmers, representing the majority of emerging producers in the SA agricultural sector.

Over a 3-year period the main objectives of the pilot programme will be to:

1. Increase the socio-economic impact of 5000+ supported farms in this programme.
2. Establish and grow businesses for youth and women in agriculture over a period of 3 years.
3. Facilitate the development of new businesses and the inclusive agricultural value chain for smallholder farmers.
4. Coordinate and support a network of farmers in three provinces (Gauteng, Western Cape and KZN), with the opportunity to expand into the Eastern Cape and Limpopo with potential partners.
5. Train and develop a coordinated agroecological extension service that can work in selected provinces and replicate training into the farming communities.
6. Increase institutional capacity of SAOSO and PGS-SA as the sector bodies that will drive the programme with selected partners and into the future.
7. Develop an advanced ICT solution for an extension service and farmer tools, and facilitate an online farmer database with selected partners.
8. Conduct targeted action-driven service and research programmes setting up a technical facility in George in the Southern Cape, with a focus around integration of appropriate technologies into smallholder agriculture, business model creation for youth and women, true cost accounting of production, carbon sequestration and soil and food testing.

Inception phase – 3 months

The objectives of the inception phase are as follows.

1. *Formation of the regional chapters in collaboration with SAOSO and PGS-SA* following correct governance procedures. Coordinate the inception meetings between key stakeholders face to face and online.

a. Introduce the regional representatives to the farmers' forums and existing support services in each province.

b. Identify potential candidates for the agri-ambassador training programme.

2. *Formation of the five working groups* for the project, capacitated with the relevant skill sets housed within SAOSO and PGS-SA.

a. The five working groups are: (i) Farmers Services; (ii) Standards and PGS; (iii) Communications and Advocacy; (iv) Value Chain; and (v) Research. Each must be capacitated with project managers, mentors, farmers and facilitators.

b. Integration with Izindaba Zokudla farmers' forum and the Youth in Agri Initiative (YAI) programme in collaboration with the University of Johannesburg; and regional expansion of the farmers' forums into the wider Gauteng and KZN provinces.

3. *Formally present the pilot programme to potential partners* and stakeholders for support and collective buy-in. Formalize partnerships in key provinces through stakeholder engagements face to face and online.

a. Legal documents and Terms of Agreements will be developed in consultation with Werksmans Attorneys.

b. Identify crucial partners in selected provinces and form partnerships.

c. Present the broader stakeholders in the network.

4. *Develop SAOSO training programme* and YAI mentorship programme for farmers, the extension service and government.

a. These training programmes must ideally be aligned to the SETA accreditation for external funding opportunities to roll out the training.

b. There is an opportunity to collaborate with the biodynamic (BD) and Rainman Landcare Foundation training programmes which are SETA accredited.

c. The conversion programme in collaboration with Agri-Skills is key to ensure that all forms of current production methodologies are included and can be converted to an agroecological food system.

5. *Software development* and integration of a consolidated ICT solutions in smallholder agriculture – #ResearchGo, Khula and Aparate.

a. Creation of the various surveys that will be conducted – baseline survey, infrastructure audit, project mapping, research and extension evaluation.

b. Streamlining of the record keeping system on the ICT solution, with accounting systems for the farmers and the extension service to use.

c. Linking the Khula online trade platform and Aparate online farmer database.

d. Developing and integrating agri-chain and other traceability and transparency software into the value chain.

6. *Planning a Communications and Advocacy Campaign* – The launch of the programme to the public using social media and other platforms to broadcast a unified message of a food sovereign SA. This should include:

a. A local campaign linked to broader topics for food system transformation. Also an online presence and public engagements hosted in collaboration with various other stakeholders.

b. A documentary on the current food system and how smallholder farmers can contribute to the socio-economic development in SA.

c. A roadshow to gather public support for the programme in collaboration with partners.

Project proposal action items for year one

The objectives for year one are as follows:

1. *Create a strong recruitment process for the agri-ambassadors* (agroecological extension service). This will look to existing success stories that are currently providing a level of agricultural service to farmers.

a. The identification of 50+ inspired candidates who are already active in the agricultural sector, to provide a solid training programme and mentorship process throughout the duration of the 1-year training and mentorship programme.

b. Ensuring that this initial batch of candidates is capacitated with the resources and business skills to develop into a sustainable service provider.

c. Develop the YAI with the University of Johannesburg and SAOSO as the programme for agri-ambassador training and business model development of youth in agriculture. This programme will research and pilot the various business models in the value chain with selected youth enterprises and partners, such as the Green Business College in Johannesburg. The successes and challenges will be monitored and reported throughout the programme.

d. Support Izindaba Zokudla (University of Johannesburg Farm School) with the establishment of regional farmers' forums in Gauteng. This platform will build the network of farmers, facilitate business development and community participation.

2. *The monitoring of working groups and regional chapters* for the project.

a. The establishment of five working groups within SAOSO that would enable the potential project to be delivered successfully. These working groups are: (i) Farmers services; (ii) Standards and PGS; (iii) Communications and advocacy; (iv) Value chain; and (v) Research.

b. The formation of regional chapters integrated with a future extension service and training programmes.

c. The formation of an effective working partnership with PGS-SA and streamline the establishment of new PGS nodes.

d. The incubation of the new PGS nodes through information and knowledge sharing and integration of technology systems that will support the new nodes to fruition.

3. *Training of the first round of 50+ agri-ambassadors* in collaboration with SAOSO and PGS-SA.

a. The training of the 50+ agri-ambassadors through a modularized programme, in conjunction with the YAI programme and Izindaba Zokudla farmers' forum to support the groups of farmers.

b. Integrate with practical work to support identified projects in each province.

c. Leadership and business development skills: strong focus to the training programme.

d. Training on the software and ICT solutions available to support them in their work.

e. Identifying upstream business opportunities and providing the support though business incubation with selected partners.

f. The identification of demonstration sites that would enable the training in production methodologies, land design, agroprocessing and technology transfer.

4. *Surveying and mapping conducted by the agri-ambassadors in Gauteng, Westerm Cape and KZN* – this will include:

a. Baseline data surveys, infrastructure audits, farmer identification, local economy surveys, project mapping, resource location, market analysis.

b. Analysis of data to inform the region-specific intervention – The database will allow for live data to be analysed and will provide important knowledge for the coordination of capacity and resources.

c. Logistics solutions will be identified and acted upon with associated partners.

d. Establishment of a research agenda and appropriate technology transfer.

5. *Support the farmer through agroecology and PGS awareness training*

a. Roll out of agri-ambassador farmer training programmes with selected service providers for support.

b. Document the training process and monitor the farmers through constant data collection and extension service.

c. Develop market linkages for farmers online and locally with agri-ambassadors through strategic partnerships within the inclusive value chain.

d. Develop farmer profiles online and implement the traceability software Agri-Chain for the produce so there is transparency and trust can develop in the market.

Project proposal action items for year two

The desired outcomes for year two will be to develop the supply and demand ratio for the 5000+ selected farmers in Gauteng, Western Cape and KZN within the inclusive value chain.

1. *Farmer support and youth development*

a. Mobilize agri-ambassadors to provide farmers with training and an extension service. With the use of the *SAOSO standards, guide PGS endorsement and third party certification* of fresh produce. Mobilize the agri-ambassadors over the 12-month period to support this process.

b. Develop market linkage and refine the ethical value distribution downstream of the value chain to the farmer and local businesses involved in the value chain.

c. Establish agroprocessing and cold chain storage for the farmers. Activate the agri-hubs, with city collaboration, to best serve the smallholder farmers and surrounding communities.

d. Develop mentorship opportunities and microbusinesses to support the inclusive value chain with a focus on Youth in Agriculture.

2. *Hold public engagements to discuss the findings of the data and mobilize the network*

a. Facilitate public engagements and public discussion panels with partners.

b. Mobilize government engagement through OSSIC.

c. Organize a festival and number of micro-gatherings in collaboration with partners.

d. Branch out into other regions through the Izindaba Zokudla farmers' forum and YAI programme in collaboration with the University of Johannesburg.

3. *Develop strong market* access through the inclusive value chain with selected partners.

a. Establish and support new and existing PGS nodes.

b. PGS-SA to play a lead role and oversight of the establishment of the nodes.

c. Facilitate a relationship with selected retail and market participants through the inclusive value chain.

4. *Develop shared services* for the smallholder farmers.

a. Support smallholder farmers with logistics throughout the value chain with selected partners.

b. Broaden access to inputs through the supply of seedlings, compost, fertility solutions and most management solutions.

c. Integrate appropriate technologies to improve production capacity of smallholder farmers.

d. Broaden access to agroprocessing opportunities and 'value adds' for the smallholder producers.

e. Facilitate agroecological farmers to share knowledge with conventional farmers on soil degradation, effective water management, biodiversity, permaculture land design; this process will develop the capacity within the network to facilitate farmer-to-farmer knowledge sharing.

Project proposal action items for year three

Year three will be the phasing out stage of the initial mentorship programme. It will entail reporting, through the monitoring and evaluation processes that were implemented over the period. Some outcomes will be:

1. *A coordinated value chain network* between smallholder farmers and market participants.

a. Supply and demand ratio will have been established between producers, retailers and consumers in the three selected provinces and the value chain will be expanding regionally.

b. Ethical pricing model will have been developed with the establishment of a farmer cooperative and market cooperative discussing pricing on certain commodities in the inclusive value chain.

c. Technology will have been successfully integrated into the supply chain and management of extension service will be streamlined with farmer support.

2. *Establishment of the export market* for smallholder farmers. This will include:

a. Identification of cash crops and commodities for the international export market, such as the UN PAGE Biopanza initiative.

b. Trade agreements with the global south and other African countries could be established in line with the current African Union 'Ecological Organic Agriculture Initiative'.

c. Integration of traceability and transparency software into the export value chain.

3. *Reporting and closure of project* including identifying further funding opportunities locally and abroad.

a. Although regular financial reporting with be built into the programme quarterly, monitoring and evaluation will take place throughout the programme.

b. Full reports will be assembled regarding the various deliverables from the associated working groups.

c. Regional chapters and PGS nodes will submit reporting documentation to SAOSO.

d. Full income and expenditure reports will be assembled.

A new, inclusive value chain can be developed building on the success of the PGS model. This value chain will include new entrants in a supported environment, where entrepreneurial endeavours will drive the development of localized economies, unlocking employment opportunities for youth and women in agriculture, and providing upstream mentorship opportunities with leaders in the sector. This paves the way for social participation in the formation of localized, regenerative economies and support for entrepreneurial activities and business creation.

Technology alone (e.g. synthetic fertilizers, agrochemicals, genetically modified seeds and irrigation) will not solve the problems of African food insecurity; it will not, as Jeffrey Sachs (2005) claims, 'End poverty in our time'. What is required is systematic capacity building of farmer institutions with farmer training using agroecological (low external input sustainable agriculture or LEISA) approaches. Together with this, it is vital to build market linkages, and to create consumer awareness of the importance of dietary diversity. A LEISA approach will assist in increasing dietary diversity, and agroecology will support improved agrobiodiversity, improved WUE and food sovereignty. Household food security will improve as small-scale farmers move to efficient subsistence farming, and the rural economies will start to develop as they move into semi-commercial farming. Government policies for sustainable rural development will have to understand these interrelated, complex truths, and if the SA Landbank is to live up to its mandate, it needs to develop supportive programmes for OA.

Indicators of Sustainability

Diverse systems have been proposed for measuring agricultural productivity and sustainability; some have proposed 'sustainable intensification' as a desirable direction. Koohafkan *et al.* (2011) point out that many systems propose measures of intensification including:

- indicators to identify areas suitable either for intensification or for ecosystem services;
- indicators to assess the performance of the intensification process; and

- situational awareness indicators to capture different dimensions of intensification.

They comment that:

> An apparent drawback of this proposal is that it identifies intensification (increase the production per area via the efficient use of inputs) as the only agricultural path for agricultural production, disregarding the diversity of other agroecological approaches that, instead of intensification, emphasize diversification, synergies and recycling.
>
> (Koohafkan *et al.*, 2011)

The founding meeting of Regeneration International in Costa Rica in May 2015 noted that both the concept of 'sustainable intensification' and the idea of 'climate smart agriculture' have often been used to promote industrial agriculture while making minimal changes in production systems. Unless the food system changes to become more regenerative and more responsive to human needs of both producers and consumers, the health of both people and planet will continue to deteriorate. Regeneration International is working through scientists and activists to improve production and the food system.

The SOAAN work took the Food and Agriculture Organization of the United Nations (FAO) indicators as a point of departure, and at the same time, FAO kept refining their indicators and their assessment tool (FAO, 2014). FAO's comprehensive guidelines, including 21 indicators grouped under 'Environment', 'Economy', 'Society' and 'Governance', now give a broad picture of sustainability. The FAO SAFA (Sustainability Assessment of Food and Agriculture Systems) guidelines (Version 3.0) constitute a very useful tool for sustainability assessment, but the process is fairly data intensive, and the IFOAM work continued, specifically focused on the organic sector. After preliminary work by IFOAM, David Gould was appointed to manage a process of consultation. This resulted in the founders making some suggestions, and the appointment of a steering committee. In 2012, there was a series of meetings in Bonn (the 'Sustainability Days') at IFOAM headquarters, before the International Sustainable Development conference, also held in Bonn. The working group continued with drafts and redrafting based on public input, and tabled several versions for comment on the IFOAM website, and this too, became a fairly complex set of measures.

Discussion of Proposed Indicators

Although the main work of SOAAN produced the well-known 'Blume' flower with 20 petals, showing 20 areas of sustainability measurement, the South Africans working at the Nelson Mandela University found this rather clumsy for farmers to use.

They amalgamated the sustainability measures into four themes ('Ecology', 'Economy', 'Culture' and 'Society'); this equates to a triple-bottom-line analytical tool, with 12 main measures, which overlap to some extent across the themes (Fig. 24.2). These farm-scale measures establish a baseline, and help the farmer to

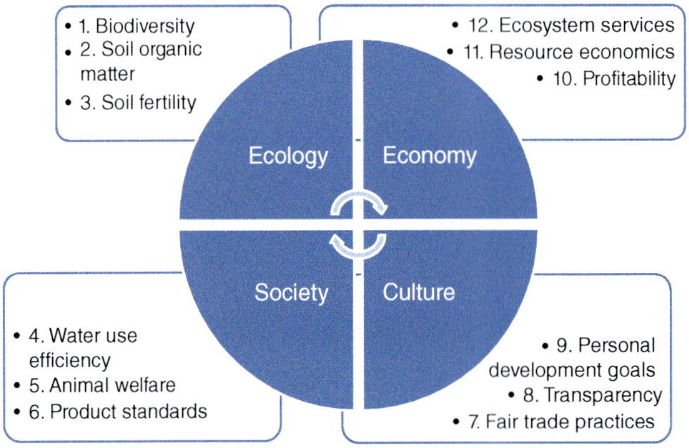

Fig. 24.2. Twelve sustainability indicators clustered in four themes. (From Auerbach, 2015.)

measure change (positive or negative), and to set targets.

The proposed indicators are all measurable, and the intention is not so much to make up a set of desirable performance standards, but more to help a farmer to measure the progress of a given farm in the direction of sustainability. This should be backed up by support for the local ecology, positive contributions to society, enrichment of culture and contribution to the local economy. Indicators which ignore what the farmer and farm family aspire to do will not result in improved sustainability. The farmer should be at the heart of the indicators (see Indicator 9 in Fig. 24.2), and if farmer aspirations and farm profitability are not adequate, the farmer will most likely leave the farm sooner or later.

More detail about these indicators is given in Box 24.1, after which, in conclusion, Table 24.1 offers a summary of a suggested programme for the organic sector in SA.

Box 24.1. Twelve measurable indicators of progress towards sustainable agriculture

1. **Biodiversity** is measured by: (i) variety (i.e. number of species); (ii) diversity (i.e. number of families); and (iii) rarity (i.e. number of endangered species).
2. **SOM** is measured by: (i) percentage of soil carbon; (ii) fraction of soil carbon which is active biomass; and (iii) an index of soil macrofauna actually present in the soil.
3. **Soil fertility** measured as change in N, P, K and pH, as an index of original soil status divided by desirable status, deducted from current soil status divided by desirable status.
4. **WUE** measured as the change in the ratio of crop production per unit of water used (current less original).
5. **Animal welfare** expressed as an index of original animal welfare index divided by desirable welfare index, subtracted from current animal welfare index divided by desirable welfare index. (This index will be calculated for each type of animal, and the proportion of each animal will be converted into numbers of large stock units and then expressed as fractions of the farm total.)
6. **Product standards** measured either as non-conformities with the local organic certification standards or, when possible, as a combination of nutrient density and toxic residue indices for each crop or animal produced.
7. **Fair trade indicators** measured as a worker satisfaction index comprising education, housing, safety, satisfaction, wages and dignity indices, together with a 'drudgery of work' index, as developed by Chand *et al.* (2015).
8. **Transparency index** measured by number of organizations with access to information about items 1–7 above; maximum score for this index is attained when all of these factors are published on the Internet (open access).
9. **Personal development goals** measured qualitatively as the extent to which the farmer feels that life is unfolding in a direction that is in line with what is desired.
10. **Profitability** is measured by expressing the farm gross margin (direct income less directly allocated variable costs) as a proportion of the farmer's desired return on investment (what the farmer would consider a satisfactory gross margin, bearing in mind all the investments of earlier generations, of the current generation, and the level of financial return which the farmer considers acceptable as a reward for all the effort of farming).
11. **Resource economics** measured as society's evaluation of whether the farm resources are being stewarded sustainably, and whether the farm is a useful resource for local people (e.g. as a local PGS or for local groups to have access to the farm for hiking, tree planting or educational activities).
12. **Ecosystem services** measured as the contributions of the farm to local water, biodiversity, amenity, and whether the farm impacts positively or negatively on local ecosystems.

A technical support service needs to be established to support the scientific work and to carry out monitoring and evaluation for the sector, as requested by farmers at all three consultative workshops during 2018. Time horizons are provided in conclusion as a summary of what needs to be done. These should be developed into a logical framework analysis for the sector, with smart indicators for each activity.

Table 24.1. Time horizons for sector planning: 1 year, 5 years, 10 years.[a]

Area of intervention	Horizon		
	1 year	5 years	10 years
Government	Organic and Agroecology Policy	Teacher training agriculture	DAFF organic extension
Farmer training	Identify 50 trainees, 35 of whom should be women	Train 500 farm women and 200 men	2000 Organic farmers
Quality management	Set up nine provincial PGS offices	Provincial PGS support for 30 PGS groups	Support for 60 PGS groups and five agrihubs
Market development	Develop organic agrihub in George	Start organic agrihub system	Five agrihubs
Food education	Recruit nine provincial food educators	Train teachers in 100 schools	Teachers in 1000 schools
Weather shocks	Identify climate-vulnerable areas	Implement water efficient gardens	30 Water wise gardens
Capacity building	Set up SAOSO offices in Gauteng and Western Cape, then roll this out so there are a total of nine such offices	Technical quality management linkages set up	10,000 Farmers linked to market apps
Technical support	Set up a soil analysis laboratory to apply results of research	1000 Samples establishing soil database for SA	3000 Samples monitoring soil fertility change
Local certification	Establish Afrisco SANAS	200 Clients	1000 Clients
Research (technical)	Restart Mandela Trials	Work with postgraduate researchers and universities	25 Research projects covering: food quality, food education, training, WUE, product markets
Research (capacity)	Set up a database for organic-sector farmers	Involve farmers with sustainability indicators	Report on progress to sustainability
Research (policy)	Set up political lobby groups for sustainable food systems	Operationalize organic and agroecology policies, local food systems	Mainstream organics as developmental agriculture

[a]DAFF, Department of Agriculture, Forestry and Fisheries; PGS, Participatory Guarantee System; SA, South Africa; SANAS, South African National Accreditation Service; SAOSO, South African Organic Sector Organisation; WUE, water use efficiency.

Conclusion

This book has examined many aspects of food systems, climate change and farmer support and development. We conclude that a grass-roots movement has started to make food systems more sustainable, and to improve food sovereignty in Africa. This movement needs support: (i) it needs volunteers; (ii) it needs funding to set up professional support systems; (iii) it needs publicity; and (iv) it needs informed consumer movements. Already GMO and poison-free zones are emerging in SA, and training, technical support and market linkages are developing. Communication and innovation are dynamic properties, and the organic sector will need to build with consumers, identifying what they want, helping to educate young people about food choices, and responding to the critiques of well-informed clients. Agroecology is the future; the Grow Food movement is growing worldwide (see the recent film of that name by Jessica Smith and Joe Rignola);

regenerative agriculture is becoming a widespread, unifying philosophy; the organic movement, as pioneer of sustainable food systems development, with standards, marketing systems and better technical support, will remain at the centre of this mass mobilization for Mother Earth. Already as a result of the research in this book and the Ecological Organic Agriculture (EOA) Initiative, the African Union has appointed Professor Raymond Auerbach to assess how EOA can be mainstreamed in 47 African countries of North, West, Central and Southern Africa, to match the progress already underway in East Africa. This assessment gathers the evidence presented in this book, summarises the situation in each of the 47 countries, and proposes a detailed way froward for each of five types of African countries. The assessment will be presented to the EOA Initiative Continental Steering Committee in Accra, Ghana in November 2019.

References

Auerbach, R. (2015) Measurable sustainability indicators for organic farms. Organics 3.0 conference, Goesan, South Korea. IFOAM, Bonn, Germany.

Auerbach, R. (2018a) African food systems. *Economia agro-alimentare/Food Economy* 20(3), 295–314. https://doi.org/10.3280/ECAG2018-002002

Auerbach, R. (2018b) Organic research and government support improve organic policy and progress in Danish, Swiss, American and African case studies. *Economia agro-alimentare/Food Economy* 20(3), 327–343. https://doi.org/10.3280/ECAG2018-002002

Chand, P., Sirohi, S. and Sirohi, S.K. (2015) Development and application of an integrated sustainability index for small-holder dairy farms in Rajasthan, India. *Ecological Indicators* 56, 23–30.

FAO (Food and Agricultural Organization of the United Nations) (2014) *Sustainability Assessment of Food and Agriculture Systems (SAFA)*. FAO, Rome. Available at: http://www.fao.org/3/a-i4113e.pdf (accessed 6 June 2018).

IFAD (International Fund for Agricultural Development) and WFP (World Food programme) (2013) *The State of Food Insecurity in the World 2013: the Multiple Dimensions of Food Security*. FAO, Rome. Available at: http://www.fao.org/docrep/018/i3434e/i3434e.pdf (accessed 9 June 2018).

Koohafkan, P., Altieri, M.A. and Holt Giminez, E. (2011) Green agriculture: foundations for biodiverse, resilient and productive agricultural systems. *International Journal of Agricultural Sustainability* 10(1), 61–75. https://doi.org/10.1080/14735903.2011.610206

Kuhn, T. (1962) *The Structure of Scientific Revolutions*. University of Chicago Press, Chicago, Illinois.

Macombe, C. (2018) Diversity of food systems for securing future food availability. *Economia agro-alimentare/Food Economy* 20(3), 344–356. https://doi.org/10.3280/ECAG2018-002002

Madonsela, T. (2017) *State of Capture*. Report 6 of 2016/17. Office of the Public Protector, Pretoria. Available at: http://saflii.org/images/329756472-State-of-Capture.pdf (accessed 8 June 2019).

National Development Plan 2030 (2012) Our Future-make it work. National Planning Commission. Department of the Presidency, Republic of South Africa.

Sachs, J. (2005) *The End of Poverty*. Penguin, London.

UNCTAD (United Nations Conference on Trade and Development) (2008) Best Practices for Organic Policy. United Nations. Available at: http://unctad.org/en/Docs/ditcted20073_en.pdf (accessed 3 June 2018).

Index